Electron Microscopy of Shale Hydrocarbon Reservoirs

Memoir 102

Edited by
Wayne K. Camp, Elizabeth Diaz, and Barry Wawak

Published by
The American Association of Petroleum Geologists

ISBN13: 978-0-89181-383-5

AAPG Editor: Michael L. Sweet
AAPG Geoscience Director: James B. Blankenship

COVER: Pseudocolor secondary electron SEM image from Ar-ion milled surface illustrating porosity development within organic matter from Eagle Ford Shale, La Salle County, Texas. Orange = calcareous coocolith fragments, blue = organic matter, black=pores, white=pyrite framboid. White rims surrounding many of the larger pores and parallel white lines are image artifacts. Image courtesy Ingrain, Inc.

This publication is available from:

The AAPG Bookstore
P.O. Box 979
Tulsa, OK U.S.A. 74101-0979
Phone: 1-918-584-2555 or 1-800-364-AAPG (U.S.A. only)
E-mail: bookstore@aapg.org
www.aapg.org

Affiliated East-West Press Private Ltd.
G-1/16 Ansari Rd., Darya Gaaj
New Delhi 110-002
India
Phone: +91-11-23279113
Fax: +91-11-23260538
E-mail: affiliate@vsnl.com

Geological Society Publishing House
Unit 7, Brassmill Enterprise Centre
Brassmill Lane, Bath BA13JN
United Kingdom
Phone: +44-1225-445046
Fax: +44-1225-442836
E-mail: sales@geolsoc.org.uk
www.geolsoc.org.uk

The American Association of Petroleum Geologists

The American Association of Petroleum Geologists Books Refereeing Procedures

The Association makes every effort to ensure that the scientific and production quality of its books matches that of its journals. Since 1937, all book proposals have been refereed by specialist reviewers as well as by the Association's Publications Committee. If the referees identify weaknesses in the proposal, these must be addressed before the proposal is accepted.

Once the book is accepted, the Association Book Editors ensure that the volume editors follow strict guidelines on refereeing and quality control. We insist that individual papers can only be accepted after satisfactory review by two independent referees. The questions on the review forms are similar to those for the AAPG Bulletin. The referees' forms and comments must be available to the Association's Book Editors on request.

Although many of the books result from meetings, the editors are expected to commission papers that were not presented at the meeting to ensure that the book provides a balanced coverage of the subject. Being accepted for presentation at the meeting does not guarantee inclusion in the book.

More information about submitting a proposal and producing a book for The American Association of Petroleum Geologists can be found on its web site: www.aapg.org.

Table of Contents

DVD MATERIAL

Acknowledgments

AAPG wishes to thank the following for their generous contributions to
Electron Microscopy of Shale Hydrocarbon Reservoirs

AAPG Memoir 102

Memoir sponsors:

bp

INGRAIN

Weatherford® LABORATORIES

Core Lab™

FEI™ Explore. Discover. Resolve.

Chapter sponsors:
Anadarko Petroleum Corporation
Statoil
Trican Well Service

Contributions are applied toward the production cost of the publication, thus directly reducing the book's purchase price and making the volume available to a larger readership.

About the Editors

Wayne K. Camp is a distinguished geological advisor with Anadarko Petroleum Corporation, The Woodlands, Texas, where he has been employed since 1980, working various domestic and international exploration and development projects. Prior to working with Anadarko, Camp was employed for two years by Phillips Petroleum Company. Camp received his B.A. degree in geology with honors from the State University College at Oneonta, New York, and his M.S. degree in geology from Colorado State University, Fort Collins, Colorado. Camp's expertise includes identifying and evaluating new exploration play opportunities and prospect generation. Camp supervised various geological and geophysical exploration and development teams at Anadarko from 1986 to 2004. His experience includes conventional and unconventional U.S. plays onshore, subsalt and deep-water plays in the Gulf of Mexico, and offshore plays in Indonesia. Camp is an active member of AAPG, the Geological Society of America, and the Houston Geological Society. He served as chair for the AAPG Unconventional Reservoirs Research Group in 2004; was cochair for the 2005 Vail Hedberg Conference on tight-gas sands; and was coeditor and contributor to the AAPG Hedberg Series No. 3 book, which was awarded the Robert H. Dott Sr. Memorial Award for best AAPG special publication in 2010. Camp also serves as an advisor for the U.S. Department of Energy Unconventional Resources Technology Advisory Committee. Camp lives in Montgomery, Texas, with his wife, Joanne, and has two daughters and two granddaughters.

Elizabeth Diaz has been the Chief Geoscientist at Ingrain, Incorporated, for the last four and a half years; she has helped with the development and deployment of digital rock physics solutions for industry. She has been pivotal in the development of Ingrain's shale product offering and works closely with clients to help solve pertinent industry problems through the application of digital rock physics technology.

Prior to Ingrain, she was an expert in the application of rock physics and quantitative seismic interpretation techniques, evaluating and risking the presence of reservoir rocks and hydrocarbons on multiple exploration and appraisal projects worldwide (Nigeria, Brazil, Libya, Kazakhstan, Yemen, Peru, Ecuador, the United States, and Colombia).

Diaz holds a B.S. in geology from Universidad Industrial de Santander and a M.S. in geophysics from Stanford University, Stanford, California. At Stanford Rock Physics Laboratory, she conducted rock physics research on the effect of glauconite on the elastic properties, porosity, and permeability of reservoir rocks.

Barry E. Wawak is presently manager of the Reservoir Geology Group, Core Laboratories Advanced Technology Center, Houston, Texas. He has spent his 30-year career applying technological solutions to reservoir characterization problems. Recent industry developments in shale resource plays have renewed his interest in applying scanning electron microscopic solutions to geological problems. Prior to joining Core Laboratories in 2000, his career included stops at Core Petrophysics, Incorporated, in Lafayette, Louisiana; Chevron Corporation in New Orleans and La Habra, California; and California Institute of Technology Jet Propulsion Laboratory in Pasadena, California. His geology education includes undergraduate classes at California State University at Fullerton and graduate studies at California State University at Los Angeles and the University of New Orleans.

Note to the Reader

This publication is a continuation of AAPG's efforts to combine the best of both digital and print publications for its members. *Memoir 102, Electron Microscopy of Shale Hydrocarbon Reservoirs*, consists of two sections.

Chapters are printed in their entirety in this volume. The DVD located in the inside back cover of this book contains all of the chapters in digital form and special files. All are in full color.

The dual digital and print format allows authors greater scope for both length of manuscripts and number of color figures and provides the reader much more versatility in using the content. For example, figures can be used directly for making slides or Microsoft Powerpoint© presentations and can be enlarged on the screen as needed for easier viewing. Individual chapters can be easily printed for later reading, and key word searching through the papers is possible. Finally, this format allows AAPG the flexibility to publish a volume that contains material that will appeal to a wider audience with additional interests.

Introduction

W. Camp, E. Diaz, and B. Wawak, eds. Electron microscopy of shale hydrocarbon reservoirs: AAPG Memoir 102, p. ix–xi.

Wayne K. Camp

Anadarko Petroleum Corporation, 1201 Lake Robbins Dr., The Woodlands, Texas, 77380, U.S.A. (e-mail: wayne.camp@anadarko.com)

Elizabeth Diaz

Ingrain, Inc., 3733 Westheimer Rd. #300, Houston, Texas, 77027, U.S.A. (e-mail: diaz@ingrainrocks.com)

Barry Wawak

Core Laboratories, 6316 Windfern Rd., Houston, Texas, 77040, U.S.A. (e-mail: barry.wawak@corelab.com)

Recent advances in horizontal drilling and hydraulic fracture stimulation technology have opened up vast new oil and natural gas sources in shale formations. Shale gas production in the United States was 5 trillion cubic feet (141 billion m^3) in 2010, representing 23% of the total U.S. dry-gas production (U.S. Energy Information Administration, 2012). Shale gas production has also been established in Canada, and initial exploration drilling has begun outside North America, most notably in Poland, Argentina, India, Australia, and China.

This so-called shale gale Yergin (2011) can also be measured by the increase in the number of active horizontal drilling rigs in the United States that surpassed the number of active vertical drilling rigs for the first time in 2009. During the first half of 2012, active vertical drilling rigs only accounted for 30% of the nearly 2000 total active United States rigs (Baker Hughes, 2012).

The rapid increase in horizontal wells targeting shale formations has demanded an improved understanding of shale hydrocarbon reservoirs by geologists, geophysicists, and engineers, particularly as shale exploration moves away from the traditional dry-gas fairways in response to declining domestic natural gas prices in search for more liquid hydrocarbon-rich regions. Many traditional reservoir analytical methods fail to accurately characterize shale reservoirs because of their complexity, small (micro- to nanometer) pore size and low (micro- and nanodarcy) permeability.

Shale, and its nonfissile counterpart mudstone, are fine-grain sedimentary rocks composed of silt- and clay-size particles. Because of their extremely fine grain size ($<$62 μm), shale and mudstones are difficult to observe and describe using conventional optical microscope techniques. Fortunately, the high-magnification capability of electron microscopy is ideally suited to study fine-scale rock features, such as texture, composition, and porosity, down to the nanometer scale. Recent advances in electron microscopy technology have resulted in improved images and analytical methods to better describe, evaluate, and understand shale hydrocarbon reservoirs.

The purpose of this memoir is to provide a practical reference for geologists, geophysicists, engineers, and students to gain a better understanding of the various state-of-the-art techniques and applications of electron microscopy for shale hydrocarbon reservoir evaluation. This reference volume is intended for both specialists and nonspecialists and includes the first published catalog of described scanning electron microscopic (SEM) images of productive shale hydrocarbon reservoirs that is hoped will serve as a useful guide for SEM petrographic interpretation and reservoir analog studies.

ORGANIZATION

Part 1: Technical Papers

This book is composed of three parts. Part 1 is a collection of 13 solicited, highly illustrated, peer-reviewed papers by leading researchers from major universities, oil and gas companies, commercial laboratories, and SEM and related equipment manufactures. The first six chapters provide the reader with an introduction to

DOI: 10.1306/13391698M1021681

modern SEM equipment and new sample preparation, imaging, and image analysis techniques.

The remaining seven chapters are case studies from several of the major active shale plays, including the Barnett, Eagle Ford, Haynesville, and Marcellus in the United States; the Horn River, Nordegg, Montney, and Duvernay in Canada; and Posidonia in Germany. These chapters describe various applications of electron microscopy to gain a better understanding of the characterization, distribution, and evolution of productive shale hydrocarbon reservoirs.

A glossary defining common terms and acronyms used in the technical chapters is included at the end of Part 1.

Part 2: Image Catalog

Part 2 is a compilation of nearly 250 described SEM images representative of known productive and potential shale hydrocarbon reservoirs provided by Anadarko Petroleum, Core Laboratories, and Ingrain Incorporated. The images are organized in alphabetical order by formation name. All of the represented formations are from the United States, reflecting the greater amount of shale drilling activity in the United States and availably of images. A map and stratigraphic chart are included showing the geographic and stratigraphic distribution of the formations represented in the image catalog. The image catalog also includes a brief discussion and representative examples of the common image artifacts to aid in the interpretation of SEM images.

Part 3: Supplemental Data and Special Displays

Part 3 is a searchable DVD containing a digital copy of the printed material, plus an appendix with supplemental data and color and special-format images, such as large-scale high-resolution and three-dimensional animation files.

DISCUSSION

Most would agree that we still have plenty to learn about shale oil and gas reservoirs, and we have only scratched the surface (literally only a few microns deep) in our understanding of the microstructural characteristics utilizing electron microscopy. The impetuous for this volume was to bring forth current research and applications and to make available examples of shale hydrocarbon reservoirs from the increasing collection of unpublished SEM images, in part to spur future research and improve our understanding of shale oil and gas plays. Some topics for future research are outlined next.

Organic Matter

Organic matter is readily identified in backscattered electron SEM images by its high contrast, reflecting its relatively low mean atomic number compared with minerals common in shale and mudstone. However, our ability to distinguish between kerogen and bitumen (see glossary at the end of Part 1 of this volume for definitions), and our ability to identify the various types of kerogen as recognized by organic petrologists, is limited in part because of differences in the scale of observation caused by differences in magnification levels between typical SEM images and optical microscopy. Other than the well-known distinctive morphology of certain palynomorphs, such as *Tasmanites*, little work has been done to establish criteria for identifying kerogen macerals in SEM imagery. New research using transmission electron microscopy and x-ray analysis as described by Bernard et al. (Chapter 7, this volume) offers promising results, but the technique has not yet gained wide acceptance because of lack of widespread specialized equipment and expertise. High-resolution, large-scale SEM images as described by Lemmens (Chapter 4, this volume) may help to bridge the gap between optical and SEM observations. Other methods, such as high-resolution, laser-induced fluorescence spectroscopy combined with electron microscopy, might also find future applications for studies of organic matter in shale hydrocarbon reservoirs.

Porosity

Pores have traditionally been observed as part of the overall rock fabric as revealed by secondary electron SEM images from broken surfaces of shale and mudstone samples (e.g., Scholle, 1979; O'Brien and Slatt, 1990); however, legacy SEM studies gave little attention to quantifying these pores until shale became economically important as hydrocarbon reservoirs (Sondergeld et al., 2010). Identification of nanometer-scale pores within organic matter by the pioneering SEM studies of argon-ion-milled surfaces of the Barnett Shale by Loucks et al. (2009) has led to a proliferation of similar studies and advancements in quantifying porosity such as described in Chapters 5–13 of this volume. Yet, questions remain regarding the impacts of various sample preparation and imaging techniques on our ability to observe and quantify porosity and how to reconcile SEM measurements made from minute areas with laboratory indirect porosity measurements from much larger core samples.

Fluid Distribution

Dean-Stark distillation extraction (Gas Research Institute, 1996) is now routinely conducted on crushed shale samples to quantify water and oil saturations. However, these measurements (similar to other quantitative laboratory petrophysical measurements) cannot provide insights regarding the distribution of fluids within shale hydrocarbon reservoirs. Pore fluid distributions may be modeled by digital rock physics; however, these models lack calibration based on native-state observations.

Environmental scanning electron microscopy (ESEM) provides the ability to observe shale specimens in their natural state and thus may potentially be used as a tool to describe and measure the distribution of fluids using freshly broken rock samples. Future studies may be designed to address several unresolved questions, for example:

1. Does water and oil saturation vary by pore type (i.e., organic matter pores vs. inorganic matter pores)?
2. Is solid bitumen removed during Dean-Stark distillation in oil-mature samples that may be erroneously calculated as liquid-filled pore space?
3. Does oil/water wettability vary within a shale hydrocarbon reservoir?
4. Does imbibed water following hydraulic fracture stimulations displace fluids in nanometer-scale pores?
5. How do fluids introduced during drilling and hydraulic fracture stimulation alter fluid saturation?
6. What microstructural changes occur in fluid-sensitive shales?

REFERENCES CITED

Baker Hughes, 2012, North American rotary rig counts—May 25, 2012: http://investor.shareholder.com/bhi/rig_counts/rc_index.cfm?showpage=na (accessed May 30, 2012).

Gas Research Institute, 1996, Development of laboratory and petrophysical techniques for evaluating shale reservoirs: Gas Research Institute Final Report, GRI-95/0496, 286 p.

Loucks, R. G., R. M. Reed, S. C. Ruppel, and D. M. Jarvie, 2009, Morphology, genesis, and distribution of nanometer-scale pores in siliceous mudstones of the Mississippian Barnett Shale: Journal of Sedimentary Research, v. 79, p. 848–861, doi: 10.2110/jsr.2009.092.

O'Brien, N. R., and R. M. Slatt, 1990, Argillaceous rock atlas: New York, Spring-Verlag, 141 p.

Scholle, P. A., 1979, A color illustrated guide to constituents, textures, cements, and porosities of sandstones and associated rocks: AAPG Memoir 28, 201 p.

Sondergeld, C. H., R. J. Ambrose, C. S. Rai, and J. Moncrieff, 2010: Micro-structural studies of gas shales: Society of Petroleum Engineers, Unconventional Gas Conference, February 23–25, Pittsburg, Pennsylvania, SPE Paper 131771, 17 p., doi: 10.2118/131771-MS.

U.S. Energy Information Administration, 2012, Annual energy outlook-early release review: http://www.eia.gov/forecasts/aeo/er/pdf/0383er(2012).pdf (accessed May 30, 2012).

Yergin, D., 2011, The quest: Energy, security, and the remaking of the modern world: New York, Penguin Press, 804 p.

1

Huang, Jason, Timothy Cavanaugh, and Boaz Nur, 2013, An introduction to SEM operational principles and geologic applications for shale hydrocarbon reservoirs, *in* W. Camp, E. Diaz, and B. Wawak, eds., Electron microscopy of shale hydrocarbon reservoirs: AAPG Memoir 102, p. 1–6.

An Introduction to SEM Operational Principles and Geologic Applications for Shale Hydrocarbon Reservoirs

Jason Huang

Carl Zeiss, 1 Corporation Way, Peabody, Massachusetts, 01960, U.S.A. (e-mail: jason.huang@zeiss.com)

Timothy Cavanaugh and Boaz Nur

3733 Westheimer Rd., Houston, Texas, 77027, U.S.A. (e-mails: cavanaugh@ingrainrocks.com; boaz@ingrainrocks.com)

ABSTRACT

Scanning electron microscopy (SEM) and its auxiliary technologies have been exploited by various industries for over 50 years. One of its main applications has been to characterize materials at very high magnification. Recent interest by the petroleum industry to better characterize and understand shale hydrocarbon reservoirs has led to an increased interest in utilizing SEM technology for shale reservoir studies. The purpose of this chapter is to review basic SEM operational principles and applications that are particularly useful for characterizing shale hydrocarbon reservoirs.

INTRODUCTION

The mechanics of the modern scanning electron microscope (SEM) system allow for various imaging and detecting techniques that can be used to study different aspects of the composition of shale samples at very high resolution. Scanning electron microscopy, unlike conventional light microscopy, produces images by recording various signals resulting from interactions of an electron beam with the sample as it is scanned in a raster pattern across the sample surface. A fine electron probe, with a spot size from a few angstroms to several hundred nanometers, is generated by focusing electrons emanating from an electron source (conventionally called the electron gun) onto the surface of the specimen using a series of electro-optical lens elements. The combination of the source and the lens elements is called the electron column.

Figure 1a shows an example schematic diagram of a state-of-the-art electron column. The column shown here is the Zeiss GEMINI™ column. The ray path of the electrons is illustrated in Figure 1b. Specifically, electrons are emitted from a very fine heated filament with a high electrostatic extraction potential. The electrons are then immediately accelerated to a desired kinetic energy and propagate down the electron column. In the GEMINI column, a very simple two-lens system is employed to focus the electron beam. The upper lens, or the condenser lens, first demagnifies the emitted electrons to some extent. Next, some of the electrons

DOI: 10.1306/13391699M1023580

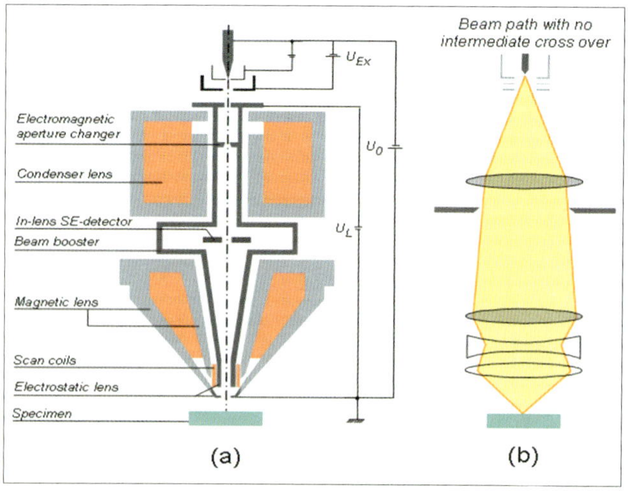

Figure 1. Schematic diagram of the GEMINI™ electron column (a) and the electron beam path in the column (b).

are cut off by an aperture. The focusing strength of the condenser lens and the size of the aperture determine the beam current in the final spot. A final objective lens assembly then focuses the beam down to the desirable spot size. Depending on manufacturer design, the lenses used in the column can be magnetic or electrostatic (or a combination of the two types). In the case of the GEMINI column, the final objective lens is a combination of magnetic and electrostatic lenses. The main advantage of such a design is that there is virtually no magnetic field on the specimen that could distort the imaging if the specimen is magnetic.

The earliest work describing the concept of SEM is that of Knoll in 1935 (Goldstein et al., 2003). In his work, the focused electron beam was generated to image a metal plate with a transmission signal. The first SEM used to examine thick specimens was described by Zworykin at the RCA Laboratories in 1942 (Goldstein et al., 2003). Everhart and Thornley improved on Zworykin's original design by employing a scintillator to convert the electrons to light, which was then transmitted by a light pipe directly to a photomultiplier. The Everhart-Thornley detector greatly improves the signal-to-noise ratio of SEM imaging. The first commercial instrument was built in 1963 by Pease at Cambridge Scientific Instruments (Goldstein et al., 2003).

Since the first commercial instruments, many advances have been made, including development of a high-brightness electron gun allowing more electrons to be focused into a smaller beam spot, an improved electron optics design that can dramatically enhance imaging resolution, and a large portfolio of detectors that capture and separate different types of information generated by the primary electron beam. A more detailed description of a detection mechanism related to imaging contrast is elaborated in the following section.

SEM CONTRAST MECHANISMS

Several types of electrons are generated as the result of the energetic bombardment of the specimen by the primary beam (Figure 2). All of these electrons carry distinct structural information about the sample and differ from one another in origin, energy, and traveling direction. For example, type I secondary electrons (SE1) emit with a high angle at a close proximity from the impact point and therefore carry high-resolution, surface-sensitive (topographic) information of the sample. Type II secondary electrons (SE2), however, are generated from a deeper and wider volume than the SE1 and reflect at a lower angle, therefore carrying intrinsically lower-resolution topographical information. Similarly, singly scattered backscattered electrons (BSE1) tend to emit at a high angle and are closely linked to compositional contrast, while multiply scattered BSE (BSE2) take off at a lower angle and are used to characterize composition and crystalline structures of a sample. Photons can also be generated from the excitation by the primary electron beam. For example, x-rays with a continuous spectrum in energy (bremsstrahlung) are generated as a result of the deceleration of the electrons. Additionally, characteristic x-ray lines

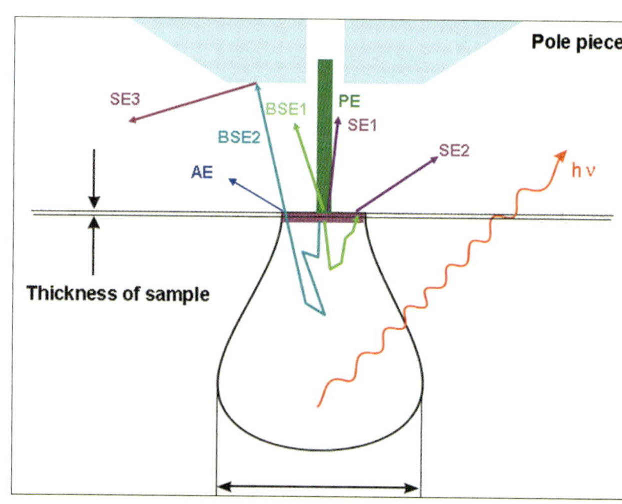

Figure 2. Profile view of a typical specimen-electron interaction volume. Note that secondary electrons (SE) originate much shallower in the sample than backscattered electrons (BSE) and therefore provide more surface-driven and less composition-influenced information.

are generated from electron excitation within specimen atoms that interact with the primary electron beam. The latter form of x-ray is particularly useful in chemical analysis with SEM.

Photons with wavelengths in the ultraviolet, visible, and infrared light regions of the electromagnetic spectrum can also be excited by the primary electrons. This phenomenon is called cathodoluminescence (CL). The generation of CL is commonly associated with the presence of certain trace impurities in minerals and therefore is also widely used in geological applications. A few examples are illustrated next to demonstrate several common contrast mechanisms useful for the geological study of shales.

SEM Image Examples

Figure 3 shows an SE1 image of a shale sample. At low landing energy (less than 1 keV), the secondary electrons collected with the in-lens detector originate from the very surface of the sample. The SE1 signal is commonly used to image surface details at the highest resolution at the expense of compositional information. On the right side of Figure 3, the SE2 image of the same area as the SE1 image on the left shows the effects of higher landing energy and deeper specimen interaction, including more compositional and less topographical information.

Figure 4a shows an SE2 image of a cross section of a shale sample that has been polished by argon-ion milling, a sample preparation technique that consists of using one or more beams of argon ions to gently polish the surface of a sample by sputtering away material, thus providing an extremely smooth surface for SEM investigation. The image was acquired with an Everhart-Thornley type secondary electron detector. The SE2 image contrast reveals both topographical and compositional information due to the greater sample interaction depth of SE2 electrons. High SE yield is scaled as lighter shades of gray, and low SE yield is scaled as dark shades of gray in SEM images. Pores and fractures appear dark in the SE2 image, reflecting lower secondary electron yield from negative depressions than from elsewhere on the sample surface. Therefore, porosity information can be readily characterized with the SE2 electrons.

The same cross section was imaged with BSE1 with the Energy-selective backscatter (EsB) detector (Figure 4b). The image contrast (grayscale variation) reflects compositional variations (mean atomic number) of the sample. For example, the midgray represents silica matrix; the darker level represents organic matter. The brighter gray level reflects higher density carbonate phases, and the brightest gray level represents pyrite. Note the greater compositional contrast provided by the BS1 image (Figure 4b) over the SE2 image (Figure 4a). Although SE2 and BS1 images are capable of providing compositional information, auxiliary techniques, such as energy-dispersive x-ray spectrometry (EDS), are required to characterize elemental composition.

Compared to BSE1, for which the contrast is modulated by mean atomic number differences, BSE2 yield depends strongly on crystalline structures such as grain orientations. Figure 5 shows a BSE2 image of

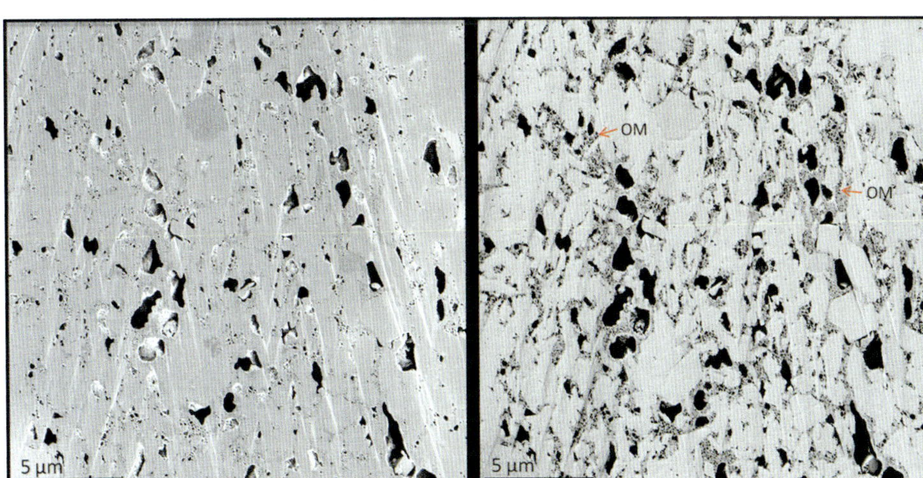

Figure 3. A shale sample imaged using SE1 signal (left) and SE2 signal (right). Surface-specific information such as pore space and surface roughness is evident in the SE1 image. The SE2 image has more compositional influence, displaying organic matter (OM) bodies that are not evident in the SE1 image.

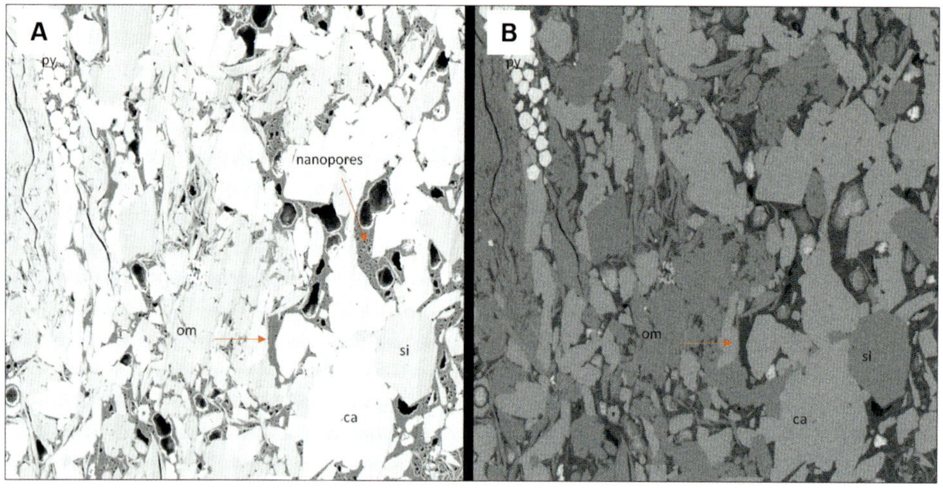

Figure 4. SE2 (a) and BSE1 (b) image of a cross section of a shale rock. Note that the contrast between carbonate (ca) and silica (si) grains is much higher in BSE1; the topographical information is greater in the SE2 image (OM-associated nanopores are not visible in BSE1).

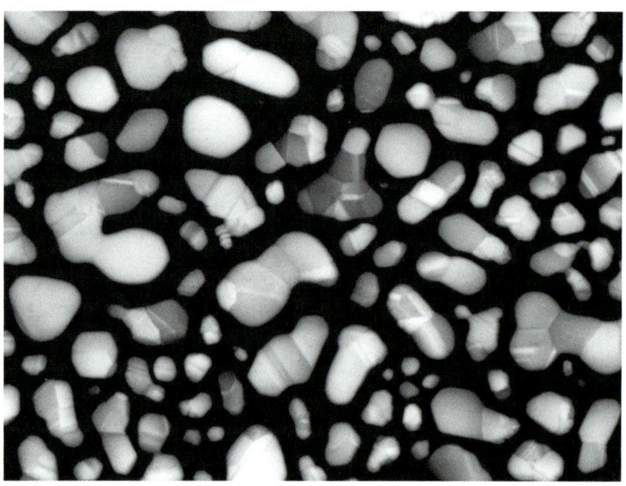

Figure 5. A BSE2 image of gold (Au) nanoparticles showing crystallographic contrast.

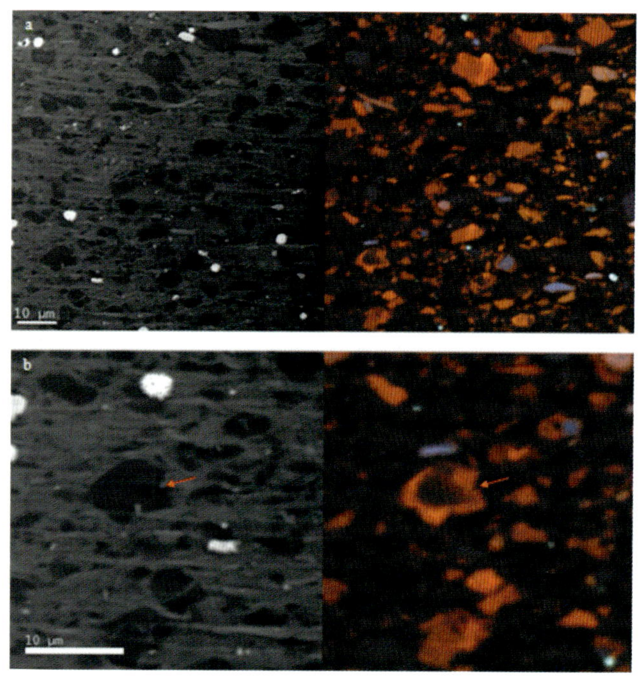

Figure 6. (a) A BSE and CL image of a polished shale sample. Orange-hued quartz grains reflect low-grade metamorphic origin (slate); blue-hued quartz grains indicate higher grade metamorphism (phyllite-schist). (b) Detail of a large quartz grain in center of image (arrow) displays multiple generations of growth in CL; this distinction in zoning is not visible in SEM.

gold (Au) nanoparticles. Contrast corresponding to different grains is revealed in the image despite all the chemically identical grains. Therefore, BSE2 electrons can be used to image crystallographic contrast in polycrystalline materials.

Another complimentary technique is the detection of CL, in which certain materials will emit photons in the form of visible light as a result of interactions between specimen electrons and primary beam electrons. Figure 6 shows an example of the effects of CL on a shale sample. The image was acquired with a dedicated CL detector. Variations in CL emission caused by mineral impurities could indicate provenance of individual quartz grains. Cathodoluminescence can also be used to differentiate between generations of quartz growth that are not distinguishable in SEM due to identical mean atomic number.

X-RAY MICROANALYSIS

Both the bremsstrahlung (continuum) and characteristics x-rays are produced by electron beam-sample interactions. The characteristic x-rays originate from the inner-shell ionization process, where an electron from the electron beam interacts with the tightly bound inner-shell electrons, ejecting an atomic electron and leaving a vacancy in that shell. The atom relaxes to its ground state through a limited set of transitions of outer-shell electrons to fill the inner-shell vacancy. The energies of the electrons in the shells are sharply defined, with values characteristic of the binding energies of the electrons; therefore, the characteristic x-rays are elemental specific, creating a powerful tool for elemental analysis in SEM. A typical x-ray spectrum is shown in Figure 7, which was acquired from a shale sample. In the spectrum, the broad, continuous background is the bremsstrahlung; the sharp peaks correspond to the elemental lines associated with specific intershell transitions.

To take advantage of the element-specific information caused by x-ray excitation, EDS (or EDX) can be performed. As the electron beam scans the sample surface pixel by pixel, a full x-ray spectrum can be acquired from each pixel. Elemental distribution can therefore be mapped using the relative peak intensity to build an image of the scanned area. This can then be interpreted to estimate mineral phase composition. Figure 8 shows a secondary electron image of a shale sample with EDS-derived mineral segmentation of the same region.

FOCUSED ION BEAM APPLICATIONS

Focused ion beam (FIB) systems also find a growing number of applications in geology (Goldstein et al., 2003). In a typical FIB-SEM system, an extraction field is applied to a gallium (Ga) liquid metal ion source to field emit Ga ions and form a Ga beam. Due to the higher atomic mass, the Ga beam not only can be used

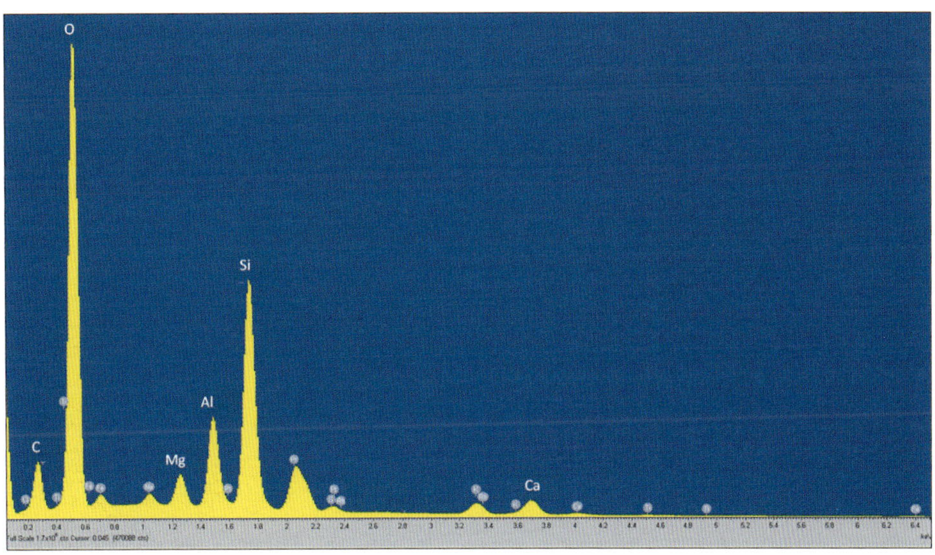

Figure 7. An example of an x-ray spectrum acquired from a shale sample. Individual peaks indicate an elevated concentration of a given element. C=carbon, O=oxygen, Mg=magnesium, Al=aluminum, Si=silicon, Ca=calcium.

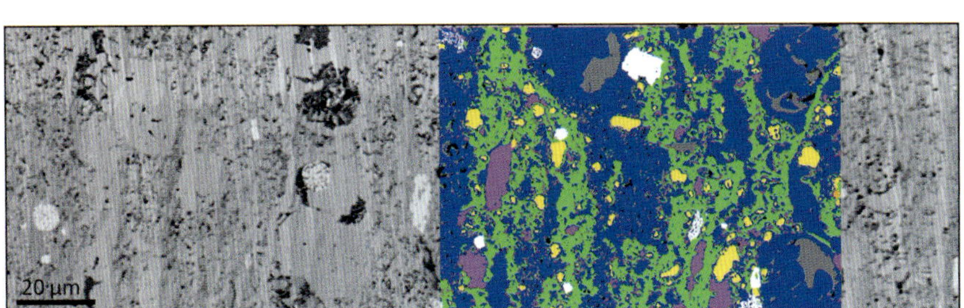

Figure 8. A secondary electron image of a shale sample with an EDS-derived mineral segmentation overlay. In the segmented region, blue = carbonate, green = clay minerals, yellow = quartz, pink = feldspar, white = pyrite, and gray = organic matter.

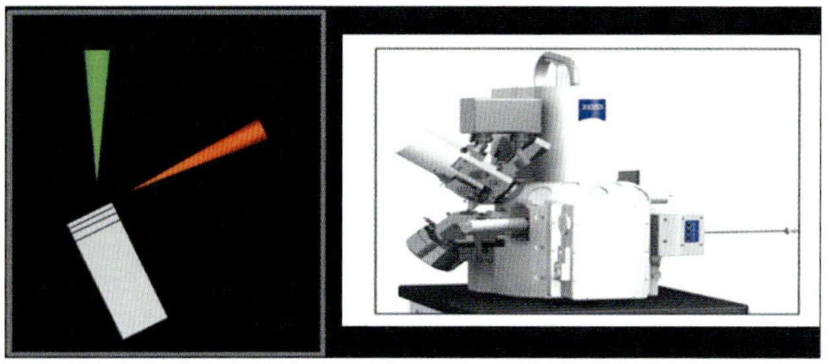

Figure 9. At left, a schematic diagram of operation of an FIB-SEM system. The FIB sputters away a thin layer of the sample at a time, while the electron beam/detector system captures an image of each newly exposed surface. At right, a picture of a commercial FIB-SEM CrossBeam™ system, Auriga.

to generate electron and ion images, but also may be used to mill samples to remove material. Figure 9a shows the schematic diagram of an FIB-SEM system where a cross section of the sample is milled by a Ga FIB beam and is imaged simultaneously by the SEM. This milling and imaging process can be automated in a serial fashion to acquire a stack of two-dimensional images, from which a 3-D image volume can be constructed from the data set. This technique is particularly useful in revealing the 3D distribution of mineral types, organic matter, porosity, and the like in shale (and other rock) samples.

CONCLUSIONS

Scanning electron microscopy provides different modes and techniques for acquiring high-quality images of shale and other rock samples. The images in this chapter demonstrate their fine resolution and their applicability for the characterization of shale reservoirs.

REFERENCE CITED

Goldstein, J., D.E. Newbury, D.C. Joy, C.E. Lyman, P. Echlin, E. Lifshin, L. Sawyer, and J.R. Michael, 2003, Scanning electron microscopy and X-ray microanalysis, 3d ed.: New York, Springer, 695 p.

Erdman, Natasha, and Nicholas Drenzek, 2013, Integrated preparation and imaging techniques for the microstructural and geochemical characterization of shale by scanning electron microscopy, *in* W. Camp, E. Diaz, and B. Wawak, eds., Electron microscopy of shale hydrocarbon reservoirs: AAPG Memoir 102, p. 7–14.

2

Integrated Preparation and Imaging Techniques for the Microstructural and Geochemical Characterization of Shale by Scanning Electron Microscopy

Natasha Erdman
JEOL USA Inc., 11 Dearborn Rd., Peabody, Massachusetts, 01960, U.S.A. (e-mail: erdman@jeol.com)

Nicholas Drenzek[1]
Statoil Research Center, Sandsliveien 90, 5020-Bergen Norway. (e-mail: ndrenzek@statoil.com)

ABSTRACT

To better understand the influence of microscale geochemical and microstructural relationships on the bulk petrophysical properties of unconventional shale systems, core samples from four producing North American formations were cross-sectioned with an argon ion polisher and imaged with a field emission scanning electron microscope (FE-SEM) using a variety of complementary detectors. We demonstrate distinct advantages of the ion-polishing technique for the preservation of the internal shale structure. Moreover, we show how such preparation affords a wider choice of imaging options for both chemical and structural characterization, such as backscatter electron observation at varying beam potentials coupled with x-ray and cathodoluminescence spectroscopic techniques.

INTRODUCTION

Gas and oil production from shale formations has become an increasing component of North American energy supply over the past ten years and may likewise rebalance the global energy portfolio in the coming decade. The fundamental microstructural and geochemical properties of shales differ markedly from traditional petroleum reservoirs, however, presenting significant technical challenges that today hinder full exploitation. Two of the more important differences lie in the small dimension and mixed wettability of their pore network (Passey et al., 2010), characteristics that themselves arise from the abundant micrometer- to nanometer-size clay mineral grains and organic matter macerals interspersed throughout more conventional carbonate or sandstone matrices. Not surprisingly, then, the application and interpretation of many conventional analysis techniques such as mercury intrusion porosimetry have proven to be more complex for

[1] *Previous address: Schlumberger-Doll Research, 1 Hampshire St., Cambridge, Massachusetts, 02139, U.S.A.*

DOI: 10.1306/13391700M1023581

Table 1. Alphabetical List of Common Acronyms Used in This Text

Acronym	Full Name
BSE	backscattered electron
CL	cathodoluminescence
CP	cross-section polisher
EA	elemental analysis
EDS	energy-dispersive spectroscopy
FE	field emission
PMT	photomultiplier tube
SDD	silicon drift detector
SE	secondary electron
SEM	scanning electron microscopy
TOC	total organic carbon
VR_O	vitrinite reflectance
VR_E	vitrinite reflectance equivalent
XRD	x-ray diffraction

such critical petrophysical data in organic matter- and clay mineral-rich shale reservoirs (e.g., Dacy, 2010).

Scanning electron microscopy (SEM; see Table 1 for a list of most of the acronyms used in this chapter), is well suited to study the complex distribution and association of organic, mineral, and pore phases at the nanoscale that govern multiphase fluid transport in shale formations. The application of on-board backscattered electron (BSE), energy dispersive spectroscopy (EDS), and cathodoluminescence (CL) detectors to ion cross-polished sample surfaces has opened an unparalleled window into such rock "fabric" (Ogura et al., 2007; Bustin and Bustin, 2008; Loucks et al., 2009; Wang and Reed, 2009; Curtis et al., 2010; Loucks et al., 2010; Milner et al., 2010; Sondergeld et al., 2010; Curtis et al., 2011a, b), which is in turn critical to describing the pore saturations, flow mechanisms, and production rates of fluids therein (O'Brien et al., 1994, 2002; Javadpour et al., 2007; Javadpour, 2009; Loucks et al., 2009; Ross and Bustin, 2009; Wang and Reed, 2009; Shabro et al., 2011; Sun et al., 2011). This chapter seeks to review the state of the art in such SEM preparation and analysis techniques for shales through a comparative case study of samples collected from several producing North American gas shale plays.

METHODS

Cores from the Marcellus, Haynesville, Woodford, and Barnett formations were collected, slabbed, subsampled into aliquots of millimeter-size fragments from 5-cm (1.9685-in.) depth intervals, split, and further pulverized for analysis (Table 2). Bulk density was measured using the Archimedes method. Values for porosity (ϕ) were acquired via gas intrusion (Handwerger et al., 2011) and total organic carbon (TOC) contents were measured by combustion elemental analysis (EA) following carbonate removal using the HCl fumigation method (Brodie et al., 2011). Thermal maturity T_{max} values were acquired by Rock-Eval pyrolysis (Behar et al., 2001), while vitrinite reflectance (VR_O), and bitumen reflectance (VR_E) measurements were obtained using standard organic petrology methods (Schieber, 2001). Mineralogy was measured by x-ray diffraction (XRD) (Ruessink and Harville, 1992).

Similar to established optical petrology techniques, several milligrams of pulverized material from the same aliquots used for bulk measurements were embedded in epoxy resin to mechanically stabilize them and polished using a broad argon ion beam cross-section polisher (CP) normal to the intended SEM imaging direction at 5 kV for 4 h. This workflow produced a typical cross-sectioned area of 1.5 × 0.5 mm with minimal smearing or deformation (Figure 1) (Erdman et al., 2006a, b; Ogura et al., 2007; Loucks et al., 2009; Sondergeld et al., 2010). Whereas any features of the native shale fabric beyond the 150-μm window were compromised through pulverization, those

Table 2. Bulk geophysical and geochemical characteristics for samples under study

Sample	ρ_B (g/cm^3)	ϕ	*TOC*	R_O (%)	T_{MAX} (°C)	*Quartz*	*Carbonate*†	*Clay*§	*Pyrite*
Marcellus	2.57	6.5	11	2.0	501	24	10	51	8
Haynesville	2.41	11.8	5	2.4*	512	27	34	29	4
Woodford	1.93	4.8	23	0.6	435	25	9	48	7
Barnett	2.47	3.9	5	0.6	441	58	38	4	1

†Calcite + dolomite + ankerite.
§Illite + smectite + kaolinite + chlorite + biotite + glauconite + muscovite.
*Solid pyrobitumen value.

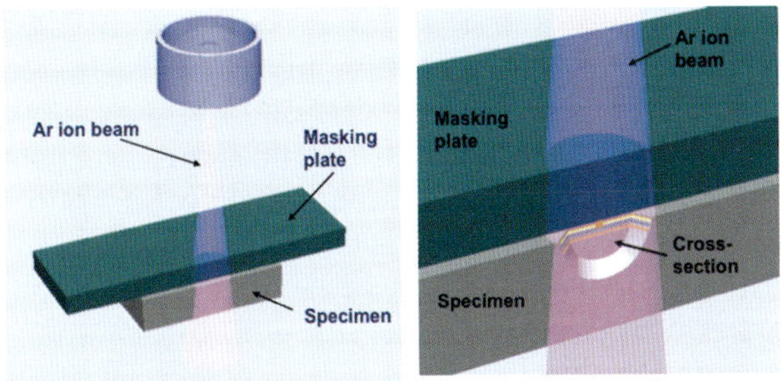

Figure 1. Principle of CP (cross-section polisher) operation. A specimen is masked so that a small portion (few tens of micrometers) protrudes from the mask. The exposed area is ablated with the argon ion beam and results in a specimen cross section with minimal artifacts.

viewed within the fragments less than 150 μm themselves should remain representative at this smaller scale. The resulting surfaces were then coated with less than a 100-Å thick elemental carbon film via vacuum evaporator to mitigate charge buildup and were examined with a Schottky field emission scanning electron microscope (FE-SEM) under high vacuum (10^{-4} Pa) using beam-accelerating voltages of 8 and 2 kV and a high-resolution, low-angle detector. Corresponding EDS elemental maps were then obtained using a 50-mm^2 silicon drift detector (SDD) at 8 kV for three frames (289 s per frame) with drift correction engaged. High-resolution CL and companion grayscale BSE images were also acquired using a FE-SEM system with a KE Centaurus CL detector containing a photomultiplier tube (PMT) set to detect photons in the 185- to 850-nm range.

RESULTS AND DISCUSSION

Cross-section polishing is an indispensable advancement for fine-scale SEM investigations of the distributions and associations of organic matter, minerals, and pores in highly consolidated geologic samples. Because major shale constituents such as organic matter, clay minerals, quartz, carbonate, and pyrite exhibit various degrees of hardness, traditional mechanical preparation (saw cutting, grinding, abrasive polishing, microtome cutting, etc.) tends to result in uneven surfaces as a result of smearing and/or pitting. Figure 2 shows an example of a Marcellus shale specimen polished with silicon carbide 1200-grit paper versus the same sample after CP preparation. The comparison clearly demonstrates that inappropriate sample preparation will confound subsequent attempts to image the exact locations of organic matter macerals and mineral grains along with both their internal and interfacial pore networks. In a study comparing native versus CP prepared surfaces, Milner et al. (2010) likewise concluded that such assessments of shale fabric are impossible without CP pretreatment.

Figure 3 shows relatively low magnification overview BSE images of CP prepared samples from the four formations investigated here. Backscatter images tend to provide better visualization of the various components of shale samples for two reasons: (1) the polished samples are very flat, thus providing minimal topography and relief-based contrast on which secondary electron (SE) images rely for interpretation, and (2) the BSE signal contrast is indicative of compositional differences—bright and dark areas indicate high and low atomic number materials, respectively. For example, kerogen and bitumen show up as dark gray regions while their chemically and spatially associated pyrite crystals are bright white, therein affording a means to readily differentiate these low and high atomic number components from the rest of the complex, intermediate atomic number mineral milieu (e.g., Figure 3). Nonetheless, the lower grayscale contrast among these other constituents and the overall compositional heterogeneity of shale specimens present a significant challenge to BSE studies of shale geochemical and microstructural relationships alone. Although the advent of EDS elemental mapping has

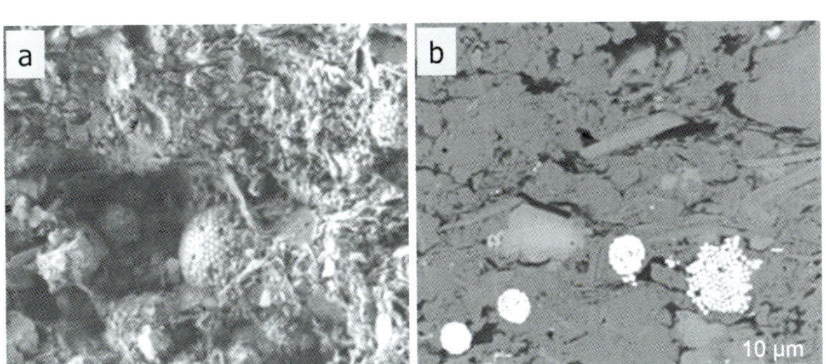

Figure 2. Field emission scanning electron microscope (FE-SEM) images of Marcellus shale specimen before (left) and after (right) ion beam milling. The image on the left was taken with an Everhart-Thornley secondary electron detector and shows a substantial amount of relief associated with mechanical abrasion. The image on the right shows a smooth, flat surface following ion milling.

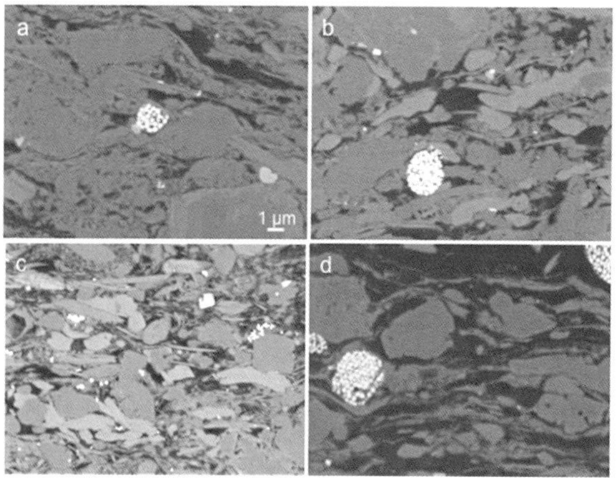

Figure 3. Backscatter field emission scanning electron microscope (FE-SEM) images of the specimens from four different formations: (a) Barnett, (b) Marcellus, (c) Haynesville, and (d) Woodford. All images were taken at 8 kV.

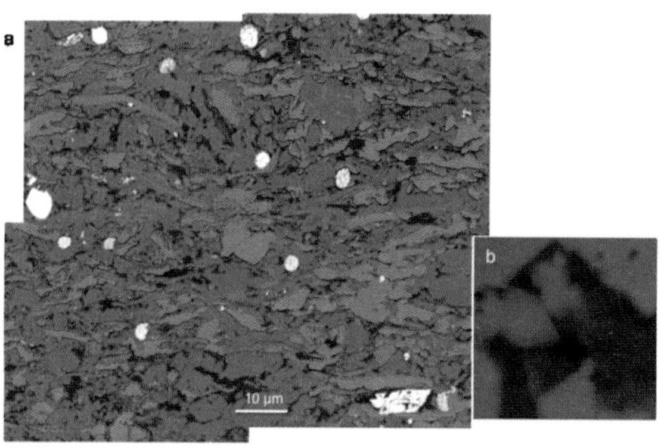

Figure 4. (a) Montage of 2 × 2 image tiles and corresponding energy-dispersive spectroscopy (EDS) maps of calcium (pink) distribution in the Marcellus shale. Both the backscattered electron (BSE) images and the EDS maps were acquired with high pixel resolution, so that the images can be later examined at higher magnification without loss of resolution. Data acquired at 8 kV using 50-mm^2 Oxford silicon drift detector (SDD). (b) A 2.5-μm^2 area zoom-in of BSE image showing organic matter (OM) and mineral porosity.

helped to further define these relationships (Figure 4), the low magnification required to yield a representative sampling area for compositional analysis obscures the corresponding nanoscale organic and mineral pore networks thought responsible for most petroleum storage and flow in shales. The user must therefore devise an approach that would combine these seemingly diverging needs into a single technique.

The acquisition of companion sets of high-pixel-resolution BSE and EDS images at low magnification is one such solution. For instance, low-magnification (e.g., ×2500) BSE images can be collected at 5000 × 4000 (5 × 4K) pixel resolution, allowing the image to be expanded to effectively ×12,500 without any significant loss of pixel resolution. The same can be done with EDS, wherein maps can be collected at up to 4 × 4K resolution. These data sets can then be integrated to simultaneously yield a broad overview of shale geochemical composition along with information on associated microstructural features. Alternatively, a series of neighboring high-magnification, high-pixel-resolution images and EDS maps can be acquired in automated fashion using current FE-SEM software and subsequently combined into a large-scale montage (see Ogura et al., 2010, for a case study using brain tissue). An example of 2 × 2 tiled BS montage images collected at 5 × 4K resolution (46 × 15 μm viewing area) for the Marcellus shale is shown in Figure 4a, along with corresponding EDS Ca Kα spectral overlays documenting the locations of calcite, dolomite, and aragonite. Figure 4b demonstrates how a 2.5-µm^2 portion of the same montage can be used to examine organic matter- and mineral-hosted porosity in more detail without any loss of resolution. This approach can be further utilized with more rigorous BSE/EDS segmentation algorithms to relate microscale compositional and structural features of shales to their macroscale petrophysical manifestations.

Using the methods described, some general microscale compositional and structural features can be delineated in each of the shale formations studied here. The siliceous Barnett, for example, displays abundant organic-matter-hosted pores with less mineral porosity, similar to previous observations (Loucks et al., 2009; Wang and Reed, 2009; Loucks et al., 2010; Milner et al., 2010; Sondergeld et al., 2010), whereas micrometer- to nanometer-size pores in the argillaceous Marcellus sample appear to be predominantly associated with organic-clay mineral interfaces as likewise reported by previous studies (Milner et al., 2010; Curtis et al., 2011a). These mixed-porosity systems are bracketed by the behavior of the calcareous Haynesville, for which the high level of thermal maturation (T_{max} = 512°C, VR_E = 2.4%) has led to a greater abundance of larger organic matter-hosted pores; and of the carbonaceous Woodford shale, in which a pore network has yet to clearly develop in its abundant yet thermally immature (T_{max} = 435°C, VR_E = 0.6%) organic matter inventory. Indeed, mineral-based pores (perhaps volumetrically dominated by diagenetic pyrite) may represent the only viable storage and migration pathway for oil and wet gas production from relatively low-maturity shale plays or sections of a play. In any case, these microscale descriptions are qualitatively consistent with bulk compositional and

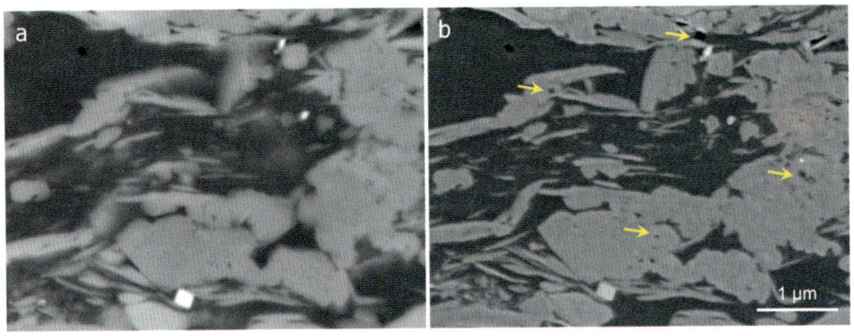

Figure 5. Comparison between (a) 8-kV and (b) 2-kV beam accelerating voltage observations of the Woodford shale specimen acquired across the same field of view. Arrows in (b) highlight pores that were previously obscured by a larger depth of investigation at 8 kV.

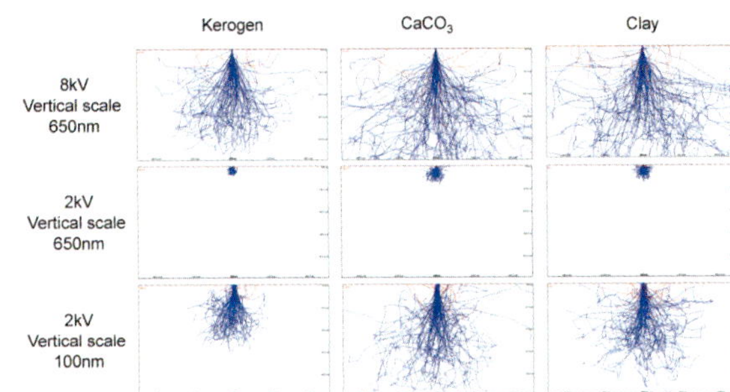

Figure 6. Monte Carlo simulations (CASINO®) of secondary (blue) and backscattered (red) electron beam-specimen interactions in three components of shale (organic matter [OM], calcite, and clay minerals) at both high (8-kV) and low (2-kV) beam accelerating voltages. Note greater interaction volume at higher voltage. $CaCO_3$ = calcium carbonate.

petrophysical properties measured on macroscale aliquots of the same material (Table 2). X-ray diffraction analysis, for example, indicates predominantly quartz, clay mineral, carbonate, and organic matter contents for the Barnett, Marcellus, Haynesville, and Woodford samples, respectively. In fact, the very high (23 wt%) *TOC* content of the Woodford as both measured by combustion EA and reflected at the SEM scale (Figure 3) is further expressed in a marked 0.5-g cm^{-3} depression in bulk density relative to its lower *TOC* counterparts (Table 2). Organic matter maturation also appears to play a dominant role in setting the comparatively high porosity of 11.8% measured for the Haynesville shale.

Given the reputed importance of organic pore networks for petroleum storage and flow in shale reservoirs (O'Brien et al., 1994, 2002; Javadpour, 2009; Loucks et al., 2009; Wang and Reed, 2009; Shabro et al., 2011), SEM has become an increasingly important tool to investigate their basic structure and distribution. Even so, their small size has typically demanded high accelerating voltages (i.e., > 15 kV) coupled with BSE detection to achieve the nanometer-size feature resolution needed for accurate characterization. Yet the large volume of interaction (up to several cubic micrometers) produced at high acceleration voltages along with the low BSE grayscale contrast between these pores and the surrounding organic matter matrix can lead to subjective segmentation workflows. New developments in FE-SEM technology (e.g., Erdman et al., 2009, 2010) allow the user to maintain single-nanometer spatial resolution while imaging the specimen and performing microanalysis at lower accelerating voltages to reduce the interaction volume and improve surface contrast. Figure 5 shows a comparison of BSE images taken at 8 and 2 kV for the same Woodford shale specimen. The arrows point to features observed at 8 kV that completely disappear at 2 kV following reduction in the beam-specimen interaction volume. This behavior can also be modeled by utilizing the Monte Carlo simulations (Hovington et al., 1997) of beam interaction with carbon, silicon, and calcium (as rough proxies for organic matter, quartz, and calcite, respectively) at 8 and 2 kV (Figure 6).

Utilization of low beam voltages also reduces radiation damage to the shale structure itself. For example, our experience demonstrated that morphologically distinct calcite fragments have been shown to be susceptible to degradation under a 5-kV beam but encumber no apparent alteration at only 2 kV. Nonetheless, companion EDS mapping of all major elemental constituents of shale (i.e., silicon, calcium, aluminum, magnesium, potassium, sodium, iron, sulfur, oxygen, and carbon) still requires the collection of x-ray spectra across energies of up to approximately 8 kV depending on whether K or L x-ray lines are used for identification and quantification. For this application, care should thus be taken to minimize the amount of time shale samples are exposed to high beam currents and accelerating voltages by employing a large, high-efficiency SDD.

Complimentary information on shale elemental composition can sometimes be obtained at lower voltages and superior resolution using SEM-mounted CL detectors. In CL, the interaction of the primary electron beam with the specimen results in the promotion

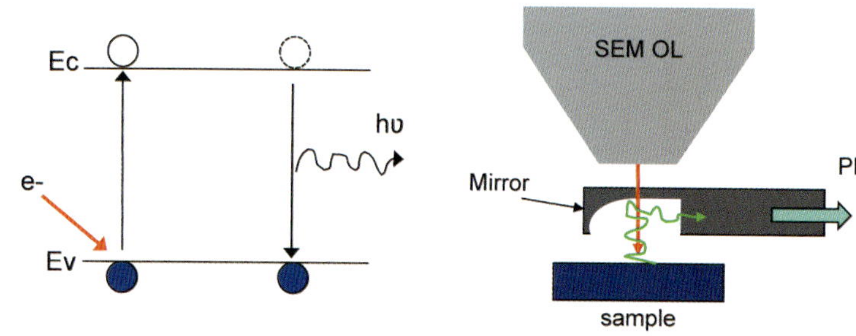

Figure 7. KE Centaurus cathodoluminescence (CL) images (left), overlaid backscattered electron (BSE) images-EDS (energy-dispersive spectroscopy) maps (center), and remaining EDS surveys (right) of the Woodford sample, all acquired at 5-kV accelerating voltage.

of valence band electrons to the conduction band. Subsequent relaxation emits photons whose energies are characteristic of particular elemental constituents or mineral defects. These photons are then collected by an elliptical mirror, separated by a monochromator (if present), and detected by a PMT to render composite grayscale or spectrally resolved compositional information across a sample surface. Similar to EDS, the use of CL can therefore aid in the differentiation between various materials of different CL characteristics but similar electron backscatter coefficient (i.e., minimal difference in backscatter image contrast) (Erdman et al., 2009, 2010; Sondergeld et al., 2010). For example, the calcite, dolomite, and quartz constituents of shale studied here have distinctly bright CL signatures relative to less optically active components like clay minerals or organic matter. Moreover, CL has been previously demonstrated to be capable of discriminating finely crystalline, low-permeability dolomites from coarser, higher-permeability morphotypes as well as primary from authigenic quartz (Boggs and Krinsley, 2006). With this in mind, combined application of CL and EDS technology (Figure 7) promises to further uncover the microscale geochemical-microstructural relationships that underpin macroscale petrophysical properties of unconventional shale systems.

SUMMARY AND PERSPECTIVE

The addition of CP preparation to standard shale SEM imaging workflows over the past several years has revolutionized the microscale characterization of shale fabric (i.e., the distribution and association of organic, mineral, and pore constituents) that is likely to strongly influence petroleum storage and flow therein. The concomitant integration of more diverse yet complimentary detection systems such as CL and EDS along with the optimization of traditional detectors (SE and BSE) has lent further insight into the fundamental geochemical controls on microstructural properties in shales, not the least of which is the development of organic-matter-hosted nanopores on thermal maturation. Although focused ion milling combined with ever-more-sophisticated image alignment/segmentation software has garnered more recent attention because of its ability to render true three-dimensional (3-D) reconstructions of shale matrices (e.g., Elgmati et al., 2011a, b), the nanometer- to micrometer-level areas of investigation to which this technique is confined (along with the high cost of acquisition) has so far limited its widespread use in shale research. Future developments should therefore focus on comparing the upscaling potential of two-dimensional versus 3-D microscale information to core or wellbore log measurements as well as integrating SEM instrumentation with dynamic experiments such as hydraulic fracturing, multiphase fluid imbibition, and pyrolytic artificial maturation.

ACKNOWLEDGMENTS

We thank the TerraTek core analysis team for providing the bulk porosity, density, mineralogy, and vitrinite reflectance data along with the SDR Petrolabs staff for additional sample preparation. Maria Mastalerz of Indiana University generously supplied the bitumen reflectance value for the Haynesville sample. Regina Campbell of JEOL is also acknowledged for help with SEM sample preparation.

REFERENCES CITED

Behar, F., V. Beaumont, and H. L. De B. Penteado, 2001, Rock-Eval 6 technology: Performances and developments: Oil & Gas Science and Technology, v. 56, no. 2, p. 111–134.

Boggs, S., and D. Krinsley, 2006, Application of cathodoluminescence imaging to the study of sedimentary rocks: Cambridge, Cambridge University Press, 176 p.

Brodie, C. R., M. J. Leng, J. S. L. Casford, C. P. Kendrick, J. M. Lloyd, Z. Yongqiang, and M. I. Bird, 2011, Evidence for bias in C and N concentrations and $\delta^{13}C$ composition of terrestrial and aquatic organic materials due to pre-analysis acid preparation methods: Chemical Geology, v. 282, p. 67–83.

Bustin, A. M. M., and R. M. Bustin, 2008, Importance of fabric on the production of gas shales: Society of Petroleum Engineers Unconventional Reservoirs Conference, February 10–12, Keystone, Colorado, SPE Paper 114167, 29 p., doi: 10.2118/114167-MS.

Curtis, M. E., R. J. Ambrose, C. H. Sondergeld, and C. S. Rai, 2010, Structural characterization of gas shales on the micro- and nano-scales: Canadian Society for Unconventional Gas/Society of Petroleum Engineers Canadian Unconventional Resources and International Petroleum Conference, October 19–21, Calgary, Alberta, SPE Paper 137693, 15 p., doi: 10.2118/137693-MS.

Curtis, M. E., R. J. Ambrose, C. H. Sondergeld, and C. S. Rai, 2011a, Investigation of the relationship between organic porosity and thermal maturity in the Marcellus shale: Society of Petroleum Engineers North American Unconventional Gas Conference and Exhibition, June 14–16, The Woodlands, Texas, SPE Paper 144370, 4 p., doi: 10.2118/144370-MS.

Curtis, M. E., R. J. Ambrose, C. H. Sondergeld, and C. S. Rai, 2011b, Transmission and scanning electron microscopy investigation of pore connectivity of gas shales on the nanoscale: Society of Petroleum Engineers North American Unconventional Gas Conference and Exhibition, June 14–16, The Woodlands, Texas, SPE Paper 144391, 10 p., doi: 10.2118/144391-MS.

Dacy, J. M., 2010, Core tests for relative permeability of unconventional gas reservoirs: Society of Petroleum Engineers Annual Technical Conference and Exhibition, September 19–22, Florence, Italy, SPE Paper 135427, 18 p., doi: 10.2118/135427-MS.

Elgmati, M., H. Zhang, B. Bai, and R. Flori, 2011a, Submicron-pore characterization of shale gas plays: Society of Petroleum Engineers North American Unconventional Gas Conference and Exhibition, June 14–16, The Woodlands, Texas, SPE Paper 144050, 19 p., doi: 10.2118/144050-MS.

Elgmati, M., M. Zobaa, H. Zhang, B. Bai, and F. Oboh-Ikuenobe, 2011b, Palynofacies analysis and submicron pore modeling of shale-gas plays: Society of Petroleum Engineers North American Unconventional Gas Conference and Exhibition, June 14–16, The Woodlands, Texas, SPE Paper 144267, 10 p., doi: 10.2118/144267-MS.

Erdman, N., R. Campbell, and S. Asahina, 2006a, Argon beam cross sectioning: Advanced Materials and Processes, v. 164, no. 6, p. 33–35.

Erdman, N., R. Campbell, and S. Asahina, 2006b, Precise SEM cross section polishing via argon beam milling: Microscopy Today, v. 14, no. 3, p. 22–25.

Erdman, N., N. Kikuchi, A. Laudate, and V. Robertson, 2009, Multispectral imaging in a FEG-SEM: Advanced Materials and Processes, v. 167, no. 9, p. 28–31.

Erdman, N., N. Kikuchi, R. Campbell, and V. E. Robertson, 2010, SEM technology advances energy research: Advanced Materials and Processes, v. 168, no. 2, p. 14–15.

Handwerger, D. A., R. Suarez-Rivera, K. I. Vaughn, and J. F. Keller, 2011, Improved petrophysical core measurements on tight shale reservoirs using retort and crushed samples: Society of Petroleum Engineers Annual Technical Conference and Exhibition, October 30–November 2, Denver, Colorado, SPE Paper 147456, 12 p., doi: 10.2118/147456-MS.

Hovington, P., D. Drouin, and R. Gauvin, 1997, CASINO: A new era of Monte Carlo code in C language for the electron beam interaction—Part I: Description of the programme: Scanning, v. 19, p. 1–14.

Javadpour, F., D. Fisher, and M. Unsworth, 2007, Nanoscale gas flow in shale sediments: Journal of Canadian Petroleum Technology, v. 46, no. 10, p. 55–61.

Javadpour, F., 2009, Nanopores and apparent permeability of gas flow in mudrocks (shales and siltstone): Journal of Canadian Petroleum Technology, v. 48, no. 8, p. 16–21.

Loucks, R. G., R. M. Reed, S. C. Ruppel, and D. M. Jarvie, 2009, Morphology, genesis, and distribution of nanometer-scale pores in siliceous mudstones of the Mississippian Barnett shale: Journal of Sedimentary Research, v. 79, p. 848–861.

Loucks, R. G., R. M. Reed, S. C. Ruppel, and U. Hammes, 2010, Preliminary classification of matrix pores in mudrocks: Gulf Coast Association of Geological Societies Transactions, v. 60, p. 435–441.

Milner, M., R. McLin, and J. Petriello, 2010, Imaging texture and porosity in mudstones and shales: Comparison of secondary and ion-milled backscatter SEM methods: Canadian Society for Unconventional Gas/Society of Petroleum Engineers Canadian Unconventional Resources and International Petroleum Conference, October 19–21, Calgary, Alberta, SPE Paper 138975, 10 p., doi: 10.2118/138975-MS.

O'Brien, N. R., R. M. Slatt, and J. Senftle, 1994, The significance of oil shale fabric in primary hydrocarbon migration: Fuel, v. 73, no. 9, p. 1518–1522.

O'Brien, N. R., M. D. Cremer, and D. G. Canales, 2002, The role of argillaceous rock fabric in primary migration of oil: Gulf Coast Association of Geological Societies Transactions, v. 52, p. 1103–1112.

Ogura, K., M. Kamidaira, S. Asahina, and N. Erdman, 2007, New methods for cross-section sample preparation using broad argon ion beam: Microscopy and Microanalysis, v. 13, p. 1518–1519.

Ogura, K., M. Yamada, O. Hirahara, M. Mita, N. Erdman, and C. Nielsen, 2010, Gigantic montages with a fully automated FE-SEM (serial sections of a mouse brain tissue): Microscopy and Microanalysis, v. 16, p. 52–53.

Passey, Q. R., W. L. Bohacs, R. Esch, R. Klimentidis, and S. Singha, 2010, From oil-prone source rock to gas-producing shale reservoir—Geologic and petrophysical characterization of unconventional shale-gas reservoirs: Society of Petroleum Engineers International Oil and Gas Conference and Exhibition in China, June 8–10, Beijing, China, SPE Paper 131350, 29 p., doi: 10.2118/131350-MS.

Ross, D. K. J., and R. M. Bustin, 2009, The importance of shale composition and pore structure upon gas storage potential of shale gas reservoirs: Marine and Petroleum Geology, v. 26, p. 916–927.

Ruessink, B. H., and D. G. Harville, 1992, Quantitative analysis of bulk mineralogy: The applicability and performance of XRD and FTIR: Society of Petroleum Engineers Formation Damage Control Symposium, February 26–27, Lafayette, Louisiana, SPE Paper 23828, 14 p., doi: 10.2118/23828-MS.

Schieber, J., 2001, A role for organic petrology in integrated studies of mudrocks: Examples from Devonian black shales of the eastern U.S.: International Journal of Coal Geology, v. 47, p. 171–187.

Shabro, V., C. Torres-Verdin, and F. Javadpour, 2011, Numerical simulation of shale-gas production: From pore-scale modeling of slip-flow, Knudsen diffusion, and Langmuir desorption to reservoir modeling of compressible fluid: Society of Petroleum Engineers North American Unconventional Gas Conference and Exhibition, June 14–16, The Woodlands, Texas, SPE Paper 144355, 11 p., doi: 10.2118/144355-MS.

Sondergeld, C. H., R. J. Ambrose, C. S. Rai, and J. Moncrieff, 2010, Micro-structural studies of gas shales: Society of Petroleum Engineers Unconventional Gas Conference, February 23–25, Pittsburgh, Pennsylvania, SPE Paper 131771, 17 p., doi: 10.2118/131771-MS.

Sun, W., J. E. Andrade, J. W. Rudnicki, and P. Eichhubl, 2011, Connecting microstructural attributes and permeability from 3D tomographic images of in situ shear-enhanced compaction bands using multiscale computations: Geophysical Research Letters, v. 38, doi: 10.1029/2011GL047683.

Wang, F. P., and R. Reed, 2009, Pore networks and fluid flow in gas shales: Society of Petroleum Engineers Annual Technical Conference and Exhibition, October 4–7, New Orleans, Louisiana, SPE Paper 124253, 8 p., doi: 10.2118/124253-MS.

3

Wayne K. Camp, Barry Wawak, 2013, Enhancing SEM grayscale images through pseudocolor conversion: Examples from Eagle Ford, Haynesville, and Marcellus shales, *in* W. Camp, E. Diaz, and B. Wawak, eds., Electron microscopy of shale hydrocarbon reservoirs: AAPG Memoir 102, p. 15–26.

Enhancing SEM Grayscale Images through Pseudocolor Conversion: Examples from Eagle Ford, Haynesville, and Marcellus Shales

Wayne K. Camp

Anadarko Petroleum Corporation, 1201 Lake Robbins Dr., The Woodlands, Texas, 77380, U.S.A. (e-mail: wayne.camp@anadarko.com)

Barry Wawak

Core Laboratories, 6316 Windfern Rd., Houston, Texas, 77040, U.S.A. (e-mail: barry.wawak@corelab.com)

ABSTRACT

The ability to visualize and interpret standard grayscale scanning electron microscopic (SEM) images is limited by image contrast, magnification level, and the capability of the human eye to distinguish between subtle shades of gray. A pseudocolor conversion technique is described using the freely available ImageJ digital image processing and analysis program that can improve the ability to interpret and communicate information contained within SEM images of shales. This technique is most suitable for backscattered electron images to enhance contrast variations and aid compositional interpretations, as illustrated by examples from the Eagle Ford, Haynesville, and Marcellus shales.

INTRODUCTION

The scanning electron microscope (SEM) produces images by detecting various signals emitted from a sample in response to an accelerated electron beam. The sample is scanned in a raster pattern (rectangular grid) across its surface. Typical signals used for SEM imaging include secondary electrons, backscattered electrons, x-rays, and photons (cathodoluminescence). Secondary electron and backscattered electron images are routinely collected for SEM analysis of shale samples because they (1) are rapid and inexpensive to obtain; (2) produce high-resolution images at sufficiently high magnification levels necessary to image clay-size particles; and (3) provide information about surface topography and composition useful for sedimentological, petrographic, and paleontological studies (Reed, 2010).

Digital SEM images are produced by assigning grayscale values to reflect signal intensity variations measured at each scanned raster location by the various signal detectors. Low-intensity values thus appear dark (black), and high-intensity values appear light (white). The SEM operator can manipulate the measured signal

DOI: 10.1306/13391701M1021681

brightness and contrast to produce a suitable image. The grayscale values representing a given signal intensity value are therefore not unique but may vary between images based on the dynamic range of signal intensity measured within the scanned area.

The ability to visualize and interpret various features in standard grayscale SEM images is limited by image contrast, magnification level, and the capability of the human eye to distinguish between subtle shades of gray. The purpose of this chapter is to illustrate the advantage of enhancing grayscale SEM images by pseudocolor conversion to improve the ability to interpret and communicate information contained within SEM images of shales.

The SEM images illustrated in this chapter were obtained from drill core samples using an FEI Quanta model 250 field emission SEM on uncoated, freshly broken (rough), and argon ion-milled (smoothed) surfaces prepared with a Leica EM TIC 020 argon ion mill. Secondary electron and backscattered electron images were obtained at low electron beam energy (10–15 keV) operating in a high-pressure vacuum chamber environment at 59–62 Pa (0.009 psi). An energy-dispersive x-ray spectrometer, which is sensitive to elements with atomic weights greater than beryllium (>9), was used to identify the mineral composition at selected target areas.

Digital images were acquired simultaneously to produce secondary electron, backscattered electron, and superimposed backscattered and secondary electron images, recorded as 8-bit grayscale images in Tagged Image File Format (TIFF). The digital grayscale images were converted to color using the ImageJ image-processing and analysis program v 1.45q (released October 4, 2011) on a Macintosh MacBook Pro laptop computer run by Mac OS X software.

Secondary Electron Images

Secondary electrons are produced principally by inelastic collisions of primary beam electrons with atoms near the sample surface, which cause the primary electrons to lose energy. By convention, detected electrons with less than 50 eV are considered secondary electrons. The number of secondary electrons that escape the surface (secondary electron yield) is a function of the orientation of the specimen surface in relation to the electron beam and secondary electron detector, varying only slightly because of compositional variations (mean atomic number), except for carbon and gold (Goldstein et al., 2003). This property is used to create highly detailed topographic images of three-dimensional (3-D) surfaces useful for analyzing the fabric, texture, and morphology of microfossils (e.g., coccoliths), grains, cements, and matrix material (Figure 1a).

Secondary electron images can also be obtained from smoothly polished or ion-milled surfaces to aid in imaging near-surface negative aberrations such as small pores (Figure 1b) and cracks, which are revealed as low-intensity (dark) areas in the secondary electron images. Because surface topography is essentially eliminated by ion milling, compositional variations can also affect secondary electron yield at the low electron beam energies typically used for SEM imaging of shale samples, most notably by organic matter and iron sulfides (Figure 1b). Higher secondary electron yields are produced from materials with higher mean atomic number (density) and are represented by brighter (whiter) grayscale values.

Grayscale secondary electron images are generally adequate for interpreting surface morphology and texture of 3-D (rough, broken) surfaces (Figure 1b). Pseudocolor conversions may aid in interpreting rugose sample surfaces in a manner similar to topography displayed by color relief maps and serve to highlight subtle compositional variations in polished or ion-milled specimens.

Backscattered Electron Images

Backscattered electrons are high-energy electrons (>50 eV) that result from multiple elastic collisions of the incident primary beam electrons with the nucleus of atoms within the sample. These electrons then exit the surface, where they are measured by a backscattered electron detector. The measured backscatter intensity is proportional to the mean atomic number of atoms in the volume sampled by the electron beam. The proportion of incident electrons exiting the sample as backscattered electrons (backscatter electron coefficient) increases with mean atomic number because there is a greater probability of elastic scattering with larger atoms (Goldstein et al., 2003). Backscattered electron images therefore provide a measure of atomic number contrast reflecting variations in elemental composition that can be useful for mineral identification.

Variations in backscatter intensity are also a function of the angle of the incident electron beam, surface roughness, applied coatings, and surface contamination (Goldstein et al., 2003). To obtain useful backscattered electron images, sample surfaces must be smoothed by careful polishing or by argon-ion or gallium-ion beam (focused ion beam) milling to eliminate backscatter signal strength variations caused by rugose surfaces.

Minerals or compounds with low mean atomic number (or low density), such as organic matter, appear dark (black), and minerals with high mean atomic number (or high density), such as pyrite, appear light (white); these are easily distinguished from each other

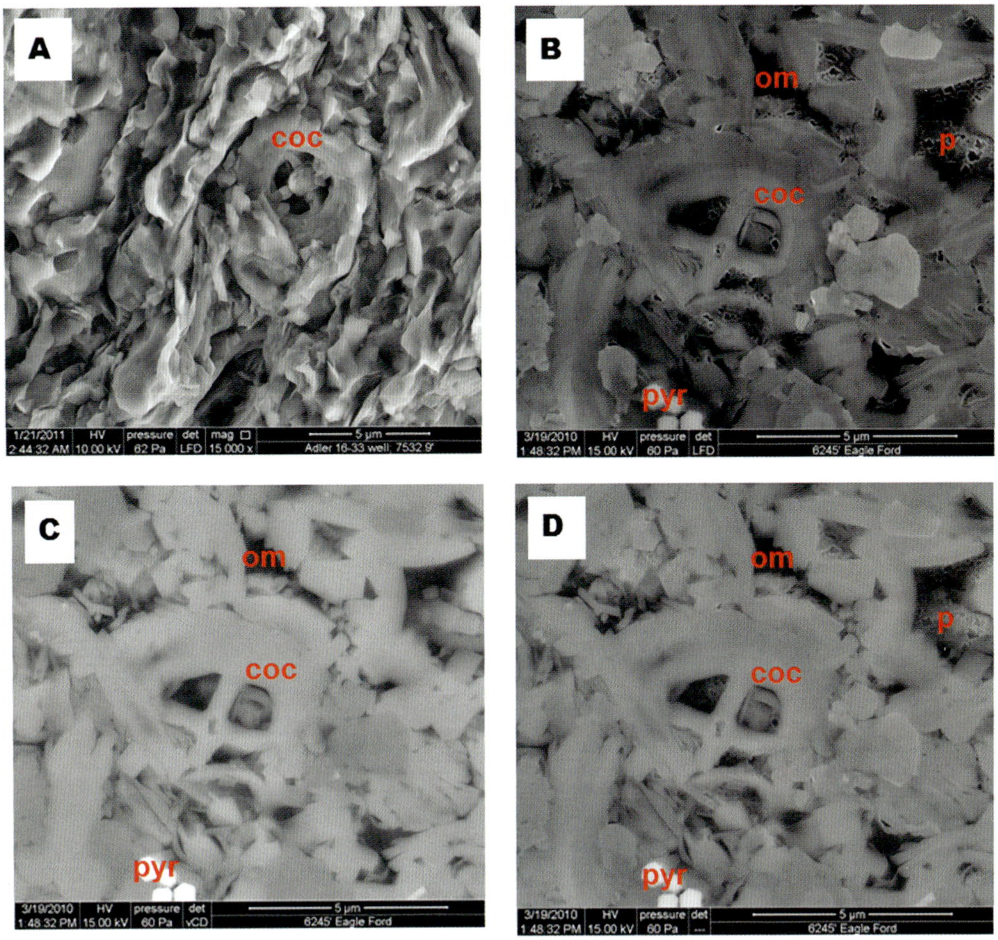

Figure 1. SEM images of a calcareous coccolith fossil at the same magnification level illustrating the different types of images produced by different surface preparations and signal detection modes: (A) secondary electron image of a freshly broken surface; (B) secondary electron image of an argon-ion-milled surface; (C) backscattered electron image of an argon-ion-milled surface; (D) superimposed secondary and backscattered electron images. Images B–D show identical areas. Coc = coccolith, om = organic matter, p = pore, pyr = pyrite.

in backscattered electron images of shales because of their high backscatter electron contrast (Figure 1c). Other minerals common in shales, such as quartz, calcite, dolomite, feldspar, and clays, exhibit low backscatter contrast, generally making it difficult to distinguish these minerals in grayscale backscattered electron images. Interpretations of mineral composition in shale in samples exhibiting low backscattered electron image contrast can be improved by pseudocolor conversion.

PSEUDOCOLOR CONVERSION

The pseudocolor conversions described in this chapter were done using the ImageJ program v 1.45q (October 4, 2011) on a Macintosh MacBook Pro laptop computer with Mac OS X operating system. ImageJ is an image-processing and analysis program originally developed by the U.S. National Institutes of Health for the medical sciences. The program is an open-source, public domain software available for free from the ImageJ Web site (http://rsbweb.nih.gov/ij/download.html). The software is written in Java and can be run on Windows, Macintosh OS X, and Linux operating platforms. Instructions for using ImageJ software for pseudocolor conversion of grayscale images is provided in the appendix of this chapter. Pseudocolor conversions may also be accomplished using commercial programs, such as Adobe Photoshop.

All secondary electron and backscattered electron SEM images are inherently grayscale images reflecting intensity variations of a single signal type. Pseudocolor images are created by assigning specific colors to each grayscale value using a color lookup table, or LUT. The differences in color thus reflect differences in SEM signal intensity instead of the actual color of the sample, hence the terms *pseudocolor* and *false color* are used.

Lookup Tables

A LUT is a table of 256 gray values (for 8-bit images) with corresponding red, green, and blue color values. When a given LUT is applied to a grayscale image, the various grayscale pixels are converted to color, creating a color image that preserves the fidelity of the grayscale image.

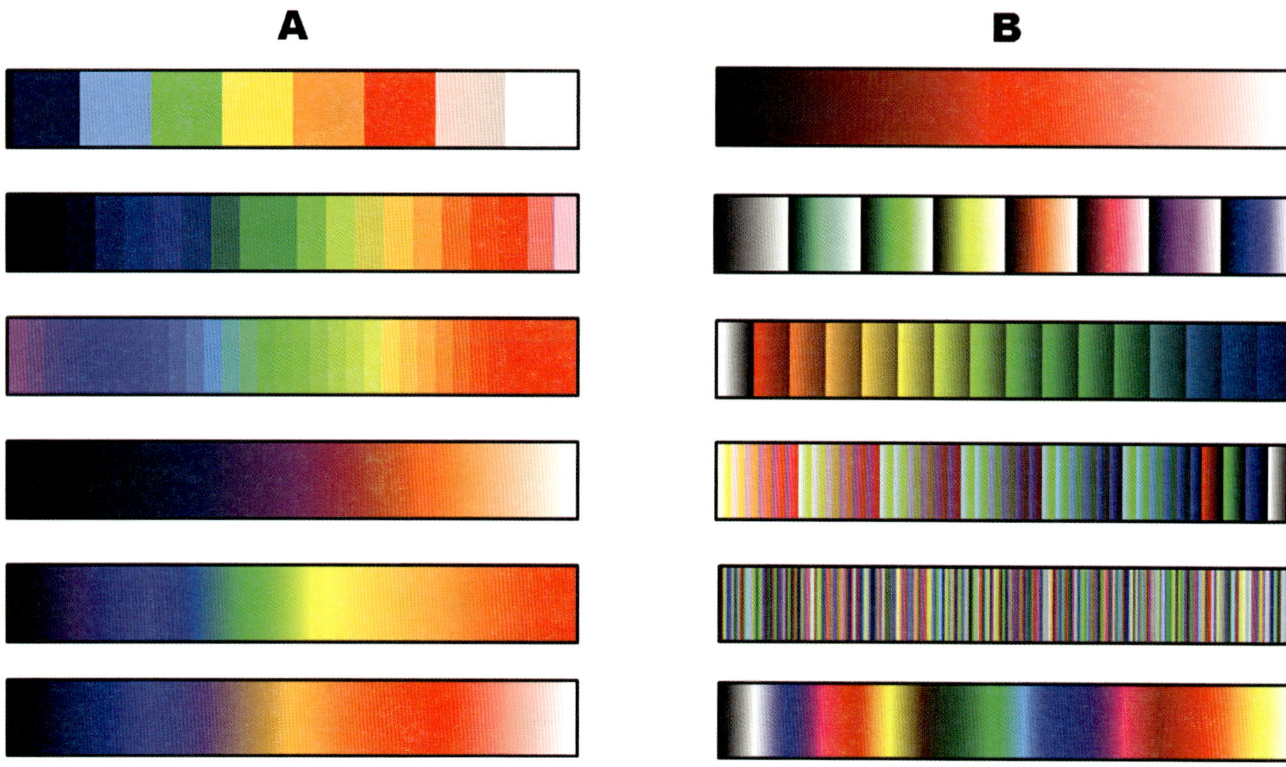

Figure 2. Examples of various predefined lookup tables (LUTs). Prismatic multicolor LUTs, such as the examples in the left column (A), are preferred for pseudocolor conversion to enhance compositional variations in scanning electron microscopic (SEM) images. Monochromatic, ramp, repeating color, and other specialized LUTs as shown in the right column (B) typically do not produce pseudocolor images useful for scientific interpretation of shale SEM images.

Several types of predefined LUTs are available for pseudocolor conversions (Figure 2). Selecting the appropriate LUT to enhance SEM grayscale images is largely a process of experimenting with various LUTs until a desired image is produced that enhances specific features of interest. The degree to which a particular LUT is able to enhance a grayscale image is limited by the LUT design (number of colors, hue, color gradient, etc.) and the contrast dynamic range of the grayscale image. Prismatic multicolor LUTs are generally suitable to enhance compositional variations exhibited by backscattered and secondary electron images (Figure 2a). Monochromatic, limited color, discrete ramp, repeating color tables, and other specialized LUTs (Figure 2b) can produce colorful images but are generally not useful in improving the ability to interpret shale SEM images.

Contrast Enhancement

The contrast of SEM images is adjusted by the SEM operator until a suitable image is obtained. The contrast will thus vary between images depending on the range of backscatter or secondary electron intensity detected within the scanned sample area. The image contrast is often limited to a narrow range of values concentrated in the midrange of the gray scale because of the similarity in backscatter intensity of minerals common in shale formations (see Table 1 and Figure 3).

Table 1. Common Shale Minerals Listed in Order of Increasing Backscatter Intensity Based on Mean Intensity (Grayscale) Values Measured from Backscattered Electron Images Used in This Study

Order	Mineral	Z	G
1	Bitumen (Graphite)	(6)	(2.23)
2	Kaolinite	10.4	2.6
3	Albite	10.7	2.62
4	Quartz	10.8	2.66
5	Plagioclase	11.9	2.76
6	Illite	–	–
7	Dolomite (nonferroan)	–	2.85
8	Calcite	12.4	2.72
9	Muscovite	11.1	2.76–2.88
10	Biotite	–	2.8–3.2
11	Apatite	14.1	3.15–3.20
12	Rutile	16.4	4.2
13	Pyrite	20.7	5

Source: Mineral mean atomic number (Z) values from Reed (2005). Specific gravity (G) values from Hurlbut and Sharp (1998).

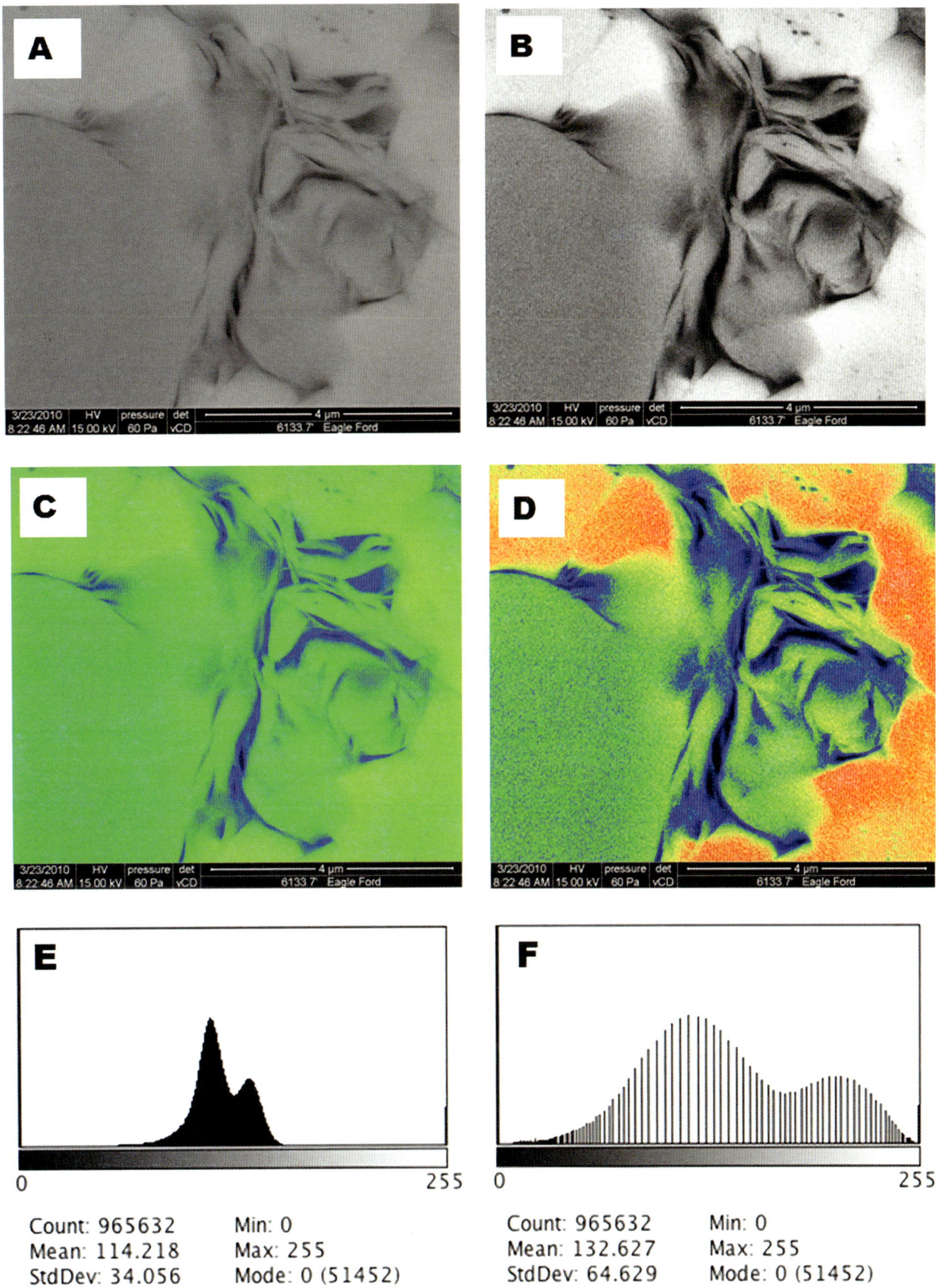

Figure 3. Comparison of contrast enhancement of grayscale and pseudocolor backscattered electron images exhibiting low dynamic range: (A) original grayscale image; (B) contrast-enhanced grayscale image; (C) pseudocolor conversion from original grayscale image; and (D) contrast-enhanced pseudocolor image. Histograms show differences in backscatter intensity distributions of original (E) and contrast-enhanced (F) images.

Pseudocolor conversions of SEM images with low contrast variation (low dynamic range) can often be improved by contrast enhancement using image intensity histogram equalization as described in the appendix.

Contrast enhancement alters the intensity values of the original grayscale image. It is possible to fundamentally alter the image when aggressive contrast adjustments are applied such that the resulting image does not accurately represent the information displayed in the original image. It is important to disclose when contrast enhancement has been applied to notify the reader that the original image fidelity has not been preserved because this could lead to erroneous interpretations (see graph, Figure 4).

PSEUDOCOLOR EXAMPLES

Secondary Electron Images

Rugose Surfaces

Figure 5 displays detailed textural characteristics typical of secondary electron images from freshly broken surfaces. Deformed clay layers surrounding a spherical pyrite framboid are clearly observed in the grayscale image (Figure 5a).

Brighter areas reflect higher secondary electron intensity measured from elevated peaks and surfaces facing the secondary electron detector (to the right). Dark, low-intensity areas are produced from depressions and surfaces directed away from the secondary

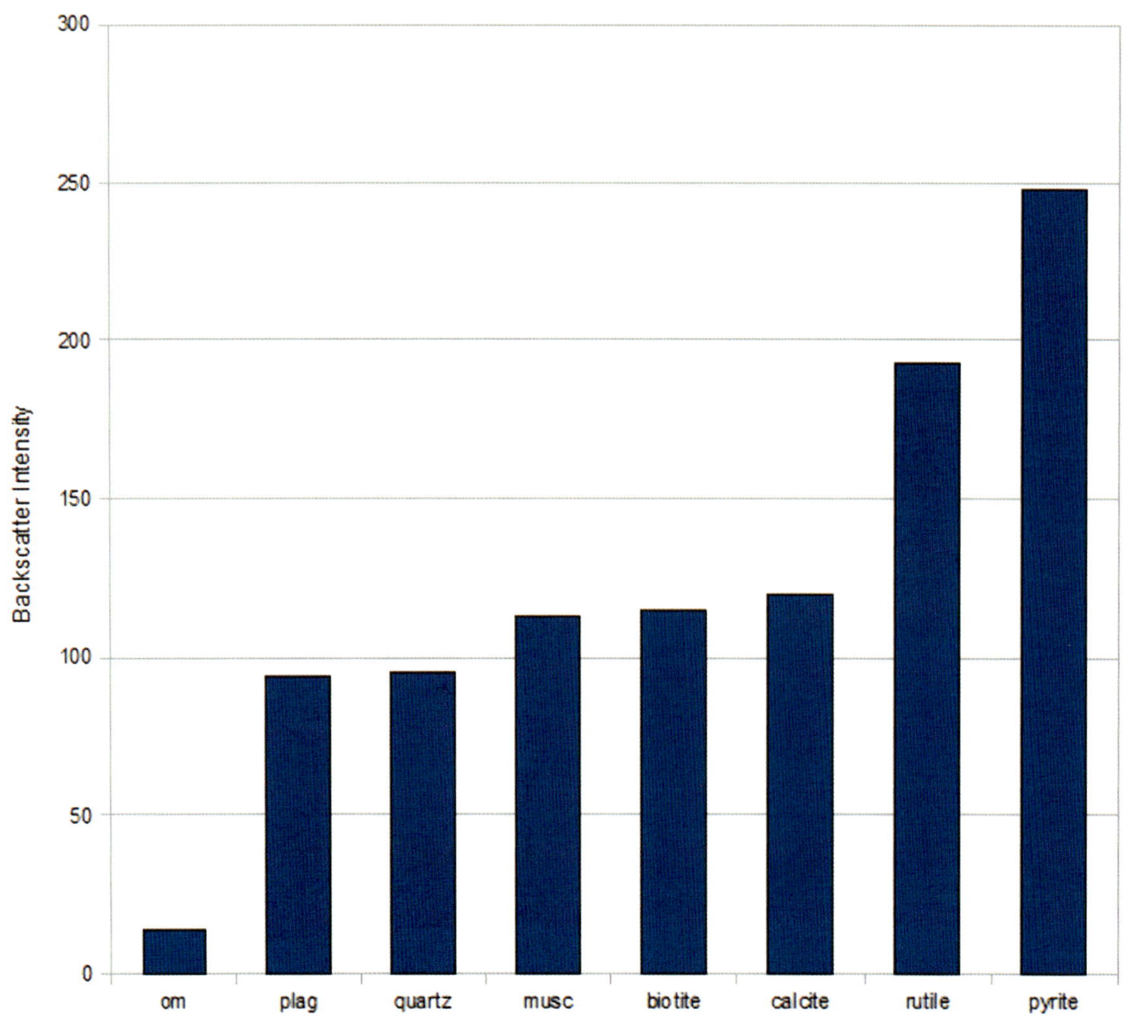

Figure 4. Bar graph showing mean backscattered electron intensity measured for various minerals present in a single scanning electron microscopy (SEM) image. Note similarity of backscatter intensity for many of the common minerals present in shale hydrocarbon reservoirs.

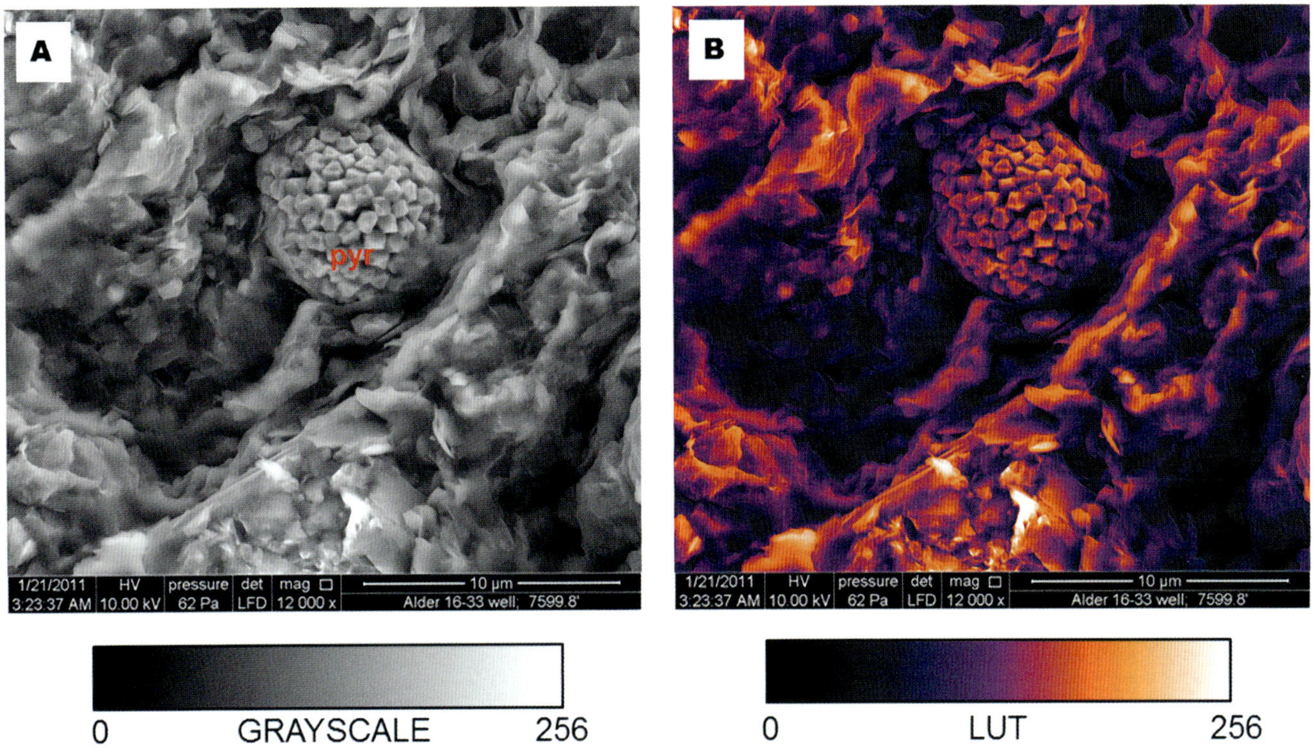

Figure 5. Secondary electron image from broken shale surface showing clay layers deformed around a pyrite framboid (pyr): (A) original grayscale image; (B) pseudocolor enhanced image.

electron detector. It is not clear if the black areas separating some of the clay grains represent natural pores or are artifacts of mechanically breaking the sample. Pore spaces between individual pyrite grains appear to be natural.

The pseudocolor image (Figure 5b) emphasizes the high signal intensity areas by bright colors (orange and white), enhancing the individual pyrite crystals forming the composite framboid. Note the secondary electron intensity is not affected by compositional variations between pyrite and clay (illite) because of the greater effect of surface rugosity on secondary electron yield.

Smoothed Surfaces

Ion beam milling produces a very flat surface by removing small amounts of material from the sample using either argon or gallium ions (focused ion beam). Because surface rugosity is essentially removed by ion beam milling, the topographical effects on secondary electron yield are minimized, allowing some compositional variations to be imaged.

Figure 6 is a secondary electron image from an argon-ion-milled surface. The variation in grayscale (Figure 6a) and color in the pseudocolor conversion (Figure 6b) is a combination of surface topography and composition. Pyrite and organic matter show compositional contrast similar to that observed in backscattered electron images. Other minerals present, including quartz, calcite, and clay (based on bulk x-ray diffraction), are less distinguishable because of their similar secondary electron intensity values.

Pores associated with organic matter appear black in grayscale images because only a low number of secondary electrons are able to escape from depressions below the argon-ion-milled surface. Light rims surrounding some of the pores (Figure 6a) are excess electron charging effects associated with sharp edges characteristic of secondary electron images obtained from uncoated nonconductive materials (Reed, 2010). Surface striations that were not removed by the argon-ion milling are also evident in the lower left portion of the grayscale and pseudocolor images as alternating bright and dark bands (Figures 6a, b).

Although some compositional variations may be observed in secondary electron images from smooth ion-milled surfaces, backscatter electron is the preferred detection mode for observing compositional variation in SEM images as described next.

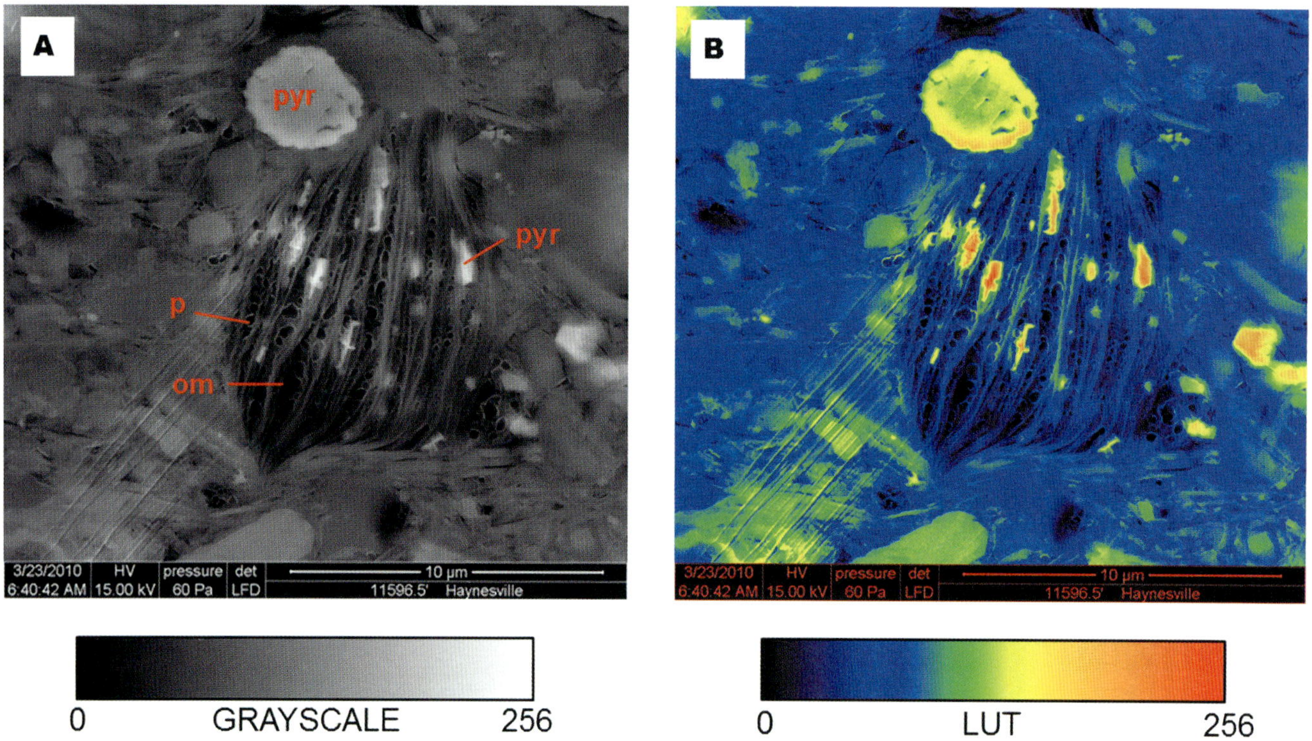

Figure 6. Secondary electron image from a smoothed, argon-ion-milled surface. Grayscale (A) and color (B) intensity variations are a result of combined surface and compositional affects. Pyr = pyrite, om = organic matter, and p = pore.

Backscattered Electron Images

Eagle Ford

Figure 7 is a backscattered electron image of an argon-ion-milled shale gas reservoir sample from the Upper Cretaceous Eagle Ford Formation, Dimmitt County, Texas. Backscatter electron intensity (grayscale brightness) increases with mean atomic number (and density). Several different minerals are indicated by variations in grayscale brightness and particle and crystal morphology (Figure 7a).

An energy-dispersive x-ray spectrometer was used for mineral identification. Pyrite and apatite have high mean atomic numbers and are exhibited in the grayscale image as white and light gray, respectively. Organic matter is easily identified by its characteristic very low backscatter intensity (black). Other identified minerals such as calcite, plagioclase, and kaolinite exhibit less grayscale contrast, making it more difficult to distinguish these minerals in the grayscale image (Figure 7a). Coccolith fragments are identified by their characteristic shape (coc in Figure 7a) and calcite composition (yellow, Figure 7b).

The pseudocolor image shown in Figure 7b has been enhanced by histogram equalization to improve the contrast dynamic range. This technique produced a noticeable color contrast between kaolinite (blue) and plagioclase (green), which normally are indistinguishable in grayscale and pseudocolor because of their similar backscatter intensity response (Table 1 and Figure 3). The LUT was also edited by replacing the high-intensity red values with white to make it easier to distinguish the small pyrite crystals from the red-orange apatite (see LUT, Figure 7b).

Small, rounded, green grains (e.g., plag? Figure 7b) exhibit the same color as the euhedral plagioclase grain (plag, Figure 7a), and thus may also have the same composition. However, the difference in grain shape suggests either different origins (detrital and authigenic) or possibly two different minerals with similar backscatter intensity. Other techniques, such as x-ray spectrometry, are required to confirm the mineral identification suggested by the pseudocolor image.

Haynesville

The backscattered electron image shown in Figure 8 is from an argon-ion-milled sample of a shale gas reservoir from the Jurassic Haynesville Formation, Harrison County, Texas. Pyrite and organic matter have high atomic number contrast and are typically easily distinguished when present in grayscale images (Figure 8a). Calcareous minerals, identified by energy-dispersive x-ray spectrometry, include calcite,

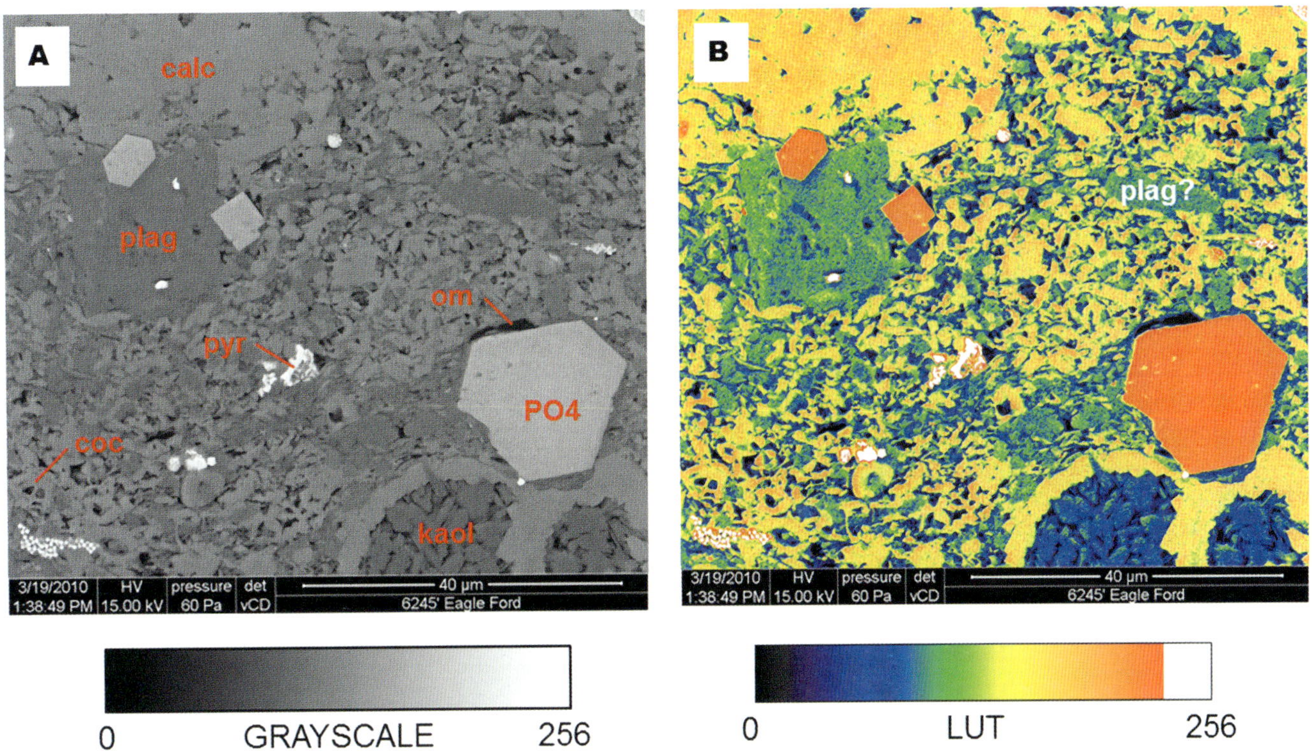

Figure 7. Backscattered electron image from argon-ion-milled surface of shale gas reservoir sample of the Eagle Ford Formation. Grayscale (A) and pseudocolor (B) intensity reveal variations in composition. Calc = calcite, coc = coccolith, kaol = kaolinite, om = organic matter, plag = plagioclase, and PO4 = calcium phosphate (apatite).

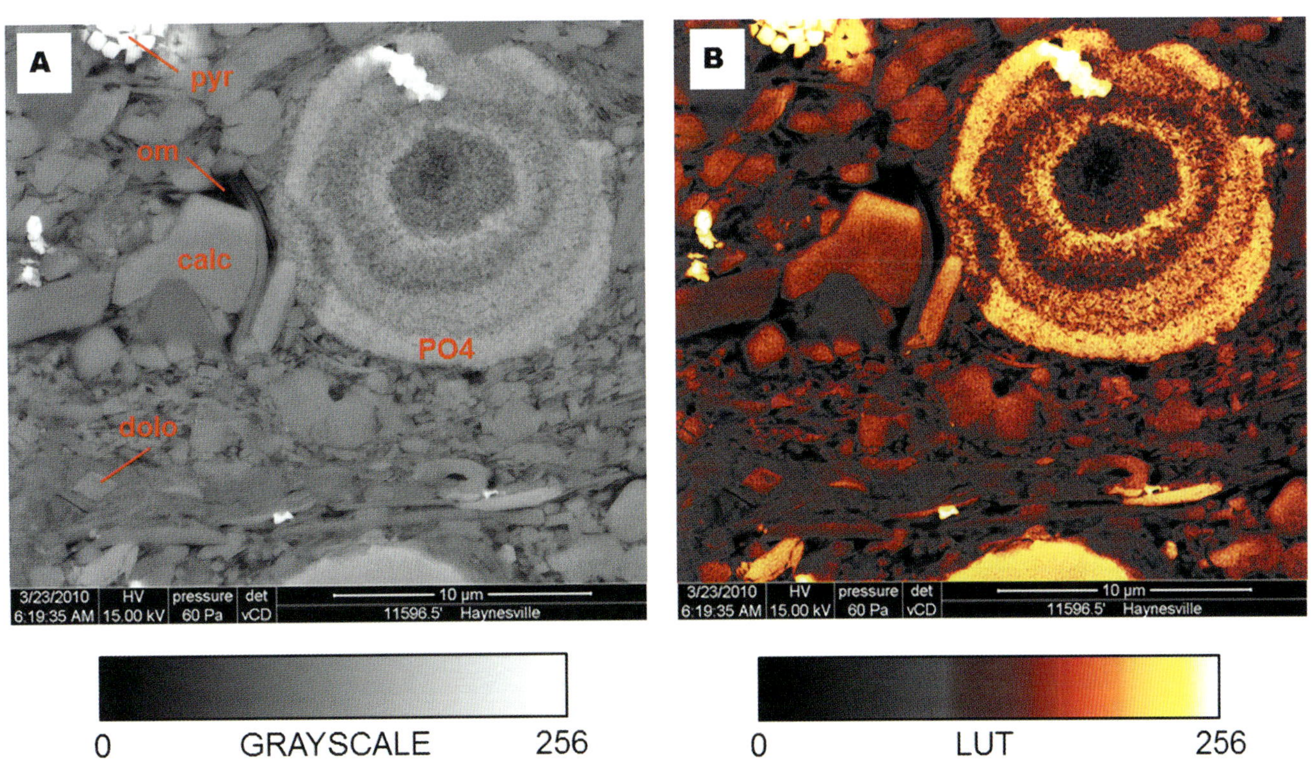

Figure 8. Backscattered electron image from argon-ion-milled surface of a shale gas reservoir sample from the Haynesville Formation. Grayscale (A) and pseudocolor (B) intensity variations reveal variations in composition. Calc = calcite, dol = dolomite, om = organic matter, and PO4 = calcium phosphate (apatite).

dolomite, and apatite and appear as light gray. Noncalcareous minerals present (quartz, illite, and feldspar) have lower backscatter intensity and appear in darker shades of gray.

The LUT selected for the pseudocolor conversion shown by Figure 8b highlights only the high-intensity values (yellow and orange) to emphasize the proportion of carbonate from noncarbonate grains. Organic matter appears black in the grayscale and pseudocolor images and is interpreted as bitumen that fills most of the available intergranular pore space. The round phosphate grain (oolite?) exhibits concentric compositional banding, evident by backscatter intensity variations in both the grayscale and pseudocolor images.

Marcellus

The backscattered electron image shown in Figure 9 is from an argon-ion-milled sample of a shale gas reservoir from the Middle Devonian Marcellus Formation, Clinton County, Pennsylvania. This sample contains a wide variety of minerals identified by energy-dispersive x-ray spectrometry. Except for pyrite and organic matter, most of the minerals present within the scanned area exhibit low grayscale contrast, making it difficult to interpret mineral identity and distribution (Figure 9a).

The variation in colors shown in the pseudocolor conversion (Figure 9b) produces an image that is less difficult to interpret. Note the difference in distinguishing between the relatively higher backscatter intensity minerals rutile and apatite and the lower grayscale minerals quartz and muscovite in the grayscale and pseudocolor images (Figure 9a, b). The pseudocolor conversion produced by this particular LUT (Figure 9b), however, was unable to differentiate between quartz (qtz) and plagioclase and between calcite and muscovite.

The ability of pseudocolor conversions to produce images useful for compositional interpretations is limited by the degree of contrast of the original image. Editing or selecting alternative LUTs does not always improve the ability to distinguish between minerals with similar backscatter intensity. However, other characteristics, such as grain morphology, may aid in interpretation, such as between calcite and muscovite in Figure 9. Images with multimineral assemblages shown in Figures 7 and 9 were useful in building the backscatter intensity ranking chart (Figure 3).

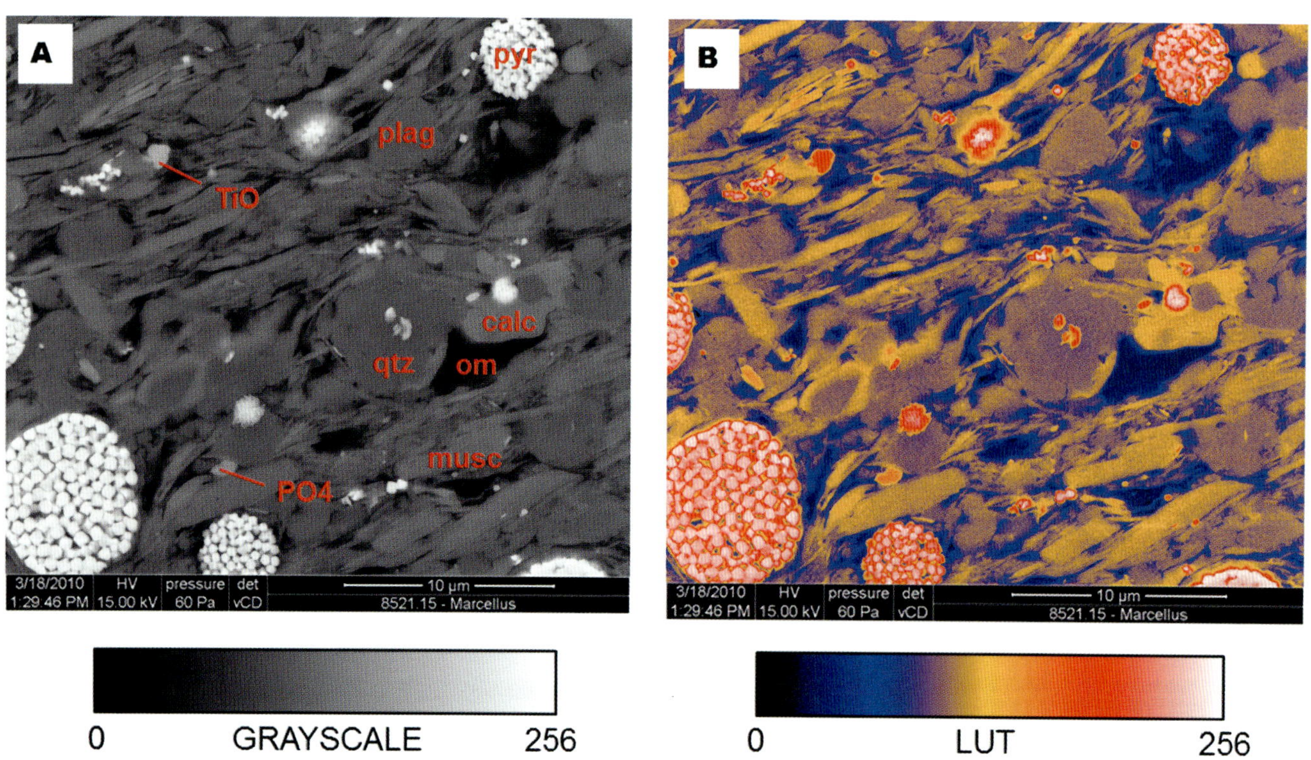

Figure 9. Backscattered electron image from argon-ion-milled surface of a shale gas reservoir sample from the Marcellus Formation. Grayscale (A) and pseudocolor (B) intensity reveal variations in composition. Calc = calcite, musc = muscovite, om = organic matter, plag = plagioclase, PO4 = calcium phosphate (apatite), and TiO = titanium oxide (rutile).

CONCLUSIONS

Pseudocolor conversions can be rapidly and inexpensively accomplished using the public domain software ImageJ. Although any grayscale SEM image can be converted to color by applying LUTs, the primary application for shale hydrocarbon reservoir studies is to facilitate compositional interpretations from backscattered electron images (either alone or superimposed with secondary electron images) acquired from ion-milled surfaces. Because grayscale values are not unique to any particular minerals or compounds, the various colors displayed by pseudocolor conversions cannot be used alone for mineral identification. Supplemental data, such as from energy-dispersive x-ray spectrometry, are required to identify minerals from their characteristic elemental compositions.

The ability of pseudocolor conversion to enhance a given grayscale image is limited by the dynamic range of the original grayscale image and the type of LUT applied. Contrast-enhancing techniques, such as histogram equalization, can improve the image dynamic range and color discrimination between minerals exhibiting subtle contrast difference. However, unlike direct LUT applications that preserve image fidelity, contrast enhancement alters the fundamental signal intensity values. Caution is urged against aggressive use of contrast and brightness adjustments, which could lead to erroneous interpretations.

ACKNOWLEDGMENTS

The senior author wishes to acknowledge Anadarko colleagues Breck Johnson, for introducing me to the ImageJ program and its potential for enhancing SEM grayscale images, and Howard White, for his efforts to implement SEM analysis for shale gas reservoir studies at Anadarko. We are also grateful to Brian Lock and William DeMis for their constructive reviews and Anadarko Petroleum and Core Laboratories for permission to publish this chapter.

APPENDIX

This section describes step-by-step instructions for creating pseudocolor conversions of grayscale SEM images using the ImageJ program. Additional information is available in the ImageJ User Guide (Ferreira and Rasband, 2011) available from the ImageJ Web site (http://rsbweb.nih.gov/ij/docs/guide/index.html).

All commands are run from the main menu bar, consisting of nine menus identified here by italic font: *File, Edit, View, Insert, Format, Table, Tools, Window,* and *Help*. Submenu commands are selected from the drop-down lists, denoted as items following the greater than symbol (>), such as *Image > Adjust > Brightness/Contrast.*

Open and Select Image

Open the ImageJ program. From the main menu bar, select *File > Open* and navigate to the location on the computer where the SEM image file of interest has been saved. Select the desired image file, then click *Open*. The selected image will open in a separate window on the computer screen.

Images illustrated in this chapter were created from and saved in TIFF. ImageJ is capable of opening other common image format files, such as Graphic Interchange Format (GIF), Joint Photographic Experts Group (JPEG), and Bitmap Image File (BMP). See the ImageJ User Guide (Ferreira and Rasband, 2011) for additional information about working with other image types and formats.

Select *Image > Type* to determine the type of image. Check *8-bit* from the drop-down menu to convert the image to an 8-bit grayscale image. Sixteen- and 32-bit may also be selected; however, color tables cannot be assigned to RGB (red, green, blue) images; these must first be converted to grayscale images. Images processed for this chapter were created as 8-bit images. If multiple image windows are open, make sure the image of interest is activated by clicking on the desired image window.

Applying Lookup Tables

The ImageJ program is preinstalled with 11 color LUTs. The color LUTs can be applied to the active grayscale image by selecting an LUT from the drop-down list following the command *Image > Lookup Tables*. Only four of the preinstalled LUTs contain multicolor spectrums suitable for pseudocolor processing of shale SEM images. These are *Fire, Ice, Spectrum,* and *Red/Green*. Selecting *Image > Lookup Tables > Invert LUT* results in four additional applicable LUTs by reversing the color table. Once a desired pseudocolor image is produced, select *Apply LUT* from the *Lookup Tables* submenu to assign the selected LUT to a grayscale image.

Because of the limited number of preinstalled LUTs, it is recommended to download additional LUTs available from the ImageJ Web site. The ZIP files luts.zip and luts2.zip and files contained within the NucMed_Image_LUTs/ file contain nearly 100 additional LUTs.

To apply one of the downloaded LUTs, use the command *File > Import > LUT*, navigate to the location on the computer where the downloaded LUT files are saved, and select the desired LUT file.

The final pseudocolor image may be saved by the command *File > Save As* and selecting the desired file format from the drop-down list (e.g., *Tiff*).

Selecting an appropriate LUT to enhance a particular grayscale image is largely a process of trying various LUTs until a suitable image is obtained that enhances the features of interest. Gradational and discrete multicolor spectrums (Figure 2) are generally suitable to enhance subtle contrast variations in images containing multiple minerals and compounds. In addition to the predefined LUTs available for use with the ImageJ software, the user may easily create additional LUTs by editing as described in the ImageJ User Guide (Ferreira and Rasband, 2011).

The LUTs used for the pseudocolor conversions illustrated in this chapter (i.e., Figures 4–9) were downloaded from the ImageJ Web site and are listed by name following the figure number:

Figure 4: 32_Colors.lut
Figure 5: ge-256.lut
Figure 6: S_Pet.lut
Figure 7: S_Pet.lut (edited)
Figure 8: smart.lut
Figure 9: Ceretec.lut

Contrast Enhancement

Pseudocolor images with low dynamic range can often be improved by contrast enhancement. The dynamic range of the grayscale image can be displayed from the command *Analyze > Histogram*.

Histograms of images with low dynamic range show a narrow distribution of intensity values along the x axis (Figure 4). To increase the image dynamic range, select *Process > Enhance Contrast*. This will open a separate window. Check the box for *Equalize Histogram*, then select *OK*. Select *Analyze > Histogram* again to observe the changes applied by histogram equalization.

Contrast enhancement using the dynamic range displayed in the entire image may produce poor results when high-contrast differences exist. Performing histogram equalization to a portion of the image with limited contrast differences may produce better results. To do this, choose an area selection tool (e.g., *Rectangular*) by clicking on the tool icon from the ImageJ toolbar. With the mouse, drag an outline within a small area of the image that exhibits a narrow range of contrast variation. Select *Process > Enhance Contrast*, ensure that the *Equalize Histogram* box is checked, and select *OK*. This will apply contrast enhancement to the whole image based on histogram equalization within the smaller selected area.

If the applied contrast enhancement does not produce satisfactory results, select *Edit > Undo* to return to the image prior to applying histogram equalization. Click anywhere outside the selected area to remove the selection. Experiment by selecting other areas with different contrast variation until the desired results are obtained.

Note histogram equalization may be applied before or after applying an LUT. Because contrast enhancement alters the grayscale values of the original image, it is recommended to evaluate pseudocolor conversions by applying various LUTs before attempting contrast enhancement.

Image Adjustments

Contrast and other image adjustments can be done from the *Image* menu by selecting one of the various applications listed in the *Adjust* submenu, for example, *Image > Adjust > Brightness/Contrast*. See instructions in the ImageJ User Guide (Ferreira and Rasband, 2011) for information and use of the available image adjustment applications. Care should be used to avoid making aggressive adjustments that could lead to erroneous interpretations from altered images.

REFERENCES CITED

Ferreira, T., and W. Rasband, 2011, ImageJ user guide, IJ 1.45m: http://rsbweb.nih.gov/ij/docs/guide/index.html (accessed October 5, 2011).

Goldstein, J. I., D. E. Newbury, D. C. Joy, C. E. Lyman, P. Echlin, E. Lifshin, L. Sawyer, and J. R. Michael, 2003, Scanning electron microscopy and x-ray microanalysis, 3d ed.: New York, Springer Science and Business Media, 689 p.

Hurlbut, C. S., Jr., and W. E. Sharp, 1998, Dana's minerals and how to study them, 4th ed.: New York, Wiley, 328 p.

Reed, S. J. B., 2010, Electron microprobe analysis and scanning electron microscopy in geology, 2d ed.: Cambridge, Cambridge University Press, 212 p.

Herman Lemmens, Dennis Richards, 2013, Multiscale imaging of shale samples in the scanning electron microscope, *in* W. Camp, E. Diaz, and B. Wawak, eds., Electron microscopy of shale hydrocarbon reservoirs: AAPG Memoir 102, p. 27–35.

4

Multiscale Imaging of Shale Samples in the Scanning Electron Microscope

Herman Lemmens
FEI, Achtseweg Noord 5, 5600 Eindhoven, The Netherlands (e-mail: herman.lemmens@fei.com)

Dennis Richards
FEI, 5350 NE Dawson Creek Dr., Hillsboro, Oregon, 97124, U.S.A. (e-mail: dennis.richards@fei.com)

ABSTRACT

Porosity in shale samples is often at such a small scale that analysis with a focused ion-beam scanning electron microscope (FIB-SEM) is needed to resolve the nanometer-size pores in three dimensions. This gain in resolution, however, comes at the expense of analysis volume (typically no more than 1000 μm^3), which may not be representative of the larger sample. This chapter describes a method of creating high-resolution, large-scale SEM images to aid in selecting more representative locations for FIB-SEM imaging and analysis than by current industry practice.

INTRODUCTION

Unconventional shale reservoirs are also largely unknown reservoirs. Every time the first high-resolution scanning electron microscopic (SEM) images of the newest shale play are presented to the interpreting scientist, a wave of excitement occurs. Each shale play seems to exhibit another intriguing network of clays, organic matter, and porosity. Typical for shales is the occurrence of porosity at the nanometer scale and very fine features in the clays, requiring high imaging magnifications. Focused ion-beam scanning electron microscopy (FIB-SEM) takes this even one step further by providing a three-dimensional (3-D) data set composed of hundreds of such high-resolution SEM images. A typical FIB-SEM data set has a spatial resolution of 10 nm for a volume of $10 \times 10 \times 10$ μm.

A most relevant question is how representative those small-volume FIB-SEM data sets are. Industry best practice is to use full-core x-ray computed tomographic (CT) analysis to define relevant core plug positions (Walls and Sinclair, 2011). A limited number of high-resolution SEM images are taken to decide what the relevant and representative features are. For a 1-in. core plug, this still means that less than 1% of the sample surface is actually imaged at a resolution able to resolve the porosity before deciding where to perform an FIB-SEM experiment.

The risk associated with the much higher resolution characterization of shale core sample material with high-resolution SEM and FIB-SEM is that the structural association with fabric is lost. Yet, fabric heterogeneity is equally important for the characterization of shale samples, and the association of nanometer-scale pore structures with fabric domains may provide important information on permeability and fluid flow at the scale of the core.

DOI: 10.1306/13391702M1023582

Tiling and stitching SEM images into mosaics enable pore-scale resolution over the entire core plug surface in one image. This allows simultaneously retrieving fabric and pore-level information without losing correlation. Furthermore, it allows choosing the FIB-SEM locations in a more representative way.

The Pore-fabric Paradox

Gas shale samples typically contain pores as small as 10 nm (or even smaller). Resolving those pores requires a pixel size of about 3 nm. A common SEM image is 4096 pixels wide. This results in a field of view of about 12 μm. Figure 1 shows such a high-resolution backscatter electron (BSE) image of a shale sample with the porosity in the organic matter clearly visible in the inset (Figure 1B). This image shows pyrite crystals (white) in a porous organic matter matrix (dark gray), but the small area imaged makes it impossible to judge whether this particular microstructure is representative for this sample. This can only be answered by studying a larger surface at the resolution required to resolve the pore space. Figure 2 shows a mosaic consisting of 12,800 SEM images, totaling 394,815 by 394,807 pixels. The high resolution in this mosaic is demonstrated in the insert (Figure 2B) of the 10-μm-wide particle filled with porous organic matter.

SAMPLE PREPARATION

Mechanical polishing is not suitable for shale samples, as this technique tends to destroy the fine structure of the pore network (e.g., smearing grain boundaries, embedment of grit into surface of sample). A noncontact polishing technique with argon ions is used to obtain a smooth surface. The traditional approach to preparing shale samples in the oil and gas petrology laboratory with the argon-ion mill uses a mask to create a cross section, but in that case the result is a parabola-shaped flat section limited to about 1 × 1 mm. Larger-scale mosaic images provide ample evidence of fabric differences in many shales on length scales larger than 1 mm.

Figure 3 shows two images from the 12 × 12 mm section in Figure 2, displaying two regions with porous and nonporous organic matter spaced more than 6 μm apart. A 1 × 1 mm section would have missed at least one of these two distinct morphologies of organic matter (Figure 3).

For the samples discussed in this chapter, the Fischione 1060 SEM Mill was used. Samples are attached to a 25-mm diameter pin-type stub using carbon tape and loaded into the 1060 SEM Mill. Under a 5° incident angle and 360° sample rotation, 5-kV ions produce a polished surface after 1.5 h.

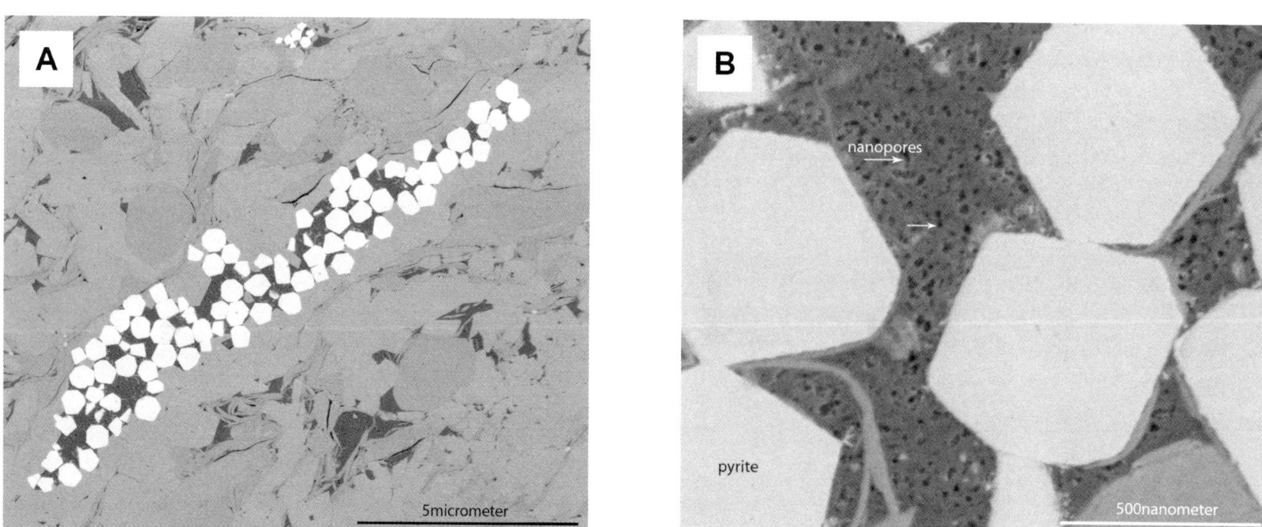

Figure 1. (A) High-resolution scanning electron microscopic (SEM) image. Horizontal field of view is 13.4 μm, effectively isolating this image from the sample fabric. (B) The porosity in the organic matter is clearly shown. Field of view is 1.5 μm.

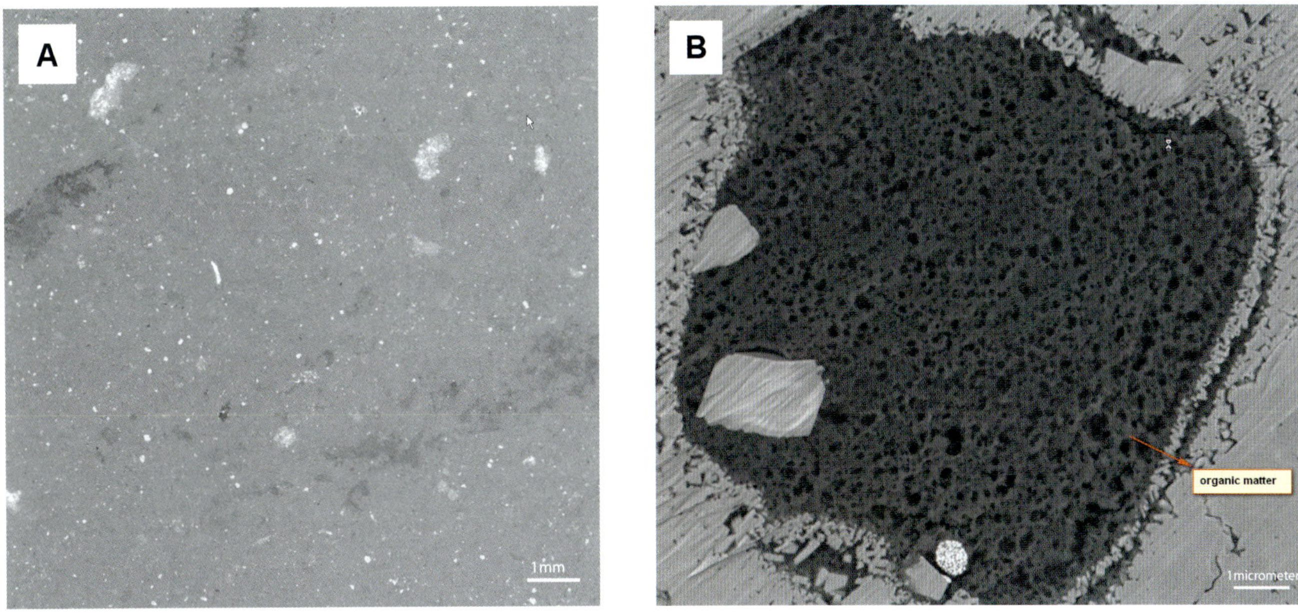

Figure 2. (A) Overview image of a 12-mm = wide shale sample consisting of 12,800 individual high-resolution scanning electron microscopic (SEM) images, totaling 156 gigapixels. (B) The pore-scale resolution obtained over this large surface (field of view is 12 μm) is shown.

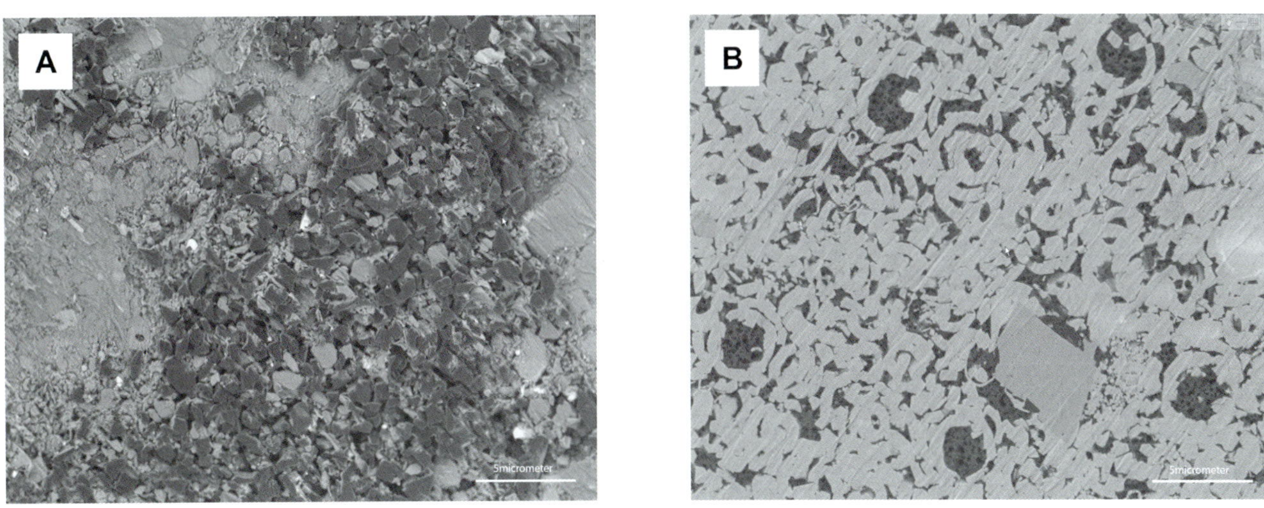

Figure 3. Two regions of the same sample, with nonporous (A) and porous (B) organic matter, spaced 6 mm apart. Field of view is 30 μm.

EXPLANATION OF IMAGING METHODS

Large-scale Imaging

Large-scale imaging uses tiling and stitching to create a high-resolution image mosaic in which the size is limited only by time constraints. Autoimaging controls (focus, brightness, contrast) contribute to an image that is truly seamless, even when it is composed of many thousands of individual SEM images. Care must be taken not to introduce image distortions at the edges that would complicate stitching. Edge distortions are most pronounced at low magnifications or in large images (over 8000 $\times$ 8000 pixels). Stitching many smaller images (typically up to 4000 $\times$ 4000 pixels) avoids edge distortions. Figure 4 shows a set of 2 $\times$ 2 individual tiles and the resulting seamless mosaic. Overlap between the tiles could be kept to a

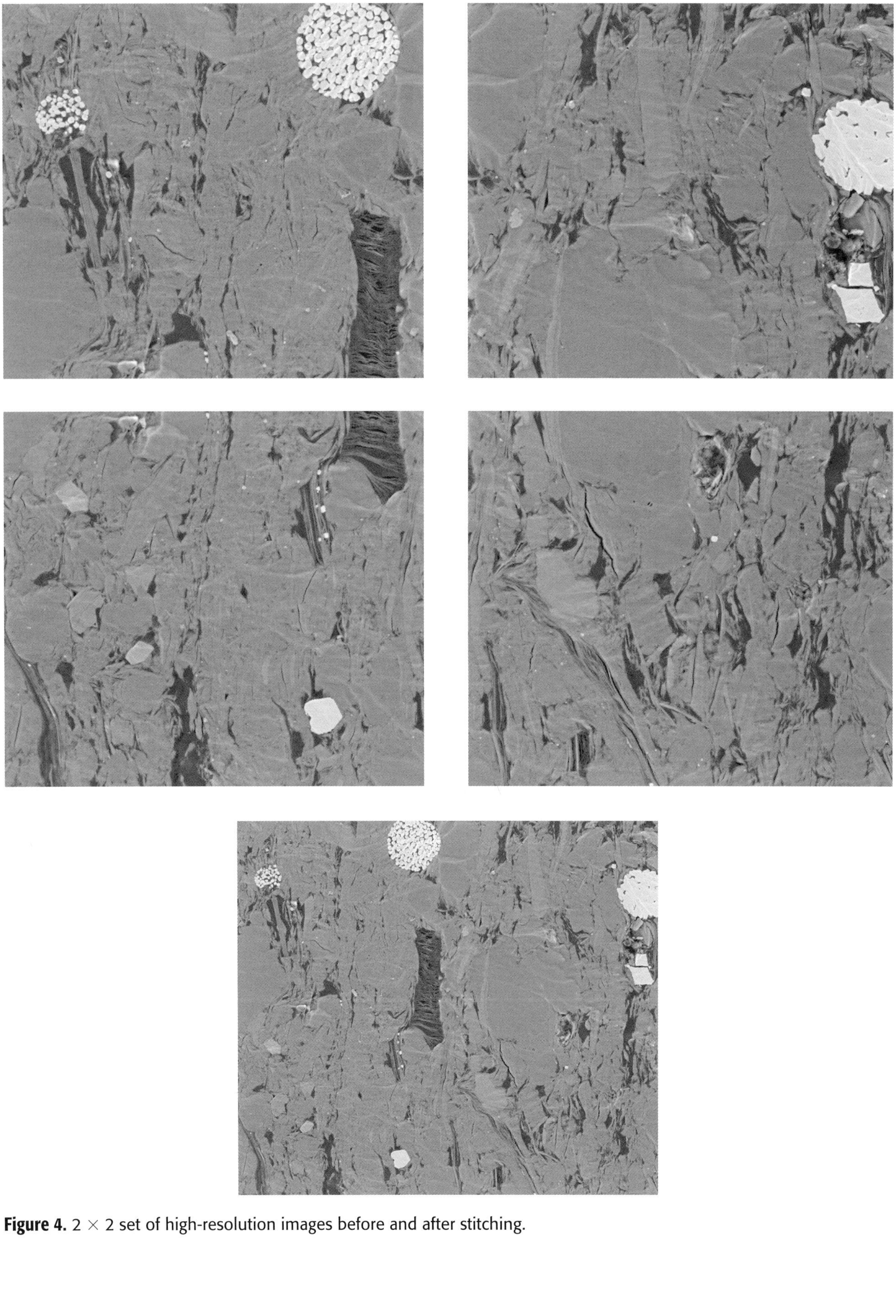

Figure 4. 2 × 2 set of high-resolution images before and after stitching.

minimum (5%) because of the high accuracy provided by piezoelectric stages. Stage accuracy also facilitates retrieving nanometer-size particles by navigating on the mosaic image.

Automated Petrography

The traditional method for fabric characterization is optical petrography on thin sections. Optical petrography generates mineralogical and textural information in a single image; however, severe resolution limitations exist on fine-grain shale samples. Automated petrography based on SEM provides a high-resolution alternative to optical petrography. Quantitative evaluation of minerals by scanning electron microscopy (QEMSCAN) uses BSE imaging and energy-dispersive x-ray (EDX) analysis to obtain the elemental composition of the phase under the electron beam (Gottlieb et al., 2000). Coupled with a library of mineral phases, automated petrography allows fast, reliable, and unattended mineral characterization along a finely spaced grid of measurement points on a two-dimensional (2-D) surface (Figure 5). In addition to mineral identification, textures such as laminations and grain size variations are revealed (Armitage et al., 2010). Although this technology was originally developed for the mining industry, it is now commonly used in the oil and gas industry. Clay standards from the Clay Minerals Society (Chantilly, VA; http:///www.clays.org) were used to build a specific clay mineral library.

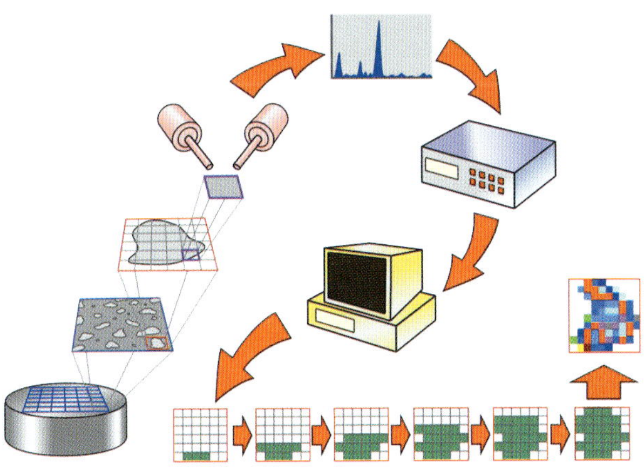

Figure 5. Schematic of the automated petrography analysis routine. From the backscatter electron image gray level, the particles are separated from the resin in case of cutting samples. For core samples, the full surface is mapped with energy-dispersive x-ray (EDX) analysis with a user-defined spacing. The EDX spectrum is matched to a spectrum from an oil-and-gas-specific mineral database.

Samples with one dominant type of clay were used to define the reference spectra, as illustrated with the illite clay standard in Figure 6. Automated petrography measurements were performed on an FEI QEMSCAN 650FEG instrument with two EDX detectors.

Focused Ion-beam Scanning Electron Microscopy

Focused ion-beam scanning electron microscopy uses a focused (gallium) ion beam (FIB) to cut parallel, evenly spaced slices in combination with SEM imaging to obtain a high-resolution 3-D data set (Figure 7). Slice thickness can be as low as 3 nm, although a thickness of 10 nm is a practicable standard developed for shale samples. Between slices, recognition marks are used for alignment of the individual slices and quality control on the slice thickness (Figure 8). The result is a set of aligned SEM images that are turned into a 3-D volume by dedicated reconstruction software. Gray-level segmentation is used to analyze spatial distributions of pores and other features (Curtis et al., 2010; Knackstedt et al., 2010).

A common issue with FIB-SEM imaging is preferential milling causing an artifact called curtaining. Shale samples are heterogeneous with phases that have very different material removal rates under the ion beam. This creates laminations in the cross section, shown as vertical bands in the SEM image, that may lead to segmentation errors. To reduce this effect without compromising milling speed and hence sample throughput, an ion beam with excellent beam profile at high beam currents is preferred.

Shale samples typically charge heavily under the electron beam, creating artifacts in the image. The Helios 650 DualBeam by FIB-SEM was used at low

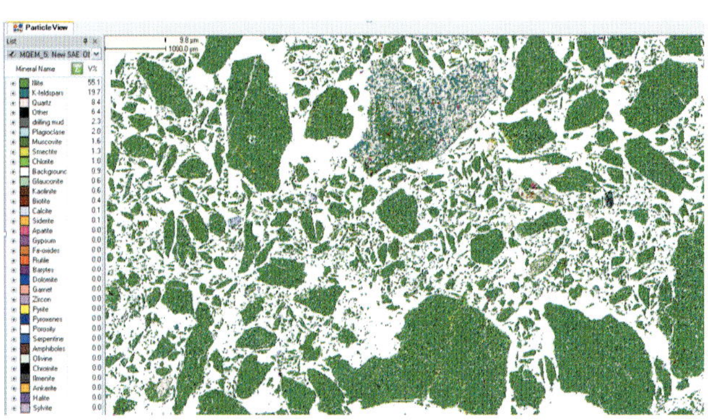

Figure 6. QEMSCAN analysis of clay standard IR6 IMt-1 Illite obtained from the Clay Minerals Society. The majority phase identified correctly is illite (green), with quartz, calcite, and K-feldspars also present.

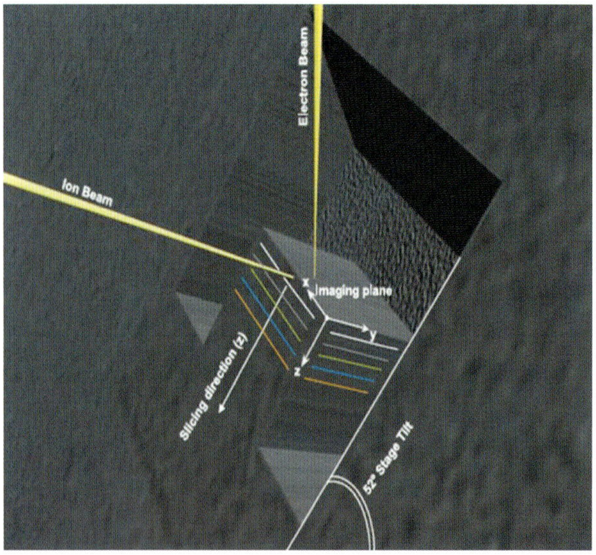

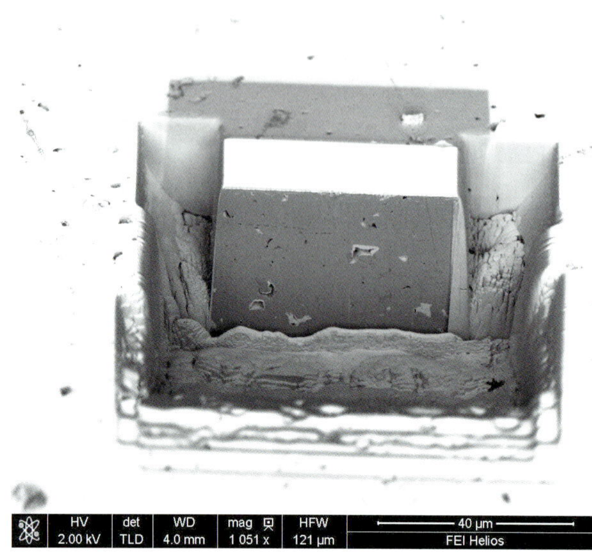

Figure 7. Sample setup in a typical focused ion-beam scanning electron microscopic (FIB-SEM) experiment. First, a trench is milled around the cube that will be analyzed. Then, FIB milling and SEM imaging are used sequentially to generate a three-dimensional data set of parallel, high-resolution images.

Source: The graphic representation is taken from L. HOLZER, et al., Journal of Microscopy Vol. 216, Part 1, October 2004, pp. 84–95

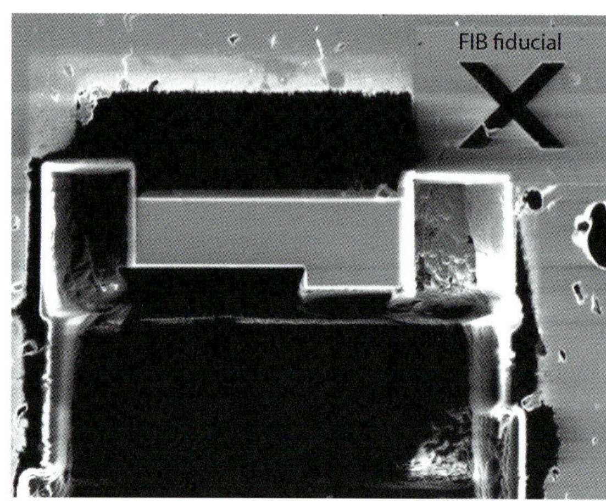

Figure 8. After the area of interest has been defined, the alignment between slices is guaranteed by calibrating on two recognition marks, one for the ion beam and one for the electron beam.

accelerating voltages to prevent charge buildup without compromising resolution or contrast. The actual accelerating voltage needed depends on the sample and probably on sample size and geometry. In the case of a Marcellus Shale sample, imaging down to 350 V was done to obtain optimal results. The FIB-SEM data sets in this study were done at 1.5 kV.

Editor's note: The Image Catalog section of this volume contains examples of some of the more common image artifacts.

RESULTS

Large-scale Imaging

An overview mosaic of a section of the sample is presented in Figure 9. This section has been chosen based on the automated petrography results on the full sample surface, as presented in Figure 11. This section consists of a clay-rich matrix between an apatite-rich inclusion at the left and an albite rich region on the right. A 10-µm-wide wide crack (black) and pyrite

Figure 9. High-resolution mosaic of a 2.2-mm-wide section.

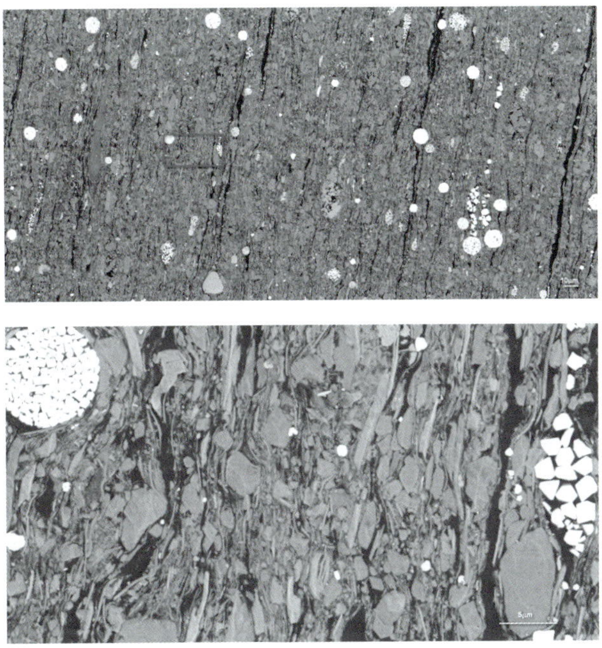

Figure 10. High-resolution details of the 2.2-mm-wide selection. The insets show the common microstructural feature in the form of elongated solid organic particles in a clay-rich matrix.

framboids (white) are clearly visible from the BSE image contrast. The mosaic software allows easy navigation and unlimited zoom-in capability over the entire mosaic. This was used to focus subsequently on a specific area considered representative for the matrix, suggested also by the automated petrography results presented later in the chapter. A high-resolution image mosaic of the specific selection is shown in Figure 10. The smaller dark linear features, aligned vertically, can be identified as non-porous organic matter. The high-resolution mosaic in Figure 9 consists of 308 SEM images, with 90-μm field of view per individual tile. The stitched image is a 3.8-gigabyte TIFF (tagged image file format) file.

Automated Petrography

Two mineral maps were obtained, with respectively 2.5- and 0.5-μm pixel spacing for the EDX

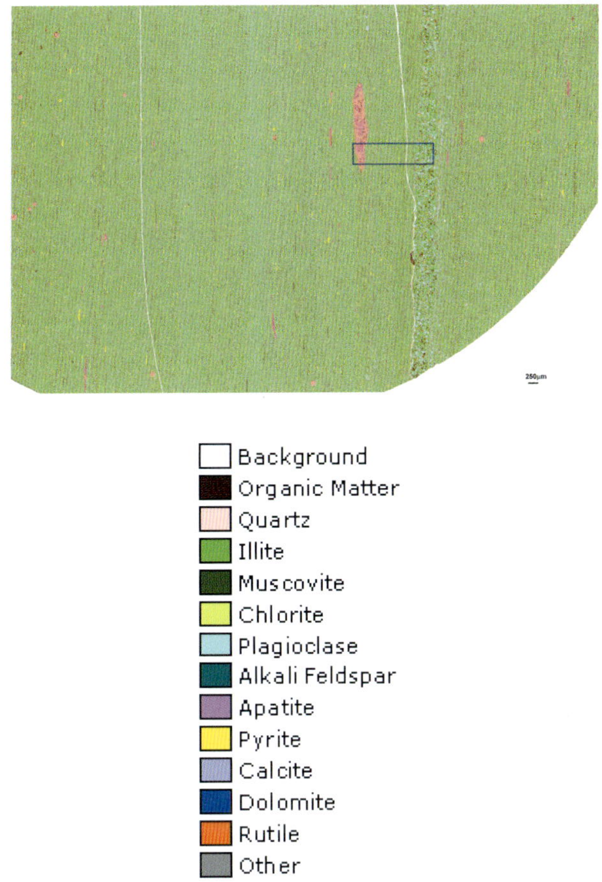

Figure 11. Petrographic image of the full surface (16 mm wide). Almost 30 million energy-dispersive x-ray (EDX) spectra were recorded in 31 hr.

Figure 12. Petrographic image of a smaller section (2.2 mm wide). Over 5 million energy-dispersive x-ray (EDX) spectra were collected in 5 hr. Note the grain size variation at either side of the fracture.

measurements (Figures 11, 12). The 2.5-μm spacing results were obtained on the full surface (16 mm wide). In total, 29,219,838 spectra were measured, analyzed, and attributed to different mineral phases in 31 hours for the full surface; and 5,220,945 spectra in five hours for the smaller section.

The spacing between individual EDX analysis points can be taken as the resolution of the mineral maps. Spacing below 0.5 μm is not recommended as the interaction volume associated with the electron-beam-induced x-rays is of this order of magnitude. For the large-scale BSE images the resolution can be much higher. Most of the images in this study were 4096 pixels wide. For a field of view of 10 μm, this results in 2.4-nm pixel size. As a rule of thumb, one-third of the feature size is used as the pixel size. For instance, to observe pores of 5-nm diameter, we aim for a 1.5-nm pixel size or smaller. In general, this gives good definition of the features of interest. Further distinction needs to be made between BSE images and secondary electron (SE) images. The advantage of backscatter images is that they are a direct representation of the mean atomic number (Z) of the material under the electron beam. Organic matter (low Z) will be dark gray, while pyrite (high Z) will be bright white. This makes image interpretation more intuitive, at least for the mineral constituents. Secondary electron images, on the other hand, show surface contrast much better, and in some cases will also have greater gray-level difference between organic matter and porosity, which facilitates the segmentation of 3-D data sets. Acceleration voltage is also important. Secondary electrons always come from the top surface (a few nanometers deep), but BSEs can come from hundreds of nanometer deep in the sample, depending on their energy (accelerating voltage) and the sample composition. We have come across a case in which the organic matter porosity was obscured by the signal from a heavier phase below in BSE imaging. With SEs, the porosity was observed readily. This effect was not observed in the current study, and BSE imaging was used throughout to make visual interpretations easier.

The higher-density mineral map in Figure 12 was taken over an area of 2.2 mm wide and yielded 5,220,945 spectra in five hours. This area overlaps with the mosaic in Figure 9. The modal mineralogy differences between the full surface and the smaller section appear to be minor (Figure 13), suggesting that the matrix is rather homogeneous over the entire surface, with the exception of the apatite-rich inclusion. The high-resolution BSE mosaics did not show any evidence of porous organic matter. All the organic matter seemed solid in the 2-D image datasets.

Focused Ion-beam Scanning Electron Microscopy

Based on the large-scale imaging and QEMSCAN results, it became clear that a prime candidate for FIB-SEM imaging is the organic matter to see how well connected that is throughout the matrix. Three hundred slices with 10-nm pixel resolution in the slices

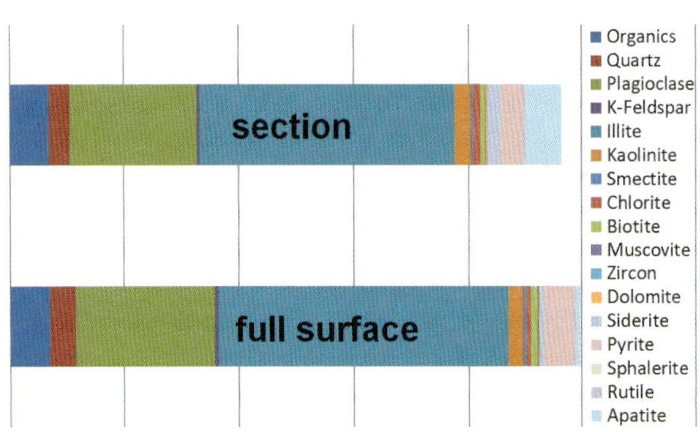

Figure 13. Modal mineralogy comparison between full surface (top) and smaller section (bottom), revealing only minor modal mineralogy differences.

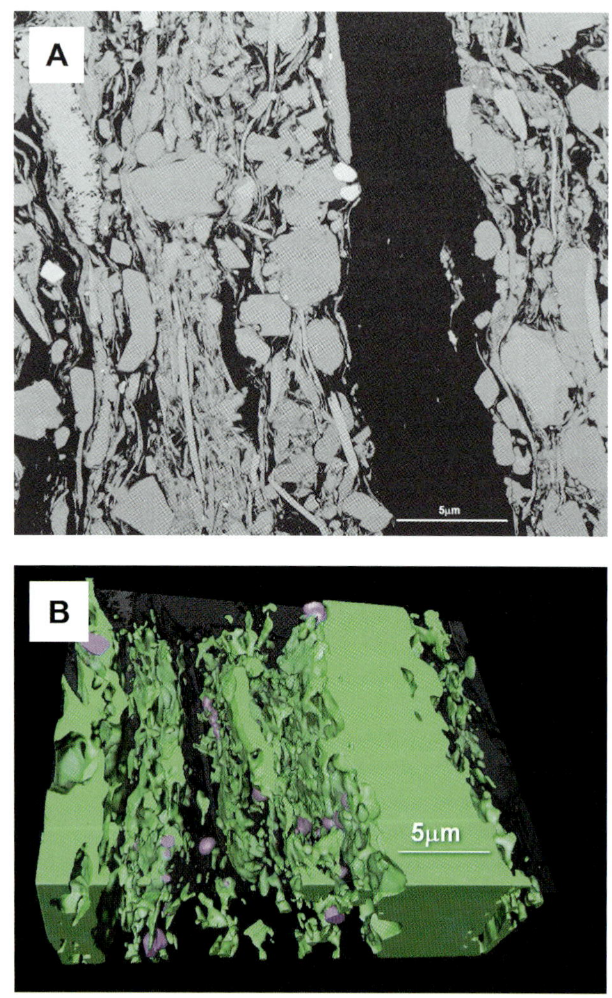

Figure 14. (A) The first slice of the focused ion-beam scanning electron microscopic (FIB-SEM) data set. Horizontal width is 20 μm. (B) The three-dimensional reconstruction on top of a two-dimensional slice shows layers of organic matter with limited connection between layers.

and 10-nm slice spacing were recorded. Figure 14 shows the first slice of this sequence and a gray-level segmented reconstruction. The organic phases are colored green and pyrite is pink. The organic matter is well connected in layers but connection between the layers is limited.

CONCLUSIONS

Large-scale, high-resolution imaging enables core-scale imaging with nanometer pore-scale resolution. The ability to see both fabric and details in one image enables identification of representative features and zooming in without losing correlation between pore and core scale.

Focused ion-beam scanning electron microscopy produces unique 3-D data sets that reveal porosity in shale at the nanometer scale. A major industry concern with this technology is the small volume analyzed and how representative this small area is of the core sample.

Mineral and textural analysis combined with pore-scale imaging at the core scale have identified elongated organic particles to be a dominant feature and hence a prime candidate for the FIB-SEM analysis.

ACKNOWLEDGMENTS

The authors appreciate the support from Paul Fischione and Rocco Cerchiara at Fischione Instruments for polishing the sample with the argon-ion mill. We thank Aukje Benedictus at FEI Australia for the QEMSCAN automated petrography measurements and David Donnet at FEI Netherlands for the Image Mosaics and the FIB-SEM work on the Helios 650 DualBeam.

REFERENCES CITED

Armitage, P. J., R. H. Worden, D. R. Faulkner, A. C. Aplin, A. R. Butcher, and J. Iliffe, 2010, Diagenetic and sedimentary controls on porosity in Lower Carboniferous fine-grained lithologies, Krechba Field, Algeria: A petrological study of a caprock to a carbon capture site: Marine and Petroleum Geology, v. 27, p. 1395–1410.

Curtis, M. E., R. J. Ambrose, C. H. Sondergeld, and C. S. Rai, 2010, Structural characterization of gas shales on the micro and nanoscales: Canadian Society for Unconventional Gas/Society of Petroleum Engineers Canadian Unconventional Resources and International Petroleum Conference, October 19–21, Calgary, Alberta, SPE Paper 137693, 15 p., doi: 10.2118/137693-MS.

Gottlieb, P., G. Wilkie, D. Sutherland, E. Ho-Tun, S. Suthers, K. Perera, B. Jenkins, S. Spencer, A. Butcher, and J. Rayner, 2000, Using quantitative electron microscopy for process mineralogy applications: Journal of the Minerals, Metals and Materials Society, v. 52, p. 24–25.

Knackstedt, M., P. Jaime, A. R. Butcher, P. W. S. K. Botha, J. Middleton, and R. Sok, 2010, Integrating reservoir characterization: 3D dynamic, petrophysical and geological description of reservoir facies: Society of Petroleum Engineers Asia Pacific Oil and Gas Conference and Exhibition, October 18–20, Brisbane, Queensland, SPE Paper 133981, 9 p., doi: 10.2118/133981-MS.

Walls, J. D., and S. W. Sinclair, 2011, Eagle Ford Shale reservoir properties from digital rock physics: European Association of Geoscientists and Engineers, First Break, v. 29, p. 97–101.

5

Roger M. Slatt, Neal R. O'Brien, 2013, Microfabrics related to porosity development, sedimentary and diagenetic processes, and composition of unconventional resource shale reservoirs as determined by conventional scanning electron microscopy, *in* W. Camp, E. Diaz, and B. Wawak, eds., Electron microscopy of shale hydrocarbon reservoirs: AAPG Memoir 102, p. 37–44.

Microfabrics Related to Porosity Development, Sedimentary and Diagenetic Processes, and Composition of Unconventional Resource Shale Reservoirs as Determined by Conventional Scanning Electron Microscopy

Roger M. Slatt

University of Oklahoma, 100 East Boyd St., Suite 810, Norman, Oklahoma, 73019, U.S.A. (e-mail: rslatt@ou.edu)

Neal R. O'Brien

State University of New York at Potsdam, 44 Pierrpont Ave., Potsdam, New York, 13676, U.S.A. (e-mail: obriennr@potsdam.edu)

ABSTRACT

This chapter describes and illustrates features of shales and mudstones at the nanometer and micrometer scales using standard scanning electron microscopy (SEM) and field emission scanning electron microscopy (FE-SEM) techniques. Microfabric observations at these scales not only provide insights into sedimentary and postdepositional processes but also offer evidence useful in understanding storage and primary migration patterns in unconventional shales. The images illustrated are suggested as references to guide future shale studies related to shale porosity and permeability. Examples are provided of various shales (Barnett, Woodford, Eagle Ford, Rhinestreet, Fayetteville, and Marcellus). Microfabric and pore features illustrated include clay flakes related to open-network floccules and clay-aligned fabric, plus other features produced by nonclay minerals. Organic matter produced by zooplankton and algae (e.g., coccolithophores, *Tasmanites*) is described because it may form organic mucus that adheres to and helps aggregate clay flakes. Organic matter is also common within porous fecal pellets. Coccolithophores, sponge spicules, and foraminifera tests contain hollow internal chambers, which provide porosity and probably permeability, even when filled with clay minerals. Conventional SEM provides a rapid and relatively inexpensive way of evaluating pores and microfabrics in shales.

INTRODUCTION

Scanning electron microscopy (SEM) is a standard technique for identifying nanometer- and micrometer-scale features of shales. Shale fabric at these scales provides information on the origin, sedimentary

DOI: 10.1306/13391703M102441

and postdepositional (diagenetic) processes, and pore networks; thus, it is of value in the exploration and exploitation for shale dry and wet gas and oil. Understanding storage and flow capacity of unconventional shales using standard SEM and field emission (FE)-SEM techniques has been recently significantly advanced (Loucks and Ruppel, 2007; Loucks et al., 2009; Milner et al., 2010; Slatt and O'Brien, 2011). Most notably, nanometer- and micrometer-size pores have been identified within the Barnett Shale using secondary electron and backscatter electron techniques in combination with argon-ion milling to identify nanoscale porosity in organic matter (Loucks et al., 2009; Milner et al., 2010), which has been considered by some (Ambrose et al., 2010; Sondergeld et al., 2010) to be the only significant pore type for gas storage. However, other types of pores in the nano- and micrometer-size range have been identified by Milner et al. (2010) and Slatt and O'Brien (2011), some of which may be at least equally important for storage in and transfer of hydrocarbons through shales. As pointed out by Milner et al. (2010), pores in the range of 10 μm (0.000393700790 in.) may not be imaged when viewed under focused ion-beam SEM of 1- to 5-μm (3.9370079e-05- to 0.00019685040-in.) shale cubes. Shale microfabrics have been documented previously by O'Brien and Slatt (1990) using standard SEM and FE-SEM techniques. This chapter discusses the role of microfabrics, as viewed under SEM and FE-SEM, in understanding unconventional shale microfabrics.

Methodology

We provide a pictorial frame of reference of various microfabric features common to shales, suggesting their usefulness to researchers using conventional electron microscopic techniques in their studies.

Our images of freshly broken surfaces were gold coated and viewed in the secondary electron mode and backscatter mode using both standard SEM and FE-SEM. The main advantage of FE-SEM is that the field emission source produces a narrower radius electron beam than a conventional tungsten filament electron gun, resulting in higher-resolution SEM images. Samples were also analyzed by energy-dispersive x-rays for elemental content and distribution. Silt-size grains may pop out during sample preparation, giving a false impression of a pore (Figure 1). These depressions resemble pores except that the bottom of the pop-out crater is easily recognized plus the clay flakes, which are wrapped around the hole once occupied by a grain, are still intact tangentially to the empty crater.

Figure 1. Quartz grains (Q) in place in shale plus an example of a pop-out cavity left by a quartz grain in shale matrix (arrow). Notice bottoms of pop-out holes are visible, and clay flakes wrap around the crater at the arrow. Fayetteville Shale (Mississippian, Arkansas).

Pore bottoms are not apparent in the open microfabric of a clay floccule.

We stress at the start of this discussion how to recognize real pore morphology at the microlevel from other features that might be identified—wrongly—as pores. Images shown in the following sections on microfabrics are interpreted based on their characteristics of real pores. Recognition of the differences in morphology is important in porosity determination.

MICROFABRICS

Flocculated Clay Microfabric

Various unconventional resource shale samples, plus other shales, are shown in this chapter. For a review of the relationships of various depositional processes and shale microfabrics, refer to the works of Bennett et al. (1991), Slatt and O'Brien (2011), and Abouelresh and Slatt (2011).

Electrolyte-rich saltwater promotes clay flake aggregation or flocculation. The floccule fabric is an open arrangement of individual, randomly oriented clay flakes (Figure 2A). Domains of stair-stepped parallel flakes may also form "the cardhouse microfabric" (Bennet et al., 1991; Slatt and O'Brien, 2011) (Figure 2B). This microfabric with open, porous, randomly oriented clay flakes is preserved as a primary fabric in many ancient, nonbioturbated, dark gray-to-black shales (O'Brien and Slatt, 1990) (Figure 3A–B). Pores are an integral part of the flocculated clay cardhouse microfabric. Bioturbation may later randomize

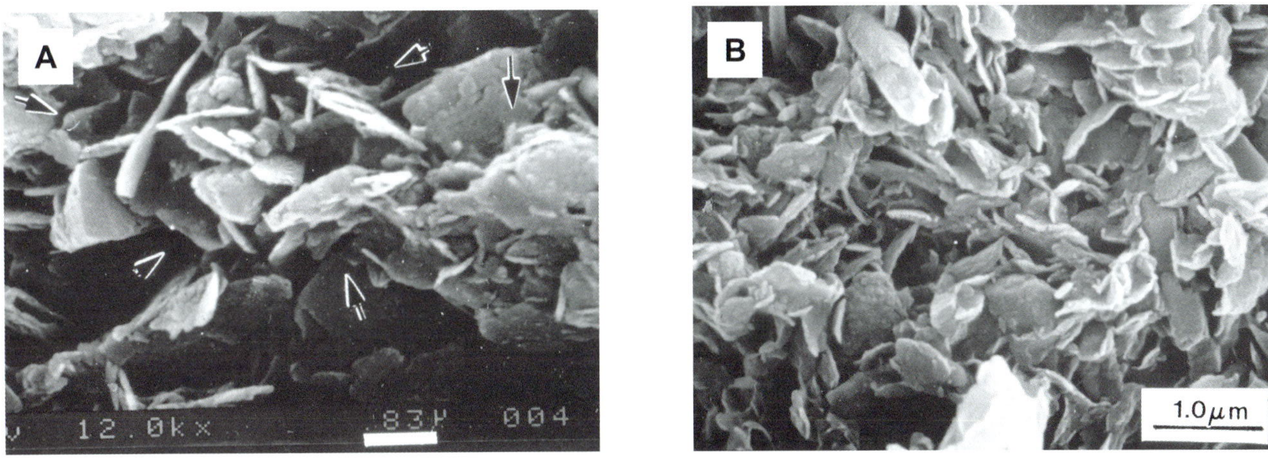

Figure 2. (A) Floccules in lake clay preserved in Pleistocene raised terrace, Great Salt Lake, Utah. Notice the cardhouse fabric of single flakes and porosity. (B) Floccules in lake clay, Pleistocene outcrop, Glens Falls, New York. Notice randomly oriented clay flakes.

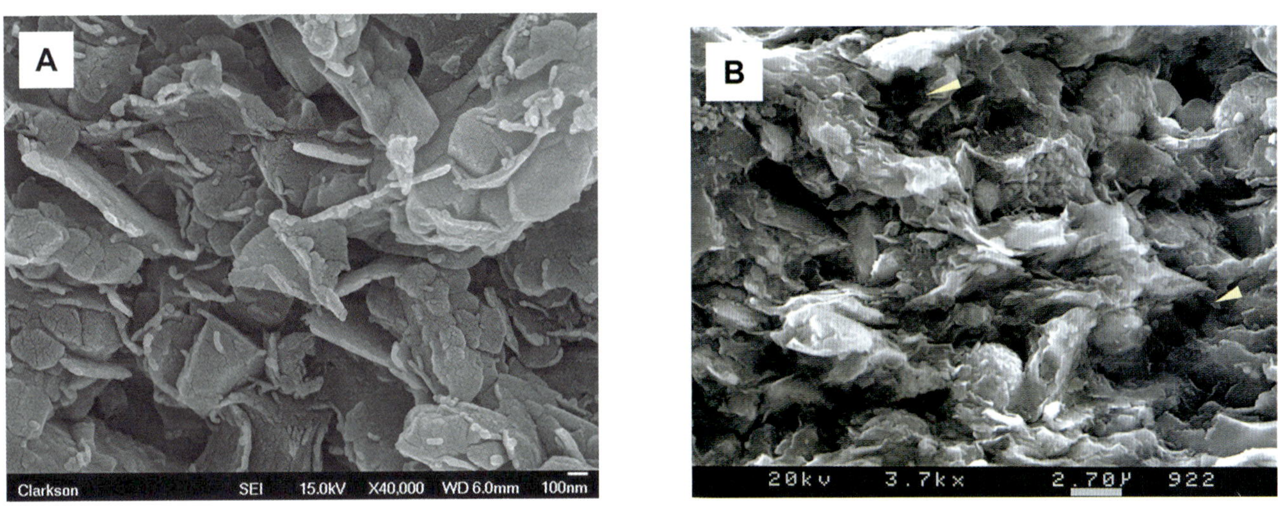

Figure 3. (A) Floccules in Eagle Ford Shale (Cretaceous, Texas). (B) Random clay flake orientation in floccules (arrows), Marcellus Shale (Devonian, New York).

the original primary fabric. Hand specimens or thin sections distinguish primary flocculated microfabric from that caused by bioturbation (O'Brien, 1987). Open porous clay floccules thus provide primary porosity (and probably permeability) in organic-rich shale (Figure 4).

Parallel-subparallel Clay Flakes

Our observations indicated that microfabric is influenced by the amount of silt within flocculated sediment. The clay flakes within silt-poor flocculated clay on burial could more easily rotate into a parallel or subparallel position (Figure 5). The resulting final microfabric produced may still provide small pores between overlapping flakes. Figure 6 illustrates the influence of increased silt content on final microfabric.

Nonclay Minerals

The amount of silt-size particles within clayey sediment may influence final microfabric and consequently porosity. Examples of the most common

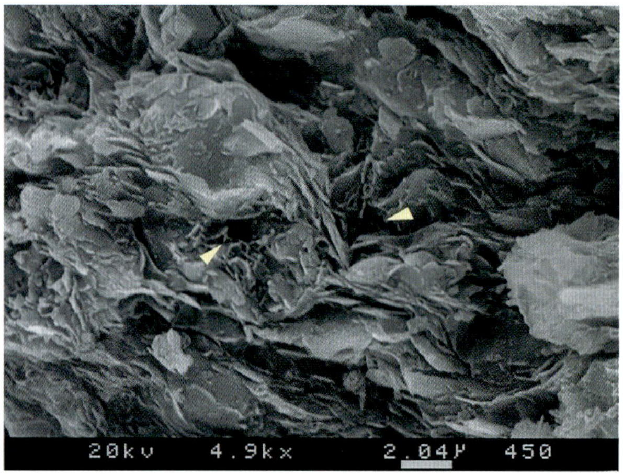

Figure 4. Pores in floccules (arrows), Woodford Shale (Devonian-Mississippian, Texas).

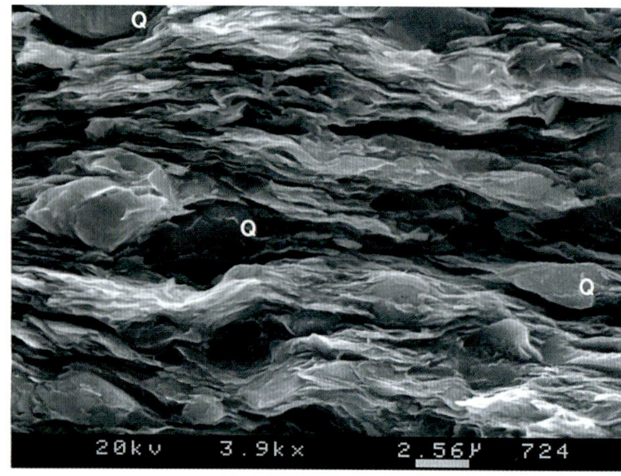

Figure 5. Parallel clay flake orientation in the Fayetteville Shale (arrows; Q = quartz grains).

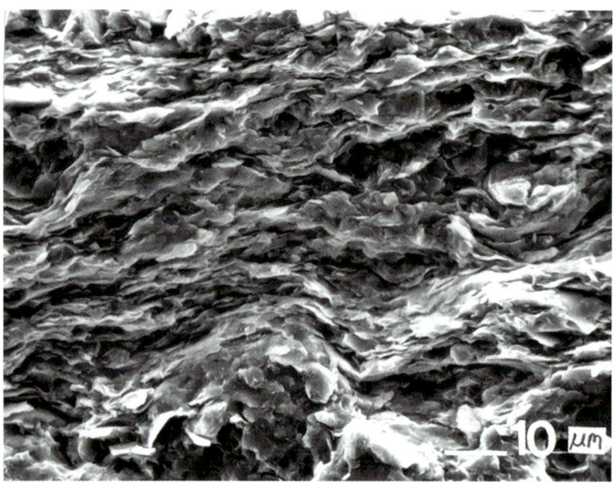

Figure 6. Subparallel clay flake orientation slightly disrupted by silt grains, Rhinestreet Shale (Devonian, New York).

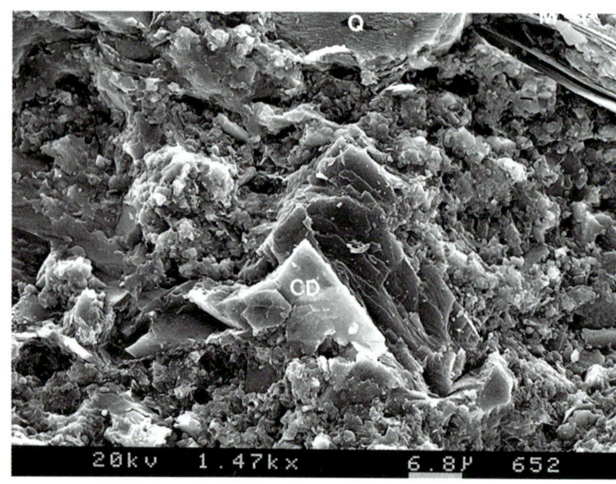

Figure 7. Examples of various types of mineral grains in a clay matrix: M = mica, Q = quartz, CD = carbonate (calcite or dolomite). Woodford Shale (Devonian-Mississippian, Texas).

Figure 8. Pores in typical framboid. Woodford Shale (Mississippian-Devonian, Oklahoma). Arrows show pores; notice clay flakes wrap around edge of framboid.

silt-size minerals are quartz, mica, carbonate-calcite/dolomite (Figure 7) and pyrite in framboids (Figure 8). Their identification should be confirmed by energy-dispersive x-ray analysis. Their presence contributes to the randomness of microfabric and thus influences porosity. However, even pyrite framboids may contain micro- or nanoscale pores (Figure 8).

Organic Matter Sources

Organic-rich shales form a significant part of the geological record and thus deserve attention, not only because of their economic potential for gas and oil generation and storage but also because of the specific sedimentary processes by which they formed.

Numerous microscopic organisms like salphs, copepods, and other zooplankton jettison their organic-rich

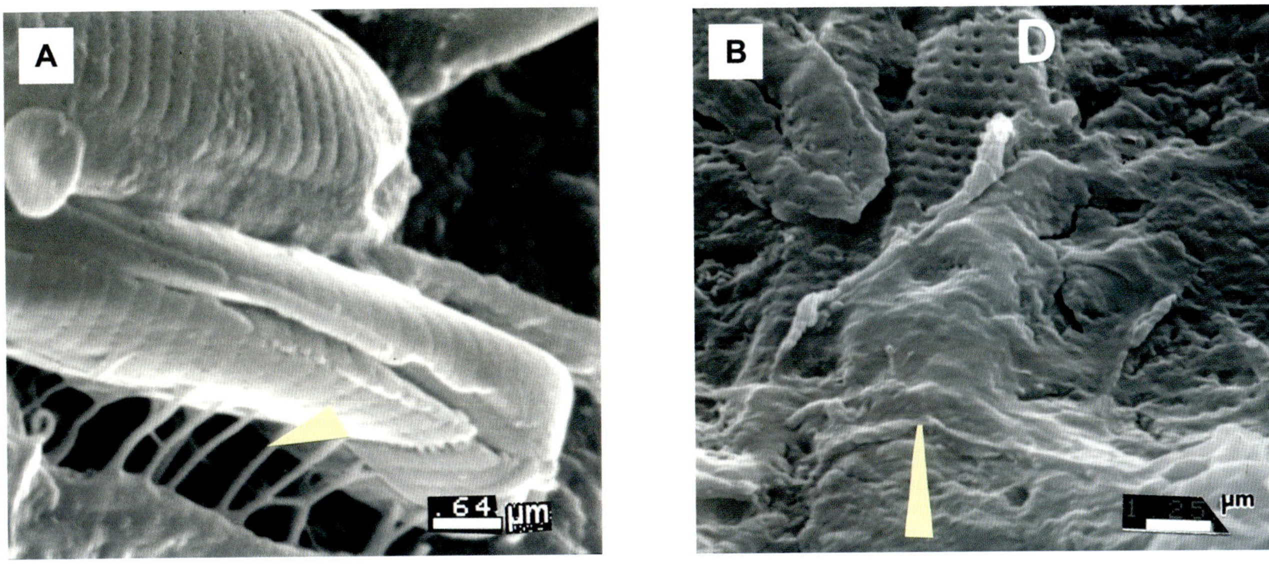

Figure 9. (A) Example of strands (arrow) of organic matter (EPS, extracellular polysaccharide; commonly called slime or mucus) extruded by diatoms grown in the laboratory. (B) A mucus EPS slimy mat (arrow) covering a diatom (D) in the Lake Florissant Shale (Oligocene Florissant Formation, Colorado).

filtering membranes when clogged. Microscopic diatoms and other phytoplankton also secrete a mucous film (Figure 9A). This sticky organic matter may entrap suspended inorganic (clay, silt) particles and other organic matter and is often referred to as "marine snow." Bennet et al. (1991) described marine snow as a "biosediment aggregate," which transports to the seafloor detrital particles trapped within an organic mass, ultimately forming organic-rich sediment. Diatoms and other zooplanktons also secrete polysaccharides called extracellular polysaccharide secretions (commonly called slime or mucus) during periodic blooms in oceans and lakes. O'Brien et al. (2008) described this mechanism as responsible for preserving fossils in the Florissant Shale (Figure 9B).

Other sources of organic matter in lake and ocean water described by Turner (2002) are fecal pellets, proposed by others as organic matter in shales (Loucks et al., 2009; Milner et al., 2010; Slatt and O'Brien, 2011). The fecal pellets preserved in shales may be a millimeter or less in diameter, suggesting a zooplanktonic source such as copepods or other minute organisms. Gowing and Silver's (1985) study of modern marine minipellets (3- to 50-µm [0.00011811024- to 0.0019685040-in.] diameter) calculated pellet carbon content was from 17.3 to 29.7% dry weight. Preservation of organic matter in pellets is consistent with their abundance in black shales in that they appear to be deposited in an anoxic environment with little to no bioturbation, where organic matter is preserved. Their presence in such an environment generally devoid of fauna is another indicator that they originate from fauna living within the overlying water column. Microfabric analysis shows that fossil fecal pellets have considerable porosity and a high carbon content: conditions necessary as both sources and storage sites for hydrocarbons (Figure 10).

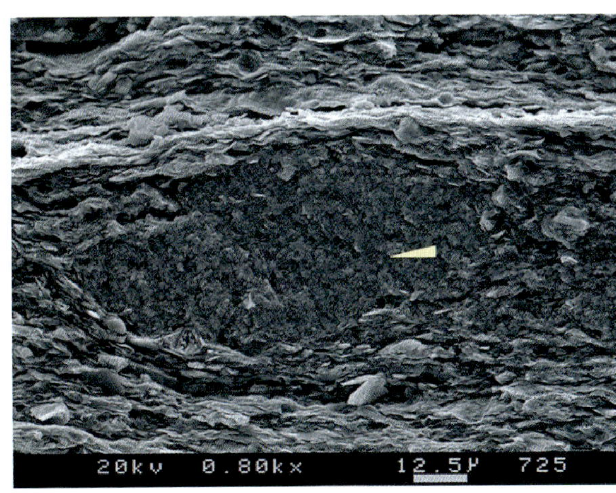

Figure 10. Fecal pellet (arrow) in Fayetteville Shale (Mississippian, Arkansas).

Certain hydrocarbon-rich source rocks (e.g., Kimmeridge and Eagle Ford formations) contain zones of calcareous algae-coccolithophores that are interbedded within organic-rich shaley intervals. Whole and broken coccolithophore particles, coccolith platelets, and hollow spines in the Eagle Ford provide abundant nano- and micropores, which provide both porosity

Figure 11. Coccoliths in the Eagle Ford Shale (Cretaceous, Texas). C = whole coccolith plate, X = coccolith plate fragments, S = cross section of a coccolithophore spine.

and permeability to store and transmit hydrocarbon molecules (Figure 11).

Pores also occur in the microfabric of sponge spicules, such as in the Barnett Shale (Figure 12A). They can be abundant in clay and silty laminae and beds. The interior chambers of many of these long, siliceous tubes may be filled with sediment or authigenic minerals, but some chambers can be hollow and provide potential storage sites for hydrocarbon molecules. The tests of foraminifera may also remain open on burial or become partially or completely filled with authigenic clays, thus providing nanometer- to micrometer-size pore spaces (Figure 12B).

Tasmanites algal cysts are common organic fragments in some shales, such as the Woodford Shale (Figure 13A). On burial, the hollow, organic cyst body may totally collapse or remain partially open, providing nanometer- and micrometer-size pores (Figure 13B).

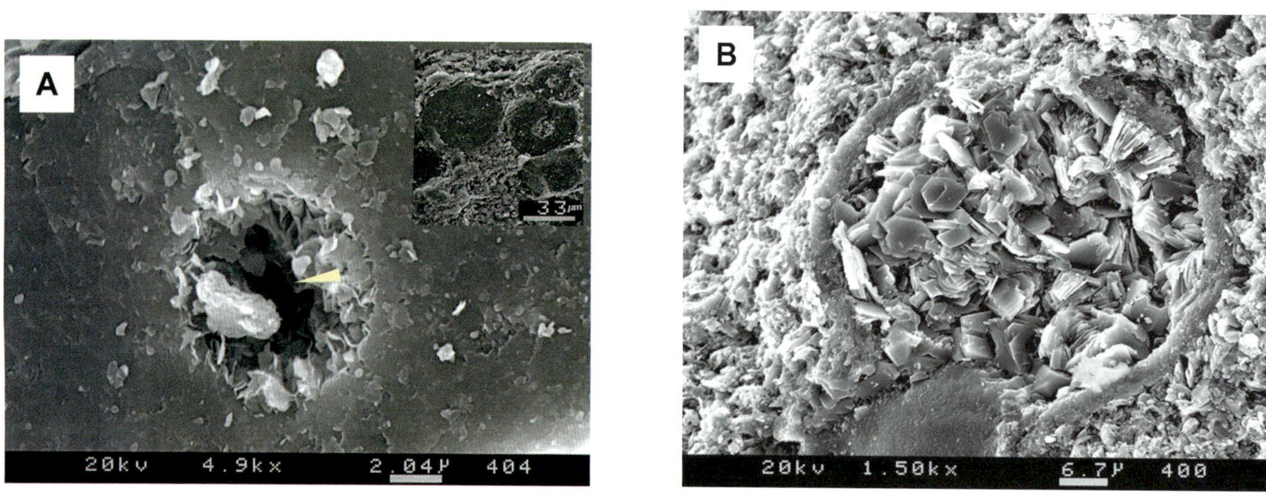

Figure 12. (A) Sponge spicule in cross section perpendicular to long axis (arrow shows cross section of a pore along sponge axis). Inset (upper right): A group of spicules in cross-sectional view. Barnett Shale (Mississippian, Texas). (B) Authigenic kaolinite filling foraminifera test. Eagle Ford Shale (Cretaceous, Texas).

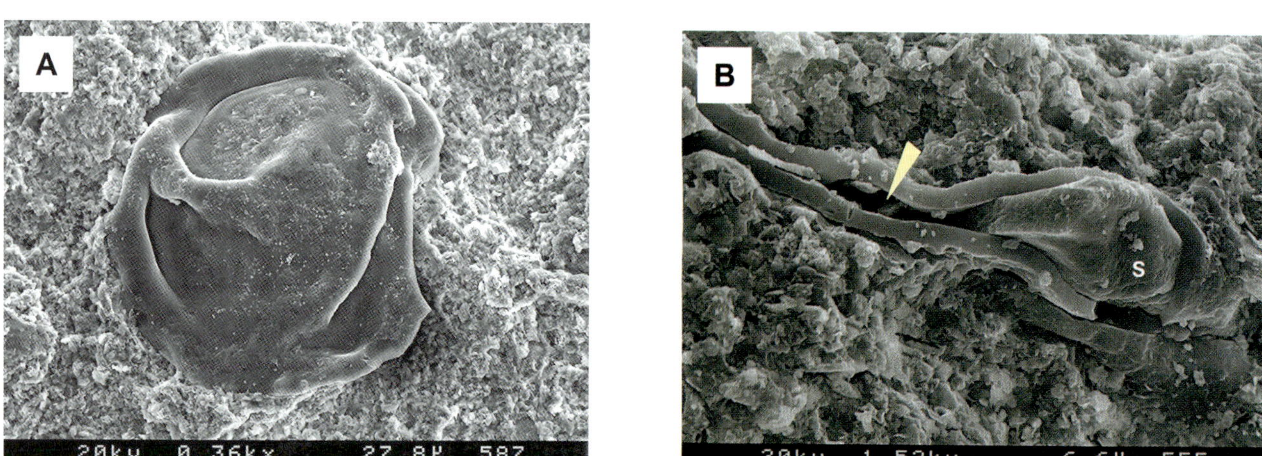

Figure 13. (A) Top view of a collapsed *Tasmanites* cyst on the bedding surface. Woodford Shale (Devonian-Mississippian, Oklahoma). (B) Partially collapsed (side view) *Tasmanites* cyst filled with authigenic SiO_2 (S). Notice pore space at arrow. Woodford Shale (Devonian-Mississippian, Oklahoma).

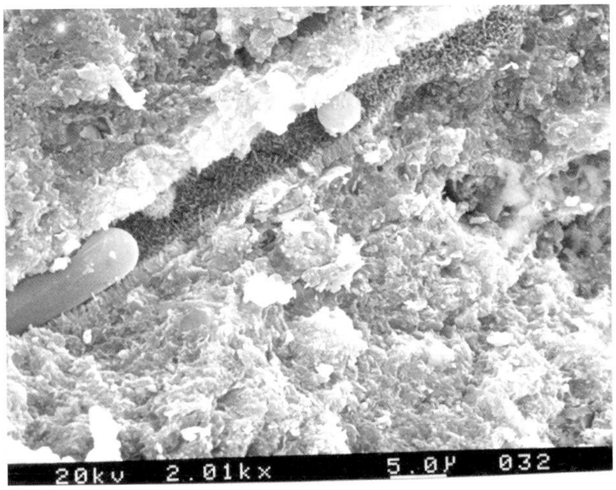

Figure 14. Oil droplets in a diatom frustule microchannel. Monterey Formation (Miocene, California).

Microfabric analysis also reveals the presence of microchannels. In the Miocene Monterey Formation, elongate tubes of centric diatom frustules are partially filled with pyrobitumen and remnants of oil droplets, which cling to the tube walls (Figure 14) (O'Brien et al., 1996). O'Brien et al. (1996) concluded that a pressure gradient produced during hydrocarbon migration propelled the droplets to aggregate and move along channels.

Other types of microchannels shown in Figures 15 A, B, C are interpreted as primary features; that is, they are not artifacts produced during sample preparation or because of unloading when cores are brought to the surface. They may also provide primary migration routes for gas and oil molecules.

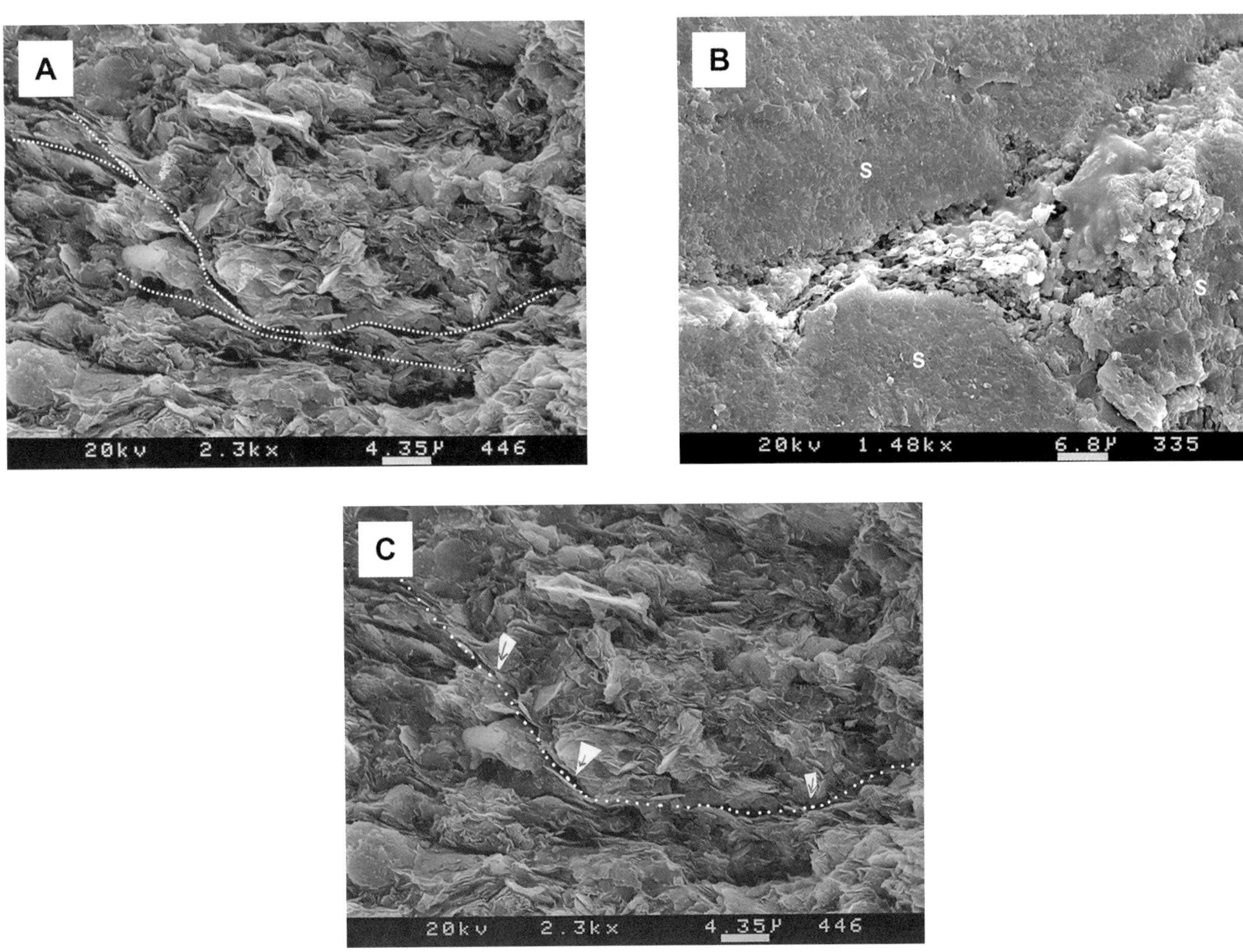

Figure 15. (A) Microchannels (dashes) within clay matrix of the Woodford Shale (Devonian-Mississippian, Texas). (B) Microchannel between sponge spicules (S) and shale matrix in the Barnett Shale (Mississippian, Texas). (C) Microchannel (dots) in the Marcellus Shale (Devonian, New York).

SUMMARY

Scanning electron microscopic microfabric analysis of rough, broken shale surfaces, as described in this chapter, provides a useful technique for characterizing unconventional resource shales that compliment other SEM techniques. This technique provides high-resolution textural (topographic) information that cannot be obtained from smoothed (argon-ion-milled) surfaces and allows for recognition and description of larger-scale features (>5 μm [0.00019685040 in.]) that cannot be evaluated by focused ion-beam SEM studies because of extremely small sample sizes. Examples provided in this chapter illustrate microfabric features common in many shales and mudstones. Such features include porous clay floccules, nonclay minerals, organic matter, and microchannels. These features may influence development and preservation of pores and thus reservoir quality (porosity and permeability) in unconventional shale reservoirs. The SEM techniques described in this chapter provide an easy, rapid, and inexpensive method for viewing and quantifying these features.

ACKNOWLEDGMENTS

Roger Slatt thanks the following students within the University of Oklahoma (OU) Institute of Reservoir Characterization for providing samples and data: Jean David Amorocho, Carlos Ceron, Andrea Micelli-Romero, Heidyl Rodriguez, Emilio Torres, Matthew Totten, and Majia Zheng. Neal O'Brien thanks Ted Champagne of Clarkson University and members of the Potsdam College (SUNY Potsdam) Shale Research Team for technical assistance. Special thanks go to Roberta Greene for the manuscript and graphics preparation. Partial funding for this research was provided by the OU-SUNY Shale Pore Research Consortium of companies (Anadarko, Black Diamond, Chesapeake, Concho, Exxon-Mobil, Marathon, Nexen, Newfield, QEP, Saudi Aramco, and Southwestern Energy), which we thank.

REFERENCES CITED

Abouelresh, M.S. and R.M. Slatt, 2011, Shale depositional processes: Example from the Paleozoic Barnett Shale, Fort Worth Basin, Texas, Cent. Europ. J. Geosciences, 12 p.

Ambrose, R. J., R. C. Hartman, M. Diaz-Campos, Y. Akkutlu, and C. H. Sonderged, 2010, New pore-scale considerations for shale gas in place calculations: Society of Petroleum Engineers Unconventional Gas Conference, February 23–25, Pittsburgh, Pennsylvania, SPE Paper 131772, 17 p., doi: 10.2118/131772-MS.

Bennett, R. H., N. R. O'Brien, and M. H. Hulbert, 1991, Determinants of clay and shale microfabric signatures: Processes and mechanisms, *in* R. H. Bennett, W. R. Brant, and M. H. Hulbert, eds., Microstructure of fine-grained sediments: From mud to shale: New York, Springer-Verlag, 582 p.

Gowing, M. M., and M. W. Silver, 1985, Mini pellets: A new and abundant size class of marine fecal pellet: Journal of Marine Research, v. 43, p. 395–418.

Loucks, R. G., and S. C. Ruppel, 2007, Mississippian Barnett Shale: Lithofacies and depositional setting of a deep-water shale-gas succession in the Fort Worth Basin, Texas: AAPG Bulletin, v. 92, p. 579–601, doi: 10.1306/11020606059.

Loucks, R. G., R. M. Reed, S. C. Ruppel, and D. M. Jarvie, 2009, Morphology, genesis, and distribution of nanometer-scale pores in siliceous mudstones of the Mississippian Barnett Shale: Journal of Sedimentary Research, v. 79, p. 848–861, doi: 10.2110/jsr.2009.092.

Milner, M., R. McLin, and J. Petriello, 2010, Imaging texture and porosity in mudstones and shales: Comparison of secondary and ion-milled backscatter SEM methods: Canadian Society for Unconventional Gas/Society of Petroleum Engineers Canadian Unconventional Resources and International Petroleum Conference, October 19–21, Calgary, Alberta, SPE Paper 138975, 5 p., doi: 10.2118/138975-MS.

O'Brien, N. R., 1987, The effects of bioturbation on the fabric of shale: Journal of Sedimentary Research, v. 57, p. 449–455.

O'Brien, N. R., and R. M. Slatt, 1990, Argillaceous rock atlas: New York: Springer-Verlag, 141 p.

O'Brien, N. R., G. Thyne, and R. M. Slatt, 1996, Morphology of hydrocarbon droplets during migration: Visual example from the Monterey Formation (Miocene), California: AAPG Bulletin, v. 80, p. 1710–1718.

O'Brien, N. R., H. W. Meyer, and I. C. Harding, 2008, The role of biofilms in fossil preservation, Florissant Formation, Colorado: GSA Special Paper 435, p. 19–31.

Slatt, R. M., and N. R. O'Brien, 2011, Pore types in the Barnett and Woodford shales: Contribution to understanding gas storage and migration pathways in fine-grained rocks: AAPG Bulletin, v. 95, p. 2017–2030.

Sondergeld, C. H., R. J. Ambrose, C. S. Rai, and J. Moncrieff, 2010, Microstructural studies of gas shale: Society of Petroleum Engineers Unconventional Gas Conference, February 23–25, Pittsburgh, Pennsylvania, 2010, SPE Paper 131771, 17 p., doi: 10.2118/131771-MS.

Turner, J. T., 2002, Zooplankton fecal pellets, marine snow, and sinking phytoplankton blooms: Aquatic Microbial Ecology, v. 27, p. 57–102, doi: 10.3357/ame027057.

6

David R. Spain, Ryan McLin, 2013, SEM characterization of shale gas reservoirs using combined secondary and backscatter electron methods: An example from the Haynesville Shale, Texas and Louisiana, *in* W. Camp, E. Diaz, and B. Wawak, eds., Electron microscopy of shale hydrocarbon reservoirs: AAPG Memoir 102, p. 45–52.

SEM Characterization of Shale Gas Reservoirs Using Combined Secondary and Backscatter Electron Methods: An Example from the Haynesville Shale, Texas and Louisiana

David R. Spain

BP America, Inc., 501 Westlake Park Blvd., Houston, Texas, 77079, U.S.A. (e-mail: david.spain@bp.com)

Ryan McLin

TerraTek, a Schlumberger Company, 1935 S. Fremont Dr., Salt Lake City, Utah, 84104, U.S.A. (e-mail: rmclin@slb.com)

ABSTRACT

Secondary electron (SE) images of rough three-dimensional shale surfaces and backscattered electron (BSE) images of smoothed argon (Ar)-ion-milled surfaces provide valuable information to the study and understanding of texture, composition, and pore systems in unconventional shale reservoirs. This chapter describes a method to acquire both traditional SE and Ar-ion-milled BSE images from the same sample by acquiring images from adjacent sample splits. This method allows side-by-side comparisons of traditional SE and Ar-ion-milled BSE images, providing much more information than can be obtained from either method alone. This dual method of imaging representative sister samples, capturing the same features at the same scale, gives powerful information that could not be effectively addressed before. The traditional scanning electron microscopic (SEM) method images secondary electrons that are ejected from the k-orbitals of the specimen atoms by inelastic scattering interactions, with beam electrons giving a large depth of field on high-relief surfaces. This methodology cannot fully assess the exact location of the dominant porosity type or the nature and distribution of the organic matter on the high-relief surface. These considerations are essential to petrographic evaluation of reservoir and completion quality in unconventional shale reservoirs. SEM samples prepared using Ar-ion milling with BSE detection allow us to effectively evaluate the nature and distribution of the organics and the porosity by looking at electron density contrast or Z-contrast. Lower-density material, such as organic matter, appears darker, and higher-density material, such as quartz, appears brighter in SEM images. Void space, such as porosity, appears black. However, this method cannot accurately determine texture or differentiate some mineral cement because

DOI: 10.1306/13391704M102434

many minerals are of similar densities; hence, they have similar Z-contrasts. Using both techniques in tandem for representative sister samples on similar sample features and textures makes up for the deficiencies of each method alone. Mudstone samples from the Haynesville Shale are characterized with this technique and illustrate the complex texture, dominant porosity types, and the nature and distribution of organic matter, including bitumen, the solid residue resulting from the thermal conversion of kerogen to hydrocarbons.

INTRODUCTION

Traditional scanning electron microscopic (SEM) studies of shales involve imaging broken rock surfaces in secondary electron (SE) mode. This method produces detailed topographic images from which pore morphology and the texture of the matrix framework may be examined. However, it is difficult to make detailed interpretations of pore types and matrix composition (including organic matter), which are essential for unconventional shale reservoir evaluation. A relatively new SEM technique has been refined for shales that involves collecting backscattered electron (BSE) and SE images from flattened surfaces prepared using Ar-ion milling. This technique allows researchers to effectively evaluate the nature and distribution of pores as well as the composition of organic matter and mineral matrix. The comparison of both traditional SE and Ar-ion BSE techniques of sister samples (Figure 1) provides a powerful tool for gathering information beyond what each method provides alone.

Backscatter electron images record compositional information as BSE intensity as a function of the mean atomic number (Z) of the specimen volume that interacts with the electron beam. High BSE intensity is scaled as light, and low BSE intensity is scaled as dark in standard grayscale digital images. Minerals and substances with high atomic number contrast, such as carbon and pyrite, are easily identified in grayscale images. However, many minerals common in shales have similar Z values (e.g., quartz and clay minerals) that are more difficult to distinguish by grayscale contrast alone. Although BSE intensity of organic matter is very low, void spaces, such as pores and fractures, exhibit even lower BSE intensity, allowing for pore identification in BSE images.

METHODS

Samples examined for this chapter were from gas-productive reservoir rocks of the Late Jurassic Haynesville Shale, spanning well locations from Caddo

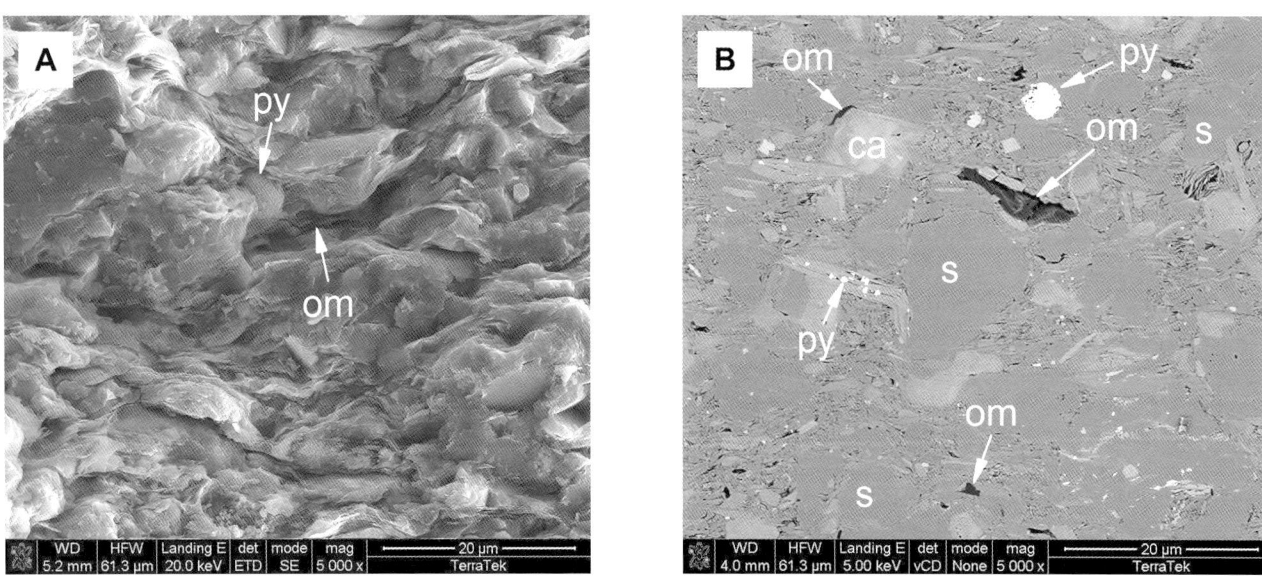

Figure 1. (A) Broken surface, secondary electron scanning electron microscopy (SEM); (B) Ar-ion-milled, backscattered electron SEM; Haynesville Shale, 10,641.4 ft (3243 m); Panola County, Texas. Low-magnification example of two representative sister samples of a shale sample taken at the same scale to illustrate similar features. The sample in (A) was prepared with the traditional method of breaking a fresh surface of the sample perpendicular to bedding and coating the sample with gold as a conductive medium. The sample was then imaged under secondary electron to capture a high depth of field. The sample in (B) was prepared with the AR-ion milling method, left uncoated, and imaged by backscattered electron to capture density contrast. ca = calcite; om = organic matter; py = pyrite; s = silt-size quartz grains. Scale = 20 μm for both images.

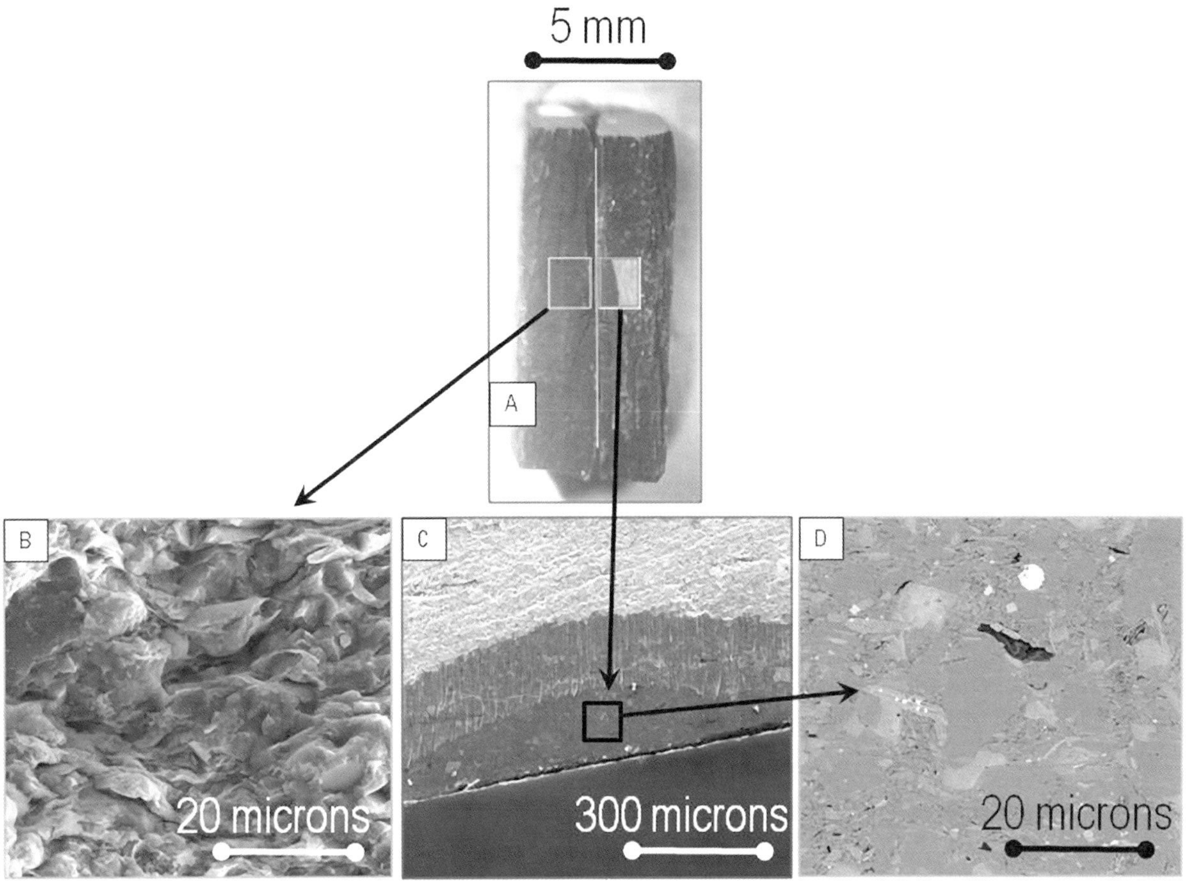

Figure 2. (A) Low-magnification binocular microscopic view; (B) broken surface, secondary electron scanning electron microscopy (SEM); (C) and (D) ion-milled, backscattered electron SEM; Haynesville Shale 10,641.4 ft (3243 m); Panola County, Texas. Representative sister samples of the same piece of shale core (A; scale = 5 mm) were split with one-half prepared the traditional way (B; scale = 20 μm [0.000787401580 in.]) and the other half prepared by Ar-ion milling (C; scale = 300 μm) and (D; scale = 20 μm).

Parish, Louisiana, to Panola County, Texas. A small, freshly broken portion of rock was taken for each sample. Each piece, measuring roughly 5 × 10 × 5 mm, was divided into adjacent halves in an orientation perpendicular to bedding so that textures and features in the two halves were as similar as possible (Figure 2A). The traditionally prepared halves were mounted on a standard aluminum SEM mount and sputter coated with a thin conductive gold coating. A polished area approximately 1000 × 500 μm was etched into the surface of each of the corresponding halves with a JEOL Ar-ion cross-section polisher (Figure 2C).

Imaging was performed using an FEI Novanano 630 scanning electron microscope equipped with an EDAX Genesis energy dispersive x-ray (EDX) spectrometer and a low-voltage, high-contrast detector that attached directly to the pole piece. The samples were then imaged at a range of magnifications to document the morphology of the rock fabric, the pore system, and the nature and distribution of organic material. The annotated, digital SEM images were arranged mainly in pairs, corresponding with similar scaled views of the traditionally prepared SEM samples and the Ar-ion beam milled sample analog (Figures 2A, D). Note that some of the magnifications are dissimilar because of the varying electron detectors and working distances used for each method; hence, it is important to scale similar features identified in each method.

RESULTS AND DISCUSSION

The detailed analysis made possible using the Ar-ion-milling method in combination with the traditional broken surface technique allows for effective identification and interpretation of many features that have direct implications on reservoir and completion quality in shale reservoirs (Spain and Anderson, 2010). For example, in Figure 3, identification of various

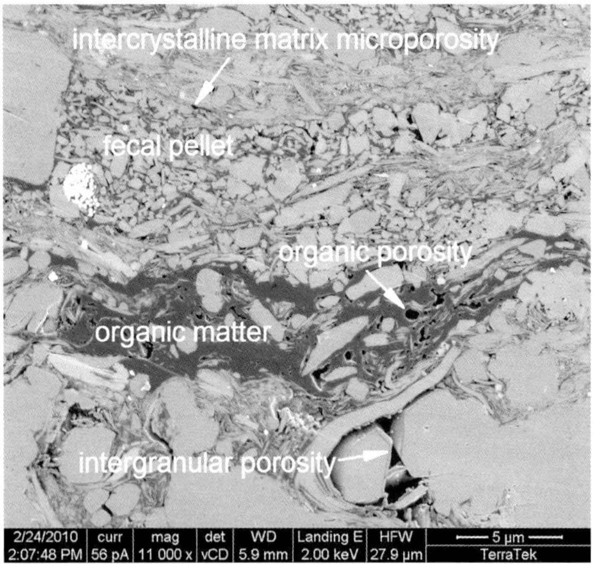

Figure 3. Ar-ion-milled, backscattered scanning electron microscopy (SEM); Haynesville Shale, 11,672.0 ft (3557.6 m); Caddo Parrish, Louisiana. This shale sample was prepared with the Ar-ion surface preparation method and imaged by backscattered density contrast. Complexities of the texture are revealed such as a fecal pellet and organic matter. The discrimination of pore types and sizes, such as intergranular porosity, organic porosity, and intercrystalline matrix microporosity is illustrated (Spain and Anderson, 2010). Scale = 5 μm.

minerals and textures from Ar-ion-milled surfaces imaged in BSE mode is improved over SE imaging alone, which reveals variations in composition (mean atomic number contrast), aiding interpretation of the textural features in the rock. It is now easier to identify certain minerals that may be sensitive to certain completion fluids (e.g., potential swelling clays; iron-rich, acid sensitive carbonates; etc.), the nature and distribution of the organic matter, and the different pore types, pore morphology, and potential connectivity of the pore system.

The advances of the Ar-ion-milling techniques and high-resolution SEM microscopy allow for description and analysis of extremely small pores tens of nanometers in diameter (Figures 4A, B), which is the upper end of most organic-matter-hosted pores reported by others (Loucks et al., 2009; Nelson, 2009; Milner et al., 2010; Schieber, 2010).

One inherent drawback of the SEM instrumentation used in the FEI Novanano 630 is in the design of the low-voltage, high-contrast detector. The detector attaches directly to the pole piece inside the SEM chamber and partially blocks the EDX detector from effectively gathering a spectral analysis. We suggest a correction for this instrumentation issue by using a JEOL JSM-7500F field emission SEM, which allows a clear view of the EDX detector under low-voltage conditions.

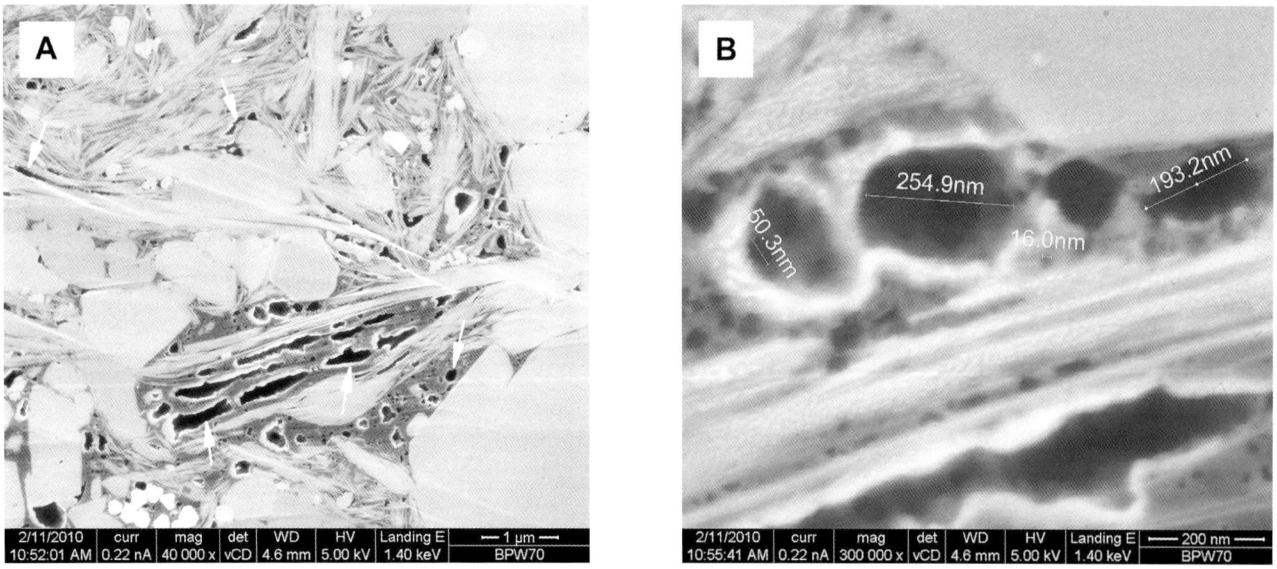

Figure 4. (A) and (B) Ion-milled, backscattered electron scanning electron microscopy (SEM); Haynesville Shale, 11,678.0 ft (3559.5 m); Caddo Parrish, Louisiana. This shale sample was prepared with the Ar-ion surface preparation method and imaged by backscattered density contrast. A low-magnification view (A; scale = 1 μm) is expanded to greater magnification (B; scale = 200 nm) to show organic-hosted micropores that are approximately 16–255 μm in diameter.

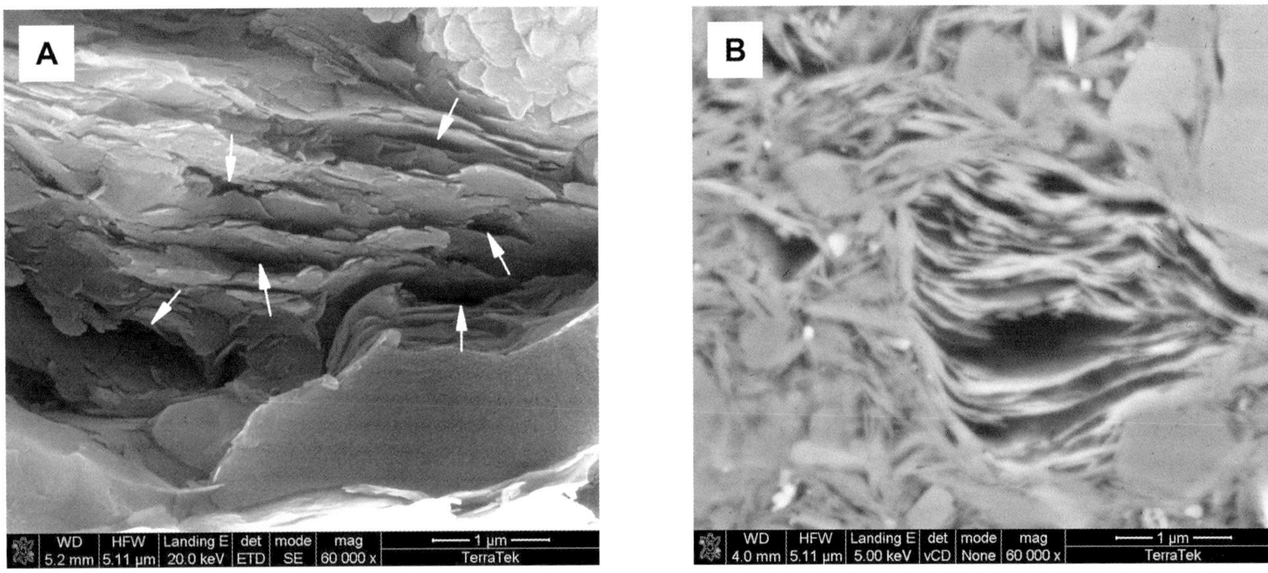

Figure 5. (A) Broken surface, secondary electron scanning electron microscopy (SEM); (B) ion-milled, backscattered electron SEM; Haynesville Shale, 10,641.4 ft (3243.5 m); Panola County, Texas. High-magnification view of sister samples illustrating matrix intercrystalline micropores located between compacted floccules of illitic clays. (A) High-resolution, three-dimensional view of clays with arrows indicating the flattened and elongated pore morphologies. (B) Two-dimensional view of matrix intercrystalline pores.

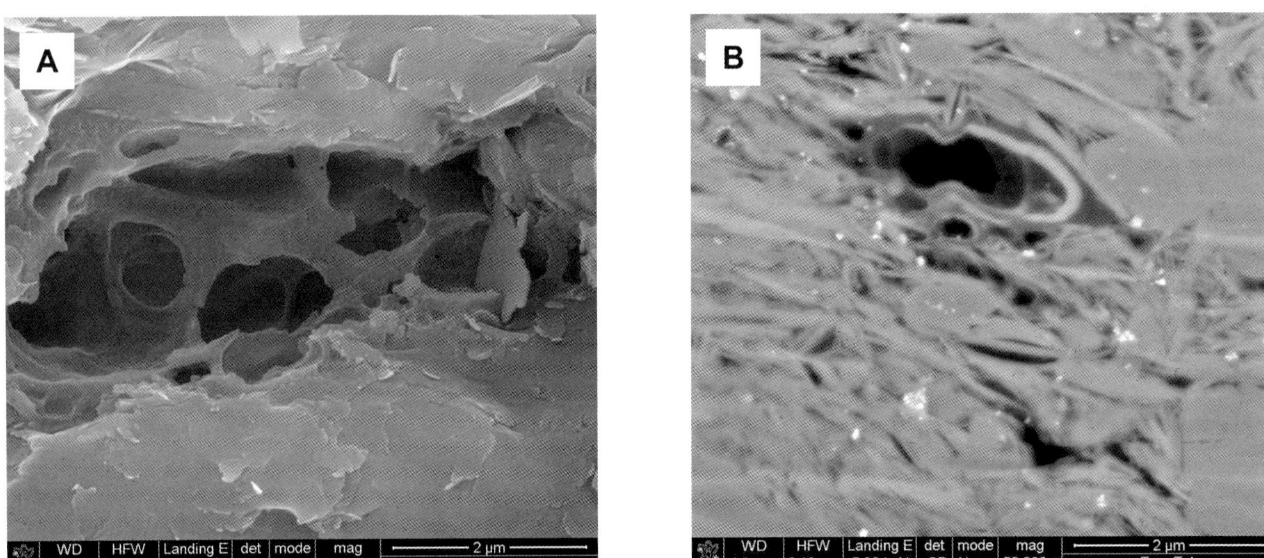

Figure 6. (A) Broken surface, secondary electron scanning electron microscopy (SEM); (B) ion-milled, backscattered electron SEM; Haynesville Shale, 10,685.9 ft (3257.1 m); Panola County, Texas. High-magnification view of sister samples illustrating pores within thermally altered organic matter. (A) High-relief view of micrometer-size voids within organic matter. (B) Two-dimensional view of a thermally degraded organic particle with a large micrometer-size void at center.

PORE TYPES

Both SE images from three-dimensional surfaces and Ar-ion-milled backscattered SEM images of matrix-supported reservoir rocks allow examination of pore types. Several shale porosity classification schemes have been proposed (e.g., Loucks et al., 2010; Milner et al., 2010; Schieber, 2010). The shale pore types described in this chapter follow that of Milner et al. (2010). Matrix intercrystalline pores are voids between particles such as clay flakes, calcite and silica microcrystals, and larger detrital particles (Figures 5A, B). Pores in organic matter range from submicron-size pores in kerogen masses to larger voids left by alteration of organic matter (Figures 6A, B; 7A, B). Intraparticle pores form connected networks between

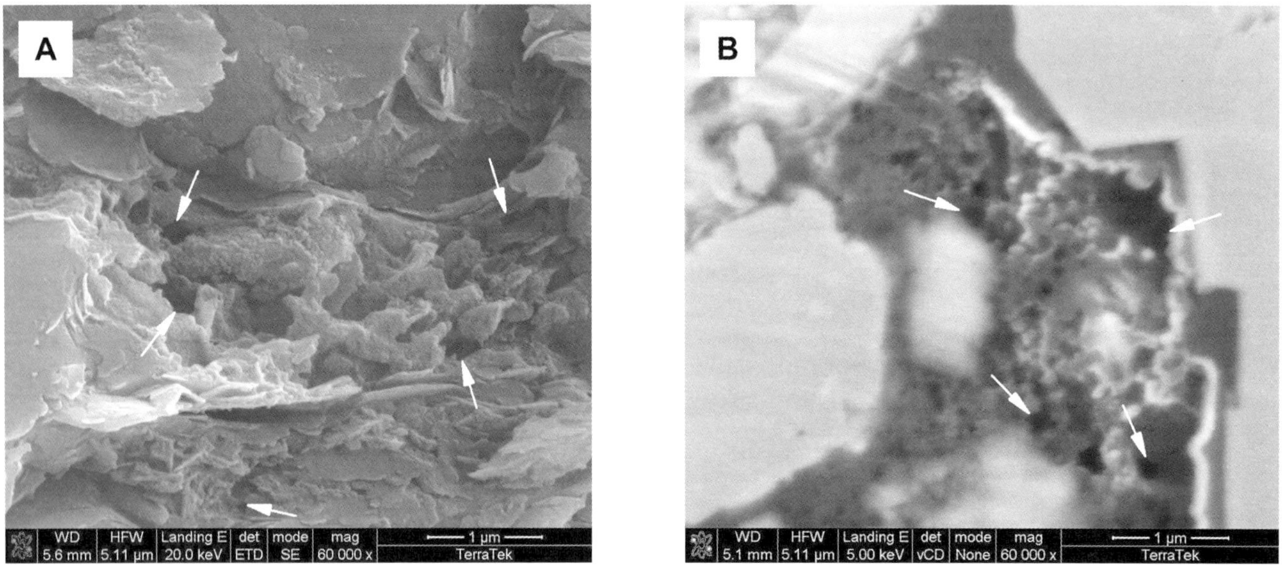

Figure 7. (A) Broken surface, secondary electron scanning electron microscopy (SEM); (B) ion-milled, backscattered electron SEM; Haynesville shale, 10,662.3 ft (3249.9 m); Panola County, Texas. High-magnification view of sister samples illustrating pores located within organic matter interpreted as thermally degraded bitumen. Figure 7A shows a three-dimensional view of submicrometer-size voids (arrows) within the bitumen (b). Figure 7B shows a two-dimensional view of a similar area of degraded bitumen illustrating rounded to ellipsoid pores (arrows) that are submicrometer in size.

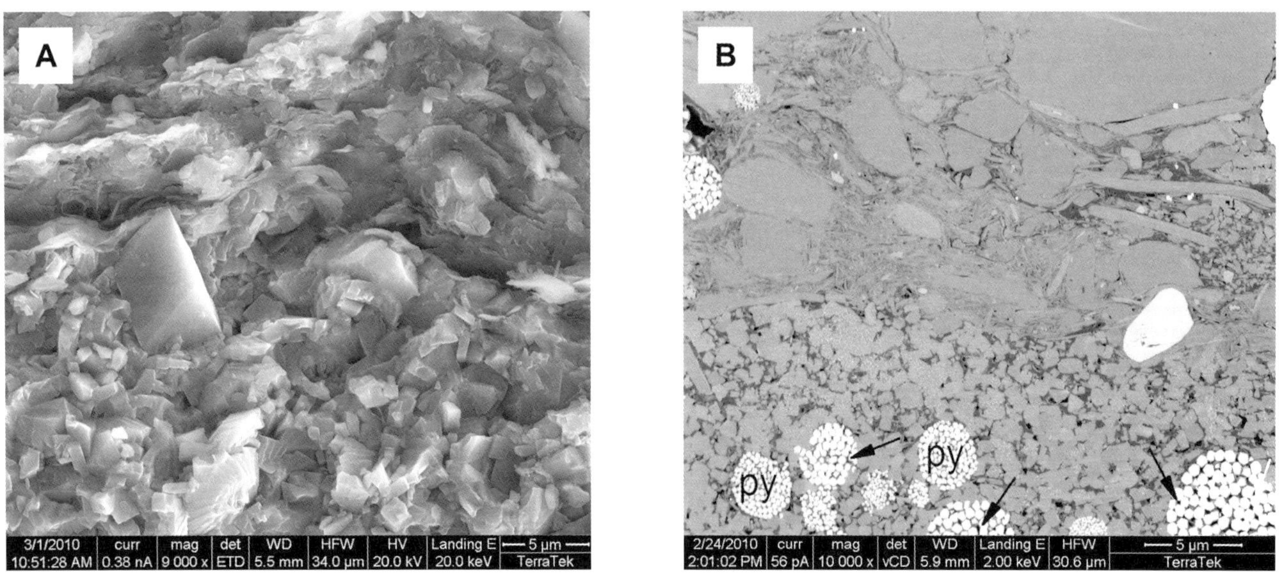

Figure 8. (A) Broken surface, secondary electron scanning electron microscopy (SEM); (B) ion-milled, backscattered electron SEM; Haynesville shale, 11,672.0 ft (3557.6 m); Caddo Parish, Louisiana. High-magnification view of sister samples illustrating intraparticle pores located within an interpreted fecal pellet. (A) High-resolution view of submicrometer-size voids between cemented coccolith plates. (B) A two-dimensional view of a similar fecal pellet showing angular submicrometer-size pores around many of the same coccolith plates. Further, minor intercrystalline micropores (arrows) are noted between large bright pyrite framboids (py).

nanofossil fragments coated with bitumen, primarily in fecal pellets composed of coccolithophorid algae plates (Figures 8A, B; 9A, B). Dissolution pores result from diagenetic alteration of grains.

Scanning electron microscopic studies by Milner et al. (2010) of selected Haynesville Shale samples suggest that matrix intercrystalline and intraparticle pores and pores in organic matter are the most common pore types observed in the Haynesville Formation. Dissolution pores were observed in only minor amounts. This study supported those observations.

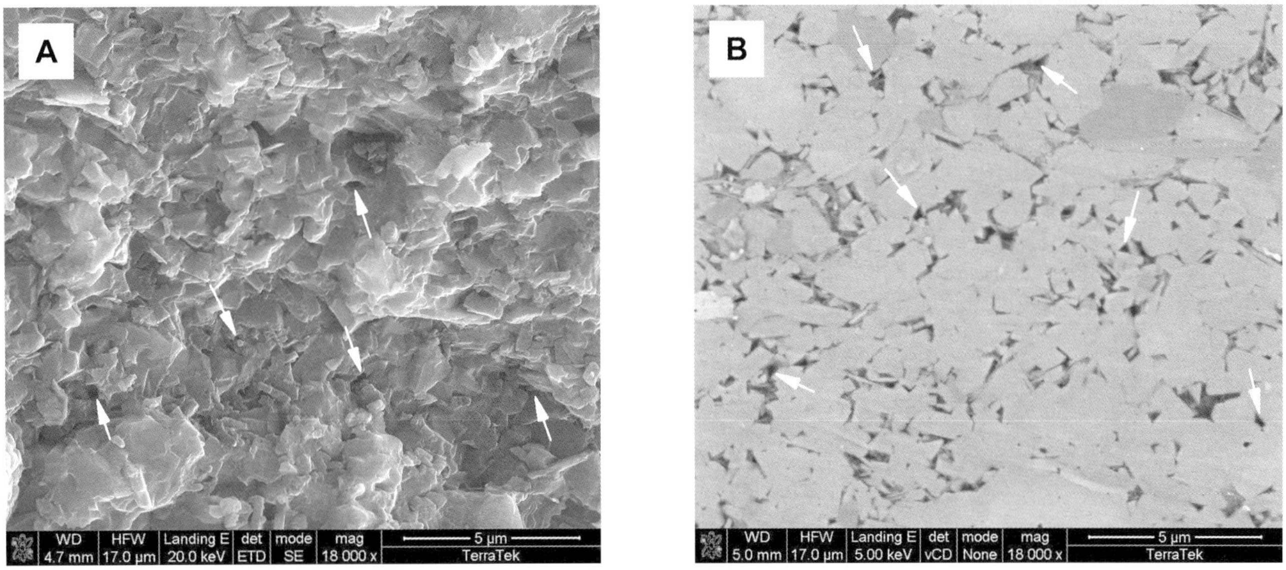

Figure 9. (A) Broken surface, secondary electron scanning electron microscopy (SEM); (B) ion-milled, backscattered electron SEM; Haynesville Shale, 10,766.2 ft (3281.5 m); Panola County, Texas. More magnified detail of fecal pellets illustrates a broken surface in (A) with angular pores identified by arrows. (B) Clear illustration of the angular pore morphologies between coccolith plate fragments. Differences in grayscale reflect variations in mean atomic number (Z); calcite is light gray (Z = 12); organic matter is black (Z − 6).

CONCLUSIONS

Traditional SEM methods of imaging three-dimensional (3-D) rock surfaces in SE mode provide highly detailed topographic images important for study of matrix microtexture (e.g., alignment of clay sheets), degree of diagenetic mineral overprint, and the nature and distribution of organic matter. Microtextural attributes are important in that they control the type of pores present and pore morphologies; however, the method takes a 3-D view of the sample surface, and true pore morphology may be obscured by sample preparation artifacts. Backscattered electron images provide additional compositional information not possible from SE images. Argon-ion milling reduces electron scattering effects associated with 3-D surfaces and thus improves BSE images to interpret the composition of mineral grains, cements, and organic matter. In addition, images from Ar-ion-milled surfaces provide a different perspective that aids in interpreting pore type and morphology and an improved ability to measure pore diameter.

Energy dispersive x-ray does not work well with low voltage and the particular detectors on BSE imaging because of the instrumentation design that partially blocks the EDX detector. Without the aid of traditionally prepared samples, the presence of organic matter can only be inferred based on distribution, mineral associations, and EDX analysis in a higher-voltage mode in BSE images, which increases surface charging and reduces detail minimizing Z-contrast. A suggested correction for this instrumentation issue is to use a JEOL JSM-7500F field emission SEM, which allows a clear view of the EDX detector under low-voltage conditions.

Pore description is critical in understanding the reservoir and completion quality of an unconventional shale gas or oil reservoir. Understanding where the dominant porosity is occurring, either the organics or the matrix, provides numerous insights into burial history and diagenesis, as well as petrophysical factors controlling fluid flow, including saturation distribution, electrical properties, methane adsorption and transport, and pore network connectivity. We can also effectively identify the presence and distribution of high-density, highly conductive minerals such as pyrite and iron-bearing chlorite, which have an impact on bulk density and wireline tool response, thus aiding our petrophysical modeling capability.

ACKNOWLEDGMENTS

We thank BP America for permission to publish these data and appreciate the support of Tim Sodergren, Sam Fluckiger, and David Handwerger at TerraTek and Bruce Abbuhl and Harris Cander at BP. We also thank Wayne Camp, Brian Lock, and Andres Mantilla for their critical review and excellent editorial feedback.

REFERENCES CITED

Loucks, R. G., R. M. Reed, S. C. Ruppel, and D. M. Jarvie, 2009, Morphology, genesis, and distribution of nanometer-scale pores in siliceous mudstone of the Mississippian Barnett Shale: Journal of Sedimentary Research, v. 79, p. 848–861.

Loucks, R. G., R. M. Reed, S. C. Ruppel, and U. Hammes, 2010, Preliminary classification of matrix pores in mudrocks: Transactions of the Gulf Coast Association of Geological Societies, v. 60, p. 435–441.

Milner, M., R. McLin, and J. Petriello, 2010, Imaging texture and porosity in mudstones and shales: Comparison of secondary and ion-milled backscatter SEM methods: Canadian Society for Unconventional Gas/Society of Petroleum Engineers Canadian Unconventional Resources and International Petroleum Conference, October 19–21, Calgary, Alberta, SPE Paper 138975, p. 1–10, doi: 10.2118/138975-MS.

Nelson, P., 2009, Pore-throat sizes in sandstones, tight sandstones, and shales: AAPG Bulletin, v. 93, p. 329–340.

Schieber, J., 2010, Common themes in the formation and preservation of intrinsic porosity in shales and mudstones—illustrated with examples across the Phanerozoic: Society of Petroleum Engineers Unconventional Gas Conference, February 23–25, 2010, Pittsburgh, Pennsylvania, SPE Paper 132370, 10 p., doi: 10.2118/132370-MS.

Spain, D. R., and G. A. Anderson, 2010, Controls on reservoir quality and productivity in the Haynesville Shale, Northwestern Gulf of Mexico Basin: Gulf Coast Association of Geological Societies Transactions, v. 60, p. 657–668.

7

Sylvain Bernard, Richard Wirth, Anja Schreiber, Leon Bowen, Andrew C. Aplin, Eliza J. Mathia, Hans-Martin Schulz, Brian Horsfield, 2013, FIB-SEM and TEM investigations of an organic-rich shale maturation series from the lower Toarcian Posidonia Shale, Germany: Nanoscale pore system and fluid-rock interactions, *in* W. Camp, E. Diaz, and B. Wawak, eds., Electron microscopy of shale hydrocarbon reservoirs: AAPG Memoir 102, p. 53–66.

FIB-SEM and TEM Investigations of an Organic-rich Shale Maturation Series from the Lower Toarcian Posidonia Shale, Germany: Nanoscale Pore System and Fluid-rock Interactions

Sylvain Bernard

Laboratoire de Minéralogie et de Cosmochimie du Museum, 61 rue Buffon, 75005, Paris, France (e-mail: sbernard@mnhn.fr)

Leon Brown

Department of Physics, Durham University, The Science Laboratories, South Road, Durham, DH1 3LE, United Kingdom (e-mail: leon.bowen@durham.ac.uk)

Richard Wirth, Anja Schreiber, Hans-Martin Schulz and Brian Horsfield

GeoForschungsZentrum Potsdam, Telegrafenberg, D-14473, Potsdam, Germany (e-mails: wirth@gfz-potsdam.de; schreib@gfz-potsdam.de; schulzhm@gfz-potsdam.de; and horsf@gfz-potsdam.de)

Andrew C. Aplin and Eliza J. Mathia

School of Civil Engineering and Geosciences, Newcastle University, Newcastle upon Tyne, NE1 7RU, United Kingdom (e-mails: andrew.aplin@newcastle.ac.uk and e.j.mathia@newcastle.ac.uk)

ABSTRACT

Although shale gas systems constitute a new target for commercial hydrocarbon production, only a little attention has been paid to the evolution of these unconventional systems with increasing thermal maturation. This study reports the characterization of samples of the Lower Toarcian (Lower Jurassic) Posidonia Shale from northern Germany at varying levels of thermal maturity (0.5–1.45%R_o [vitrinite reflectance]). Observations were made using an original combination of focused ion beam–scanning electron microscopy (FIB-SEM) and transmission electron microscopy (TEM). We document the formation of microfracture-filling bitumen in close association with kerogen residues with increasing maturity. Porosity evolves from mostly submicrometric interparticle pores in immature samples to intramineral and intraorganic pores (irregular-shape pores of about 1 to 200 nm occurring within the macromolecular structure of pyrobitumen masses) in overmature (gas mature) samples. This intraorganic nanoporosity has most likely come about by the exsolution of gaseous hydrocarbon and been hydrocarbon wet

DOI: 10.1306/13391705M1023583

during the thermal maturation processes. The mineralogical assemblage of the investigated samples strongly evolves with increasing thermal maturity. The formation of most of the mineral phases within the oil and gas mature samples is interpreted as resulting from the percolation of sulfate-rich evaporite-derived brines at temperatures of about 140 to 180°C. Although FIB-SEM and TEM images are small compared to field size, the present study emphasizes the need for nanoscale imaging to better constrain hydrocarbon generation processes in gas shale systems.

INTRODUCTION

Context

The volume of gas generated within shale gas systems is largely caused by secondary cracking of oil, which largely depends on oil retention capabilities of the system (i.e., on the specific adsorption capabilities of organic and inorganic constituents, their structural relationships, and the pore and fracture networks) (e.g., J. B. Curtis, 2002; Jenkins and Boyer, 2008; Nelson, 2009; Ross and Bustin, 2009; Ambrose et al., 2010; M. E. Curtis et al., 2010, 2012; Passey et al., 2010; Sondergeld et al., 2010a, b; Slatt and O'Brien, 2011). Diagenesis may significantly affect shale porosity and permeability (Pearson, 1999; Ross and Bustin, 2009; Yang and Aplin, 2010), which may range over more than an order of magnitude (e.g., Mondol et al., 2008; Mehmani et al., 2011). Volume changes during conversion of kerogen to hydrocarbons may induce porosity creation (e.g., Loucks et al., 2009). At a given porosity, slight variations of clay contents may increase mudstone permeability by an order of magnitude (e.g., Schneider et al., 2011). Many published pore size distributions have been determined using mercury-injection porosimetry (e.g., Dewhurst et al., 1999; Pearson, 1999; Hildenbrand and Urai, 2003). Although this technique provides an appropriate measure of permeability and pore throat size distribution, it does not provide a measure of total porosity, which is required for volumetric reserve estimates.

Finely elucidating the pore structure in relation to maturation level while precisely unraveling organic/inorganic relationships appears necessary to better understand gas generation processes occurring within shale gas systems and better estimate performance and producibility of such strategic resources. The fine-scale chemical and structural variability of shale gas reservoirs has thus recently started to be documented (e.g., Loucks et al., 2009, 2010; M. E. Curtis et al., 2010, 2011a, b, 2012; Passey et al., 2010; Heath et al., 2011; Slatt and O'Brien, 2011; Walls and Sinclair, 2011; Bernard et al., 2012a, b). The present study reports focused ion beam–scanning electron microscopic (FIB-SEM) and transmission electron microscopic (TEM) observations performed down to the nanometer scale on a wide range of thermal maturation levels (0.5–1.45%R_o [vitrinite reflectance]) of Lower Toarcian (Lower Jurassic) Posidonia Shale samples from northern Germany.

Previous Studies

Confocal laser scanning microscopy and conventional scanning electron microscope (SEM) are inadequate to characterize submicrometer grains and pores from broken or mechanically polished shale samples. Argon-ion beam milling can be used to overcome this difficulty and obtain sufficiently flat samples to allow high-magnification SEM to be performed and permeability pathways to be investigated at a high spatial resolution (Desbois et al., 2009, 2010; Loucks et al., 2009). Atomic force microscopy has also been successfully used to characterize mudrock nanopores (Javadpour, 2009; Yao et al., 2011). The recent advancements of FIB-SEM systems have offered a new alternative for investigating the three-dimensional (3-D) submicrometric fabric of a variety of samples (Krueger, 1999; Sivel et al., 2004; Tomutsa et al., 2007; Drobne et al., 2008; Haswell et al., 2008; De Winter et al., 2009; Hekking et al., 2009; Schiffbauer and Xiao, 2009; Leser et al., 2010; Park et al., 2010). This technique has thus recently been adopted for organic geochemical studies (e.g., Desbois et al., 2009, 2010; M. E. Curtis et al., 2010, 2011a, b, 2012; Sondergeld et al., 2010a, b; Bera et al., 2011; Elgmati et al., 2011a, b; Heath et al., 2011; Walls and Sinclair, 2011). The FIB-SEM technique can also be used to extract ultrathin (<100 nm thick) sections of sample across areas of interest, thus providing suitable samples for TEM, which offers a unique combination of chemical and structural information with unsurpassed spatial resolution (Heaney et al., 2001; Wirth, 2004, 2009; Benzerara et al., 2005; Bernard et al., 2007, 2010a).

Geologic Setting and Sample Location

The samples investigated in this study come from the Lias ε level of the Lower Toarcian (Lower Jurassic) Posidonia Shale from the Hils half-graben in the lower Saxony Basin, northern Germany (see map in Mann et al., 1990). These black shales were deposited during the Lower Toarcian (Lower Jurassic) in an epicontinental sea of moderate depth where water stratification leading to oxygen-depleted bottom waters could easily develop (e.g., Littke et al., 1991). Core samples were obtained from three boreholes with varying thermal maturity levels documented by vitrinite reflectance: Wickensen

($0.50\%R_o$), Harderode (0.85%Ro), and Haddessen ($1.45\%R_o$). Pyrolysis screening and optical microscopic investigations revealed the homogeneous nature of the Posidonia Shale from each of these three wells in terms of organic matter type and maturity. The structural and mineralogical differences observed between the boreholes were interpreted as resulting from thermal maturation processes (e.g., Bernard et al., 2012a).

ANALYTICAL PROCEDURES

Focused Ion Beam–scanning Electron Microscopy

The dual-beam FEI Helios Nanolab 600 microscope operating at the Durham GJ Russell Microscopy Facility has been used to perform FIB-SEM investigations. This microscope is equipped with both a gallium and an electron beam. The standard angle between the ion and electron column is 52°. Conventional polished thin sections were scanned and specific areas chosen for analysis. Prior to milling, a 3 to 4 μm protective ribbon of platinum was applied to protect the area of interest from excessive beam damage and noticeable curtaining effects (Ishitani et al., 2004). Trenches with maximum dimensions of 15 μm wide and 5 μm deep were cut normal to bedding with a gallium beam operating at 30 kV and 3.3 nA beam current. Final cleaning used lower energy (from 16 to 8 kV). Backscattered electron (BSE) images were then obtained using a 1.5 kV beam.

Ultrathin Section Extraction by Focused Ion Beam Milling

Ultrathin foils of samples, typically about 15 × 5 × 0.10 μm, were prepared using the FIB single-beam device (FEI FIB 200 TEM) operating at GFZ Potsdam following the procedure described by Heaney et al. (2001) and Wirth (2004, 2009). This extraction procedure maintains textural integrity, even in the case of loosely consolidated materials (Langford, 2006; Drobne et al., 2007) and prevents shrinkage and micrometer-to-nanometer pore deformation, even in the case of highly sensitive materials (e.g., Smith et al., 2001; Thompson et al., 2006). Milling at low-gallium-ion currents prevents common artifacts like local

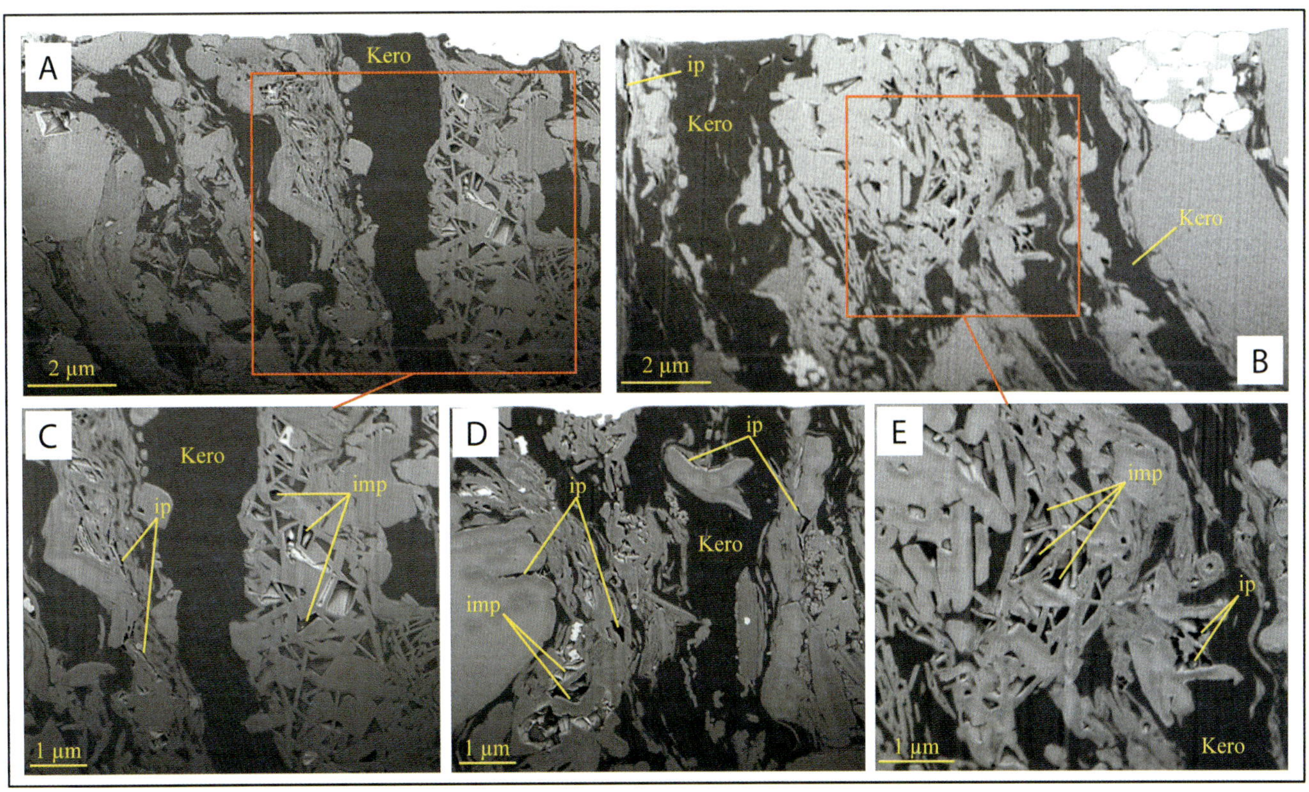

Figure 1. (A)–(E) Focused ion beam–scanning electron microscopy (FIB-SEM) images of an immature Lower Toarcian (Lower Jurassic) Posidonia Shale core sample from the Wickensen well ($0.5\%R_o$) in backscattered electron (BSE) mode. Organic matter appears dark, mineral grains various lighter shades of gray, and pyrite white. Wide continuous layers of nonporous organic matter (Kero), interparticle pores (ip), and intramineral pores within carbonate shells of microfossils (imp) are the main observable features. (C), (E) Higher-magnification images of the areas of (A) and (B) shown by squares.

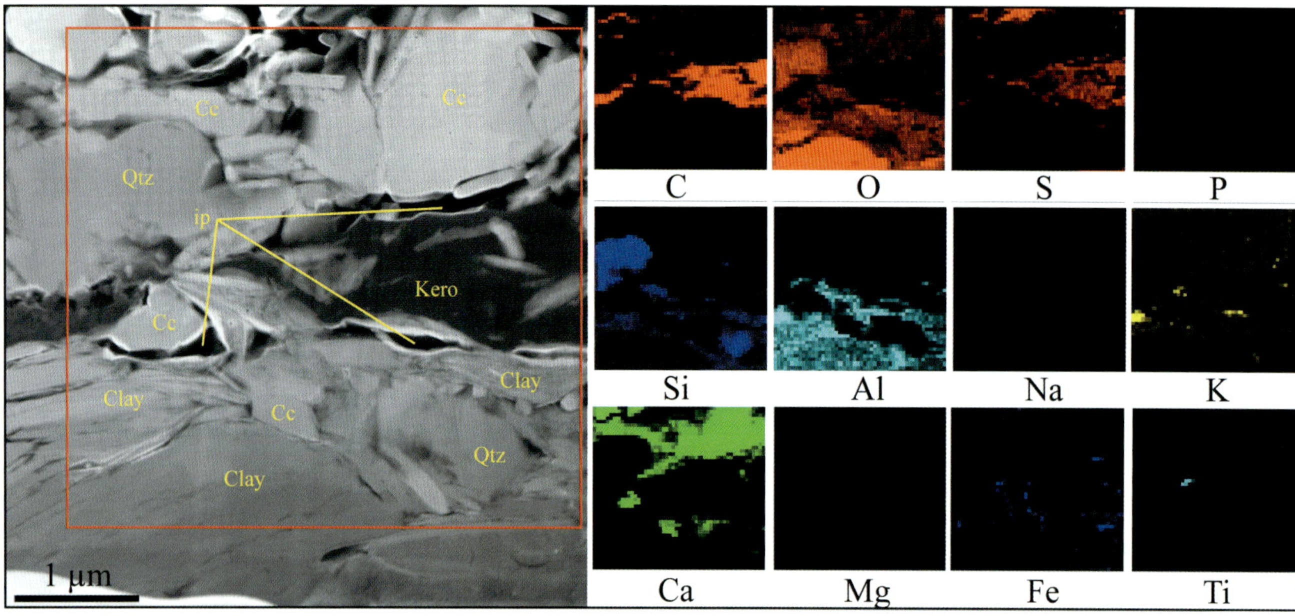

Figure 2. Left, high-magnification scanning transmission electron microscopy (STEM) image (high-angle annular dark-field [HAADF] mode) of a focused ion beam (FIB) foil extracted from an immature Posidonia Shale sample from the Wickensen well (0.5%R_o) (Z-contrast). Organic matter appears dark; silicates and carbonates appear bright. Square denotes area of energy-dispersive x-ray spectroscopy (EDXS) elemental maps (right): carbon (C), oxygen (O), sulfur (S), phosphorus (P), silicon (Si), aluminum (Al), sodium (Na), potassium (K), calcium (Ca), magnesium (Mg), iron (Fe), and titanium (Ti). Calcium carbonates (Cc), quartz (Qtz), and iron-rich clays (Clay) are the main minerals associated with sulfur-rich immature organic matter (Kero). A few interparticle pores (ip) can be observed.

gallium implantation, mixing of components, creation of vacancies or interstitials, creation of amorphous layers, local composition changes, or redeposition of the sputtered material on the sample surface (Rubanov and Munroe, 2004; Obst et al., 2005; Langford, 2006; Drobne et al., 2007; Mayer et al., 2007). In addition, milling at low gallium-ion currents prevents significant changes in the speciation of complex carbon-based polymer as evidenced from scanning transmission x-ray microscopy (STXM)-based carbon x-ray absorption near edge structure (C-XANES) data (Bernard et al., 2009; Bassim et al., 2012). For the present study, two ultrathin foils were extracted from ten samples from each well.

Transmission Electron Microscopy

Transmission electron microscopy was performed on FIB foils at the GFZ Potsdam using a TECNAI F20 XTWIN TEM operated at 200 kV with a field emission gun as the electron source. The TEM is equipped with a Gatan Tridiem™ energy filter, an EDAX Genesis™ x-ray analyzer, and a Fishione high-angle annular dark-field detector (HAADF), allowing image acquisition in the scanning transmission electron microscopic (STEM) mode (Z-contrast). Energy-dispersive x-ray spectroscopy (EDXS) element maps are always displayed as background subtracted intensity maps.

OBSERVATIONS

Thermally Immature Wickensen Samples (0.50%R_o)

Organic Matter

Organic matter represents about 25–30% of the Wickensen sample volume based on FIB-SEM images (Figure 1), consistent with reported total organic content (TOC) values of about 12 wt.% at this maturity level (Bernard et al., 2012a). This sulfur-rich (as revealed by EDXS) immature organic matter mainly forms continuous and nonporous layers that are a few micrometers wide (Figures 1, 2).

Porosity

The porosity is limited to nanometer-scale interparticle pores localized around kerogen grains, associated with clays, micas, or carbonate shells of microfossils (Figure 1). No clear interconnected porosity (permeability pathways) can be observed.

Mineralogy

Biogenic calcium carbonates and phosphates, detrital quartz, aluminum-rich silicates, and iron-rich clays are the main constituents of this immature shale based on EDXS elemental maps (Figure 2). Only a few pyrite crystals and some rare titanium oxides are present, sparsely disseminated within the matrix (Figures 1, 2).

Thermally Mature (Oil Mature) Harderode Samples (0.85%R_0)

Organic Matter

Organic matter represents about 10–12% of the sample volume based on FIB-SEM images (Figure 3), consistent with reported TOC values of about 6 wt.% at this maturity level (Bernard et al., 2012a). Some rare organic layers exhibiting no internal porosity can be observed and interpreted as mature kerogen (Figure 3). Most of the intergranular areas appear to be filled by organic matter exhibiting no internal porosity (Figure 3). Following Bernard et al. (2012a), this (now solid) organic matter, which likely originally existed as a fluid that migrated and filled the available pore spaces, is interpreted as sulfur-free bitumen.

Porosity

Clay particles and pyrite framboids display significant internal porosity (Figure 3). Abundant fragmentary

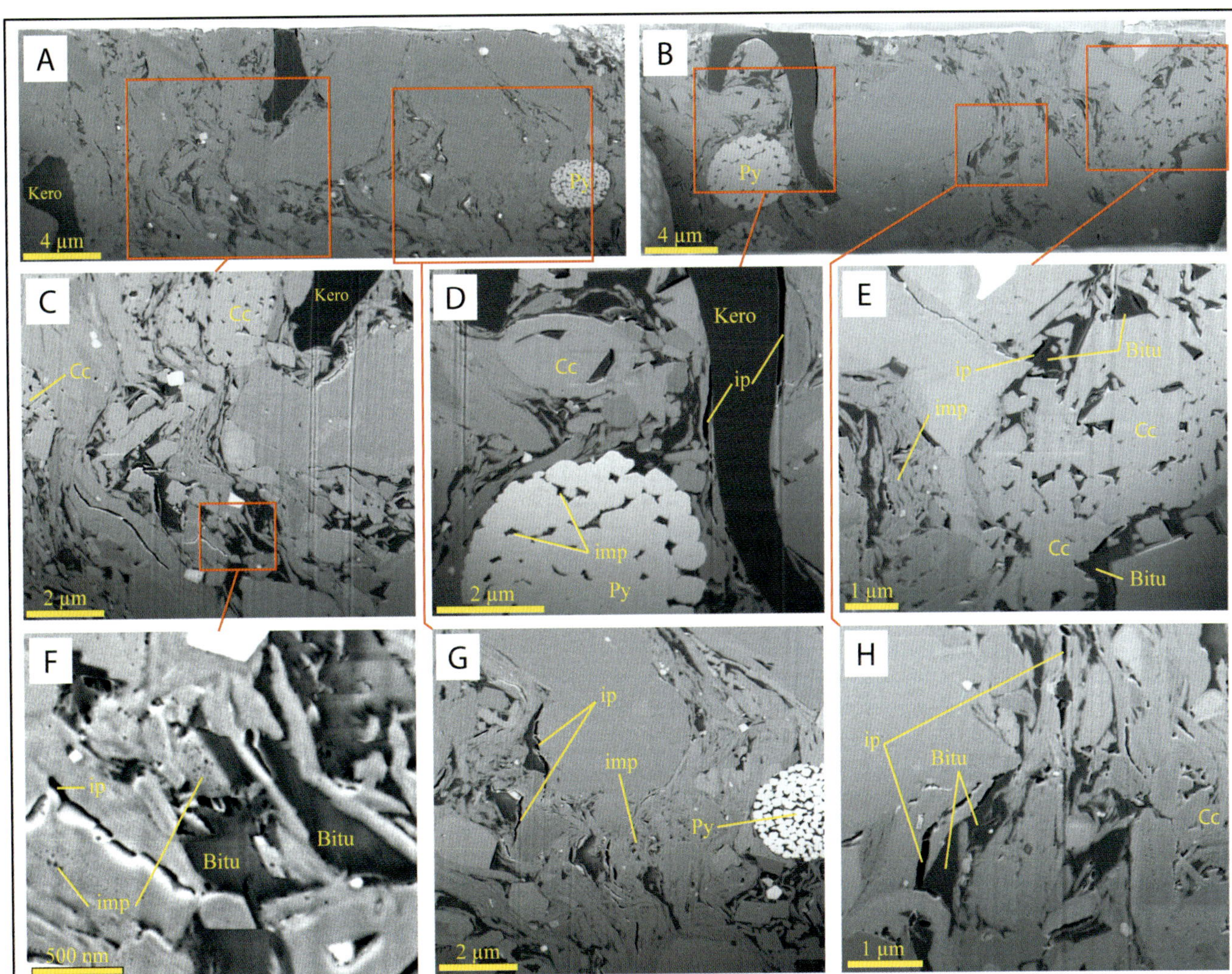

Figure 3. (A)–(H) Focused ion beam–scanning electron microscopy (FIB-SEM) images of a mature Lower Toarcian (Lower Jurassic) Posidonia Shale core sample from the Harderode well (0.85%R_0) in backscattered electron (BSE) mode. Organic matter appears dark, mineral grains various lighter shades of gray, and pyrite white. Layers of nonporous organic matter (Kero) and interparticle-pore-filling nonporous organic matter (likely bitumen) (Bitu) appear dark, pyrite framboids (Py) white, and carbonate cements within fragmentary microfossils (Cc) light gray. Elongated interparticle pores (ip) can be observed mostly around interparticle-pore-filling bitumen (Bitu); intramineral pores (imp) can be seen within pyrite framboids and clays. (C)–(E), (G), (H) Higher-magnification images of the areas of (A) and (B) shown by squares. (F) Higher-magnification image of the area of (C) shown by a square.

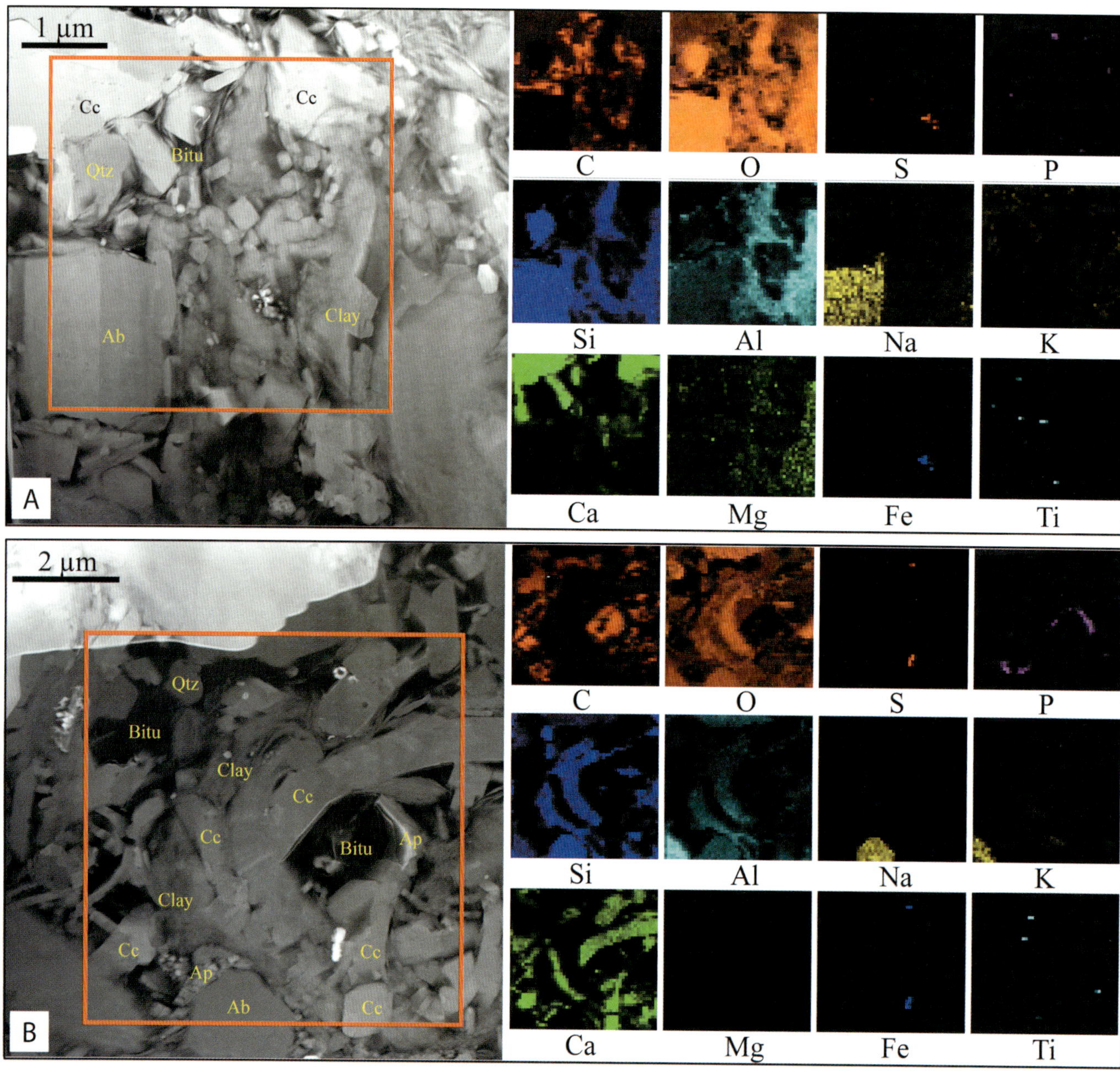

Figure 4. (A), (B) Left, high-magnification scanning transmission electron microscopy (STEM) image (high-angle annular dark-field [HAADF] mode) of two focused ion beam (FIB) foils extracted from a mature Posidonia Shale sample from the Harderode well (0.85%R_o) (Z-contrast). Organic matter appears dark; silicates and carbonates appear bright. Square denotes area of energy-dispersive x-ray spectroscopy (EDXS) elemental maps (right): carbon (C), oxygen (O), sulfur (S), phosphorus (P), silicon (Si), aluminum (Al), sodium (Na), potassium (K), calcium (Ca), magnesium (Mg), iron (Fe), and titanium (Ti). In addition to calcium carbonates (Cc), quartz (Qtz), and clays (Clay), interparticle-pore-filling bitumen (Bitu) masses are associated with albite (Ab) and apatite (Ap) crystals. Note lack of sulfur in organics and lack of iron in clays.

microfossils appear to have experienced cementation, limiting the porosity associated with this material.

Mineralogy

The mineral matrix of the Harderode samples consists of calcium carbonates, detrital and euhedral quartz crystals, euhedral and framboidal pyrites, aluminum-rich silicates, potassium-rich clays, as well as nanometric titanium oxides sparsely disseminated within the matrix (Figures 3, 4). Absent from the immature Wickensen matrix, some albite crystals containing almost no potassium and calcium and displaying a very restricted compositional and structural variability and small apatite crystals are disseminated within the Harderode matrix. The appearance of abundant pyrite framboids seems to have been concomitant to the loss of sulfur from the nonporous kerogen (Figure 4).

Thermally Overmature (Gas Mature) Haddessen Samples (1.45%R_o)

Organic Matter

Organic matter represents about 12 to 14% of the investigated sample volume, consistent with reported TOC values of about 6.5 wt.% at this maturity level (Bernard et al., 2012a). In contrast to Wickensen and Harderode samples, Haddessen organic matter is distributed in a patchy pattern and fills most of the intergranular and grain contact regions (Figure 5). The Haddessen organic matter does not contain any significant sulfur (Figure 6).

Porosity

Some organic matter exhibits irregular-shaped roughly circular pores of about 1 to 200 nm (Figure 5). Similar nanoporous organic matter has been previously identified as pyrobitumen compounds according to STXM-based XANES measurements (Bernard et al., 2012a, b). These intraorganic nanopores constituted the most obvious porosity of these overmature samples, although some rare elongated (compacted) interparticle pores could be observed.

Mineralogy

The Haddessen mineral assemblage appears similar to the Harderode samples: calcium carbonates and quartz crystals, isolated euhedral and framboidal pyrites, aluminum-rich silicates, potassium-rich clays, authigenic pure albite crystals, well-crystallized apatite, and titanium oxides (Figures 5, 6).

INTERPRETATIONS

Porosity Evolution Model

Although various porosity classification schemes exist (e.g., Desbois et al., 2010; Loucks et al., 2010; Heath et al., 2011), we discuss the evolution of the Posidonia Shale pore structure by differentiating between interparticle, intramineral, and intraorganic pores. Such classification allows relating the porosity evolution to mechanical and chemical diagenesis.

Interparticle Pores

Within the investigated Posidonia Shale samples, interparticle pores occur around large inorganic or organic grains as well as dispersed within the matrix at all levels of maturity. Our observations suggest that they are highly sensitive to burial: These pores display an elongated spheroidal shape within the immature Wickensen samples (Figure 1C–E), exhibit a pronounced ellipsoidal shape within the mature Harderode samples (Figure 3D–H), and are almost totally closed within the overmature Haddessen samples (Figure 5B, C, E). Such evolution is most likely related to compaction and cementation during burial, causing this early-formed interparticle porosity to collapse (Kim et al., 1999; Mondol et al., 2008; Milliken and Reed, 2010; Mehmani et al., 2011).

Intramineral Pores

Within the immature Wickensen samples, these pores observed within mineral (inorganic) grain boundaries are generally polyhedral with straight margins and mostly associated with carbonate shells of microfossils (Figure 1C–E). With increasing maturity (Harderode and Haddessen samples), the pores associated to microfossils are progressively filled by diagenetic carbonate cements (Figures 3C–E; 5B). The number and size of intramineral pores associated with pyrite framboids increase from Wickensen (immature) to Harderode (mature), potentially because of the diagenetic formation of the pyrite framboids. Most of these pores are filled with bitumen in the Haddessen (overmature) samples, indicating that they were once sufficiently interconnected to allow for bitumen to migrate. Haddessen samples also exhibit nanometer-scale intramineral pores within carbonate and clay minerals, not observed in the less mature samples. The appearance of these secondary pores is most likely related to organic matter thermal decarboxylation. Such thermal degradation has likely produced organic acids that partially dissolved carbonates and clays (Moore, 1997). Intramineral pores are thus more sensitive to chemical diagenesis than to mechanical diagenesis compared with the interparticle pore group.

Intraorganic Pores

Similar to what has been reported for the Barnett Shale (Loucks et al., 2009, 2010; Passey et al., 2010; Slatt and O'Brien, 2011; M. E. Curtis et al., 2012; Bernard et al., 2012b); the Marcellus Shale (M. E. Curtis et al., 2012); the Woodford Shale (Curtis et al., 2012); the Horn River Shale (M. E. Curtis et al., 2012); the Eagle Ford Shale (Walls and Sinclair, 2011; M. E. Curtis et al., 2012); the Gothic Shale (Heath et al., 2011); the Tuscaloosa Shale (Heath et al., 2011); the Haynesville

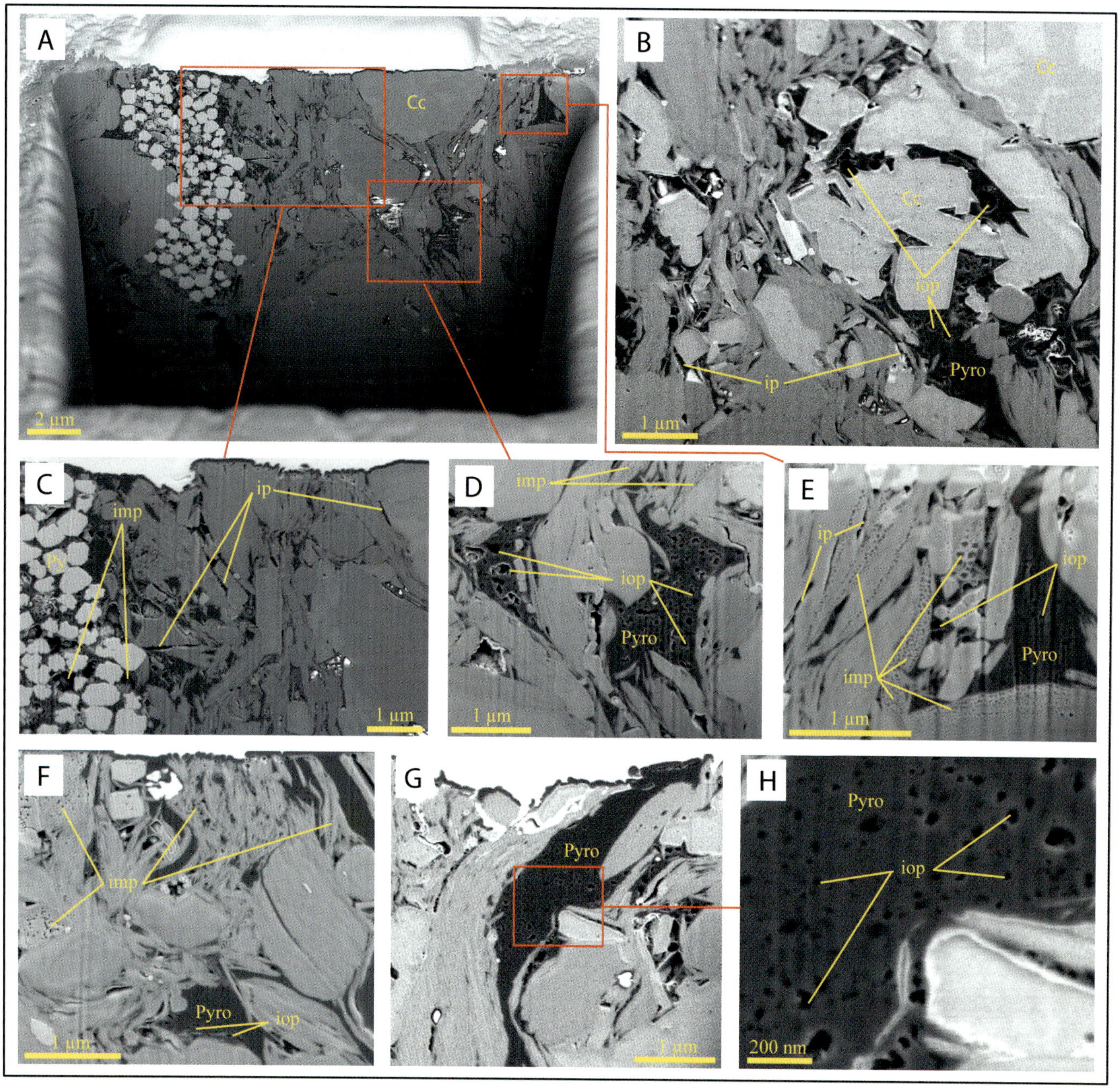

Figure 5. (A)–(H) Focused ion beam–scanning electron microscopy (FIB-SEM) images of an overmature Lower Toarcian (Lower Jurassic) Posidonia Shale core sample from the Haddessen well (1.45%R_o) in backscattered electron (BSE) mode. One can note the absence of continuous lamina of organic material; the presence of a network of irregular-shape, roughly circular nanopores (iop) of about 1 to 200 nm within pyrobitumen masses (Pyro) appearing dark; and the occurrence of intramineral pores (imp) within pyrite framboids (white) and clay packets (gray) and some rare elongated interparticle pores (ip). (C)–(E) Higher-magnification images of the areas of (A) shown by squares. (H) Higher-magnification image of the area of (G) shown by a square.

Formation (Elgmati et al., 2011b; M. E. Curtis et al., 2012); and the Utica Shale (Elgmati et al., 2011a), pores of about 1 to 200 nm exhibiting a slightly irregular ellipsoidal shape have been observed within some organic masses of the investigated Haddessen samples (Figures 5D–H). No doubt exists that this particular pore type has been acquired during increasing thermal maturation. As previously suggested by Loucks et al. (2009), this intraorganic porosity is most likely a result of the exsolution of hydrocarbon gas during the thermal cracking of retained oil. Consistently, the Haddessen nanoporous organic matter has been previously identified as pyrobitumen residues formed as a result of diagenetic degradation of Posidonia organic matter (Bernard et al., 2012a). Although intraorganic nanopores do not necessarily dominate in total volumetric contribution, the 3-D arrangement of this nanoporous organic matter has likely great influence on gas storage capacity and permeability of

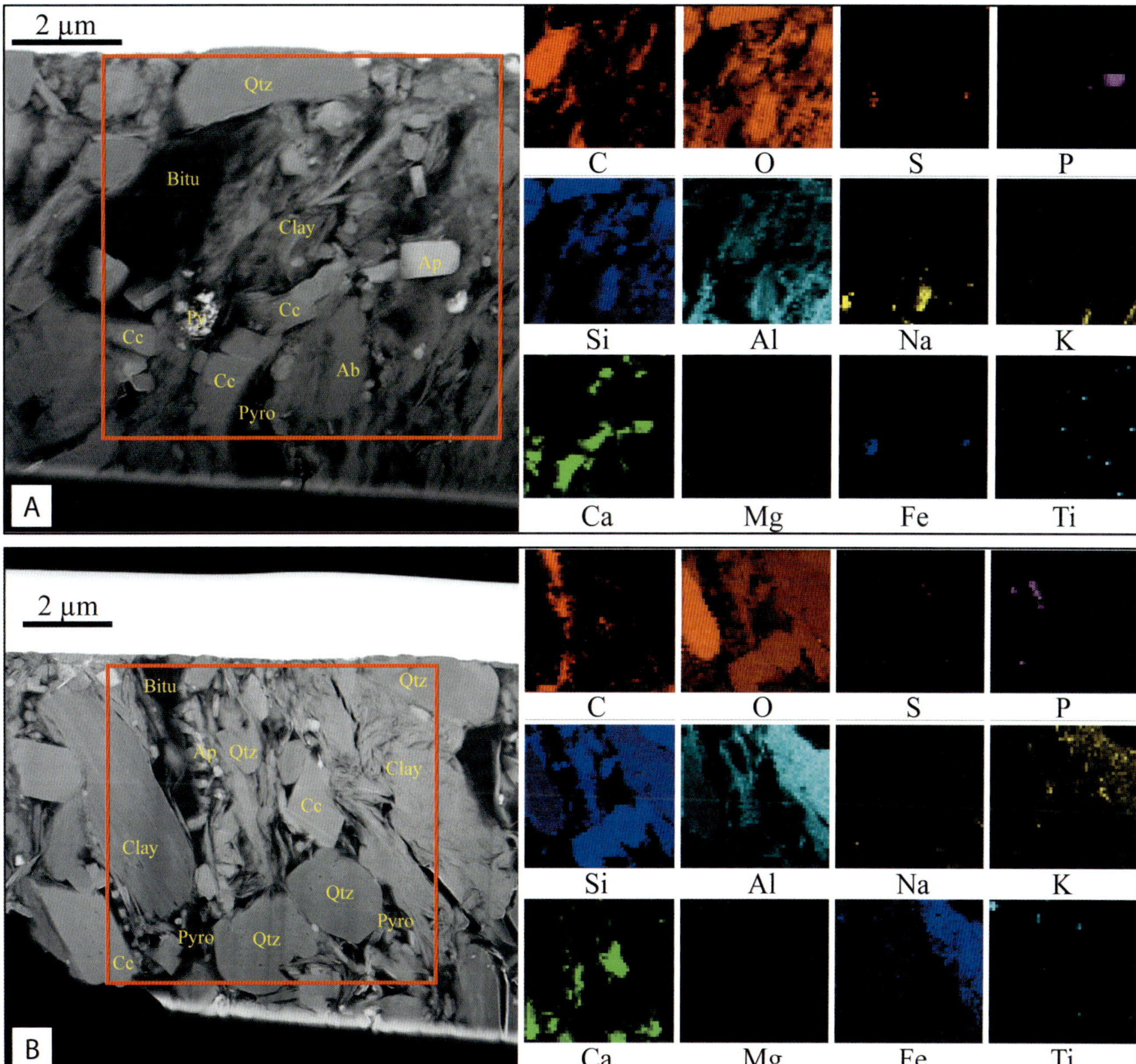

Figure 6. (A), (B) Left, high-magnification scanning transmission electron microscopy (STEM) image (high-angle annular dark-field [HAADF] mode) of two focused ion beam (FIB) foils extracted from a mature Posidonia Shale sample from the Haddessen well (1.45%R_o) (Z-contrast). Organic matter appears dark; silicates and carbonates appear bright. Square denotes area of energy-dispersive x-ray spectroscopy (EDXS) elemental maps (right): carbon (C), oxygen (O), sulfur (S), phosphorus (P), silicon (Si), aluminum (Al), sodium (Na), potassium (K), calcium (Ca), magnesium (Mg), iron (Fe), and titanium (Ti). In addition to calcium carbonates (Cc), quartz (Qtz), and clays (Clay), interparticle-pore-filling bitumen (Bitu) and pyrobitumen masses (Pyro) are associated with small pyrite framboids (Py), albite (Ab), and apatite (Ap) crystals. Note lack of sulfur in organic particles and lack of iron in clays.

such overmature shales (e.g., Ambrose et al., 2010; M. E. Curtis et al., 2010, 2012; Passey et al., 2010; Slatt and O'Brien, 2011).

Mineralogy Evolution Model: Fluid-rock Interactions

Temperature Conditions

Microthermometric measurements performed on hydrocarbon-bearing fluid inclusions observed in calcite-filling fractures have suggested that the Harderode samples experienced a temperature of about 150°C (Jochum et al., 1995). A maximum temperature of about 170–180°C has been qualitatively estimated based on the characteristics of the Raman spectrum of Haddessen organic matter (Bernard et al., 2010b), in good agreement with the temperature of 170°C that would be anticipated based on kinetic analysis of secondary gas generation (Dieckmann et al., 1998). The trend in maturity level from Wickensen to Haddessen has previously been interpreted as resulting from accelerated coalification caused by lateral heat transfer from a large intrusive body of Turonian age, the "Vlotho Massif" (e.g., Rullkötter et al., 1988; Jochum et al., 1995). Alternatively, Petmecky et al. (1999) proposed that a stable high-temperature regime coupled with ascending brine fluids was the ultimate cause for the maturation anomalies in the southern Lower Saxony Basin.

Diagenetic Carbonate Cements and Pyrite Framboids

The precipitation of diagenetic carbonate cements and pyrite framboids might have been triggered in the sediments during very early diagenesis by anaerobic bacterial sulfate reduction that may have produced hydrogen sulfide and generated local high alkalinity (e.g., Berner, 1968, 1984; Brock et al., 2006; Bernard et al., 2010a). However, although very common within the Harderode (mature) and Haddessen (overmature) samples, euhedral and framboidal pyrites, as well as carbonate cements associated with fossils, were almost absent from the Wickensen (immature) samples, suggesting a formation during advanced diagenesis. The important degree of Wickensen kerogen sulfurization evidenced from TEM-based EDXS data likely results from bacterial sulfate reduction processes having occurred at the early stages of sediment burial (e.g., Bernard et al., 2012a). Reactive iron is usually considered to outcompete organic matter for sulfur incorporation because of faster formation of pyrite (e.g., Meyers, 2007). Accordingly, the retarded formation of pyrite framboids within the Posidonia sediments would result from the originally low concentration of reactive iron (e.g., Bernard et al., 2012a). The advanced diagenetic euhedral and framboidal pyrites as well as carbonate cements associated with fossils observed within the Harderode and Haddessen samples might have precipitated as a result of thermochemical sulfate reduction processes that likely occurred during increasing maturation. Such processes can occur at relatively low temperatures of about 100–150°C (Krouse et al., 1988; Goldstein and Aizenshtat, 1994), that is, consistent with the temperature experienced by the Harderode and Haddessen samples. The abiotic reduction of sulfate in the presence of bitumen compounds in a reservoir leads to the production of CO_2 and H_2S (e.g., Goldstein and Aizenshtat, 1994; Kelemen et al., 2010). Within such systems, in addition to the formation of highly aromatic solid bitumen residues (Kelemen et al., 2010), the precipitation of pyrite, either euhedral or framboidal (Spötl et al., 1999; Ohfuji and Rickard, 2005), and calcitic or dolomitic cements (Machel, 1987; Ghazban and Al-Aasm, 2010) is relatively common. The sources of iron and calcium most likely originated from the local dissolution of the Posidonia calcium- and iron-rich clays during hydrocarbon generation processes (consistently, calcium- and iron-rich clays are common in Wickensen samples but absent from the Harderode and Haddessen samples). The supply of aqueous sulfate to the reaction may have derived either from the thermal degradation of the originally sulfur-rich kerogen or from ascending evaporate-derived brines.

Authigenic Apatite, Albite Crystals, and Titanium Globules

The euhedral apatite and albite crystals and the titanium globules found within the mature Harderode and overmature Haddessen samples are interpreted as authigenic minerals. Apatite crystals may have precipitated directly from phosphorous-rich mineralizing fluids (e.g., Ruttenberg and Berner, 1993; Davis et al., 2011). Percolation of hydrothermal brines in the presence of organic compounds has been shown to favor the solubility of titanium (Rapp et al., 2010). The titanium globules observed within the Harderode and Haddessen samples might thus have formed from interactions with ascending brines. From the presence of nanometric halite inclusions within the potassium- and calcium-free pure albite crystals sparsely disseminated in the Harderode and Haddessen samples, Bernard et al. (2010b, 2012a) proposed the participation of evaporite-derived brines in their crystallization, following Kastner and Siever (1979) and Spötl et al. (1999). Although albitization of detrital feldspars may occur in shales at relatively shallow depths (from about 1 km [~0.6 mi]; e.g., Milliken, 1992), brine-related albite authigenesis has been reported to occur in shales at advanced temperatures (140 to 200°C; e.g., Spötl et al., 1999), that is, consistent with the temperature experienced by the Harderode and Haddessen samples.

Fluid-rock Interactions

The formation of most of the mineral phases observed within the mature and overmature Posidonia Shale samples has likely involved the percolation of sulfate-rich evaporite-derived brines at temperatures of about 140 to 180°C. Isotopic and fluid inclusion microthermometry measurements would be necessary to estimate these temperatures precisely. The origin of these brines cannot be precisely determined. Multiple intercalated evaporite deposits older than the Lower Toarcian (Lower Jurassic) Posidonia Shale exist in the Lower Saxony Basin: the Late Permian "Zechstein" evaporites, the Upper Buntsandstein "Röt" sulfate and halite series, as well as the Middle Muschelkalk evaporite deposits (e.g., Petmecky et al., 1999). Diapir formation and development from Late Triassic to Early Cretaceous in the Lower Saxony Basin (Scheck et al., 2003; Mohr et al., 2005) would have favored the upward migration of high-temperature brines. These brines would have mostly circulated within the weakest units of the Mesozoic cover during the Late Triassic to Jurassic extension and during the Late Cretaceous to Early Cenozoic compression of the basin (Scheck et al., 2003; Mohr et al., 2005). The fractures induced by hydrocarbon generation processes (e.g., Lash et al., 2004) likely served as migration paths and facilitated evaporate-derived brine access and migration. In turn, such brine percolation may have accelerated organic matter maturation. Therefore, the high maturity levels of the Lower Toarcian (Lower Jurassic) Posidonia Shale would not solely result from high heat flows related to magmatic intrusions or deep burial.

CONCLUSIONS

We used FIB-SEM to document the submicrometer-scale pore structure of organic-rich shale samples over a maturation gradient (0.5–1.45%R_0). Three families of pores have been identified within the investigated samples: interparticle, intramineral, and intraorganic pores, the last having only been observed within the gas-mature Haddessen samples. While interparticle pores appear to be sensitive to both mechanical and chemical diagenesis (i.e., to compaction and cementation), intramineral and intraorganic pores seem more sensitive to chemical diagenesis. The number and volume of intramineral pores appear to be controlled by authigenesis and dissolution/precipitation processes, while intraorganic pores only appear once the gas window has been reached (>1.45%R_0) within newly formed pyrobitumen compounds (Bernard et al., 2012a).

In parallel, nanometer-scale mineralogical heterogeneities have been documented using TEM at all levels of thermal maturity. Chemical rock-fluid interactions have been inferred from the detection of diagenetic and authigenic minerals, namely, carbonate cements, pyrite framboids, apatite, and albite crystals. The formation of these mineral phases identified within the mature and overmature samples has likely involved thermochemical sulfate reduction processes during the percolation of sulfate-rich, evaporite-derived brines at temperatures of about 140 to 180°C. Such processes have undoubtedly influenced gas generation. By providing fine-scale evidence of the evolution of both the nanoscale pore structure and the mineral assemblage of gas shales as a response to the thermal evolution of the macromolecular structure of kerogen and (pyro)bitumen, the observations reported here constitute a new step toward better constraining natural shale gas maturation processes.

REFERENCES CITED

Ambrose, R. J., R. C. Hartman, M. Diaz-Campos, I. Y. Akkutlu, and C. H. Sondergeld, 2010, New pore-scale considerations for shale gas in place calculations: Proceedings of the North American Unconventional Gas Conference and Exhibition, Society of Petroleum Engineers, SPE Paper 131772, 17 p., doi: 10.2118/131772-MS.

Bassim, N. D., B. T. de Gregorio, A. L. D. Kylcoyne, K. Scott, T. Chou, S. Wirick, G. D. Cody, and R. M. Stroud, 2012, Minimizing damage during FIB sample preparation of soft materials: Journal of Microscopy, v. 245, no. 3, p. 288–301.

Benzerara, K., N. Menguy, F. Guyot, and C. Vanni, 2005, TEM study of a silicate-carbonate-microbe interface prepared by focused ion beam milling: Geochimica et Cosmochimica Acta, v. 69, p. 1413–1422.

Bera, B., M. K. Sushanta, and V. Douglas, 2011, Understanding the micro structure of Berea Sandstone by the simultaneous use of micro-computed tomography (micro-CT) and focused ion beam-scanning electron microscopy (FIB-SEM): Micron, v. 42, no. 5, p. 412–418.

Bernard, S., K. Benzerara, O. Beyssac, N. Menguy, F. Guyot, G. E. Brown Jr., and B. Goffé, 2007, Exceptional preservation of plant fossils in high-pressure metamorphic rocks: Earth and Planetary Science Letters, v. 262, no. 1–2, p. 257–272.

Bernard, S., K. Benzerara, O. Beyssac, G. E. Brown Jr., L. G. Stamm, and P. Duringer, 2009, Ultrastructural and chemical study of modern and fossil sporoderms by scanning transmission x-ray microscopy (STXM): Review of Palaeobotany and Palynology, v. 156, no. 1–2, p. 248–261.

Bernard, S., K. Benzerara, O. Beyssac, and G. E. Brown Jr., 2010a, Multiscale characterization of pyritized plant tissues in blueschist facies metamorphic rocks: Geochimica et Cosmochimica Acta, v. 74, p. 5054–5068.

Bernard, S., B. Horsfield, H. M. Schulz, A. Schreiber, R. Wirth, T. T. A. Vu, F. Perssen, S. Könitzer, H. Volk, N. Sherwood, and D. Fuentes, 2010b, Multi-scale detection of organic and inorganic signatures provides insights into gas shale properties and evolution: Chemie der Erde-Geochemistry, v. 70, no. S3, p. 119–133.

Bernard, S., B. Horsfield, H. M. Schulz, R. Wirth, A. Schreiber, and N. Sherwood, 2012a, Geochemical evolution of organic-rich shales with increasing maturity: A STXM and TEM study of the Posidonia Shale (Lower Toarcian, northern Germany): Marine and Petroleum Geology, v. 31, no. 1, p. 70–89.

Bernard, S., R. Wirth, A. Schreiber, H. M. Schulz, and B. Horsfield, 2012b, Formation of nanoporous pyrobitumen residues during maturation of the Barnett Shale (Fort Worth Basin): International Journal of Coal Geology, v. 103, p. 3–11.

Berner, R. A., 1968, Calcium carbonate concretions formed by the decomposition of organic matter: Science, v. 159, p. 195–197.

Berner, R. A., 1984, Sedimentary pyrite formation: An update: Geochimica et Cosmochimica Acta, v. 48, no. 4, p. 605–615.

Brock, F., R. J. Parkes, and D. E. G. Briggs, 2006, Experimental pyrite formation associated with decay of plant material: Palaios, v. 21, no. 5, p. 499–506.

Curtis, J. B., 2002, Fractured shale-gas systems: AAPG Bulletin, v. 86, no. 11, p. 1921–1938.

Curtis, M. E., C. H. Sondergeld, R. J. Ambrose, and C. S. Rai, 2010, Structural characterization of gas shales on the micro- and nano-scales: Proceedings of the Canadian Unconventional Resources and International Petroleum Conference, 'Society of Petroleum Engineers', SPE Paper 137693, 15 p., doi: 10.2118/137693-MS.

Curtis, M. E., R. J. Ambrose, C. H. Sondergeld, and C. S. Rai, 2011a, Investigation of the relationship between organic porosity and thermal maturity in the Marcellus Shale: Proceedings of the North American Unconventional Gas Conference and Exhibition, Society of Petroleum Engineers, SPE Paper 144370, 4 p., doi:10.2118/144370-MS.

Curtis, M. E., R. J. Ambrose, C. H. Sondergeld, and C. S. Rai, 2011b, Transmission and scanning electron microscopy investigation of pore connectivity of gas shales on the nanoscale: Proceedings of the North American Unconventional Gas Conference and Exhibition, Society of Petroleum Engineers, SPE Paper 144391, 10 p., doi:10.2118/144391-MS.

Curtis, M. E., R. J. Ambrose, C. H. Sondergeld, and C. S. Rai, 2012, Microstructural investigation of gas shales in two and three dimensions using nanometer-scale resolution imaging: AAPG Bulletin, v. 96, no. 4, p. 665–677.

Davis, W. J., Q. Gall, C. W. Jefferson, and R. H. Rainbird, 2011, Fluorapatite in the Paleoproterozoic Thelon Basin: Structural-stratigraphic context, in situ ion microprobe U-Pb ages, and fluid-flow history: GSA Bulletin, v. 123, no. 5–6, p. 1056–1073.

Desbois, G., J. L. Urai, and P. A. Kukla, 2009, Morphology of the pore space in claystones—evidence from BIB/FIB ion beam sectioning and cryo-SEM observations: Earth Discussion, v. 4, p. 1–19.

Desbois, G., J. L. Urai, M. E. Houben, and Y. Sholokhova, 2010, Typology, morphology and connectivity of pore space in claystones from reference site for research using BIB, FIB and cryo-SEM methods: EPJ Web of Conferences, doi: 10.1051/epjconf/20100622005.

Dewhurst, D. N., Y. Yang, and A. C. Aplin, 1999, Permeability and fluid flow in natural mudstones: GSL, Special Publications, v. 158, p. 23–43.

De Winter, D. A. M., C. T. W. M. Schneijdenberg, M. N. Lebbink, B. Lich, A. J. Verkleij, M. R. Drury, and B. M. Humbel, 2009, Tomography of insulating biological and geological materials using focused ion beam (FIB) sectioning and low-kV BSE imaging: Journal of Microscopy, v. 233, no. 3, p. 372–383.

Dieckmann, V., H. J. Schenk, B. Horsfield, and D. H. Welte, 1998, Kinetics of petroleum generation and cracking by programmed-temperature closed-system pyrolysis of Toarcian Shales: Fuel, v. 77, no. 1–2, p. 23–31.

Drobne, D., M. Milani, V. Leser, and F. Tatti, 2007, Surface damage induced by FIB milling and imaging of biological samples is controllable: Microscopy Research and Technique, v. 70, no. 10, p. 895–903.

Drobne, D., M. Milani, V. Leser, F. Tatti, A. Zrimec, N. Znidarsic, R. Kostanjsek, and J. Strus, 2008, Imaging of intracellular spherical lamellar structures and tissue gross morphology by a focused ion beam/scanning electron microscope (FIB/SEM): Ultramicroscopy, v. 108, no. 7, p. 663–670.

Elgmati, M., H. Zhang, B. Bai, and R. Flori, 2011a, Submicron-pore characterization of shale gas plays: Proceedings of the North American Unconventional Gas Conference and Exhibition, Society of Petroleum Engineers, SPE Paper 144050, 19 p., doi: 10.2118/144050-MS

Elgmati, M., M. Zobaa, H. Zhang, B. Bai, and F. Oboh-Ikuenobe, 2011b, Palynofacies analysis and submicron-pore modeling of shale-gas plays: Proceedings of the North American Unconventional Gas Conference and Exhibition, Society of Petroleum Engineers, SPE Paper 144267, 10 p., doi: 10.2118/144267-MS.

Ghazban, F., and I. S. Al-Aasm, 2010, Hydrocarbon-induced diagenetic dolomite and pyrite formation associated with the Hormoz Island Salt Dome, Offshore Iran: Journal of Petroleum Geology, v. 33, no. 2, p. 183–196.

Goldstein, T. P., and Z. Aizenshtat, 1994, Thermochemical sulfate reduction—A Review: Journal of Thermal Analysis, v. 42, no. 1, p. 241–290.

Haswell, R., U. Zeile, and K. Mensch, 2008, Van Gogh's painting grounds: An examination of barium sulphate extender using analytical electron microscopy—SEM/FIB/TEM/EDX: Microchimica Acta, v. 161, no. 3–4, p. 363–369.

Heaney, P. J., E. P. Vicenzi, L. A. Giannuzzi, and K. J. T. Livi, 2001, Focused ion beam milling: A method of site-specific sample extraction for microanalysis of earth and planetary materials: American Mineralogist, v. 86, no. 9, p. 1094–1099.

Heath, J. E., T. A. Dewers, B. J. O. L. McPherson, R. Petrusak, T. C. Chidsey, A. J. Rinehart, and P. S. Mozley, 2011, Pore networks in continental and marine mudstones: Characteristics and controls on sealing behavior: Geosphere, v. 7, no. 2, p. 429–454.

Hekking, L. H. P., M. N. Lebbink, D. A. M. De Winter, C. T. W. M. Schneijdenberg, C. M. Brand, B. M. Humbel, A. J. Verkleij, and J. A. Post, 2009, Focused ion beam-scanning electron microscope: Exploring large volumes of atherosclerotic tissue: Journal of Microscopy, v. 235, no. 3, p. 336–347.

Hildenbrand, A., and J. L. Urai, 2003, Investigation of the morphology of pore space in mudstones—first results: Marine and Petroleum Geology, v. 20, no. 10, p. 1185–1200.

Ishitani, T., K. Umemura, T. Ohnishi, T. Yaguchi, and T. Kamino, 2004, Improvements in performance of focused ion beam cross-sectioning: Aspects of ion-sample interaction: Journal of Electron Microscopy, v. 53, no. 5, p. 443–449.

Javadpour, F., 2009, Nanopores and apparent permeability of gas flow in mudrocks (shales and siltstone): Journal of Canadian Petroleum Technology, v. 48, no. 8, p. 16–21.

Jenkins, C., and C. Boyer, 2008, Coalbed and shale-gas reservoirs: Journal of Petroleum Technology, v. 60, no. 2, p. 92–99.

Jochum, J., G. Friedrich, D. Leythaeuser, R. Littke, and B. Ropertz, 1995, Hydrocarbon-bearing fluid inclusions in calcite-filled horizontal fractures from mature Posidonia Shale (Hils syncline, NW Germany): Ore Geology Reviews, v. 9, p. 363–370.

Kastner, M., and R. Siever, 1979, Low temperature feldspars in sedimentary rocks: American Journal of Science, v. 279, no. 4, p. 435–479.

Kelemen, S. R., C. C. Walters, P. J. Kwiatek, H. Freund, M. Afeworki, M. Sansone, W. A. Lamberti, R. J. Pottrof, H. G. Machel, K. E. Peters, and T. Bolin, 2010, Characterization of solid bitumens originating from thermal chemical alteration and thermochemical sulfate reduction: Geochimica et Cosmochimica Acta, v. 74, p. 5305–5332.

Kim, J. W., W. R. Bryant, J. S. Watkins, and T. T. Tieh, 1999, Electron microscopic observations of shale diagenesis, offshore Louisiana, USA, Gulf of Mexico: Geo-marine Letters, v. 18, no. 3, p. 234–240.

Krouse, H. R., C. A. Viau, L. S. Eliuk, A. Ueda, and S. Halas, 1988, Chemical and isotopic evidence of thermochemical sulfate reduction by light-hydrocarbon gases in deep carbonate reservoirs: Nature, v. 333, no. 6172, p. 415–419.

Krueger, R., 1999, Dual-column (FIB-SEM) wafer applications: Micron, v. 30, no. 3, p. 221–226.

Langford, R.M., 2006, Focused ion beams techniques for nanomaterials characterization: Microscopy Research and Technique, v. 69, no. 7, p. 538–549.

Lash, G., S. Loewy, and T. Engelder, 2004, Preferential jointing of Upper Devonian black shale, Appalachian Plateau, USA: Evidence supporting hydrocarbon generation as a joint-driving mechanism, *in* J. W. Cosgrove and T. Engelder, eds., The initiation, propagation, and arrest of joints and other fractures: GSL Special Publications, v. 231, p. 129–151.

Leser, V., M. Milani, F. Tatti, Z. P. Tkalec, J. Strus, and D. Drobne, 2010, Focused ion beam (FIB)/scanning electron microscopy (SEM) in tissue structural research: Protoplasma, v. 246, no. 1–4, p. 41–48.

Littke, R., D. Leythaeuser, J. Rullkotter, and D. R. Baker, 1991, Keys to the depositional history of the Posidonia Shale (Toarcian) in the Hils syncline, northern Germany: GSL Special Publications, v. 58, no. 1, p. 311–333.

Loucks, R. G., R. M. Reed, S. C. Ruppel, and D. M. Jarvie, 2009, Morphology, genesis, and distribution of nanometer-scale pores in siliceous mudstones of the Mississippian Barnett Shale: Journal of Sedimentary Research, v. 79, no. 12, p. 848–861.

Loucks, R. G., R. M. Reed, S. C. Ruppel, and U. Hammes, 2010, Preliminary classification of matrix pores in mudrocks: Gulf Coast Association of Geological Societies Transactions, v. 60, p. 435–441.

Machel, H. G., 1987, Saddle dolomite as a by-product of chemical compaction and thermochemical sulfate reduction: Geology, v. 15, no. 10, p. 936–940.

Mann, U., S. Düppenbecker, A. Langen, B. Ropertz, and D. H. Welte, 1990, Pore network evolution of the lower Toarcian Posidonia Shale during petroleum generation and expulsion—A multidisciplinary approach: Zentralblatt für Geologie und Paläontologie Teil I, v. 8, p. 1051–1071.

Mayer, J., L. A. Giannuzzi, T. Kamino, and J. Michael, 2007, TEM sample preparation and FIB-induced damage: MRS Bulletin, v. 32, no. 5, p. 400–407.

Mehmani, A., A. Tokan-Lawal, M. Prodanović, and A. P. Sheppard, 2011, The effect of microporosity on transport properties in tight reservoirs: Proceedings of the North American Unconventional Gas Conference and Exhibition, Society of Petroleum Engineers, SPE Paper 144383, 14 p., doi: 10.2118/144384-MS.

Meyers, S. R., 2007, Production and preservation of organic matter: The significance of Iron: Paleoceanography, v. 22, no. 4, p. PA4211.

Milliken, K. L., 1992, Chemical behavior of detrital feldspars in mudrocks versus sandstones, Frio Formation (Oligocene), South Texas: Journal of Sedimentary Petrology, v. 62, p. 790–801.

Milliken, K. L., and R. M. Reed, 2010, Multiple causes of diagenetic fabric anisotropy in weakly consolidated mud, Nankai accretionary prism, IODP Expedition 316: Journal of Structural Geology, v. 32, p. 1887–1898.

Mohr, M., P. A. Kukla, J. L. Urai, and G. Bresser, 2005, Multiphase salt tectonic evolution in NW Germany: Seismic interpretation and retro-deformation: International Journal of Earth Sciences, v. 94, no. 5–6, p. 917–940.

Mondol, N. H., K. Bjorlykke, and J. Jahren, 2008, Experimental compaction of clays: Relationship between permeability and petrophysical properties in mudstones: Petroleum Geoscience, v. 14, no. 4, p. 319–337.

Moore, C. H., 1997, Carbonate diagenesis and porosity: Developments in Sedimentology 46, Elsevier, p. 1–338.

Nelson, P. H., 2009, Pore throat sizes in sandstones, tight sandstones, and shales: AAPG Bulletin, v. 93, p. 1–13.

Obst, M., P. Gasser, D. Mavrocordatos, and M. Dittrich, 2005, TEM-specimen preparation of cell/mineral interfaces by focused ion beam milling: American Mineralogist, v. 90, no. 8–9, p. 1270–1277.

Ohfuji, H., and D. Rickard, 2005, Experimental syntheses of framboids—a review: Earth-Science Reviews, v. 71, no. 3–4, p. 147–170.

Park, S. E., A. Schertel, and S. S. Han, 2010, Altered structure of synapse in a brain of the Swedish mutation which causes early-onset Alzheimer's disease: A FIB-SEM microscopy study: Journal of Neurogenetics, v. 24, no. 1, p. 38–39.

Passey, Q. R., K. M. Bohacs, W. L. Esch, R. Klimentidis, and S. Sinha, 2010, From oil-prone source rock to gas-producing shale reservoir—Geologic and petrophysical characterization of unconventional shale gas reservoirs: Proceedings of the North American Unconventional Gas Conference and Exhibition, Society of Petroleum Engineers, SPE Paper 131350, 29 p., doi: 10.2118/131350-MS.

Pearson, F. J., 1999, What is the porosity of a mudrock?: GSL, Special Publications, v. 158, p. 9–21.

Petmecky, S., L. Meier, H. Reiser, and R. Littke, 1999, High thermal maturity in the Lower Saxony Basin: Intrusion or deep burial?: Tectonophysics, v. 304, no. 4, p. 317–344.

Rapp, J. F., S. Klemme, I. B. Butler, and S. L. Harley, 2010, Extremely high solubility of rutile in chloride and fluoride-bearing metamorphic fluids: An experimental investigation: Geology, v. 38, no. 4, p. 323–326.

Ross, D. J., and R. M. Bustin, 2009, The importance of shale composition and pore structure upon gas storage potential of shale gas reservoirs: Marine and Petroleum Geology, v. 26, no. 6, p. 916–927.

Rubanov, S., and P. R. Munroe, 2004, FIB-induced damage in silicon: Journal of Microscopy, v. 214, no. 3, p. 213–221.

Rullkötter, J., et al., 1988, Organic matter maturation under the influence of a deep intrusive heat source: A natural experiment for quantitation of hydrocarbon generation and expulsion from a petroleum source rock (Toarcian shale, northern Germany): Organic Geochemistry, v. 13, no. 4–6, p. 847–856.

Ruttenberg, K. C., and R. A. Berner, 1993, Authigenic apatite formation and burial in sediments from non-upwelling, continental margin environments: Geochimica et Cosmochimica Acta, v. 57, p. 991–1007.

Scheck, M., U. Bayer, and B. Lewerenz, 2003, Salt movements in the Northeast German Basin and its relation to major post-Permian tectonic phases-results from 3D structural modelling, backstripping and reflection seismic data: Tectonophysics, v. 361, no. 3–4, p. 277–299.

Schiffbauer, J. D., and S. Xiao, 2009, Novel application of focused ion beam electron microscopy (FIB-EM) in preparation and analysis of microfossil ultrastructures: A new view of complexity in early eukaryotic organisms: Palaios, v. 24, no. 9, p. 616–626.

Schneider, J., P. B. Flemings, R. J. Day-Stirrat, and J. T. Germaine, 2011, Insights into pore-scale controls on mudstone permeability through resedimentation experiments: Geology, v. 39, p. 1011–1014.

Sivel, V. G. M., J. Van Den Brand, W. R. Wang, H. Mohdadi, F. D. Tichelaar, P. F. A. Alkemade, and H. W. Zandbergen, 2004, Application of the dual-beam FIB/SEM to metals research: Journal of Microscopy, v. 214, no. 3, p. 237–245.

Slatt, R. M., and N. R. O'Brien, 2011, Pore types in the Barnett and Woodford gas shales: Contribution to understanding gas storage and migration pathways in fine-grained rocks: AAPG Bulletin, v. 95, no. 12, p. 2017–2030.

Smith, A. J., P. R. Munroe, T. Tran, and M.S. Wainwright, 2001, FIB preparation of a sensitive porous catalyst for TEM elemental mapping at high magnifications: Journal of Materials Science, v. 36, no. 14, p. 3519–3524.

Sondergeld, C. H., R. J. Ambrose, C. S. Rai, and J. Moncrieff, 2010a, Micro-structural studies of gas shales: Proceedings of the SPE Unconventional Gas Conference, Society of Petroleum Engineers, SPE Paper 131771, 17 p., doi: 10.2118/131771-MS.

Sondergeld, C. H., K. E. Newsham, J. T. Comiski, M. C. Rice, and C. S. Rai, 2010b, Petrophysical considerations in evaluating and producing shale gas: Proceedings of the SPE Unconventional Gas Conference, Society of Petroleum Engineers, SPE Paper 131768, 34 p., doi: 10.2118/131768-MS.

Spötl, C., F. J. Longstaffe, K. Ramseyer, and B. Rudinger, 1999, Authigenic albite in carbonate rocks—a tracer for deep-burial brine migration? Sedimentology, v. 46, no. 4, p. 649–666.

Thompson, L. E., P. M. Rice, E. Delenia, V. Y. Lee, P. J. Brock, T. P. Magbitang, G. Dubois, W. Volksen, R. D. Miller, and H. C. Kim, 2006, Imaging thin films of nanoporous low-K dielectrics: Comparison between ultramicrotomy and focused ion beam preparations for transmission electron microscopy: Microscopy and Microanalysis, v. 12, p. 156–159.

Tomutsa, L., D. B. Silin, and V. Radmilovic, 2007, Analysis of chalk petrophysical properties by means of submicron-scale pore imaging and modeling: SPE Reservoir Evaluation and Engineering, v. 10, no. 3, p. 285–293.

Walls, J. D., and S. W. Sinclair, 2011, Eagle Ford shale reservoir properties from digital rock physics: First Break, v. 29, p. 97–101.

Wirth, R., 2004, Focused ion beam (FIB): A novel technology for advanced application of micro- and nanoanalysis in geosciences and applied mineralogy: European Journal of Mineralogy, v. 16, no. 6, p. 863–876.

Wirth, R., 2009, Focused ion beam (FIB) combined with SEM and TEM: Advanced analytical tools for studies of chemical composition, microstructure and crystal structure in geomaterials on a nanometre scale: Chemical Geology, v. 261, no. 3–4, p. 217–229.

Yang, Y., and A. C. Aplin, 2010, A permeability-porosity relationship for mudstones: Marine and Petroleum Geology, v. 27, no. 8, p. 1692–1697.

Yao, S., K. Jiao, K. Zhang, W. X. Hu, H. Ding, M. C. Li, and W. M. Pei, 2011, An atomic force microscopy study of coal nanopore structure: Chinese Science Bulletin, v. 56, no. 25, p. 2706–2712.

Tian Dong, Nicholas B. Harris, 2013, Pore size distribution and morphology in the Horn River Shale, Middle and Upper Devonian, northeastern British Columbia, Canada, *in* W. Camp, E. Diaz, and B. Wawak, eds., Electron microscopy of shale hydrocarbon reservoirs: AAPG Memoir 102, p. 67–79.

8

Pore Size Distribution and Morphology in the Horn River Shale, Middle and Upper Devonian, Northeastern British Columbia, Canada

Tian Dong and Nicholas B. Harris

University of Alberta, 126 Earth Sciences Building, 116 St. and 85 Ave., Edmonton, Alberta, Canada, T6G 2E3 (e-mails: td2@ualberta.ca, nharris@ualberta.ca)

ABSTRACT

Porosity and pore size distributions are key attributes for characterizing shale gas storage capacity and flow in shale. In this study of shale samples from the Horn River Formation, analysis of field emission scanning electron microscopic (SEM) images of argon-ion-milled samples, nitrogen adsorption and desorption experiments, and mercury injection capillary pressure were applied to characterize pore volumes, pore morphology, and pore sizes. Helium porosity measurements range from 2.3 to 6.6%. Porosity estimated from analysis of SEM images was significantly lower, suggesting that this approach fails to resolve a significant pore volume represented by small pores. Based on SEM image analysis, most observed pores were intraparticle pores within organic matter and less commonly within inorganic matrix. Interparticle pores were observed in pyrite framboids, between grains, and surrounding grain rims. Microfractures noted in some images may influence both gas storage and transport. Estimates of pore size distributions also depend on the analytical method. Image analysis of secondary electron SEM images indicated the dominant pore size range is 10 to 40 nm, whereas nitrogen adsorption/desorption isotherm (BET) analysis indicated a much smaller dominant pore size range of 3 to 5 nm. Mercury injection capillary pressure results demonstrated a dominant pore throat size ranging from 3 to 8 nm, consistent with the nitrogen adsorption/desorption isotherm analysis. The discrepancy between SEM and nitrogen adsorption/desorption results also suggests that SEM images obtained with field emission SEM failed to resolve an important class of pores.

DOI: 10.1306/13391706M1023584

INTRODUCTION

Organic-rich shale and mudstone have long been recognized as primary sources of oil and natural gas. At elevated thermal maturities, organic-rich shale formations may contain large volumes of natural gas such as those now exploited for gas production in North America. Significant examples include the Barnett Shale in the Fort Worth Basin (Bowker, 2007), the Fayetteville Shale in the Arkoma Basin (Mason, 2012), the Haynesville Shale in the Gulf of Mexico Basin (Hammes and Carr, 2009), the Marcellus Shale in the Appalachian Basin (Boyce, 2010), and the Horn River Formation in British Columbia (Reynolds and Munn, 2010).

Understanding pore structure within shale reservoirs is critical to determining gas storage capacity and understanding gas flow. This has been a difficult problem to address, largely because of the fine-grain nature of the rocks and small pore sizes, much smaller than conventional sandstone and carbonate reservoirs. Standard techniques for determining permeability are not applicable because flow through shale and mudstone plugs cannot be achieved under normal laboratory conditions in a reasonable amount of time, although pulse decay techniques appear to yield good results. Mercury injection methods for determining pore throat size require very high pressure, not possible with some laboratory setups. Finally, standard scanning electron microscopic (SEM) imaging of pore systems had not revealed true pore sizes, largely because of damage to sample surfaces during sample preparation (Ogura et al., 2007).

Recently, argon-ion milling has been applied to shale samples to create ultrasmooth polished surfaces to better image pores in shale and mudstone samples. Initial research on samples prepared with this technique, imaged with field emission or focused ion beam SEM instruments (Loucks et al., 2009; Wang et al., 2009; Curtis et al., 2011a, b), demonstrated that a considerable amount of the porosity in shale occurs in the nanometer size range and that in some formations (i.e., Barnett Shale), most of the porosity develops in organic matter within thermally mature samples (Loucks et al., 2009).

In granular porous media, permeability depends on the abundance and size of pores and pore throats (Yang and Aplin, 1998; Dewhurst et al., 1999). Published absolute permeability measurements from a variety of shale reservoirs and analytical methods typically fall in the nanodarcy range. Kwon et al. (2004) reported a range from 0.3 to 300 nd measured on illite-rich Paleocene to Eocene Wilcox mudstones from West Baton Rouge Parish, Louisiana, in the Gulf of Mexico sedimentary basin. Permeability was measured parallel to bedding through a pulse decay technique.

In an analysis of shale samples between 4700 and 5600 m (15419.948 and 18372.703 ft) deep in wells from the Scotian shelf, Katsube et al. (1991) reported permeability values ranging from 0.1 to 16 nd perpendicular to bedding, using a transient decay technique at a range of confining pressures; permeability decreased over this range with increasing pressure. They also used a high-pressure mercury injection system to determine pore sizes in the same samples and determined a range of pore throat sizes from 3 nm to 10 μm.

More recently, Al-Bazali et al. (2009) applied pressure transmission tests to three different shales (Cretaceous Pierre Shale from the Western Interior Sedimentary Basin and two others unspecified) with a variety of nonwetting fluids to measure pore size distribution; they reported permeability values ranging from 0.5 to 6.5 nd and average pore throat radii of 20 to 32 nm.

Schettler et al. (1989) used gas adsorption/desorption experiments to investigate the physical structure of Devonian shale and reported an average internal surface area of 8.6 m^2/g (92.5696260 ft/g) and pore sizes ranging from 3.4 to 5.5 nm. These data are consistent with those of Curtis et al. (2012) in an analysis of samples from the Horn River shale reservoir based on sequential SEM images, which showed that the most common pores had radii in the range of 3 to 6 nm.

Sondergeld et al. (2010) applied a mercury injection capillary pressure test to investigate the pore throat size and obtained a dominant pore throat size range from 2 to 8 nm in a data set mostly from the Barnett Shale. Curtis et al. (2011b) also suggested that many pore throats are less than 2 nm, based on high-pressure mercury injection capillary pressure results. They noted that pore throats in this size range approach the resolution limits of SEM and tested a higher-resolution technique, transmission electron microscopy, to address this issue.

This chapter presents porosity data from helium porosimetry and SEM image analysis and pore size distribution from SEM images, nitrogen adsorption/desorption isotherms, and mercury injection capillary pressure. The focus of this study was the Horn River Formation, a Middle to Upper Devonian sequence consisting largely of mudstone and shale, deposited in the Horn River Basin of northeastern British Columbia and southwestern Northwest Territories (Figure 1). The Horn River Shale units have been the focus of significant shale gas exploration efforts during the past several years (Smith, 2011).

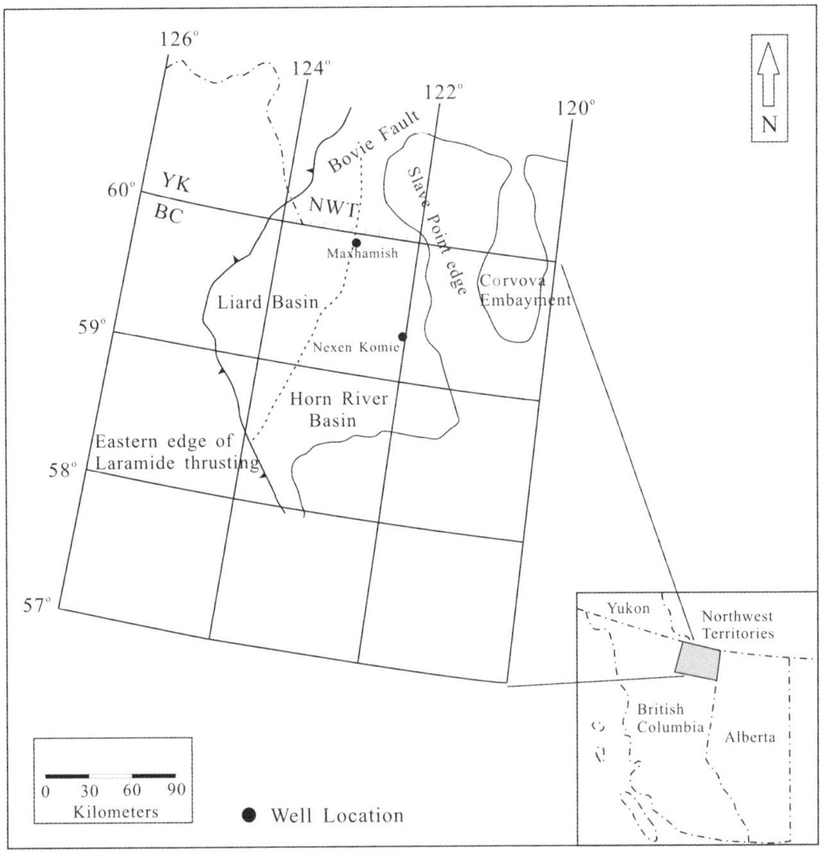

Figure 1. Map of Horn River Basin and adjacent areas (modified after Ross and Bustin, 2008) (30 km = 18.6 mi).

Shale units of the Horn River Formation are basinal equivalents of coeval carbonate platform deposits of the Middle Devonian Upper Keg River, Sulphur Point, and Slave Point formations; the carbonate platforms bordered the Horn River Basin to the east and south. Basinal mudstones compose the Evie, Otter Park, and Muskwa members of the Horn River Formation (Figure 2). The Evie shale consists of dark gray-to-black, high-gamma-ray, organic-rich, pyritic, variably calcareous, and siliceous shale. The Otter Park Shale consists of medium-to-dark-gray calcareous shale with lower gamma-ray and resistivity well log response than the Evie and Muskwa units (Figure 3). The Muskwa consists of gray-to-black, organic-rich, pyritic, siliceous shale and is characterized on well logs by high-gamma-ray and high-resistivity readings (McPhail et al., 2008).

METHODOLOGY

Our work integrated data obtained from well files at the British Columbia Ministry of Energy and Mines (BC-MEM) in Victoria, with analyses performed at the University of Alberta. Data obtained from the BC-MEM included SEM images, mineralogical and total organic carbon (TOC) content analyses. Work performed at the University of Alberta included helium porosimetry,

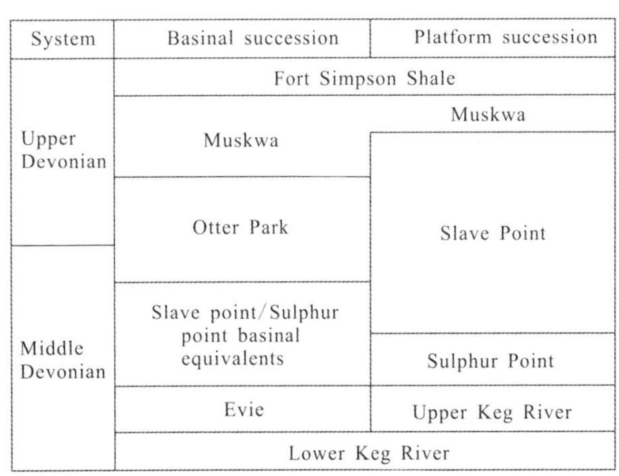

Figure 2. Middle and basal Upper Devonian stratigraphy of the Horn River Basin and Cordova embayment (McPhail et al., 2008).

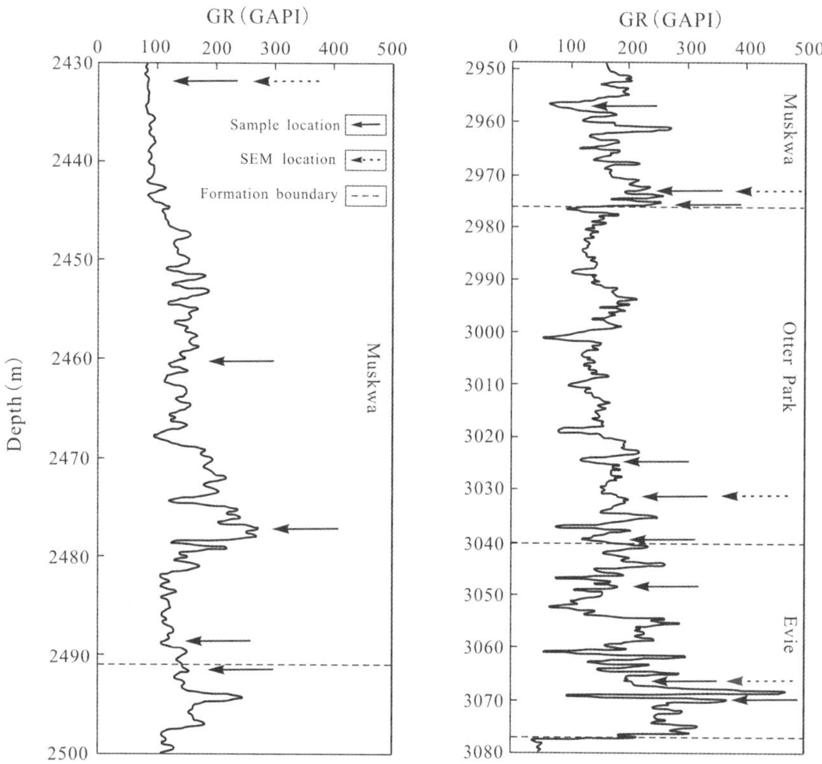

Figure 3. Location of samples from the Nexen Komie and Maxhamish wells. The Nexen Komie well log (left) and the Maxhamish well log (right) (10 m = 32.8 ft).

Table 1. Surface Area, Pore Volume, and Porosity of Horn River Shale

Area	Well Name	Depth (m)	Formation	TOC (wt.%)	Surface Area (m^2/g)	Pore Volume (cm^3/g)	Porosity (%)
Horn River	Nexen Komie	2431.83	Muskwa	0.44	10.09	0.015	3.61
Horn River	Nexen Komie	2460.14	Muskwa	2.64	11.01	0.01434	3.19
Horn River	Nexen Komie	2477.12	Muskwa	4.76	15.26	0.01483	3.20
Horn River	Nexen Komie	2488.66	Muskwa	1.91	11.84	0.01533	3.50
Horn River	Nexen Komie	2491.43	Otter Park	2.44	12.54	0.01479	3.47
Horn River	Maxhamish	2956.91	Muskwa	6.36	8.04	0.009726	2.27
Horn River	Maxhamish	2972.49	Otter Park	6.365	19.42	0.02443	5.37
Horn River	Maxhamish	2975.39	Otter Park	2.15	9.18	0.01291	2.90
Horn River	Maxhamish	3024.87	Otter Park	2.61	7.73	0.01054	2.59
Horn River	Maxhamish	3031.23	Otter Park	4.63	16.70	0.02214	5.15
Horn River	Maxhamish	3039.17	Otter Park	2.87	16.74	0.01636	3.93
Horn River	Maxhamish	3048.15	Evie	5.66	6.06	0.01039	2.49
Horn River	Maxhamish	3066.51	Evie	4.05	26.33	0.02627	6.16
Horn River	Maxhamish	3069.78	Evie	5.89	15.13	0.01571	

nitrogen adsorption/desorption experiments on core samples, and quantitative analysis of SEM images.

Samples and data from two wells in the Horn River Basin, the Nexen Komie D-94-A/94-O-8 and EOG Maxhamish D-012-L/094-O-15, were analyzed. Samples were chosen to represent different parts of the Horn River stratigraphy and a range of TOC contents. Sample depths ranged from 2431.83 to 3069.78 m (7978.4449 to 10071.457 ft) (Figure 3; Table1).

Field emission SEM images on four argon-ion-milled samples were obtained from well files at the BC-MEM. Images of samples from the Maxhamish well were acquired by TerraTek in 2006 as part of a tight rock analysis for the operator, EOG Resources.

Total organic carbon (TOC) contents were also analyzed by TerraTek using LECO combustion and are compiled in Table 1. Prior to instrument analysis, dried samples reacted sufficiently with hydrochloric acid in order to remove inorganic carbon and then were combusted in an oxygen atmosphere using a LECO Carbon Analyzer and any carbon present was converted to CO_2. Scanning electron microscopic images of samples from the Nexen Komie well were acquired by Weatherford for Devon Canada. Total organic carbon contents for the Nexen Komie well were determined by CBM Solutions using LECO combustion and measurement of the evolved carbon dioxide.

Pore sizes and pore size distributions were determined from SEM images using software to outline and measure all the individual pores and fractures in areas of interest. Pore diameters and aspect ratios were determined for each pore, and average and median diameters were computed for groups of pores. Pores with diameters less than 10 nm may be undercounted because of scale and resolution limitations of the archived SEM images. Because true porosity is a volume measurement, some error may also be introduced in extrapolating from two-dimensional (2-D) pore space area measurements to three-dimensional (3-D) pores and porosities (Loucks et al. 2009).

The porosity of 13 samples (Table 1) was analyzed by a multipycnometer helium porosimeter at the University of Alberta. The samples included splits of, or were adjacent to, core samples used for SEM analysis (Figure 3). Samples were crushed and sieved to 0.5- to 1-mm diameter. Impurities were removed by washing, and then the samples were dried in an oven for 24 hours.

Samples for nitrogen adsorption/desorption experiments were splits of samples used for helium porosimetry. Adsorption and desorption experiments were conducted on a Quantachrome Autosorb-1 instrument. Samples were crushed and sieved into fractions of 0.5–1 mm, desiccated for one day, and put into a sample cell for degassing (typically 5 hr). The Autosorb-1 measures the quantity of gas adsorbed onto or desorbed from a solid surface at some equilibrium vapor pressure by the static volumetric method. The analysis is performed by admitting or removing a known quantity of adsorbate gas into or out of a sample cell containing the solid adsorbent maintained at a constant temperature below the critical temperature of the gas. As adsorption or desorption occurs, the pressure in the sample cell changes until equilibrium is established. The quantity of gas adsorbed or desorbed at the equilibrium pressure is the difference between the amount of gas admitted or removed and the amount required to fill the space around the adsorbent (void space). When the quantity of adsorbate on a surface is measured over a wide range of relative pressures at constant temperature, the result is an adsorption isotherm; the opposite process is a desorption isotherm (Quantachrome Instruments, 2005).

The BET method (Brunauer et al., 1938; Schettler et al., 1989) provides a direct measure of the surface area of sample materials. The BET equation relates the weight of adsorbed gas to pressure:

$$1/W((P_0/P) - 1) = 1/W_mC + (C - 1)(P/P_0)/W_mC$$

where W_m is the weight of adsorbate constituting a monolayer of surface coverage, W is the weight of gas adsorbed at a relative pressure P/P_0, and the term C is the BET constant. The total surface area S_t of the sample can be expressed as

$$S_t = W_m\, N\, A_{cs}/M$$

where N is the Avogadro number (6.023×10^{23} molecules/mol), and M is the molecular weight of the adsorbate. The cross-sectional area A_{cs} for nitrogen is 16.2 $Å^2$.

Barrett, Joyner, Halenda (BJH) theory (Barret et al., 1951; Gregg and Sing, 1982; Bustin et al., 2008) can be used in combination with the Kelvin equation to obtain mesopore size distributions (pore diameter 2–50 nm) from desorption isotherms, assuming cylindrical geometry.

Two samples were tested by mercury injection porosimetry analysis to corroborate our analysis results of pore size. Analysis was carried out on a Micromeritics AutoPore IV 9510, with a maximum pressure in our experiments of approximately 410 MPa.

RESULTS

TOC and Thermal Maturity

The TOC content of the samples averaged 3.6 wt.%, ranging from 0.44 to 6.36 wt.%, which is generally consistent with published analyses by Ross and Bustin (2008). Thermal maturity in the Horn River Formation is high, with vitrinite reflectance between 1.6%R_o and 2.5%R_o (Ross and Bustin, 2008, 2009). Most T_{max} values from Rock-Eval analysis of samples from the Maxhamish well cluster between 600 and 610°C, consistent with the work of Ross and Bustin (2008). The T_{max} values in this range are equivalent to vitrinite reflectance of 3.65 to 3.80%R_o, according to the formula published by Jarvie et al. (2001); however, S_2 values are extremely low, rarely exceeding 0.20 mg/g, which makes precise determination of T_{max} unreliable.

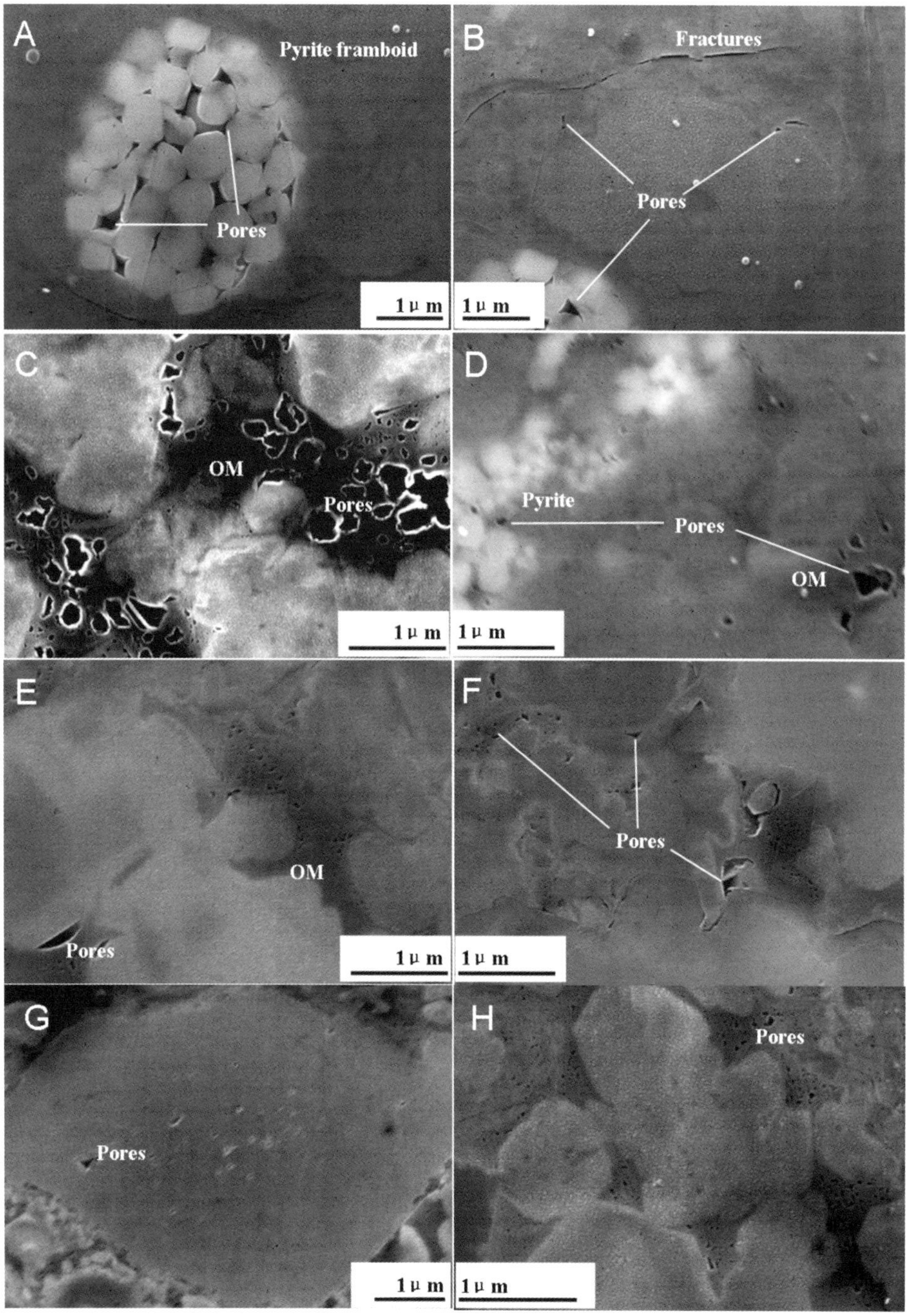

Figure 4. Secondary electron scanning electron microscopic (SEM) images of pores in shales, depth of (A), (B) was 2431.8 m (7978.4 ft); depth of (C), (D) was 2972.5 m (9752.2 ft); depth of (E), (F) was 3031.25 m (3031.25 ft); depth of (G), (H) was 3066.5 m (10060.6 ft).

Porosity Measurements

Helium-measured porosity in the samples ranged from 2.27 to 6.16% (Table 1). The different members of the Horn River are characterized by slight differences in porosity, although the data set is too small for a rigorous comparison. The two Evie samples averaged 4.3% porosity, with a considerable range in porosity. Five Muskwa samples averaged 3.2% porosity, with small variation in measured porosity. Six Otter Park samples averaged 3.9% porosity, with a large range in measured porosity. Our measurements were typically 10 to 20% less than the CMB-Solutions-reported effective porosity or the TerraTek-reported total porosity for samples at that depth. No correlation is evident between our helium porosity data and TOC data from the CBM Solutions and TerraTek reports (Table 1).

Table 2. Pore Size and Porosity Calculations in Horn River Shales

Well Name	Depth (m)	Minimum Length (nm)	Mean Length (nm)	Median Length (nm)	Maximum Length (nm)	Porosity (%)	Counts
Nexen Komie	2431.8	16.85	80.33	72.46	187.75	0.78	68
Maxhamish	2972.5	7.99	80.76	46.68	1890.46	2.31	1194
Maxhamish	3031.25	8.68	26.69	25.07	49.97	0.62	611
Maxhamish	3066.5	6.64	29.24	24.47	362.33	0.52	979

Source: SEM images were quantitatively analyzed using Image-Pro Plus software.

A systematic discrepancy exists between porosity determined from helium porosimetry (Table 1) and from SEM image analysis (Table 2), with the latter substantially lower than the former. For example, porosity determined for the Nexen Komie sample at 2431.8 m (7978.4 ft) by SEM image analysis was 0.78%, compared with the helium porosity value of 3.61%. This observation is consistent with results of Curtis et al. (2012), who described a 3- to 20-fold discrepancy in SEM versus helium porosity for a number of shale formations. Helium-measured porosity should be more reliable than determinations from SEM analysis because of two problems inherent with the latter: the resolution of small pores and conversion from 2-D pore diameters to 3-D pore volumes.

Porosity Characteristics from Image Analysis

Images of four core samples from two wells were analyzed using the methods described. Quantitative results of SEM image analysis are reported in Table 2. Several types of pores and microfractures are recognized (Figure 4). Intraparticle pores were found predominantly within organic matter (Figures 4C–F, H) and less commonly in inorganic particles (Figure 4G). Interparticle pores were present in pyrite framboids (Figures 4A, B), surrounding quartz grains (Figure 4B), and between organic matter and mineral grains (Figure 4F). In some images, microfractures containing porosity have been observed (Figure 4B).

The morphology of pores varied widely. Intraparticle organic pores and intraparticle inorganic pores were primarily rounded and ellipsoidal (Figure 4C). Interparticle pores are mostly irregular, polyhedral, or ellipsoidal (Figures 4A, B, F).

Pore sizes (Figure 5) measured from SEM images displayed a unimodal distribution with a maximum size of more than 1000 nm and a minimum size of less than 10 nm, probably close to 2 nm. The mode of the pore size distribution in these samples ranged from 10 to 40 nm.

Nitrogen Adsorption/Desorption Experiments

Results of the nitrogen adsorption and desorption experiments are shown in Table 1. The specific surface area of the samples ranged from 6.06 m^2/g to 26.33 m^2/g with an average of 13.29 m^2/g; pore volume (pore space within sample) ranged from 0.009726 to 0.02627 cm^3/g with an average of 0.015912 cm^3/g. The large surface area in conjunction with small pore volume implies that the average pore diameter must be small. In samples from the Nexen Komie well, surface area, pore volume, and porosity were fairly consistent; however, a great variation existed of these parameters in samples from the Maxhamish well.

Pore sizes (Figure 6) calculated from nitrogen desorption curves showed a unimodal distribution ranging from less than 2 nm to more than 100 nm. The peaks for these pore size distribution curves are typically between 3 and 5 nm. This result clearly differs from SEM observations, in which pores range dominantly from 10 to 40 nm.

The shape of the adsorption/desorption curves in a single experimental run, termed a hysteresis loop, is informative about the dominant shape of pores. De Boer (1958) identified five types of hysteresis loops and correlated them with various pore shapes. In general, the hysteresis loops in our experiments were quite similar (Figure 7) and can be classified as type B hysteresis, which is associated with slit-shaped pores.

Mercury Injection Capillary Pressure

Figure 8 shows pore throat size distributions of two samples obtained from mercury injection

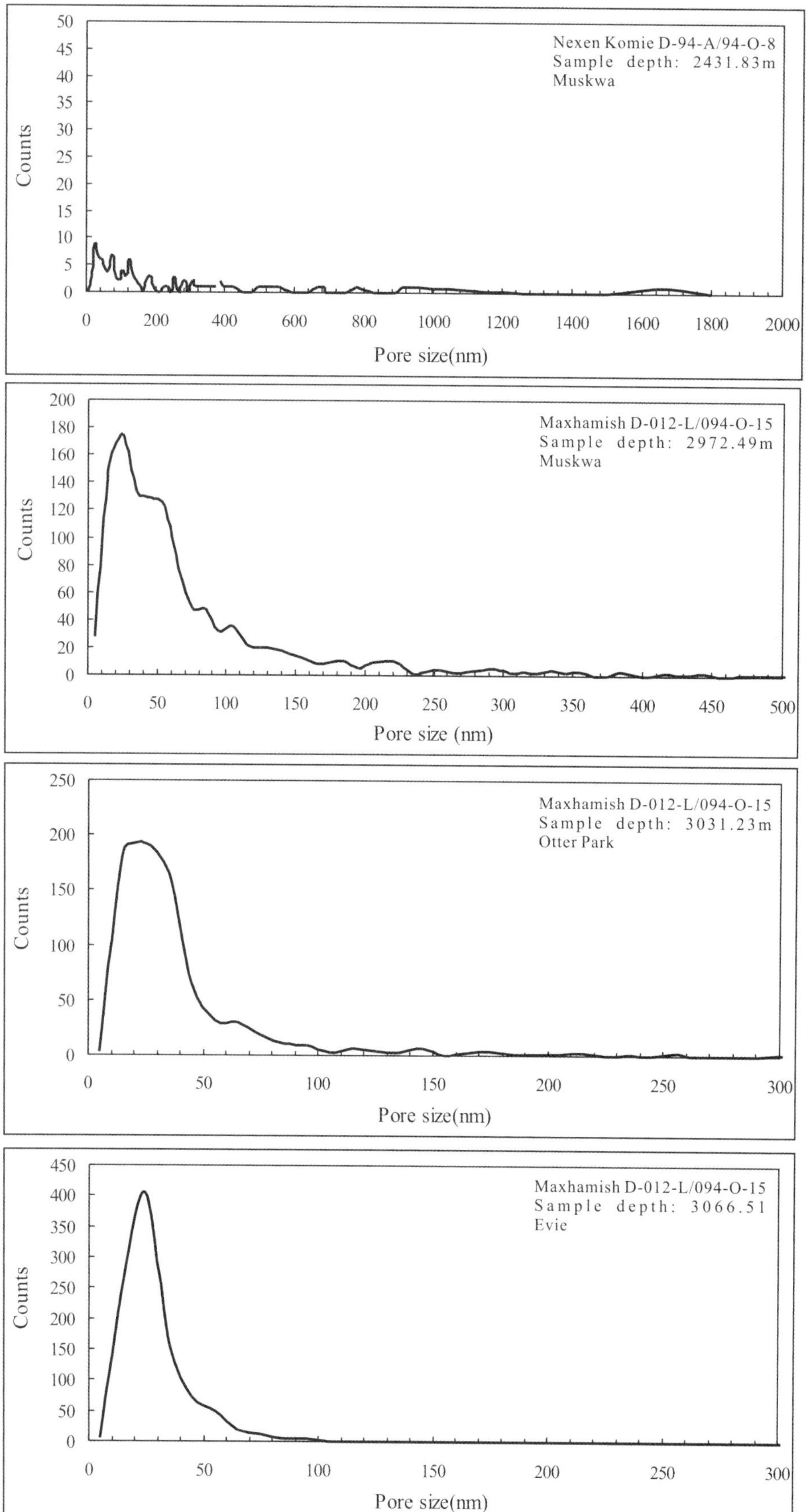

Figure 5. Pore size distribution of shale samples determined from analysis of scanning electron microscopic (SEM) images of argon-ion-milled samples from the Muskwa and Otter Park members of the Horn River Formation.

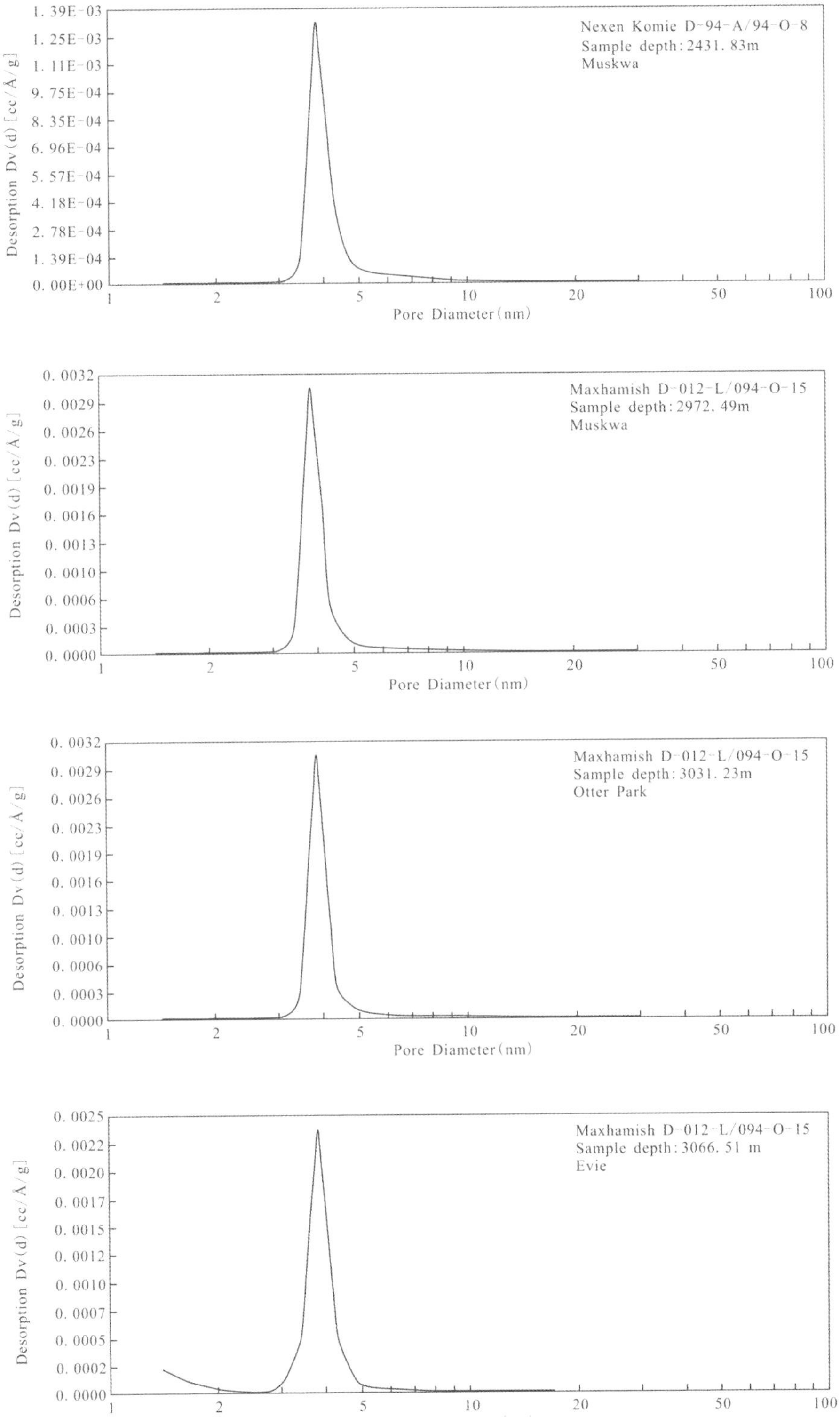

Figure 6. Pore size distribution of shale samples determined from N_2 adsorption and desorption experiments from the Muskwa, Otter Park, and Evie members of the Horn River Formation.

capillary pressure. A peak ranging from 3 to 8 nm dominates in the pore size, which was consistent with both our nitrogen adsorption/desorption result and other shale formations (Sondergeld et al., 2010; Sakhaee-Pour and Bryant, 2011).

DISCUSSION

The recent application of argon-ion-milling preparation to SEM imaging of shale samples has introduced a new perspective on porosity development in shale.

One important outcome of this analysis has been the awareness that a considerable fraction of shale porosity is localized in the organic fraction, perhaps resulting from organic maturation and cracking of kerogen to hydrocarbons (Loucks et al., 2009; Sondergeld et al., 2010; Curtis et al., 2012).

Our initial studies of porosity in the Horn River Shale compared porosity determined by analysis of SEM images and helium porosimetry. While our helium porosity results were consistent with reports of porosity in other shale formations, the SEM images appeared to undercount porosity by 60 to 90% in samples for which we can make direct comparisons between the two methods.

This undercount appears to take place in the small pores. Whereas pore size distributions based on SEM image analysis indicated modal pore sizes on the order of 10 to 40 nm, measurements based on nitrogen adsorption/desorption isotherms had modes in the range of 3 to 5 nm. The adsorption/desorption data were also broadly similar to the mercury injection results, although the two techniques measured different properties, the former measuring pore radii and the latter pore throat dimensions.

The discrepancy between the SEM and nitrogen adsorption/desorption pore size determinations may not be significant for estimating or predicting permeability, assuming shale permeability is substantially

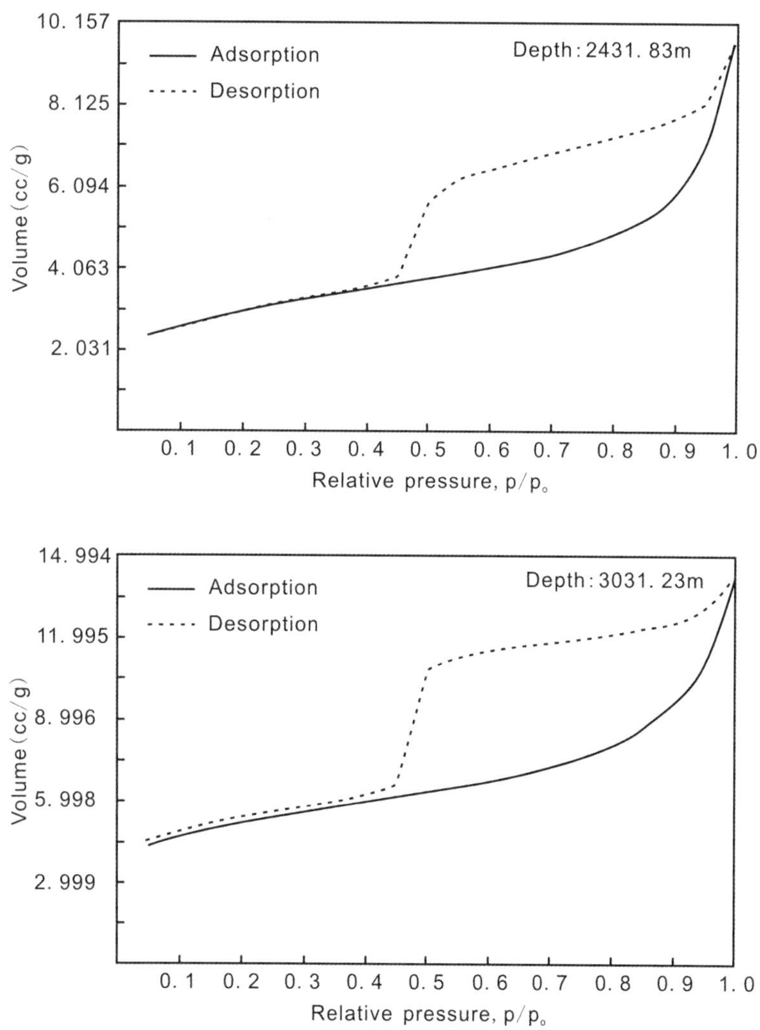

Figure 7. Hysteresis loops from nitrogen adsorption-desorption experiments on samples from the Muskwa and Otter Park members of the Horn River Formation. The shape of the hysteresis loops is classified as type B, indicative of slit-shaped pores.

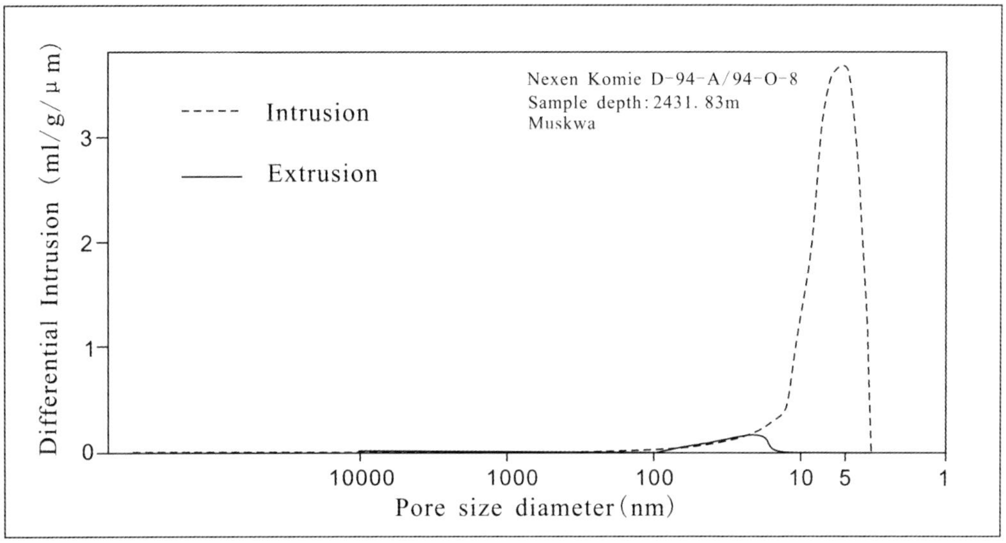

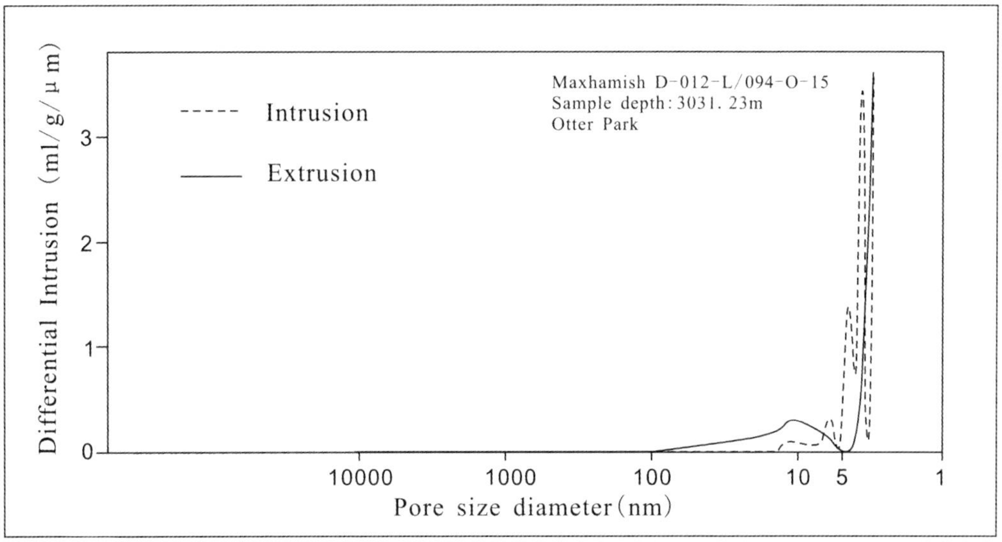

Figure 8. Pore size distribution of Horn River Formation shale samples determined from mercury injection capillary pressure. (2431.83 m and 3013.23 m = 7978.44 ft and 9885.92 ft)

governed by the largest pores and pore throats (Yang and Aplin, 1998). For this purpose, pore size estimates based on SEM image analysis may provide valuable information about size and where larger pores are localized (e.g., within organic matter) that can be used to develop models for permeability and gas flow.

Our results suggest that SEM-image-based analysis underestimates pore volume and thus should not be used for volumetric estimates, such as in-place or recoverable reserves. Clearly, this topic is understudied at present, and a more complete understanding depends both on gathering information, building larger databases, and on applying higher-resolution imaging methods such as atomic force microscopy and transmission electron microscopy.

CONCLUSIONS

Our studies showed that pores in Horn River Shale are dominantly nanometer in scale, typically ranging from 2 to 1000 nm. Most pores are associated with organic matter. Less-common pore types include interparticle pores present in pyrite framboids, interparticle pores between grains or surrounding grain rims, and intraparticle pores within mineral grains. Microfractures present in some images may be significant for both gas storage and transport.

Most pore diameters calculated from field emission SEM images fall within the 10- to 40-nm range, substantially larger than the 3- to 5-nm mode determined by nitrogen adsorption and desorption experiments. The difference in results from the two techniques suggests that the fraction of small pores (<10 nm) is significantly underestimated by SEM image analysis because of (1) limited resolution of the available images and (2) smaller sample area (volume) evaluated. This interpretation was also supported by a comparison of total porosity estimates by helium porosimetry and SEM image analysis; results from image analysis were lower in comparison to helium porosity values by factors of two to ten in splits of the same or adjacent samples. The conclusion was also supported by mercury injection capillary pressure results, which showed a dominant pore throat size distribution ranging from 3 to 8 nm.

We suggest that if the fraction of very small (<10 nm) pores is substantially undercounted in SEM image analysis, there will be little impact on models for permeability structure in shale because permeability is likely governed by the larger pores. However, models for porosity and storage capacity may be strongly impacted by underrepresentation of small pores in SEM images.

ACKNOWLEDGMENTS

We thank the British Columbia Ministry of Energy and Mines for funding this initial research on the Horn River shales and, in particular, Warren Walsh of the BC-MEM for arranging access to data files at the ministry in Victoria. We also thank Drs. Nancy Zhang and Eric Flaim of the Integrated Nanosystems Research Facility at the University of Alberta for training and guidance in use of the Autosorb-1 instrument for measurement of specific surface area. We thank Patrick Monahan and Carl Sondergeld for their helpful reviews.

REFERENCES CITED

Al-Bazali, T. M., J. Zhang, and M. E. Chenevert, 2009, Estimating the reservoir hydrocarbon capacity through measurement of the minimum capillary entry pressure of shale cap-rocks: Society of Petroleum Engineers Annual Technical Conference, October 4–7, New Orleans, SPE Paper 121450, 14 p., doi: 10.2118/121450-MS.

Barret, E. P., L. G. Joyner, and P. P. Halenda, 1951, The determination of pore volume and area distribution in porous substances. I. Computation from nitrogen isotherms: The Bureau of Chemistry and Soils and George Washington University, v. 73, p. 373–380.

Bowker, K. A., 2007, Barnett shale gas production, Fort Worth Basin: Issues and discussion: AAPG Bulletin, v. 91, p. 523–533.

Boyce, M. L., 2010, Sub-surface stratigraphy and petrophysical analysis of the middle Devonian interval of the central Appalachian Basin, west Virginia and southwest Pennsylvania: Ph.D. thesis, West Virginia University, Morgantown, 159 p.

Brunauer, B. S., P. H. Emmett, and E. Teller, 1938, Adsorption of gases in multimolecular layers: The Bureau of Chemistry and Soils and George Washington University, v. 60, p. 309–319.

Bustin, R. M., A. M. M. Bustin, X. Cui, D. J. K. Ross, and V. S. Murthy Pathi, 2008, Impact of shale properties on pore structure and storage characteristics: Society of Petroleum Engineers Shale Gas Production Conference, November 16–18, Fort Worth, Texas, SPE Paper 119892, 28 p., doi: 10.2118/119892-MS.

Curtis, M. E., R. J. Ambrose, C. H. Sondergeld, and C. S. Rai, 2011a, Investigation of the relationship between organic porosity and thermal maturity in the Marcellus shale: Society of Petroleum Engineers North American Unconventional Gas Conference and Exhibition, June 14–16, The Woodlands, Texas, SPE Paper 144370, 4 p., doi: 10.2118/144370-MS.

Curtis, M. E., R. J. Ambrose, C. H. Sondergeld, and C. S. Rai, 2011b, Transmission and scanning electron microscopy investigation of pore connectivity of gas shales on the nanoscale: Society of Petroleum Engineers North American Unconventional Gas Conference and Exhibition, June 14–16, The Woodlands, Texas, SPE Paper 144391, 10 p., doi: 10.2118/144391-MS.

Curtis, M. E., C. H. Sondergeld, R. J. Ambrose, and C. S. Rai, 2012, Microstructural investigation of gas shales in two and three dimensions using nanometer-scale resolution imaging: AAPG Bulletin, v. 96, p. 665–677.

De Boer, J. H., 1958, The shape of capillaries, *in* D. H. Evertt and R. H. Ottewill, eds., The structure and properties of porous materials: London, Butterworths, p. 68–94.

Dewhurst, D. N., Y. Yang, and A. C. Aplin, 1999, Permeability and fluid flow in natural mudstones, *in* A. C. Aplin, A. J. Fleet, and J. H. S. Macquaker, eds., Muds and mudstones: Physical and fluid flow properties, Geological Society Special Publication 158, p. 23–43.

Gregg, S. J., and K. S. W. Sing, 1982, Adsorption, surface area and porosity, 2d: London, Academic Press, 303 p.

Hammes, U., and D. L. Carr, 2009, Sequence stratigraphy, depositional environments, and production fairways of the Haynesville shale-gas play in East Texas: AAPG Annual Convention, Search and Discovery Article 110084, p. 36.

Jarvie, D.M., B. Claxton, B. Henk, and J. Breyer, 2001, Oil and shale gas from the Barnett Shale, Ft. Worth Basin, Texas (abs.): AAPG National Convention, June 3–6, Denver, Colorado.

Katsube, T. J., B. S. Mudford, and M. E. Best, 1991, Petrophysical characteristics of shales from the Scotian shelf: Geophysics, v. 56, p. 1681–1689.

Kwon, O., A. K. Kronenberg, A. F. Gangi, B. Johnson, and B. E. Herbert, 2004, Permeability of illite-bearing shale: 1. Anisotropy and effects of clay content and loading: Journal of Geophysical Research, v. 109, B10205, 19 p.

Loucks, R. G., R. M. Reed, R. M. S. C. Ruppel, and D. M. Jarvie, 2009, Morphology, genesis, and distribution of nanometer-scale pores in siliceous mudstones of the Mississippian Barnett Shale: Journal of Sedimentary Research, v. 79, p. 848–861.

Mason, J. E., 2012, Well production profiles for the Fayetteville shale gas play: Oil and Gas Journal (article accepted for publication), 14 p.

McPhail, S., W. Walsh, C. Lee, and P. A. Monahan, 2008, Shale units of the Horn River Formation, Horn River Basin and Cordova embayment, northeastern British Columbia (abs.): Canadian Society of Petroleum Geologists and Canadian Well Logging Society Convention, May 12–14, Calgary, Alberta, 14 p.

Ogura, K., M. Kamidaira, S. Asahina, and N. Erdman, 2007, New methods for cross-section sample preparation using broad argon ion beam: Microscopy and Microanalysis, v. 13 (Supplement 2), p. 1518–1519.

Quantachrome Instruments, 2005, Autosorb-1 AS1 Win Version 1.51, Operation Manual, 127 p.

Reynolds, M. M., and D. L. Munn, 2010, Development update for an emerging shale gas giant field—Horn River Basin, British Columbia, Canada: Society of Petroleum Engineers Unconventional Gas Conference, February 23–25, Pittsburgh, Pennsylvania, SPE Paper 130103, 17 p., doi: 10.2118/130103-MS.

Ross, D. J. K., and R. M. Bustin, 2008, Characterizing the shale gas resource potential of Devonian-Mississippian strata in the western Canada sedimentary basin: Application of an integrated formation evaluation: AAPG Bulletin, v. 92, p. 87–125.

Ross, D. J. K., and R. M. Bustin, 2009, The importance of shale composition and pore structure upon gas storage potential of shale gas reservoirs: Marine and Petroleum Geology, v. 26, p. 916–927.

Sakhaee-Pour, A., and S. L. Bryant, 2011, Gas permeability of shale: Society of Petroleum Engineers Annual Technical Conference and Exhibition, October 30–November 2, Denver, Colorado, SPE Paper 146944, 14 p., doi: 10.2118/146944-MS.

Schettler, P. D., C. R. Parmely, and W. J. Lee, 1989, Gas storage and transport in Devonian shales: SPE Formation Evaluation, p. 371–376.

Smith, K. G., 2011, The Horn River Basin: The opportunity and challenges of a giant shale gas field: 2011 Canadian Society of Petroleum Geologists and Canadian Well Logging Society Convention, May 9-13, Calgary, Alberta, 4 p.

Sondergeld, C. H., R. J. Ambrose, C. S. Rai, and J. Moncrieff, 2010, Micro-structural studies of gas shales: Society of Petroleum Engineers Unconventional Gas Conference, February 23–25, Pittsburgh, Pennsylvania, SPE Paper 131771, 17 p., doi: 10.2118/131771-MS.

Wang, F. P., R. M. Reed, and K. G. Jackson, 2009, Pore networks and fluid flow in gas shales: 2009 Society of Petroleum Engineers Annual Technical Conference and Exhibition, October 4–7, New Orleans, SPE Paper 124253, 8 p., doi: 10.2118/124253-MS.

Yang, Y., and A. C. Aplin, 1998, Influence of lithology and compaction on the pore size distribution and modeled permeability of some mudstones from the Norwegian margin: Marine and Petroleum Geology, v. 15, p. 163–175.

9

Raphael A. J. Wüst, Brent R. Nassichuk, R. Marc Bustin, 2013, Porosity characterization of various organic-rich shales from the Western Canadian Sedimentary Basin, Alberta and British Columbia, Canada, *in* W. Camp, E. Diaz, and B. Wawak, eds., Electron microscopy of shale hydrocarbon reservoirs: AAPG Memoir 102, p. 81–100.

Porosity Characterization of Various Organic-rich Shales from the Western Canadian Sedimentary Basin, Alberta and British Columbia, Canada

Raphael A. J. Wüst

Trican Geological Solutions Ltd., 621-37 NE Calgary, Canada T2E 2MI and James Cook University, 4811 Townsville, Australia (e-mail: rwust@trican.ca)

Brent R. Nassichuk

Trican Geological Solutions Ltd., 621-37 NE Calgary, Canada T2E 2MI (e-mail: bnassichuk@trican.ca)

R. Marc Bustin

University of British Columbia, 2329 W. Mall, Vancouver, British Columbia V6T 1Z4, Canada (e-mail: bustin@mail.ubc.ca)

ABSTRACT

Scanning electron microscopy (SEM) of shales from three unconventional gas/liquid plays (Nordegg, Montney, Duvernay) of the Western Canadian Sedimentary Basin were combined with routine analytical investigations (x-ray diffraction, source rock analyses, mercury porosimetry, and petrography) to characterize mineral composition, mineral assemblage, morphology, organic content, and porosity. The investigations demonstrated that despite marked diagenetic differences between these shales, some common textural and pore characteristics occurred in all samples. The study showed that SEM morphological investigations of unconventional shale reservoirs provided important information about mineral aggregates, cementation, and clay mineral distribution, which allows interpretations about diagenetic history. Combining petrographic analyses with SEM is critical in sediments with pronounced cementation and mineral overgrowth. Conventional secondary electron SEM studies of untreated samples have several advantages over focused ion beam (FIB) milled surfaces as morphological characteristics of larger areas can be evaluated. The study showed that the organic material may not be fully identified, but organic fragments and some bituminous material can still be recognized. In general, abundant large pores (200 to > 2000 nm) can be observed under SEM, but other techniques, such as the mercury porosimetry, failed to identify these larger pores in these shales. The main morphological porosity types in these shales were intergranular (between particles), dissolution, residual growth, fracture, and phyllosilicate pores. The high number of 5- to 20-nm pores as determined by mercury porosimetry could not be confirmed by SEM, which may have been because of resolution issues. Hence, future porosity characterization of gas shales needs to investigate these problems of discrepancy between analytical and imaging technologies.

DOI: 10.1306/13391707M1023585

INTRODUCTION

Characterizing textural parameters of shale hydrocarbon reservoirs, including porosity and permeability, is critical to understand the potential for flow and fluid transport during production. Pore space influences the amount of hydrocarbons stored within the rocks; permeability governs the flow through the reservoir. In tight reservoir rocks, including shales and mudstones, matrix permeability is often small (in the range of 10^{-3} to 10^{-7} md), but artificial stimulation and hydraulic fracturing increase fracture density and thus reduce the flow length through the tight rock to the fracture. In tight-rock resource plays, porosity characterization through empirical and visual means is instrumental for any exploration strategy.

Shale and mudstones are fine-grain (particle size <63 μm) sedimentary rocks that can contain marked amounts of organic matter (up to 30 wt.%) (Riediger et al., 1990). Rocks with high organic content often represent ideal source rocks and have become common targets in current exploration trends. Unconventional coal and shale gas has reshaped the energy future of our planet, and reserves for at least the next 135 years are suggested (Conti et al., 2011). In Alberta alone, government estimates suggest that at least 850 tcf of shale gas may be present. Shale plays in Western Canada (see Figure 1) have become valuable and important sources for gas, condensate, and oil.

This study investigated three organic-rich shale successions of the Western Canadian Sedimentary Basin (Figure 1) that have markedly different compositional characteristics. A total of six samples (two of each formation/unit) have been investigated from the Devonian Duvernay Formation, the Lower Triassic Montney Formation, and the Lower Jurassic Nordegg Member of the Fernie Formation.

The Lower Triassic Montney Formation in British Columbia and Alberta is a thick (>200–300 m [656.1–984.2 ft] in areas), fine-grain clastic deposit that has been classified as a gas shale play, although the formation consists of tight sandstone and siltstone interbedded with shale. Initial focus on the Montney Formation centered on the conventional targets; however, recent research and developments in horizontal drilling and hydraulic fracture stimulation have allowed for the exploitation of the unconventional source rock horizons. On the other hand, some other thin (<40 m [131.2 ft]) shale plays, such as the Devonian Duvernay Formation and the Lower Jurassic Nordegg Member of the Fernie Formation, have only recently attracted attention because of their potential for wet gas and other hydrocarbon liquids as gas prices have plummeted to new lows over the last 10 years.

The focus of this research was to characterize both the intrinsic and fracture-related pores of these rocks. Previous gas shale studies in the United States have characterized four different pore types, which range from nano- to micrometer sizes (Loucks et al., 2009; Milner et al., 2010; Schieber, 2010), although these ignored the large micro- to submillimeter-size pores within some of the shale reservoirs. The four types previously defined are (1) phyllosilicate framework (voids between clays and grain aggregates or cements); (2) organic matter (pores associated with organic matter, mostly kerogen and bitumen); (3) carbonate dissolution (pores resulting from diagenetic alteration of certain grains, mainly carbonate minerals); and (4) intraparticle pores. According to Milner et al. (2010), the intraparticle porosity forms networks between nanofossil fragments coated with amorphous kerogen and occurs primarily in fecal pellets.

GEOLOGIC SETTINGS

An overview of the formations in western Alberta/eastern British Columbia and sample depth with geophysical logs is provided in Figure 2. Samples were selected from interpreted organic-rich intervals based on geophysical log characteristics (high gamma ray and resistivity; Figure 2).

Nordegg

The Lower Jurassic Nordegg Member is composed of 25–45 m (78.7–147.6 ft) of organic-rich marine sediments (quartz and/or carbonate rich) that were deposited in a restricted basinal environment adjacent to reefs and carbonate platforms (Riediger et al., 1990). The Nordegg Member often consists of three lithological facies: (1) a lower phosphatic organic-rich shale; (2) a silty or carbonate-rich mudstone; and (3) an upper phosphatic, organic-rich shale. In areas of low thermal maturity, total organic content values of the Nordegg are up to 23 wt.% (Riediger and Bloch, 1995). The high total organic carbon (TOC) contents reflect high algal productivity with reduced bottom-water mixing. Nordegg adsorbed gas capacities range from 0.05 cm^3/g (1.6 scf/t) to over 2 cm^3/g (64.1 scf/t) in organic-rich zones (Ross and Bustin, 2007).

Montney

The Montney Formation in northeastern British Columbia and adjacent parts of Alberta consists

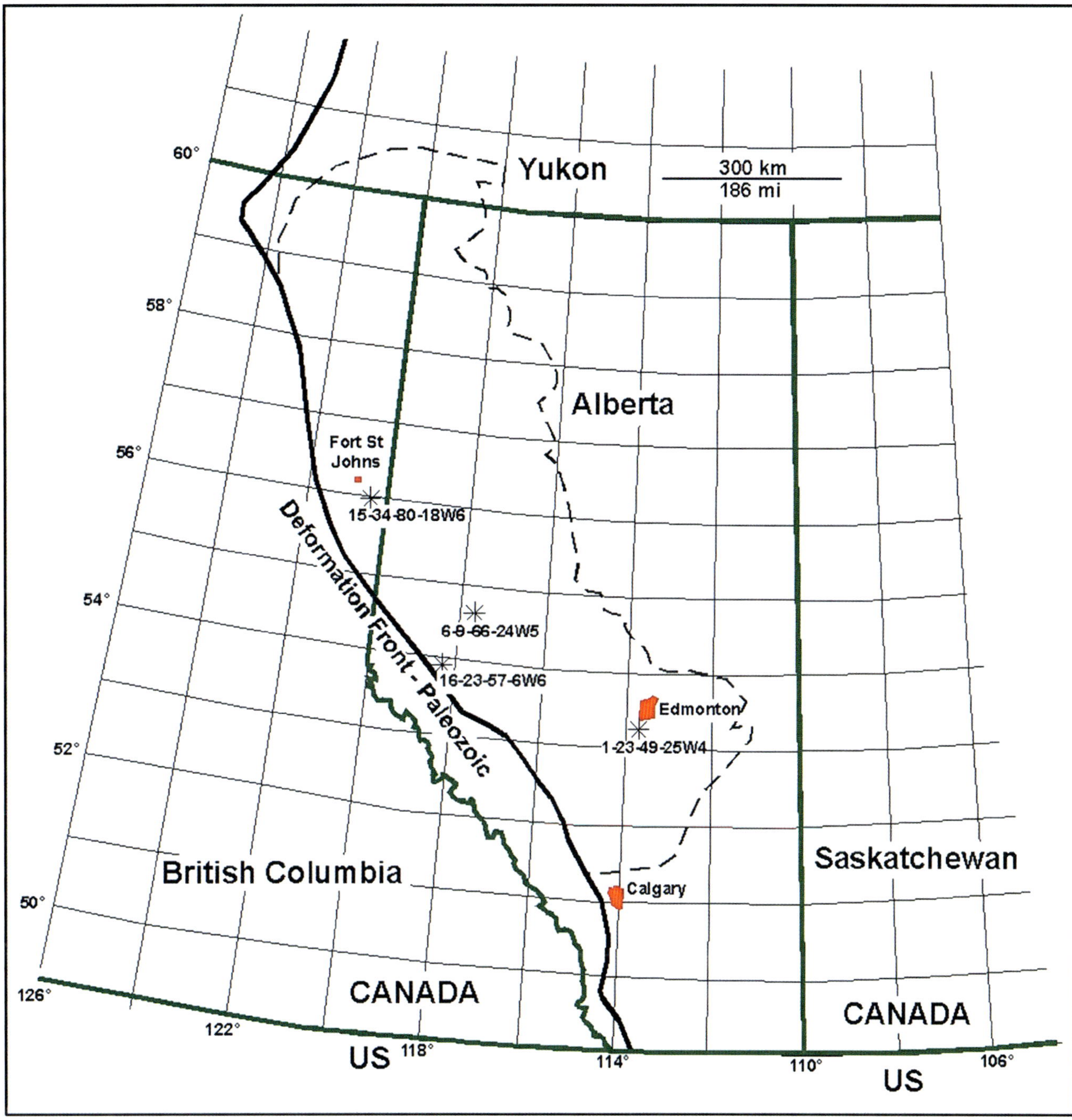

Figure 1. Overview map of the Western Canadian Basin showing the deep shale extent of the Devonian Duvernay Formation (stippled line) that is bounded in the east by the deformation front (black line) (after Edwards et al., 1994). Location of the two wells from the Nordegg Member, one situated in the oil window (Upper Nordegg 6-9-66-24W5) and the other one from the dry-gas window (Lower Nordegg 16-23-57-6W6), of the well of the Montney Formation (15-34-80-18W6; Upper Montney) and the well of the Duvernay Formation (1-23-49-25W4) were investigated.

mostly of relatively deep-water, mid- to distal-shelf, and slope shales, siltstones, and rare limestones, some of which are tempestites and/or turbidites (Edwards et al., 1994). The deposition occurred on the western margin of the North American craton in a variety of settings under arid, midlatitudinal conditions. The sediments are dominantly siltstones that contain abundant immature feldspars and few clay minerals shed from a low-gradient continental interior.

Duvernay

The Duvernay interval is a unique depositional unit within the Woodbend Group (Stoakes and Creany, 1985; Fowler and Stasiuk, 2001). The Duvernay Formation is characterized by extensive basinal deposits and synchronous with the middle stage of Leduc reef growth (Figure 2). The Duvernay Formation consists of dark brown bituminous shale with up to 17 wt.% TOC (Riediger et al., 1990; Stasiuk and Fowler, 2004) and limestone. The high organic carbon content reflects a major change in the stratification and oxygenation of basinal waters during the maximum transgressive stage of the Woodbend. Adjacent to the Leduc reef complexes, the Duvernay is thick and intermixed with reef-derived detritus and occurs especially within protected Leduc embayments. Basinward from the reefs, Duvernay basin-fill generally thins markedly.

METHODS

Samples of the Lower Triassic Montney Formation were collected from the British Columbia core repository at Charlie Lake, British Columbia; samples from the Devonian Duvernay Formation and the Nordegg Member were collected from the Energy Resources Conservation Board in Calgary, Alberta. Rock samples were cut and prepared for petrographic microscopy and scanning electron microscopic (SEM) analyses. Subsamples were also analyzed for mineralogy, thermal maturity, and TOC as well as mercury

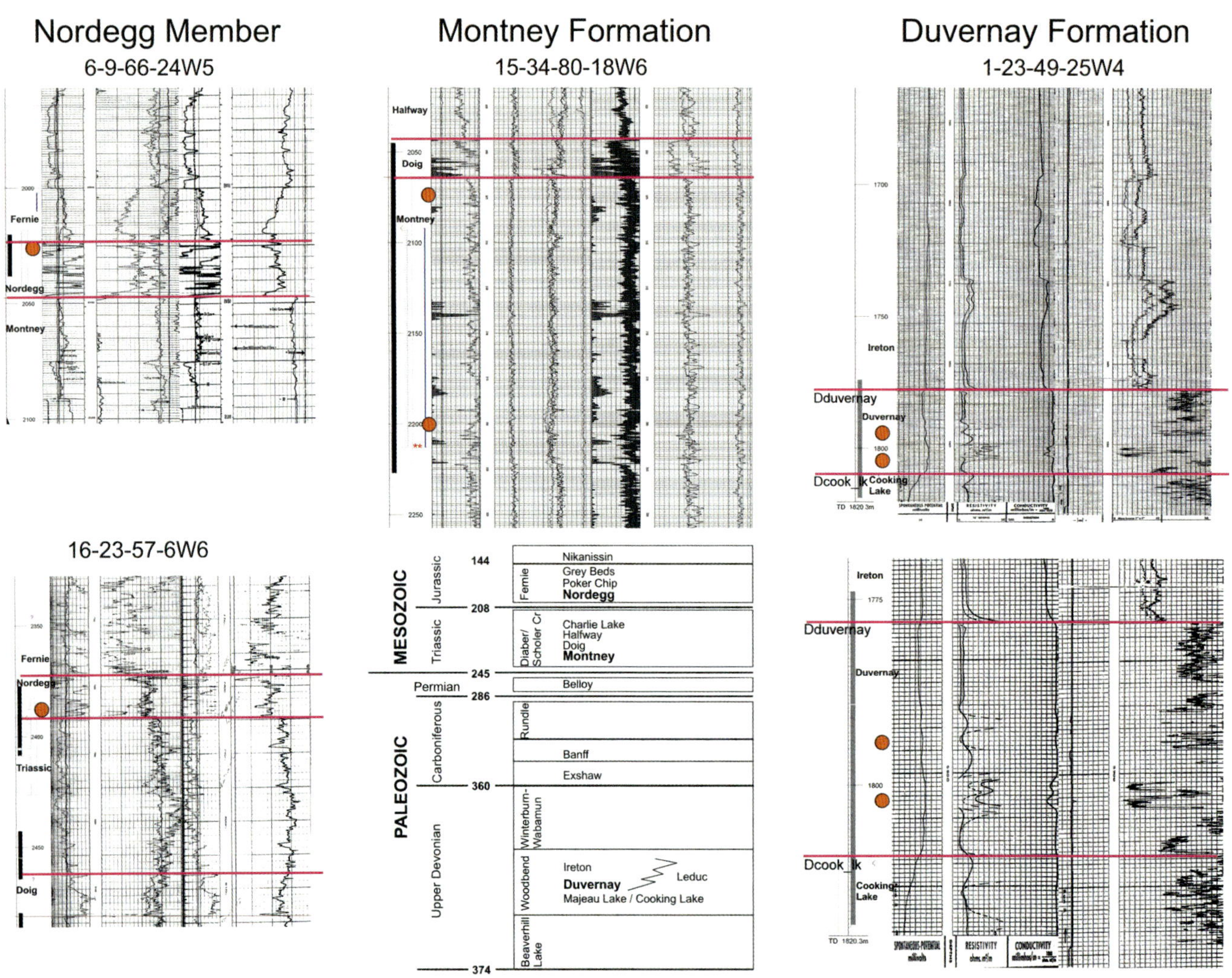

Figure 2. Geophysical logs and sample depths (dots) of the three formations/member (four wells) investigated in this study. A schematic overview of the formations in the Western Canadian Sedimentary Basin of eastern central Alberta and western British Columbia is also given.

porosimetry (mercury injection capillary pressure, MICP) analyses.

Geochemical analyses were conducted using a Source Rock Analyzer™ (SRA), a similar instrument to Rock-Eval, to determine the TOC and degree of organic thermal maturity based on T_{max} values. In the Western Canadian Sedimentary Basin, the T_{max} values of rocks are commonly used to infer thermal maturity because it is a rapid and inexpensive technique, and many organic-rich marine shales lack true vitrinite macerals required for vitrinite reflectance measurements. The TOC content in shale reflects the potential for sorbed gas capacity. The SRA is equipped with a flame ionization detector and two infrared detectors. X-ray diffraction (XRD) analyses were carried out to determine and quantify the mineralogical composition.

A Bruker D4 Endeavour XRD instrument with a LynxEye Detector was used with a cobalt x-ray tube. The samples were analyzed between 4 and 70° 2theta at 40 kV and 30 mA. The XRD trace data were analyzed using Rietveld methods (Jade 9) to quantify the mineralogy. Data quality assurance was achieved using internal standards (shale mixtures).

A Micromeritics AutoPore IV Mercury Porosimeter 9500 was used to measure mercury porosimetry by MICP, up to 60,000 psi. Mercury porosimetry routinely is applied over a capillary diameter range from 0.003 to 360 μm.

An Olympus BX51 research microscope with a DP72 Digital Camera System and an EXFO (x-Cite Series 120Q) fluorescence illuminator was used for the petrographic analysis of the samples. Photomicrographs were taken at ×40, ×100, ×200, and ×400. All thin sections were prepared using an ultraviolet (UV)-fluorescent epoxy (Rhodamine B) and polished to 30-μm thickness. Half of the thin section was then stained with a dual-carbonate stain (Alizarin Red S to distinguish between calcite and dolomite; potassium ferricyanide to distinguish carbonate minerals containing iron).

A JEOL JSM-6610 SEM was used to determine surface characteristics down to nanometer scale of all samples. Two to four subsamples of each core sample were picked and mounted on stubs and gold coated. Micrometer-scale elemental analysis was completed using an Oxford Instruments INCAx-act EDS (energy dispersive spectroscopy) system containing a silicon drift detector (SDD). Most images were taken under high-vacuum mode using SE (secondary electron) at 25 kV, spot size 25–40, and working distance between 11 and 13 mm.

RESULTS

Table 1 shows the following information about each formation: depth, mineralogical composition, mercury

Table 1. Well Identifier, Formation/Unit Names, General Maturity Level, Sample Depths, Mineralogical Composition (XRD), Skeletal Density, Average Pore Diameter, Total Porosity, Average Permeability (All Data from Mercury Porosimetry), Thermal Maturity, S1–S3, and TOC Values (SRA) of the Six Samples Analyzed Quartz (Qtz), albite (Ab, microcline (Mc), orthoclase (Or), calcite (Cal), dolomite (Dol), ankerite (Ank), Illite/mica (Ill/Mca), chlorite (Chl), kaolinite (Kln), pyrite (Py), apatite (Ap).

Method						*XRD and Rietveld*											
UWI	Fm/ Mbr	Unit	Maturity	Sample	Depth (m)	Qtz	Feldspars			Carbonates			Clays			Py	Ap
							Ab	Mc	Or	Cal	Dol	Ank	Ill/Mca	Chl	Kln		
100/06-09-066-24W5/00	NORDEGG	U	Oil	2026.24	2026.24	34.0	0.5			34.6	25.4		3.7			1.3	0.6
100/16-23-057-06W6/00		L	Dry gas	7838 Top	2389.02	29.2	0.8			48.0	5.0		11.3			4.2	1.5
				7838 Btm	2389.02	23.5	2.3			62.5	1.3		6.4			3.0	1.1
100/15-34-080-18W6/00	MONTNEY	U	Oil	25	2073.65	38.4	8.7	6.5		13.1	8.6	8.4	8.2		0.3	7.8	
			Dry gas	152	2200.05	43.0	10.0	1.3	5.6	10.0	5.4	4.8	13.4	4.4	0.1	1.8	
100/01-23-049-25W4/00	DUVERNAY	n/a	Oil	5886	1794.05	36.2	5.0		4.3	32.8		3.2	12.3		1.3	3.5	1.5
				5912	1801.98	17.8	4.8		4.7	36.9		1.8	20.1		1.2	11.7	1.0

Method		*MICP*					*SRA*										
UWI	Sample	ρ_{sk} (g/ml)	Avg d_p in nm	Total P in %	Avg k mdarcy	Stem vol %	Tmax °C	S1 (mg/g)	S2 (mg/g)	S3 (mg/g)	PC (%)	PI	S2/S3	S1/TOC	TOC %	HI	OI
100/06-09-066-24W5/00	2026.24	2.47	19.4	6.00	1.9×10^{-3}	58	433.6	9.28	26.41	0.16	2.96	0.26	165.1	1.69	5.50	480	3
100/16-23-057-06W6/00	7838 Top	2.48	15.1	1.80	16×10^{-2}	22	595.0	0.16	1.07	0.15	0.10	0.13	7.1	0.02	7.24	15	2
	7838 Btm						593.8	0.12	0.86	0.08	0.08	0.12	10.8	0.02	6.48	13	1
100/15-34-080-18W6/00	25	2.64	21.5	4.06	1×10^{-4}	45	431.8	1.65	0.73	0.14	0.20	0.69	5.2	1.11	1.49	49	9
	152	2.55	11.1	1.49	16×10^{-2}	19	467.2	4.00	2.36	0.06	0.53	0.63	39.3	1.34	2.99	79	2
100/01-23-049-25W4/00	5886	2.33	11.9	1.09	64×10^{-2}	15	434.2	4.04	56.35	0.08	5.01	0.07	704.4	0.44	9.14	617	1
	5912	2.32	10.5	1.26	13×10^{-2}	10	438.4	4.48	57.97	0.2	5.18	0.07	289.9	0.44	10.28	564	2

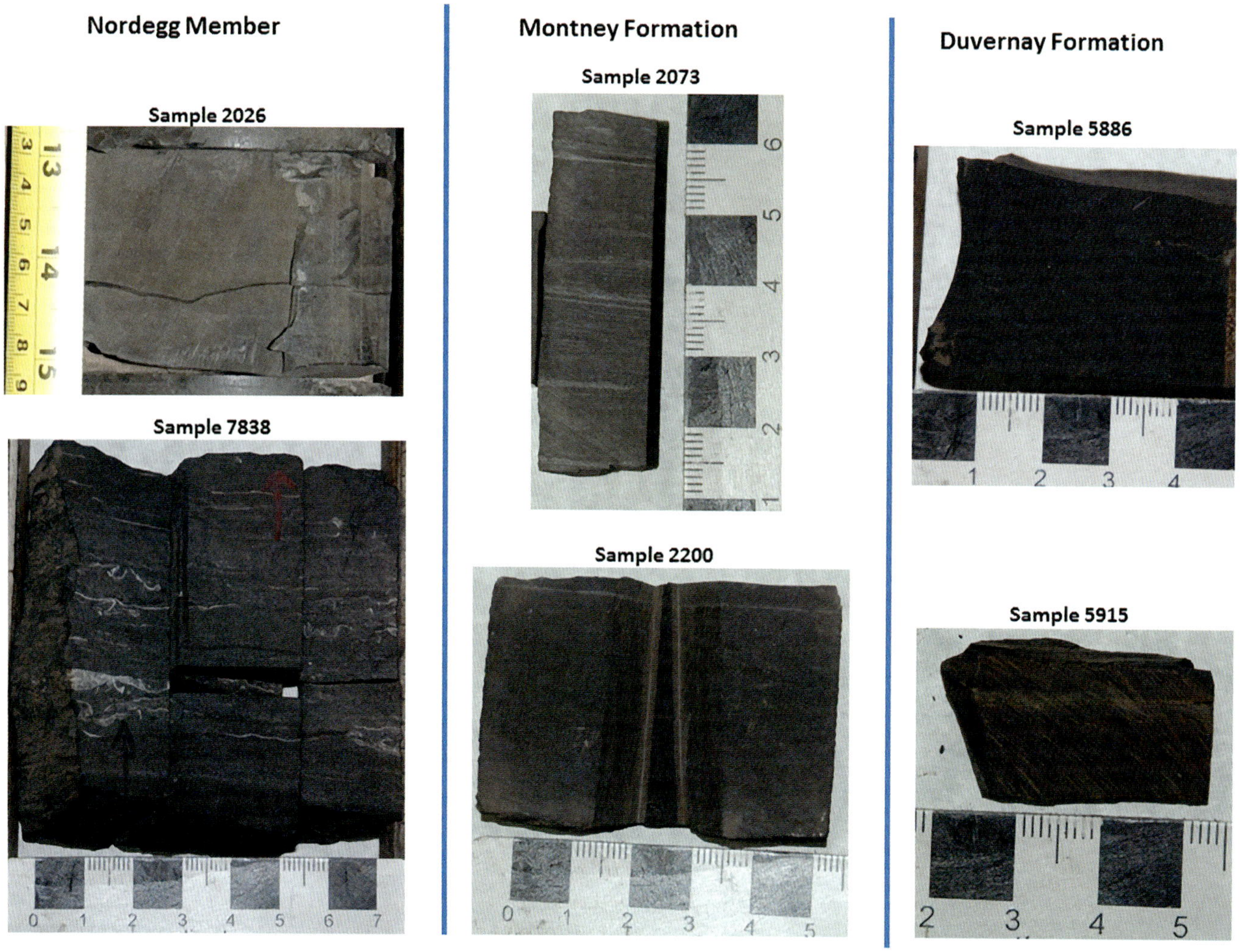

Figure 3. Hand specimens of the six samples analyzed for this study. Top orientation toward top of page. Scale in centimeters. See Figure 2 for formation details.

porosimetry (porosity and permeability), thermal maturity, and TOC analyses. Core photographs are shown in Figure 3. The data highlights the difference between the three formations. Duvernay samples often contain slightly higher clay contents than the other formations; the Montney samples often have elevated feldspar and dolomite contents. Petrographic descriptions and thin section photomicrographs are available in the Appendix of this chapter in the accompanying CD-ROM of this volume.

Lower Jurassic Nordegg Member (Fernie Formation)

Two wells from the Nordegg Member were selected; one is situated in the oil window (upper Nordegg 6-9-66-24W5), and the other one is from the dry-gas window (lower Nordegg 16-23-57-6W6) (Table 1). Both samples have 50–60% carbonates and about 30–35% quartz with small amounts of clays, pyrite, apatite, and feldspars. The samples contain abundant radiolarian and microfossil debris (Appendix A9; Figures 4, 5), and TOC is high (5.5–7.2 wt.%). The sample of the lower Nordegg contains some layers with abundant shell material, and two XRD and SRA analyses were performed (Table 1) to capture and determine the variability in carbonate content. Mercury porosimetry determined that mean pore diameters were 19 nm (upper Nordegg) and 15 nm (lower Nordegg); total porosities were 6% (upper) and 1.8% (lower). The upper Nordegg Member sample (2026) had a TOC of 5.5 wt.% and T_{max} of about 434°C (mature), and the lower Nordegg sample (7838) had a TOC of 6.5–7.2 wt.% and T_{max} of 595°C (overmature) (located in close proximity of the deformed belt line; Figure 1).

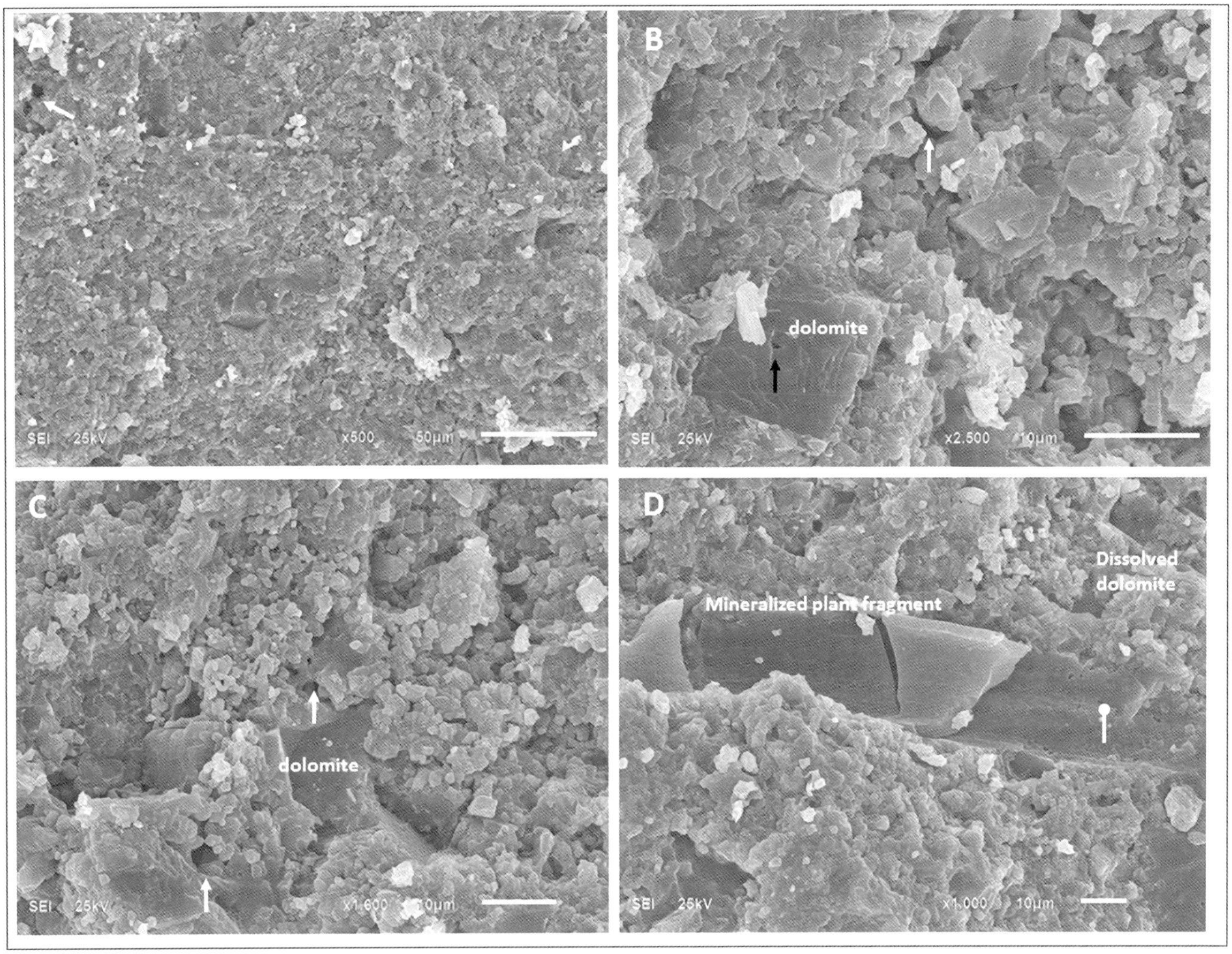

Figure 4. Scanning electron microscopic (SEM) photomicrograph of the Upper Nordegg (06-09-066-24W5; sample 2026.24 m [6647.7690 ft]) showing an overview (A) of the fine-grain nature of this shale with some idiomorphic dolomite (B), (C) and mineralized (Si-rich) plant fragments (D). All arrows refer to various pores (dissolution/residual growth, black; intergranular/fracture, white; organic matter, white with round tip).

Scanning electron microscopic images showed that these rocks were very fine-grain (<15 μm) sediments with high organic matter content (Figures 4–7). The organics not only coat mineral matter but also infill intercrystalline pore space (bituminous residue). In addition to radiolaria, sample 2026 showed other abundant microfossils (coccolith fragments; see Appendix 1). These aggregates of microfossils formed a complex framework of abundant macropores. Neoformed diagenetic minerals were also abundant, including dolomite and quartz, and some of the carbonates had dissolution pores. Sample 7838 showed a pronounced alignment of minerals and clays along the bedding planes, with associated fractures (Figures 6, 7), which is interpreted as a result of higher pressure caused by greater burial depth (current depth 363 m [1190.9 ft]) and tectonic stress near the deformed belt. Clay minerals (and possibly organic matter) showed draping features around grains, indicating differential compaction.

Lower Triassic Montney Formation

The Montney Formation was investigated in well 15-34-80-18W6 (upper Montney) at depths of 2073 and 2200 m (6801 and 7218 ft) (Table 1). The samples had about 38–43% quartz, 15–17% feldspars, 20–28% carbonates (about 10% calcite), 8–17% clays, and 2–8% pyrite. The samples were dominated by fine-sand to

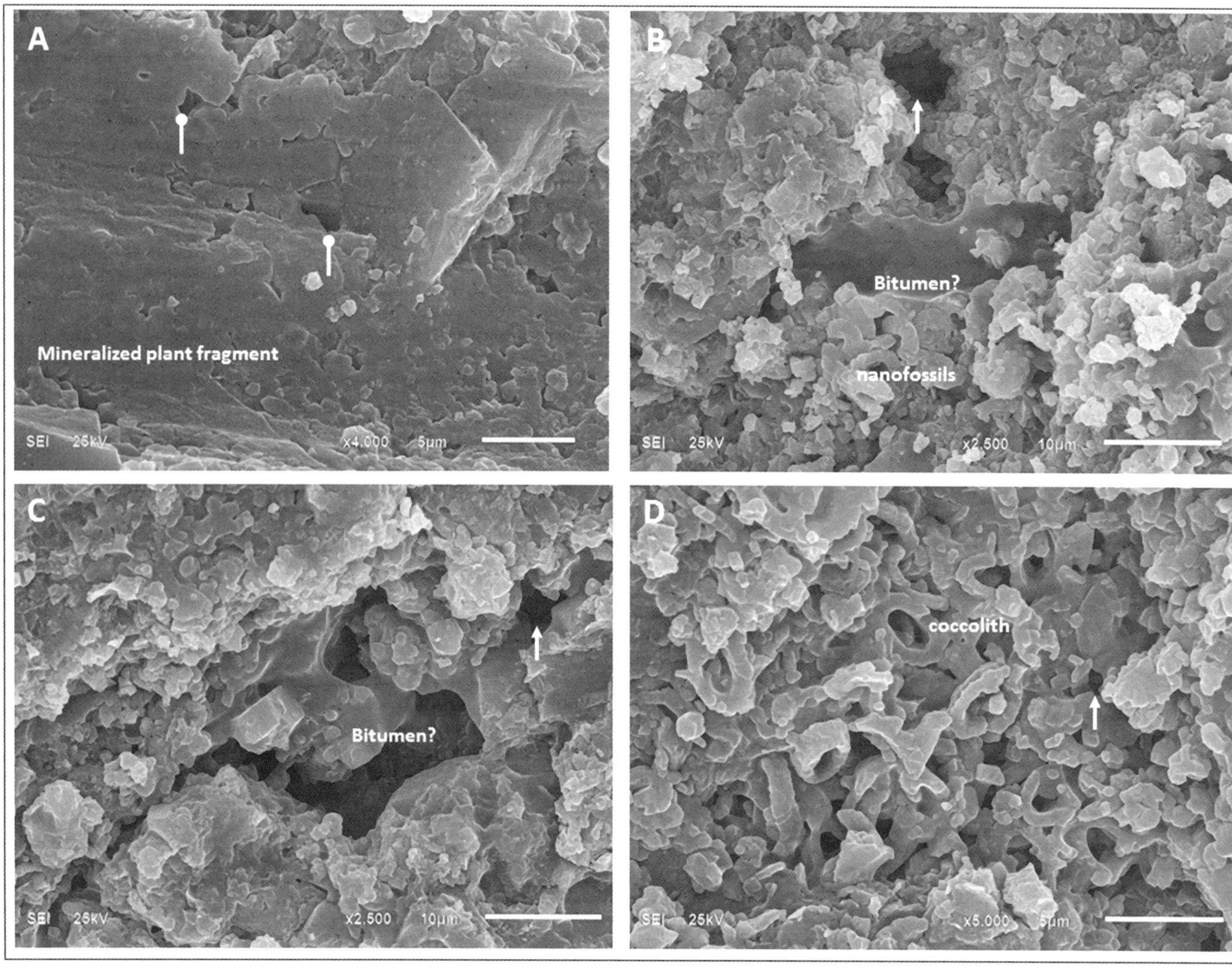

Figure 5. Scanning electron microscopic (SEM) photomicrographs of the Upper Nordegg (06-09-066-24W5; sample 2026.24 m [6647.7690 ft]) showing details of the mineralized plant fragment with some large pores (A), abundant nannofossils (B), (D) and bituminous material coating some of the mineral matter (TOC was 5.5 wt.%).

silt-size detritus (feldspars, quartz, mica; see Appendix A9-2 and A9-3) with carbonate cements and moderate organic material (TOC 1.5–3 wt.%). Mercury-porosimetry-determined average pore diameters were 22 nm (2073) and 11 nm (2200); total porosities were 4% (2073) and 1.5% (2200). Sample 2073 had a TOC of 1.5 wt.% and T_{max} of about 432°C (mature); sample 2200 had a TOC of 3 wt.% and T_{max} of 467°C (overmature). The lower porosity and pore sizes in sample 2200 appeared to be related to greater depth of maximum burial, as reflected by the T_{max} measurements.

Scanning electron microscopic analysis of the upper sample (2073) showed the presence of abundant coarse angular minerals (mainly secondary carbonates), with some quartz, feldspars, and mica sheets (Figures 8, 9). Primary minerals often showed clay mineral coatings, some of which had grown during diagenesis as they consist of fibrous structures. Dissolution, abandoned growth (interpreted as caused by pore-fluid changes or lattice poisoning; see Figures 9), and phyllosilicate pores were commonly observed. The deeper sample (2200) contained more fracture pores than the upper sample and showed quartz overgrowth and clay coatings (Figures 10, 11).

Upper Devonian Duvernay Formation

The Duvernay Formation was investigated in well 1-23-49-25W4 at depths of 5886 ft (1794 m) and

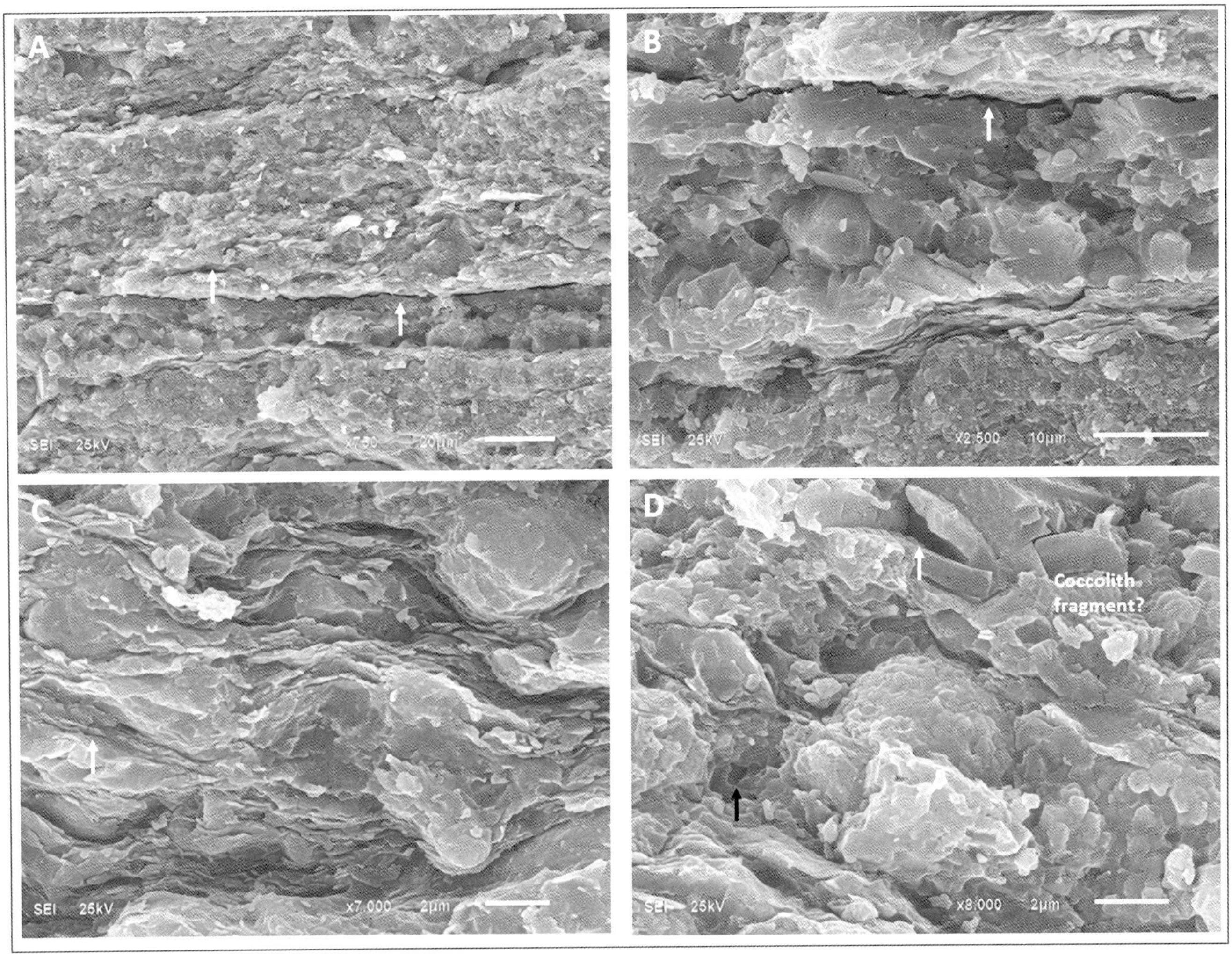

Figure 6. Scanning electron microscopic (SEM) photomicrographs of the Lower Nordegg (16-23-057-06W6; sample 7838 ft [2389.0224 m]) showing an overview of the laminated sample (A), details of the bedding-parallel fractures (B), and clay fabric deformed during burial diagenesis (C). This well has been subject to higher-stress regimes as it is situated in close proximity to the deformed belt. Higher levels of diagenesis may have led to recrystallization of the delicate nannofossils as only possible fragments were observed in this sample (D).

5912 ft (1802 m) (Table 1). The samples had about 18–36% quartz, about 10% feldspars, 36–39% carbonates (about 33–37% calcite), 14–21% clays, and pyrite (3–12%). They were dominated by dark, organic-rich shales (Appendix A9-4) with TOC up to 13 wt.% (Riediger et al., 1990; Stasiuk and Fowler, 2004) and lime mudstones. Mercury-porosimetry-determined average pore diameters were 12 nm (5886) and 11 nm (5912); total porosities were 1.1% (5886) and 1.3% (5912). Sample 5886 had a TOC of 9.1 wt.% and T_{max} of about 434°C (mature); sample 5912 had a TOC of 10.3 wt.% and T_{max} of 438°C (mature).

Scanning electron microscopic images showed recrystallization with some quartz and carbonate cements forming smooth surfaces; other grains were often coated with clay minerals (Figures 12–15). Secondary cements had replaced and infilled microfossils and pore space along the bedding planes. Some pores were the result of interrupted growth (residual growth pores), which was likely caused by chemical changes of the fluid composition during growth of the minerals (Figures 12, 13, 15). Both dissolution and abandoned growth pores are common and are up to 2 µm in diameter. In general, these rocks appeared fairly tight with few interconnected pores. Pore spaces were typically elongated and were up to 15–20 µm in length and 1–2 µm wide. The deeper sample contained higher clay mineral contents, and phyllosilicate pore

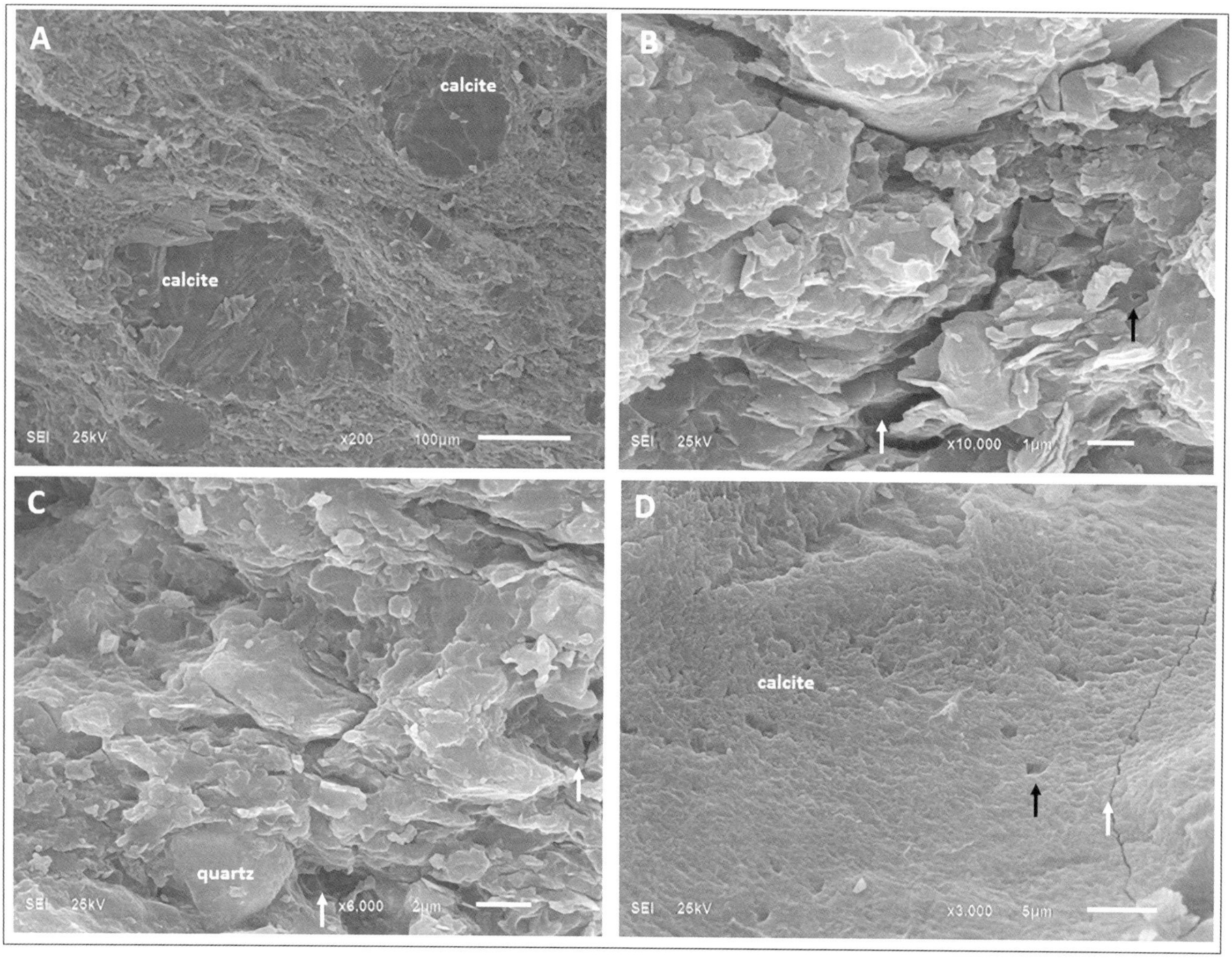

Figure 7. Scanning electron microscopic (SEM) photomicrographs of the Lower Nordegg (16-23-057-06W6; sample 7838 ft [2389.0224 m]) showing an overview of the sparry calcite-filled radiolarian (A), various fractures (B), (D), and the fine-grain nature of the sediment (see quartz grain embedded in clay-organic matter).

space was common. The clays also showed differential compaction around competent grains (Figure 15).

DISCUSSION

The focus of this study was to investigate grain morphology and porosity of three different shale hydrocarbon reservoirs from the Western Canadian Sedimentary Basin. The formations investigated were the Lower Jurassic Nordegg Member (Fernie Formation), the Lower Triassic Montney Formation, and the Upper Devonian Duvernay Formation. Scanning electron microscopy was complemented by petrographic, XRD, SRA, and mercury porosimetry. The SEM samples were prepared from fresh broken rock material, and the surfaces were coated with gold. This sample preparation technique had the advantage over argon-ion and focused ion beam (FIB) milling in that it preserved the rock fabric and allowed for investigations of a larger (more representative) sample area. Ion-milled samples appeared to have an advantage, however, in imaging nanometer-scale pores in organic matter (e.g., Ambrose et al., 2010; Milner et al., 2010), and FIB-SEM studies provided valuable 3-D images to evaluate porosity and permeability distributions (Keller et al., 2011). However, care needs to be taken with ion-milled samples as energy level and vacuum conditions can easily introduce new pore space in organic material because of desiccation and shrinkage. In addition, sputter coating may also introduce artificial porosity. Here, we focused on analyzing

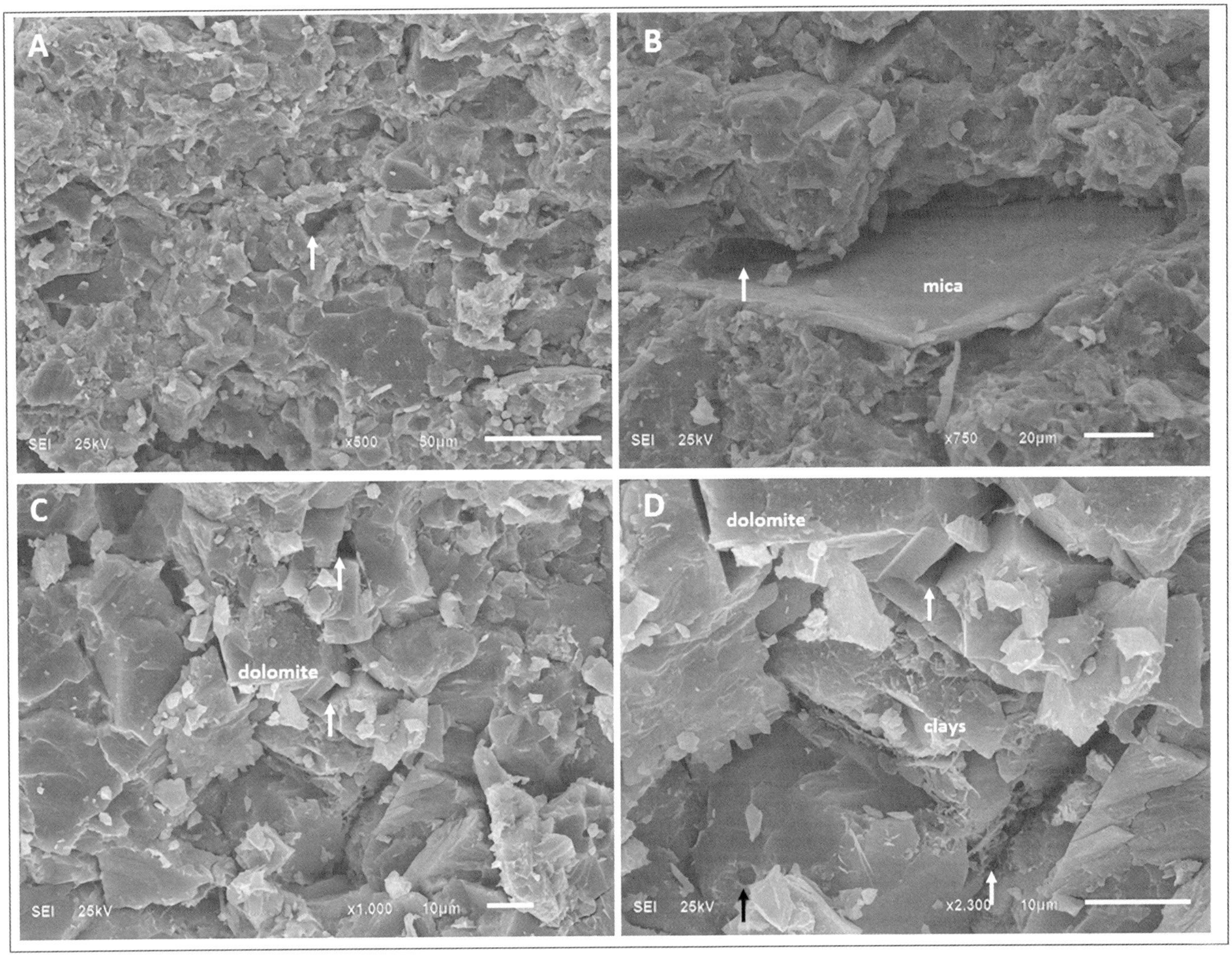

Figure 8. Scanning electron microscopic (SEM) photomicrographs of the Upper Montney (15-34-080-18W6; 2073.65 m) showing an overview of the sample (A), embedded mica sheets (B), and abundant carbonate (mainly dolomite) showing overgrowth features on sample surfaces (C), (D). Note also some of the filamentous clays on the grain surfaces with high pore space.

fresh broken surfaces to gather information about mineral and microfossil topography, relief and grain shapes, organics, pore shapes and characteristics, as well as mineral growth, overgrowth, and clay coating features.

Mineral Aggregates and Special Features

The SEM images of the three different formations showed characteristic microstructures. Both the Nordegg Member and Duvernay Formation samples contained significant amounts of biogenic silica and calcite with little detrital material. Overgrowth cements were recognized by their smooth surfaces. The diagenetic cements were confirmed using petrographic analysis. Radiolaria in the Nordegg samples that had their original shell replaced by calcite were common together with abundant microfossils (coccoliths) (Figure 5; Figure A1), but these could not be recognized in the deeper sample, likely because of recrystallization (Figures 6, 7). Mineral grains from the Nordegg samples were often less than 10 μm with only diagenetic minerals (carbonate, quartz, pyrite) forming larger crystals. Commonly, clay minerals mask all other detrital mineral matter. Biogenic silica and calcite clearly contributed much of the material in these rocks (Figures 4–7), and authigenic mineral matter was a common constituent in these shales (Figures 12–14). The Montney shale samples were

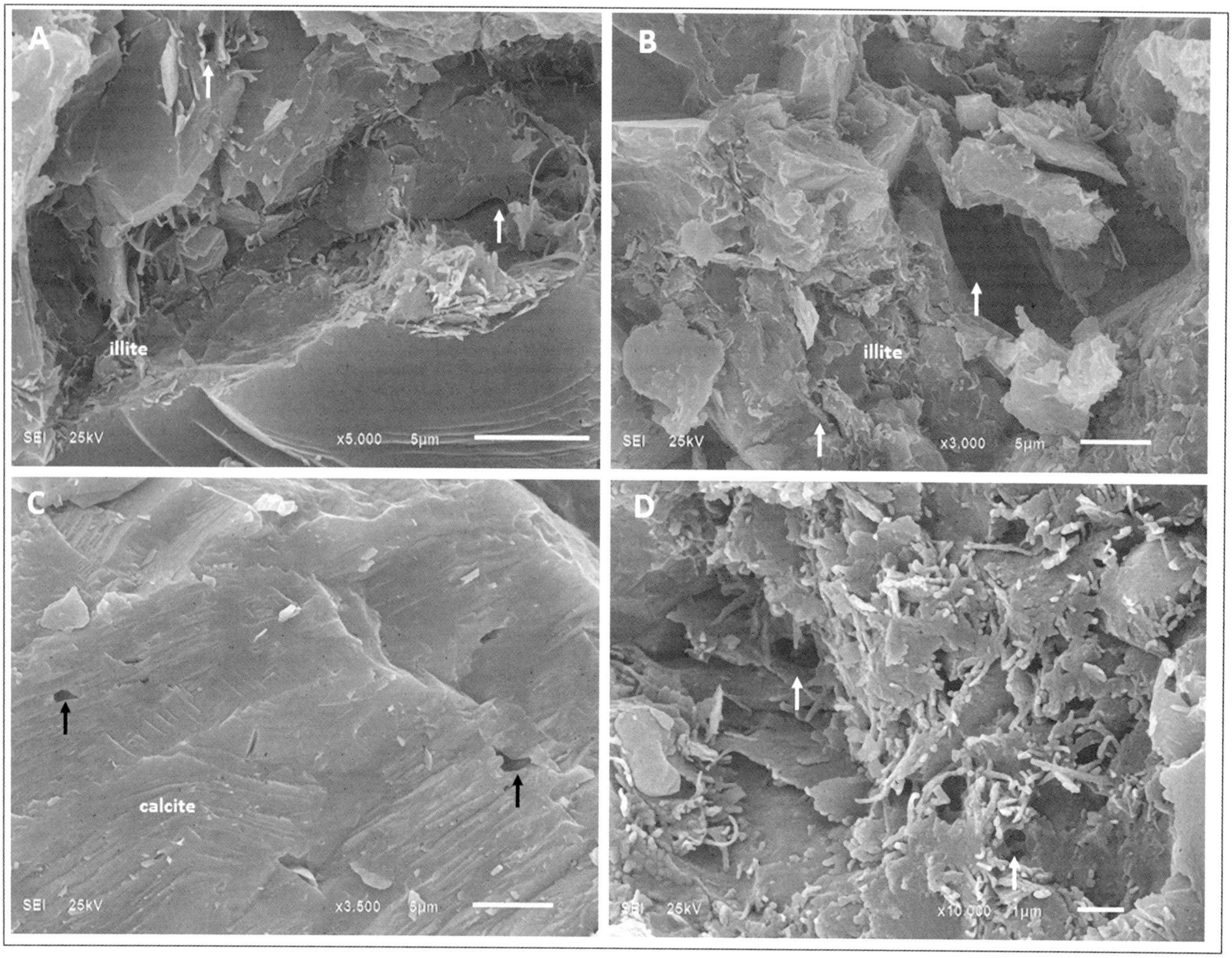

Figure 9. Scanning electron microscopic (SEM) photomicrographs of the Upper Montney (15-34-080-18W6; 2073.65 m [6803.3137 ft]) showing details of filamentous clay minerals (A), (B), (D), which appear to have had unrestricted space to grow. Large intergranular pores (arrows) and dissolution/residual growth pores (C) can also be observed.

markedly different and were dominated by detrital minerals (quartz, feldspars, mica) (Figures 8–11; Figures A2, A3) and early diagenetic carbonates (dolomite/calcite). Most of these detrital grains and early diagenetic carbonates showed late diagenetic overgrowth features and mineral grains composed of quartz, carbonates, or clay minerals. The cements within the sandy siltstones of the Montney Formation (TOC 1.5–3 wt. %) had occluded much of the original porosity (Figures 8, 12; Figures A2, A3), resulting in the low-permeability nature of the Montney Formation.

Pore Space and Pore Shapes

The advent of shale as an unconventional reservoir play has resulted in numerous studies focusing on porosity characterizations (e.g., Curtis et al., 2010; Schieber, 2010; Sondergeld et al., 2010; Curtis et al., 2011). Methods included electron imaging (transmission electron microscopy [TEM], SEM, FIB-SEM), optical petrography, and analytical methods (mercury porosimetry, helium porosimetry, gas adsorption, etc.). The most common pore types observed were interparticle, dissolution/growth, and fracture porosity. Although rare, phyllosilicate framework pores, similar to those described by Schieber (2010), were observed in all six samples of this study (e.g., Figures 6, 9–11, 15) and were interpreted as the result of diagenetic clay formation and transformation (Scotchman, 1989; Burley and MacQuaker, 1992; Michalopoulos and Aller, 2004). Porous kerogen was not observed (except for the mineralized plant fragment in Figures 4, 5) despite the high TOC contents, but this may have been caused by the sample preparation technique.

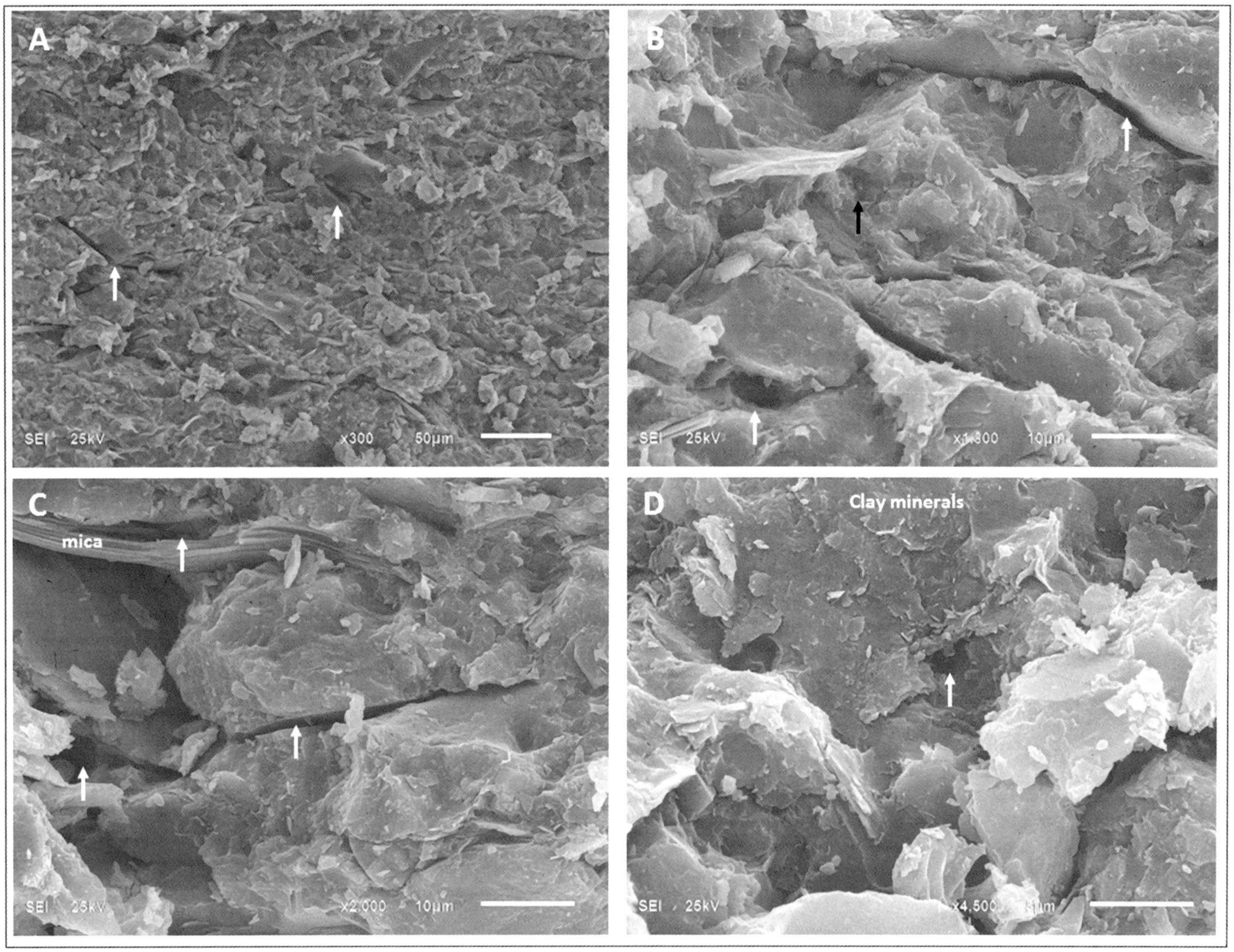

Figure 10. Scanning electron microscopic (SEM) photomicrographs of the Upper Montney (15-34-080-18W6; 2200.05 m [7218 ft]) showing an overview of this slightly finer-grain sample (A). Note that the surface was covered by secondary cements (overgrowth; see petrographic images) and some clay coatings, which make grain identification difficult. Large pores (>1–2 μm) were common, and some short (~5–60 μm) thin fracture porosity was also observed (B)–(D). Clay coatings (thin platelets) were common (D).

The pore sizes observed in SEM images ranged between 200 and 2000 nm, but mercury porosimetry data showed only a few pore sizes in that range (Figure 16). The most abundant pore sizes measured by mercury porosimetry ranged between 4 and 100 nm. The pore size distribution difference between SEM image analysis and mercury porosimetry is a common problem that has been observed in many other studies (Curtis et al., 2010; Sondergeld et al., 2010). It was suggested that the larger pores in conventional SEM analysis may be caused by grain plucking during sample preparation (Sondergeld et al., 2010). The morphological analyses of this study indicated, however, that grain plucking was relatively rare and could easily be identified and excluded in pore size measurements. Abundant large (>500 nm) dissolution pores were common (e.g., Figure 16) and yet not recognized by mercury porosimetry. During mercury injection, only connected pores were measured; thus, it may be possible that some of the larger pores observed in SEM might not have been connected to be accessed by mercury during the measurements. In addition, the principle of mercury porosimetry is based on the assumption of cylindrical pores (Webb and Orr, 1997); thus, larger pore spaces with narrow pore throats may result in incorrect pore size distribution data. In addition, the pore shapes observed in SEM images were not cylindrical and were often long, narrow pores with limited aperture. With imaging techniques, pores are typically measured by the longest dimension, whereas mercury porosimetry measures the entry (the smallest) dimension.

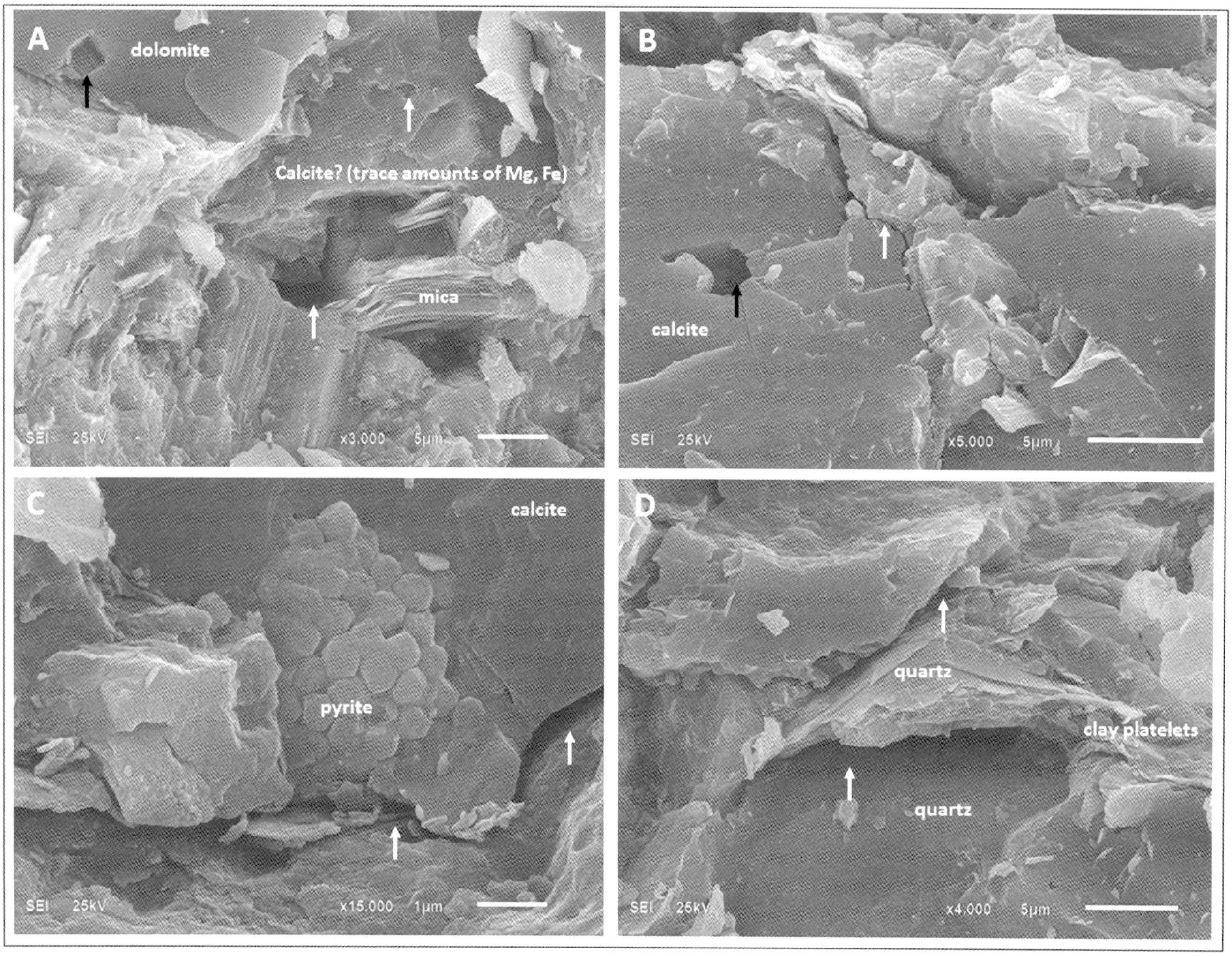

Figure 11. Scanning electron microscopic (SEM) photomicrographs of the Upper Montney (15-34-080-18W6; 2200.05 m [7218.0118 ft]) showing details of pores (arrows) between different grains (A), dissolution (A), residual growth (B), and fractures porosity (B), (C), framboidal pyrite embedded in carbonate cements and some clays (C), and quartz grain with quartz overgrowth in cross section (D). Note that some of the carbonates in (A) and (B) appear to have thin clay sheet inclusions.

This phenomenon requires further attention even though our average pore diameter data for this study determined by the mercury method (Table 1) were in line with other shale studies from the United States (e.g., Devonian shales in the Appalachian basin) using similar techniques (Nelson, 2009).

Dissolution and/or residual growth (i.e., interrupted crystal growth) pore space in both carbonate and quartz grains imply changes in pore water chemistry. Pore water chemistry changes during burial diagenesis of organic-rich sediments commonly occur because of decarboxylation of organic matter (kerogen) (Hunt, 1996) and the formation of organic acids (including carboxylic). The presence of organic acids could contribute to carbonate dissolution (e.g., Figures 9, 11, 15). In low-permeability mudstones, dissolution of carbonates will lead to chemical buffering, and carbonate cements may form. For dissolution of silicates such as biogenic silica, quartz, or feldspars, pH levels are often thought to be alkaline (>8.5) or extremely acidic (Alexander et al. 1954; Lewin, 1961), conditions that can be achieved during burial diagenesis. Extreme acid fluid conditions would dissolve most of the carbonates, which then would lead to a buffering of the pore fluid system. In sedimentary deposits, Schott and Oelkers (1995) observed that the hydrolysis of mixed oxide silicates is a multistep process that involves breaking of several bonds and structure/chemistry of the mineral's near-surface region. Thus, several minerals could contribute to aqueous silica from the conversion of isolated silica units through the hydrolysis of modifier

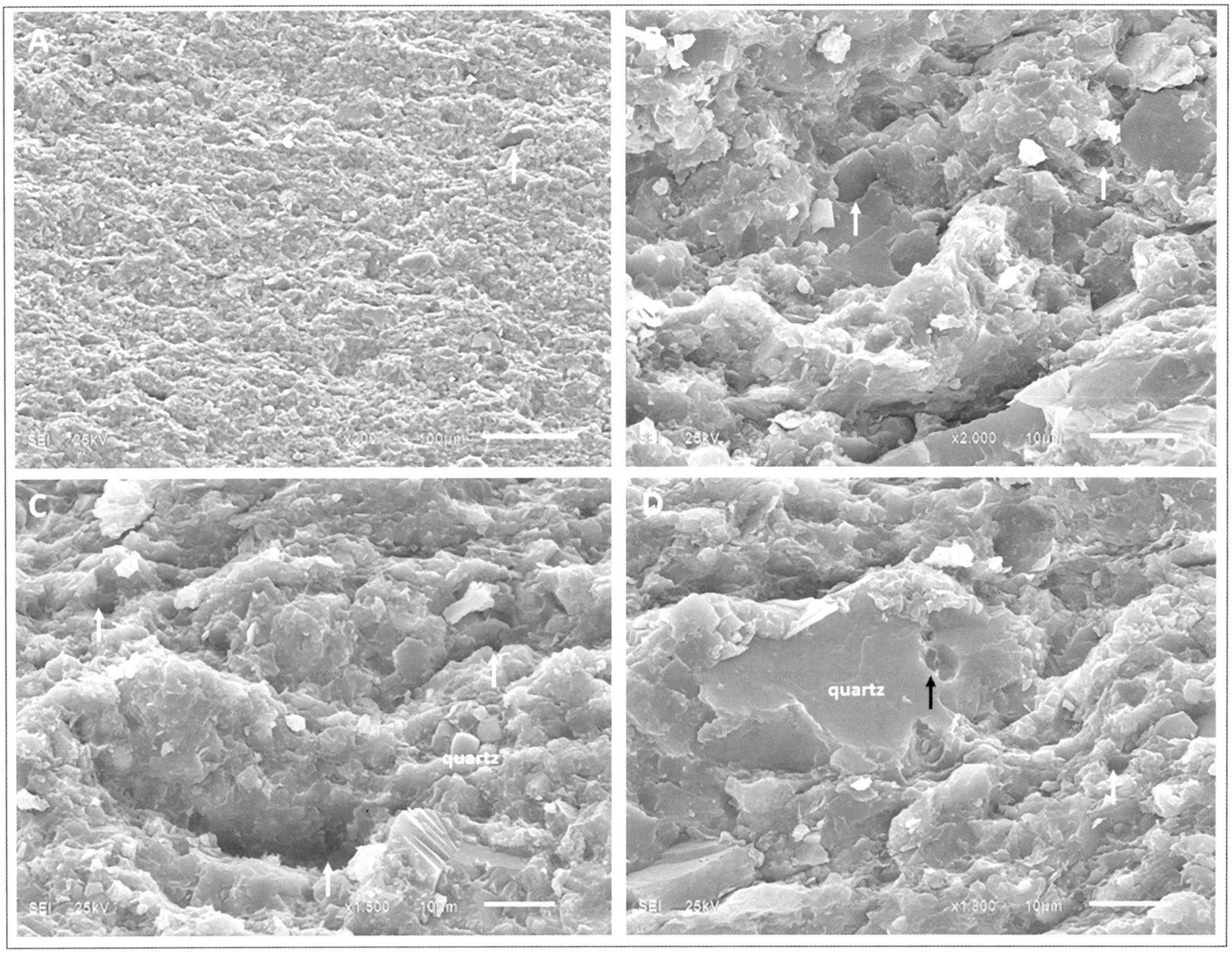

Figure 12. Scanning electron microscopic (SEM) photomicrographs of the Duvernay (1-23-049-25W4; 5886 ft [1794 m]) showing its fine-grain nature (A). Various pores (arrows) (A)–(D) and clay coatings (B)–(D) with some quartz cements (C), (D) can be observed.

cation-oxygen bonds. Hence, surface or partial dissolution of silicate minerals, including clays, quartz, and feldspars (Figures 9, 13), followed by precipitation of silicates (quartz, clays) (Figures 12, 14) may be more common in these shales than thought. Alternatively, some of these pores represent residual growth forms or inhibition of crystal growth. Folk (1974), Dowty (1976), Lahann (1978), and others discussed various mechanisms. There, Mg^{2+} ions are suggested to induce calcite growth inhibition because of accumulated lattice strain; thus, these ions introduce a poisoning of the crystal growth perpendicular to the calcite c axis (Folk, 1974). Lahann (1978) suggested that both surface charge and colloid chemical changes induce crystal growth inhibition; Dowty (1976) suggested that the template fraction (fraction of total bond energy) dictates growth of mineral surfaces. In our study, both calcite and quartz showed pore spaces that resembled residual (inhibition) growth (Figures 9, 11–13), and it is likely that pore fluid changes contribute to these shapes.

Clay Minerals

The six samples from the three formations had clay contents between 4 and 21% (Table 1) and were composed of illite/mica and occasionally small amounts of chlorite and kaolinite. Scanning electron microscopic images showed that most clay minerals often formed fibrous or platelet aggregates on the surface of other minerals (Figures 9, 10, 13). Under elevated burial pressure, alignments of clay minerals were also observed (Figures 6, 15). The morphology of the

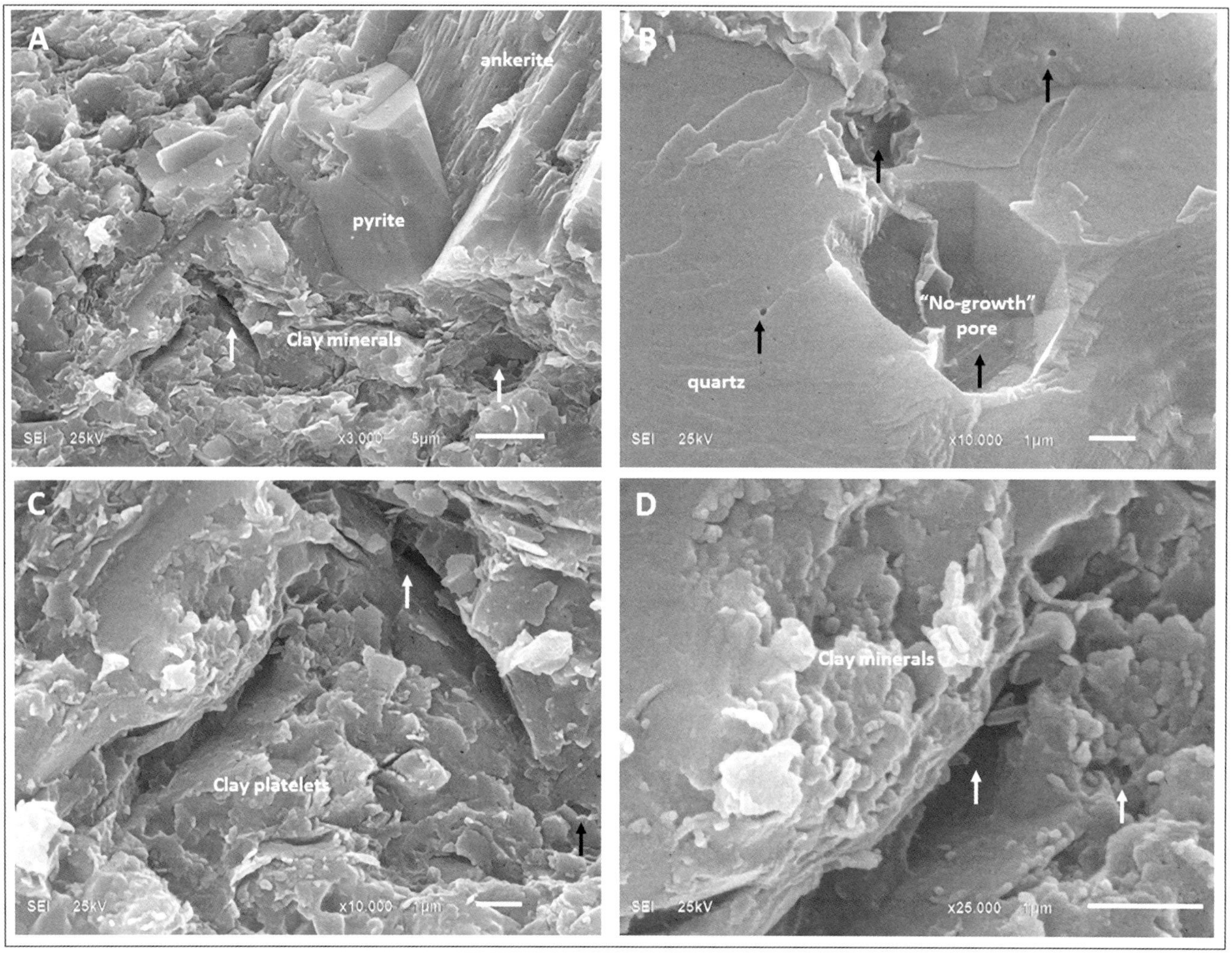

Figure 13. Scanning electron microscopic (SEM) photomicrographs of the Duvernay (1-23-049-25W4; 5886 ft [1794 m]) showing different pore types (arrows) (A)–(D) in these samples. Fracture (A), (C), dissolution and growth (B), and intergranular porosity (C), (D) were common. Note also the clay platelets that covered the surface of minerals.

clays suggests that the clay minerals (apart from the mica) were mostly diagenetic and formed likely during a later stage of diagenesis. However, we suggest that this is the result of rearrangement of existing clay aggregates instead of precipitation and neoformation from fluids. Such diagenetic clay aggregates are common in sedimentary rocks (e.g., Scotchman, 1989). The scope of this study could not determine if the growth of the clays was related to transformation schemes (smectite-illite, kaolinite-illite) or simply the rearrangement of existing illite/mica and kaolinite/chlorite clay minerals. However, in the Montney Formation samples, fibrous illite clay minerals (Figure 9) appeared to be authigenic, and large mica sheets (elongated length up to 100 µm) were detrital (Figures 8, 10, 11).

CONCLUSIONS

Conventional SEM analyses of shale hydrocarbon reservoirs from three different formations of the Western Canadian Sedimentary Basin were combined with routine analytical investigations (XRD, SRA, mercury porosimetry, petrography) to characterize mineral composition and assemblage, morphology, organic contents, and porosity. The investigations demonstrated that despite marked diagenetic differences between these shales, some common texture and pore characteristics were observed. The main findings of this study are as follows:

1. Scanning electron microscopic morphological studies of unconventional shales provide

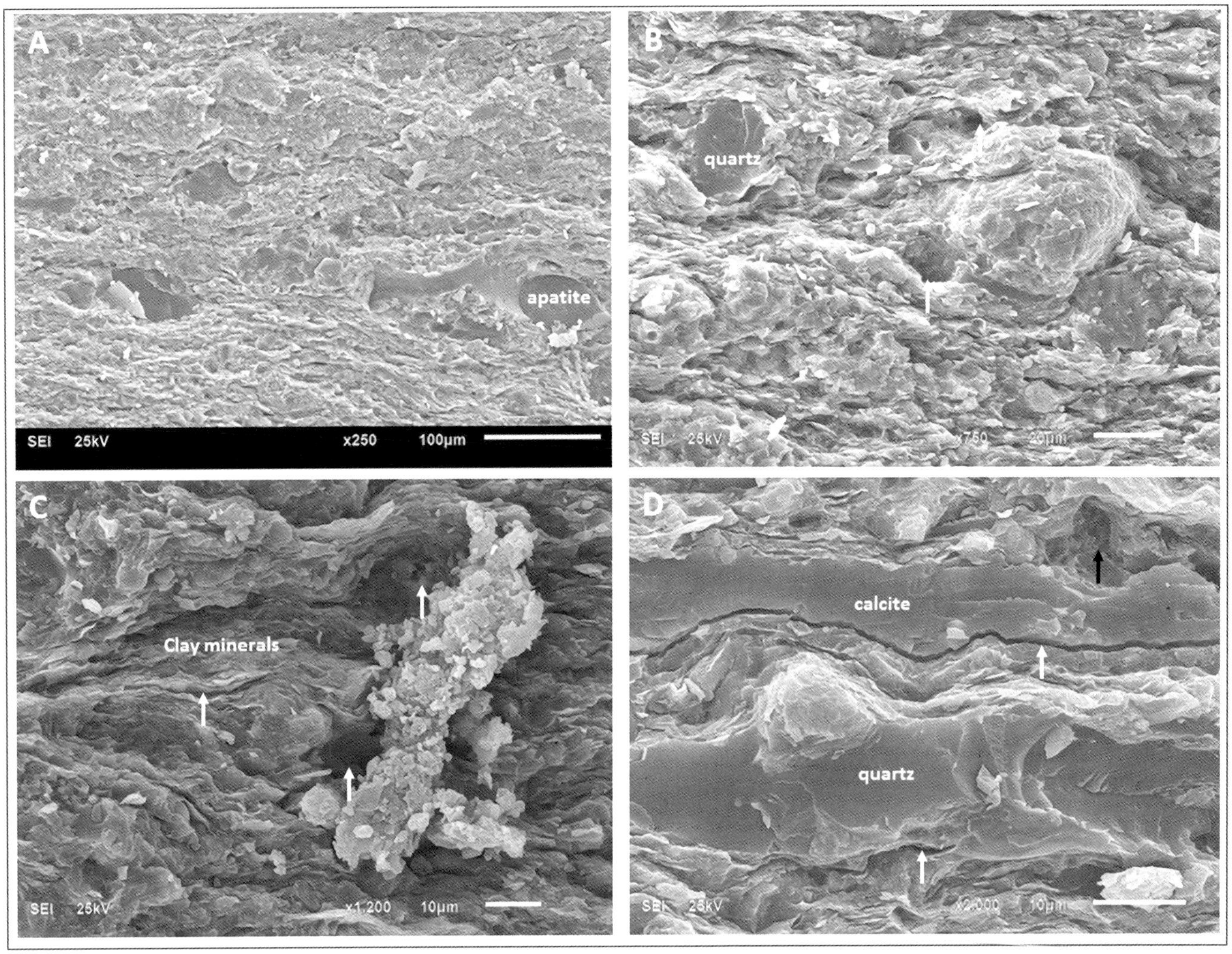

Figure 14. Scanning electron microscopic (SEM) photomicrographs of the Duvernay (1-23-049-25W4; 5912 ft [1802 m]) showing an overview of this shale (A). Note the deformed textures (clays) and pore space associated (B), (C); some grains showed competent habits (formed prior to late compaction). Along one former fracture, quartz and calcite cements were observed (bedding parallel) with some fracture porosity (D). In thin section, organic matter is often enriched along such features (former fractures; see Appendix A9-4).

important information about mineral aggregates, cementation, and clay mineral distribution, which allows interpretations of diagenetic history.

2. Scanning electron microscopic combined with petrographic analyses provides critical complementary information in particular for sediments in which cementation and mineral overgrowth are pronounced.
3. Scanning electron microscopic analysis of organic-rich shales may not be able to fully characterize the organic material, but organic fragments and some bituminous material can still be recognized. Organic matter porosity has been modeled in other SEM studies using argon- and gallium-ion (FIB) ion-milled samples.
4. Scanning electron microscopic investigations of three shale types from the Western Canadian Sedimentary Basin showed that large pores (>2000 nm) were present and had various shapes, which may be the primary reason why mercury porosimetry fails to identify the larger pores in these shales as MICP methods identify pore throat sizes.
5. The main types of pores in these shales were intergranular, dissolution, residual growth (growth inhibition), fracture, and phyllosilicate porosity. Mercury porosimetry identified abundant 5- to 20-nm pores (i.e., pore throats) that were not identified in the samples using SEM.

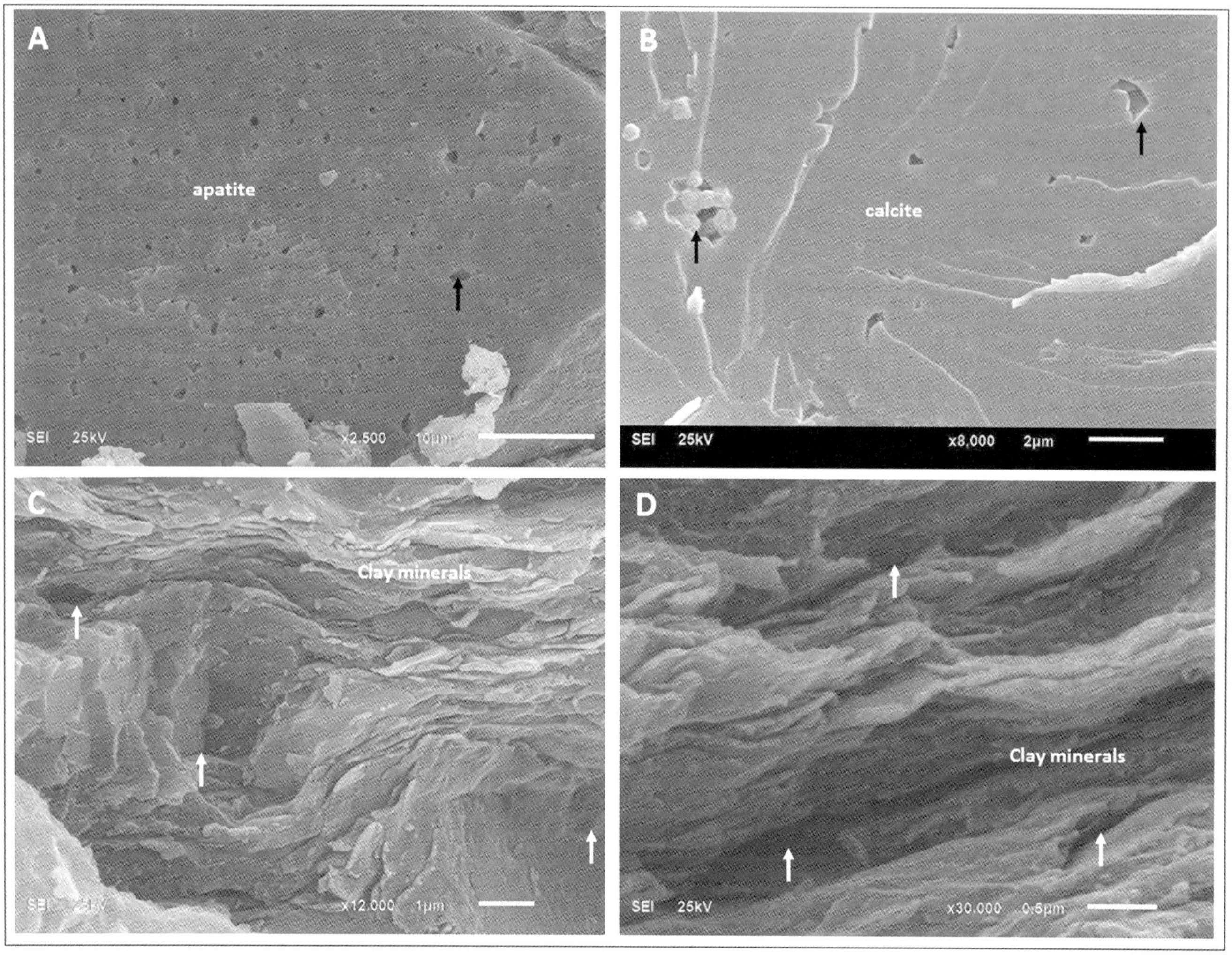

Figure 15. Scanning electron microscopic (SEM) photomicrograph of the Duvernay (1-23-049-25W4; 5912 ft [1802 m]) showing details of an apatite grain (A), some dissolution and residual growth features (B). Note that some of these pores (B) were infilled with secondary authigenic crystals (here quartz, but also calcite and dolomite), illustrating pore fluid chemistry changes during diagenesis. Most clay minerals show an elongate texture parallel to the bedding planes (C), (D).

ACKNOWLEDGMENTS

We would like to thank N. Minions, R. Monzon, F. Wang, and C. Twemlow for lab support and Wayne Camp, Patrick Monahan, and Craig Hall for their comments and suggestions that improved this manuscript. Financial support for Bustin was received from Geoscience British Columbia, Canada.

REFERENCES CITED

Alexander, G. B., W. M. Heston, and R. K. Iler, 1954, The solubility of amorphous silica in water: Journal of Physical Chemistry, v. 58, no. 6, p. 453–455, doi: 10.1021/j150516a002.

Ambrose, R. J., R. C. Hartman, M. Diaz-Campos, Y. Akkutlu, and C. H. Sondergeld, 2010, New pore-scale considerations for shale gas in place calculations: Society of Petroleum Engineers Unconventional Gas Conference, February 23–25, Pittsburgh, Pennsylvania, SPE Paper 131772, 17 p., doi: 10.2118/131772-MS.

Burley, S. D., and J. H. S. MacQuaker, 1992, Authigenic clays, diagenentic sequences and conceptual diagenetic models in contrasting basin-margin and basin-center North-Sea Jurassic sandstones and mudstones, *in* D. W. Houseknecht and E. D. Pittman, eds., Origin, diagenesis, and petrophysics of clay minerals in sandstones: SEPM Special Publication 47, p. 81–110.

Conti, J. J., P. D. Holtberg, J. A. Beamon, A. M. Schaal, J. C. Ayoub, and J. T. Turnure, 2011, Annual energy outlook 2011 with projections to 2035: U.S. Energy Information Administration, Washington D.C., 235 p.

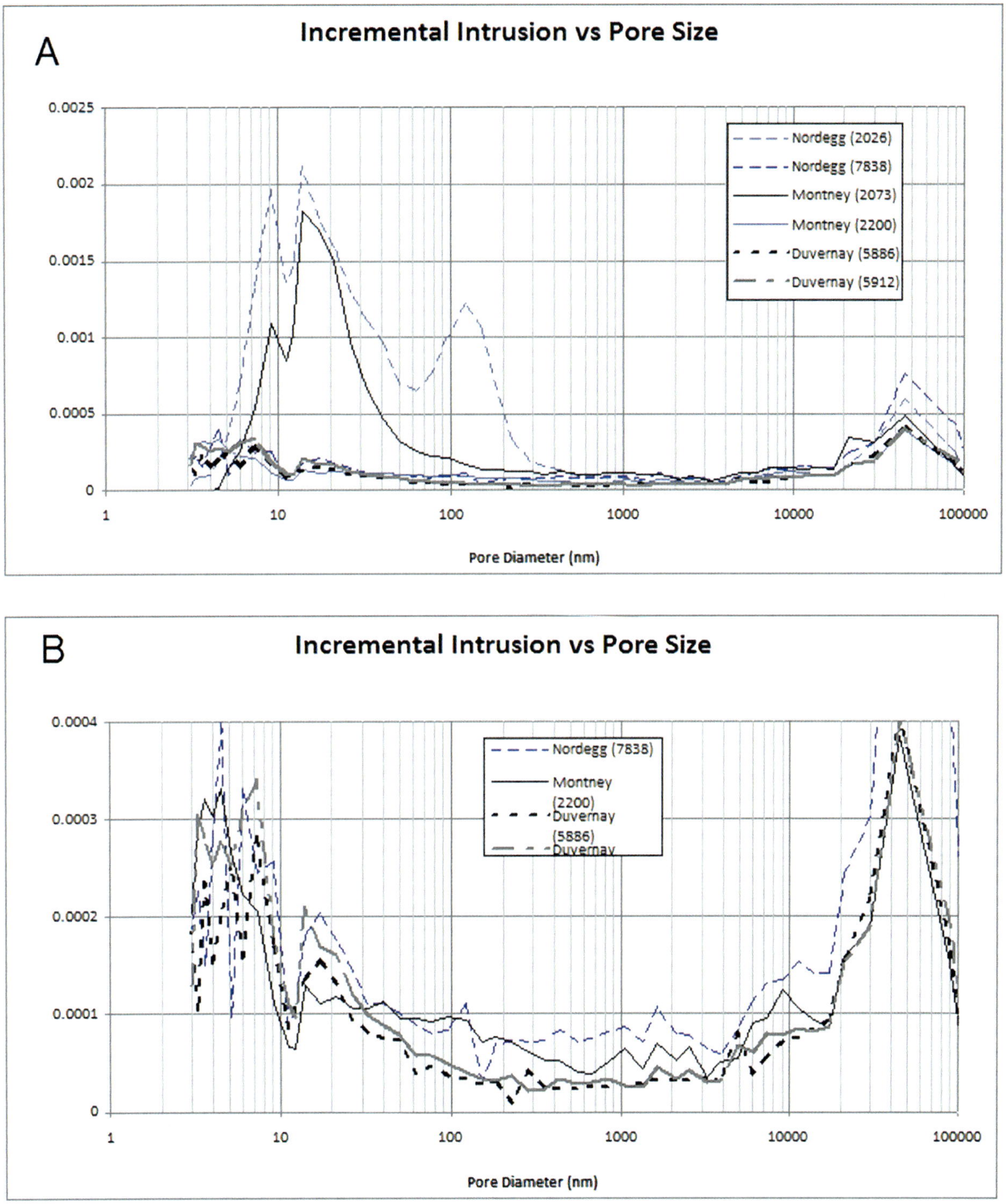

Figure 16. (A, top): Incremental pore volume versus pore diameter derived from the mercury porosimetry showing all, and the four samples with lower incremental pore volume (B, bottom). Note that small pore diameters between 200 and 2000 nm were documented in the results, which is in contrast to the scanning electron microscopic (SEM) observations.

Curtis, M. E., R. J. Ambrose, C. H. Sondergeld, and C. S. Rai, 2010, Structural characterization of gas shales on the micro- and nano-scales: Canadian Society for Unconventional Gas/Society of Petroleum Engineers Canadian Unconventional Resources and International Petroleum Conference, October 19–21, Calgary, Alberta, SPE Paper 137693, 15 p., doi: 10.2118/137693-MS.

Curtis, M. E., R. J. Ambrose, C. H. Sondergeld, and C. S. Rai, 2011, Transmission and scanning electron microscopy investigation of pore connectivity of gas shales on the nano-scale: Society of Petroleum Engineers North American Unconventional Gas Conference, June 14–16, The Woodlands, Texas, SPE Paper 144391, 10 p., doi: 10.2118/144391-MS.

Dowty, E., 1976, Crystal structure and crystal growth: 1. The influence of internal structure on morphology: American Mineralogist, v. 61, p. 448–459.

Edwards, D. E., J. E. Barclay, D. W. Gibson, G. E. Kvill, and E. Halton, 1994, Triassic strata of the Western Canadian Sedimentary Basin, *in* G. Mossop and I. Shetsen, eds., Geological atlas of the Western Canada Sedimentary Basin: Canadian Society of Petroleum Geologists and Alberta Research Council, Special Report 4, p. 259–275 http://www.ags.go.ab.ca/publications/wcsb_atlas/a_ch16/ch_16.html (accessed January 2, 2011).

Folk, R. L., 1974, The natural history of crystalline calcium carbonate: Effect of magnesium content and salinity: Journal of Sedimentary Petrology, v. 44, p. 40–53.

Fowler, M. G., and L. D. Stasiuk, 2001, Devonian hydrocarbon source rocks and their derived oils in the Western Canada Sedimentary Basin: Bulletin of Canadian Petroleum Geology, v. 49, no. 1, p. 117–148.

Hunt, J. M., 1996, Petroleum geochemistry and geology: New York, Freeman, 743 p.

Keller, L. M., L. Holzer, R. Wepf, P. Gasser, B. Munch, and P. Marschall, 2011, On the application of focused ion beam nanotomography in characterizing the 3D pore space geometry of Opalinus clay: Physics and Chemistry of the Earth, Parts A/B/C, v. 36, no. 17–18, p. 1539–1544.

Lahann, R. W., 1978, A chemical model for calcite crystal growth and morphology control: Journal of Sedimentary Petrology, v. 48, no. 1, p. 337–344.

Lewin, J., 1961, The dissolution of silica from diatom walls: Geochimica et Cosmochimica Acta, v. 21, no. 3–4, p. 182–198.

Loucks, R. G., R. M. Reed, S. C. Ruppel, and D. M. Jarvie, 2009, Morphology, genesis, and distribution of nanometer-scale pores in silicious mudstones of the Mississippian Barnett Shale: Journal of Sedimentary Research, v. 79, p. 848–861.

Michalopoulos, P., and R. C. Aller, 2004, Early diagenesis of biogenic silica in the Amazon delta: Alteration, authigenic clay formation, and storage: Geochimica et Cosmochimica Acta, v. 68, no. 5, p. 1061–1085.

Milner, M., R. McLin, and J. Petriello, 2010, Imaging texture and porosity in mudstones and shales: Comparison of secondary and ion-milled backscatter SEM methods: Canadian Society for Unconventional Gas/Society of Petroleum Engineers Canadian Unconventional Resources and International Petroleum Conference, October 19–21, Calgary, Alberta, SPE Paper 138975, 5 p., doi: 10.2118/138975-MS.

Nelson, P. H., 2009, Pore-throat sizes in sandstones, tight sandstones, and shales: AAPG Bulletin, v. 93, no. 3, p. 329–340.

Riediger, C. L., M. G. Fowler, L. R. Snowdon, F. Goodarzi, and P. W. Brooks, 1990, Source rock analysis of the Lower Jurassic "Nordegg Member" and oil-source rock correlations, northwestern British Columbia: Bulletin of Canadian Petroleum Geology, v. 38A, p. 236–249.

Riediger, C. L., and J. D. Bloch, 1995, Depositional and diagenetic controls on source-rock characteristics of the Lower Jurassic "Nordegg Member," Western Canada: Journal of Sedimentary Research, v. A65, no. 1, p. 112–126.

Ross, D. J. K., and R. M. Bustin, 2007, Shale gas potential of the Lower Jurassic Gordondale Member, Northeastern B.C., Canada: Bulletin of Canadian Petroleum Geology, v. 55/1, p. 51–75.

Schieber, J., 2010, Common themes in the formation and preservation of intrinsic porosity in shales and mudstones—illustrated with examples across the Phanerozoic: Society of Petroleum Engineers Unconventional Gas Conference, February 23–25, Pittsburgh, Pennsylvania, USA, SPE Paper 132370, 10 p., doi: 10.2118/132370-MS.

Schott, J., and E. H. Oelkers, 1995, Dissolution and crystallization rates of silicate minerals as a function of chemical affinity: Pure and Applied Chemistry, v. 67, p. 903–910.

Scotchman, I. C., 1989, Diagenesis of the Kimmeridge Clay Formation, onshore UK: Journal of the Geological Society, v. 146, no. 2, p. 285–303.

Sondergeld, C. H., R. J. Ambrose, and C. S. Rai, 2010, Microstructural studies of gas shales: Society of Petroleum Engineers Unconventional Gas Conference, February 23–25, Pittsburgh, Pennsylvania, SPE Paper 131771, 17 p., doi: 10.2118/131771-MS.

Stasiuk, L. D., and M. G. Fowler, 2004, Organic facies in Devonian and Mississippian strata of Western Canada Sedimentary Basin: Relation to kerogen type, paleoenvironment, and paleogeography: Bulletin of Canadian Petroleum Geology, v. 52, no. 3, p. 234–255.

Stoakes, F. A., and S. Creaney, 1985, Sedimentology of a carbonate source rock: The Duvernay Formation of Alberta, Canada, *in* M. W. Longman, K. W. Shanley, R. F. Lindsay, and D. E. Eby, eds., Rocky Mountain Carbonate Reservoirs: SEPM Special Publication 7, p. 343–375.

Webb, P. A., and C. Orr Jr., 1997, Analytical methods in fine particle technology: Norcross, Georgia, Micrometrics Instrument Corporation Publishers, 325 p.

10

Debra S. Jennings, Jonathan Antia, 2013, Petrographic characterization of the Eagle Ford Shale, South Texas: Mineralogy, common constituents, and distribution of nanometer-scale pore types, *in* W. Camp, E. Diaz, and B. Wawak, eds., Electron microscopy of shale hydrocarbon reservoirs: AAPG Memoir 102, p. 101–113.

Petrographic Characterization of the Eagle Ford Shale, South Texas: Mineralogy, Common Constituents, and Distribution of Nanometer-scale Pore Types

Debra S. Jennings and Jonathan Antia

Core Laboratories, 6316 Windfern Rd., Houston, Texas, 77040
(e-mails: debra.jennings@corelab.com and jonathan.antia@corelab.com)

ABSTRACT

Pore system and matrix constituents of the Eagle Ford Shale are described from 86 representative samples from southwestern and eastern Texas. Argon-ion-milled samples were imaged using a field emission scanning electron microscope (FE-SEM) to document matrix constituents and pores. Samples were chosen based on lithology, x-ray diffraction mineralogy, matrix porosity, and effective permeability measurements. The Eagle Ford Shale is characterized by a complex pore system, reflecting variations in depositional environment, composition, and diagenesis, resulting in heterogeneous reservoir properties. Interparticle pores are the most common nonorganic pore type and typically occur between coccolith fragments. Limestone intervals contain less organic matter and may be recrystallized. The porosity and effective permeability associated with intercrystalline and interparticle pores within limestone intervals is strongly controlled by the degree of recrystallization. As the degree of recrystallization increases, porosity and permeability decrease. Organic nanopores, when present, are most abundant and display the best-developed pore systems within fecal pellets exhibiting open fabric and in organic-matter-filled interparticle pores between coccolith fragments in the matrix. Organic nanopores are also observed in organic-filled foraminifer chambers and commonly occur in conjunction with intracrystalline pores between authigenic clays, which form localized complex pore systems. Although there appears to be a relationship between the abundance of organic nanopores, predominant pore type, pore connectivity, and effective permeability, common vertical and lateral variability in the presence and abundance of organic porosity, even in similar microfacies, suggests that the parameters controlling organic nanopore development are not fully understood and deserve additional detailed studies. Regional differences in pore abundance observed along the Texas Eagle Ford Shale trend are the result of variations in depositional environments and clastic (clay) input. In general, the Eagle Ford in southwestern Texas has higher quality matrix properties because of the higher abundance of interparticle and intraparticle pores present in marlstones compared to lower reservoir quality associated by the more clay-rich Eagle Ford and Maness shales in east Texas.

DOI: 10.1306/13391708M1023586

INTRODUCTION

The pore system and mineralogy of the Eagle Ford Shale, as with other shales, cannot be easily observed using conventional petrographic techniques because of the small size of the pores and grains. In addition, mineralogy indicated by x-ray diffraction (XRD) analyses commonly appears inconsistent with data visible in thin-section photomicrographs, particularly with respect to quartz, kaolinite, and calcite abundances. Although sparse preliminary Eagle Ford pore-type data are scattered throughout current literature (e.g., Burton et al., 1987; Cusack et al., 2010; Dawson and Almon, 2010; Lock et al., 2010; Walls and Sinclair, 2011), no detailed studies have been published. The purpose of this study was to document the primary constituents, pore types, and distribution of pore types in the Eagle Ford Formation in eastern and southwestern Texas.

METHODOLOGY

Eighty-six subsurface samples from 13 wells in southwestern and eastern Texas (Figure 1) were selected based on mineralogy (volume percentage calculated from XRD weight percent data, bulk density, and organic content); total organic carbon (TOC) content; and porosity and effective permeability measured from crushed core, as received samples (sensu, Luffel et al., 1992). Samples of marls and silty shales that displayed the highest, lowest, and middle range TOC, porosity, and permeability, as well as a range of microfabrics (identified in thin-section micrographs) and a range of vitrinite reflectance values (0.6–1.89%R_o) were selected for preliminary scanning electron microscopic (SEM) analysis. In addition, two representative wells, one from eastern Texas and one from southwestern Texas, were heavily sampled (Figure 1). Sampling protocol for the two representative wells included selecting (1) representative marls with similar rock properties; (2) limestone samples that displayed similar fabric, mineralogy, and TOC, but with distinctly different porosity and effective permeability data; and (3) samples from each of the representative depositional facies.

All samples were cut perpendicular to bedding prepared using argon-ion beam milling to obtain flat surfaces suitable for very high magnification imaging. Backscattered electron and secondary electron images were collected from uncoated samples using a Quanta FEG 250 field emission scanning electron microscope (FE-SEM) fitted with an energy dispersive x-ray spectrometric (EDS) detector. Argon-ion-milled samples were imaged under low vacuum (60 Pa) and low voltage (10–15 kV). The mineralogical composition of constituents was verified with EDS data collected from representative areas on each sample.

Because high-quality, argon-ion-milled samples have submicrometer-scale topography, backscattered electron images reveal subtle, high-resolution chemical differences that are especially useful in differentiating the contacts between organic matter within pores that have low atomic mass (displayed as black) and the mineral constituents that delineate the pore walls, which have a relatively higher atomic mass (displayed as white to gray). Nanoscale authigenic constituents are also more confidently identified. Comparatively, secondary electrons are very sensitive to topographic and spatial resolution, making them particularly valuable for imaging nanometer- to micrometer-scale open pores. Given that pores in shales may be open or filled with hydrocarbons, it is crucial that shale samples be evaluated with both backscattered electron and secondary electron images to comprehensively characterize pore types and constituents present in any given sample. All samples in this study were analyzed with both backscattered electron and secondary electron detectors.

FEI software was used to measure pore body diameters with a calibrated accuracy of ±1 nm. A range of pore body sizes on each sample were measured from pore wall to pore wall across the widest part of individual pores and across a field of view at ×60,000, ×30,000, and ×15,000 on both backscattered and secondary electron images. Average pore body diameter was calculated from at least ten pores collected at each magnification.

GEOLOGIC SETTING

The Eagle Ford Formation was deposited in lower shoreface/inner shelf to distal slope/basin settings (100- to 300-m [328- to 984 ft] depth) at the southern end of the Western Interior Sea of North America during the Cenomanian-Turonian (Sageman and Arthur, 1994; Arthur and Sageman, 2004). Estimated paleolatitude is about 30°N (Sageman and Arthur, 1994). Paleopedological and paleobiogeographic data indicate that the climate was warm and humid subtropical (Mack, 1992). The prevailing wind direction trended from west-southwest to east-northeast (Figure 1).

Episodic delivery of volcaniclastic material by winds and overland flow to the basin, combined with circulating currents from the north, transported reworked volcaniclastic material into the Maverick and Gulf Coast basins and significantly contributed to an influx of iron- and phosphorus-rich nutrients (Arthur and Sageman, 2004). As sea level rose, inshore

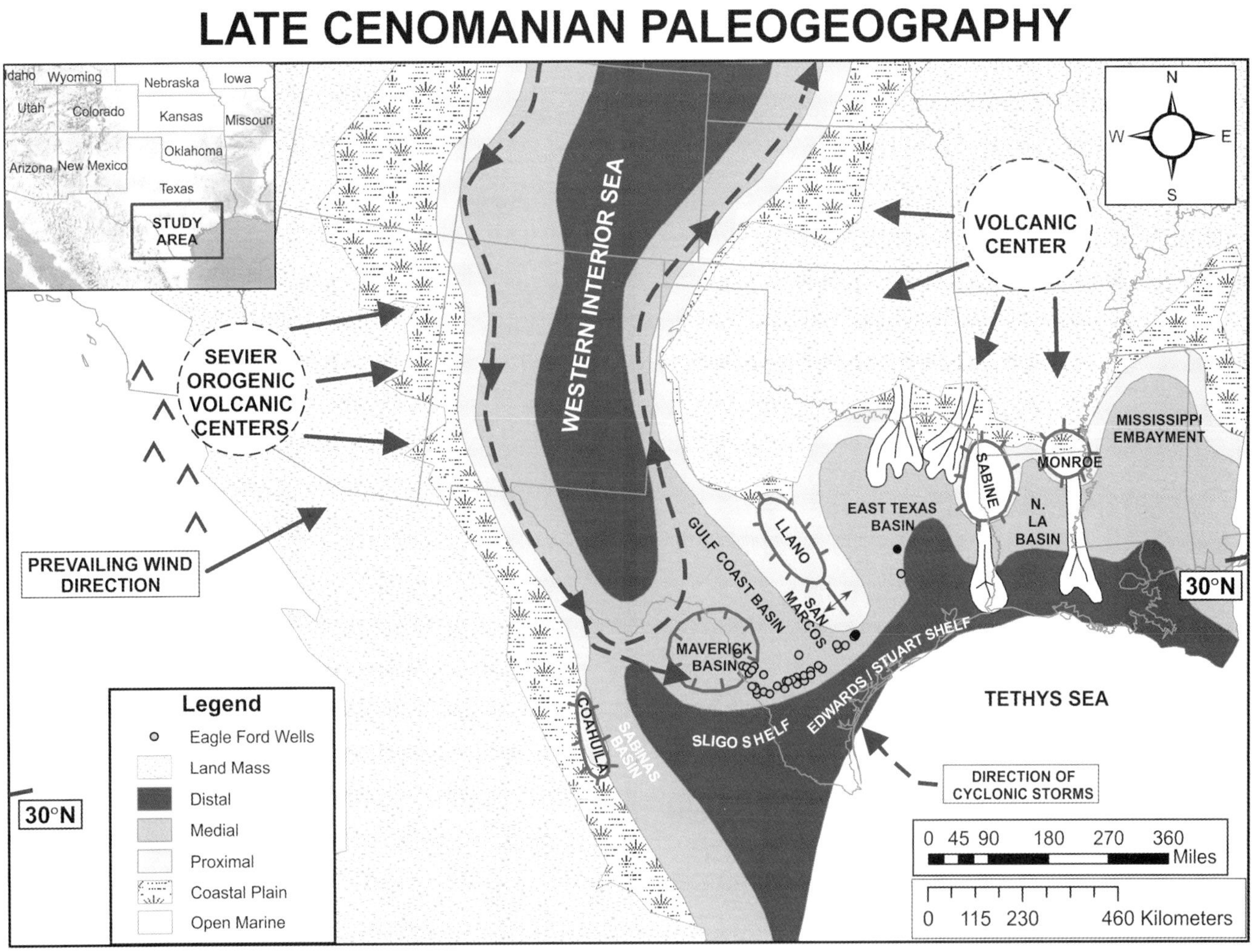

Figure 1. Paleogeography and study area map (compiled from Sageman and Arthur, 1994; Arthur and Sageman, 2004; Donovan and Staerker, 2010). The study area included the southwestern Gulf Coast and the Maverick basins in the southwestern and the eastern Texas basin. Circles locate wells used in this study. The location of the two wells that were heavily sampled is indicated by black circles. Dashed lines with arrows indicate interpreted marine current directions and direction of Cenomanian-Turonian cyclonic storms. Volcanic centers are located along the Sevier orogenic front and northeast of the eastern Gulf Coast basin and west of the Western Interior Sea. Black arrows radiating away from the volcanic centers indicate the likely direction of transported continental sediments.

sediment trapping restricted the transport of clastic sediments to the basin, freshwater delivery increased, and weathering-derived phosphorous was delivered to the basin (Arthur and Sageman, 2004; D. D. Adams et al., 2010). The subsequent enhanced nutrient supply to warm, clear, marine water to the south greatly increased organic productivity in the well-oxygenated, upper water column and subsequently increased the delivery of abundant organic matter to the seafloor. Anoxic conditions below the oxygen minimum zone resulted in preservation of a significant amount of organic matter during burial (Arthur and Sageman, 2004).

Eagle Ford deposits in the study area received freshwater from highlands on either side of the Western Interior Sea that mixed with saline marine waters from the Tethys Sea to the south, resulting in comparatively low sulfate concentrations in this portion of the Western Interior Sea (Slingerland et al., 1996; D. D. Adams et al., 2010). Influx of freshwater likely also contributed to a persistently stratified water column, favorable for organic matter preservation (Arthur and Sageman, 2004). Unlike other areas in the Western Interior Sea, the Gulf Coast and Maverick basins were less restricted, received much less siliciclastic material, and experienced better communication with open ocean water. The San Marcos Arch restricted the influx of sediments transported from eastern highlands from reaching the southwestern Gulf Coast and Maverick basins. Siliciclastic sediment supply to the eastern Texas basin was much greater than in the southwestern Gulf Coast and Maverick basins, resulting in less marine organic matter production and increased dilution of organic carbon (A. E. Adams and Carr, 2010). A relatively higher influx of freshwater likely resulted in restricted biologic productivity (foraminifera and coccoliths) and a better-oxygenated and less-stratified water column, which in turn decreased the preservation potential of organic carbon.

Overall, a wide range of microfacies have been documented in the Gulf Coast Basin wells (Dawson, 1997, 2000). Eagle Ford deposits in the Gulf Coast Basin are predominantly composed of dark gray, faintly and thinly laminated, locally burrowed marls and organic-rich marls, interbedded with predominantly recrystallized, foram-rich wackestones and packstones (Dawson and Almon, 2010). Planktonic foraminifera and coccoliths are abundant in all lithologies. Altered, bioturbated to graded, reworked ash beds are locally common. Facies generally include (1) proximal facies, recrystallized limestone and partially recrystallized marls; (2) medial facies, interbedded limestones and marls with ripple laminations, small-scale cross bedding, and internal truncation surfaces; and (3) distal facies, massive-to-laminated, clay-rich marls with sparse, massive-to-faintly laminated limestones. Intervals that display the thickest marls typically contain microburrows and fecal pellets composed of foraminifera and coccolith fragments.

Comparatively, the Eagle Ford in the eastern Texas basin wells is predominantly composed of faintly laminated, calcareous shales and organic-rich, calcareous shales; laminated, dolomitic shale; and minor laminated marls (Figure 2). Thin, reworked ash beds are locally present. Facies in eastern Texas are interpreted as prodelta to delta deposits. In general, microfacies in eastern Texas wells are more siliceous and contain significantly less calcareous fossils and organic material (Arthur and Sageman, 2004). Graded ash beds tend to be less abundant as well. Most samples are laminated, calcareous-to-noncalcareous, silty shale. Massive fabrics are rare. The Maness Shale underlies the Eagle Ford Shale in eastern Texas and is predominantly composed of silty, slightly calcareous or dolomitic, massive, and thinly laminated shales that are interpreted as prodelta deposits. Detrital quartz silt, mica, plagioclase grains, and skeletal fragments are the most abundant nonmatrix constituents (Figure 3F).

COMPOSITIONAL AND RESERVOIR PROPERTY DATA

On average, the Eagle Ford samples in the southwestern Gulf Coast Basin were highly calcareous (56.2 vol. %) with moderate clay content (13.9 vol. %). Volume percent was calculated using bulk density, grain density, total organic carbon (TOC), and XRD data. The XRD also revealed (in decreasing volume percent) quartz (15.1%), kerogen (8.3%), plagioclase (3.5%), dolomite (1.9%), and pyrite (1.8%). Quartz was commonly present as scattered silt-size grains, as a replacement mineral (foraminifer test walls and coccolith fragments), and as an authigenic matrix constituent (Figure 3). The high volume of calcite in southwestern Gulf Coast samples was caused by the presence of abundant coccolith fragments and foraminifera in the matrix (Figures 3A, B). Mixed-layer illite-smectite, illite-mica, and kaolinite were the predominant clay phases (8.0%, 4.3%, and 2.6% clay average volume percent, respectively). Kaolinite commonly occurs as an authigenic mineral in ash beds and infilling foram chambers (Figure 3C) and as an authigenic matrix constituent (Figure 3D). Minor chlorite was also locally present as a pseudomorph after kaolinite in foraminifera chambers (Figure 3C). Most wells were within the oil-to-dry-gas zone based on measured vitrinite reflectance (Cusack et al., 2010). In general, southwestern Gulf Coast Basin samples had a wide range of TOC values (0.03–10.5 wt.%). Permeability averaged 7.41E-06 md, and total porosity averaged 8.6%.

In contrast, Eagle Ford samples in the eastern Texas basin contained a much higher average clay content (36 vol. %) and lower carbonate (calcite plus dolomite) content (44.1 vol. %). Average mineral compositions as measured by XRD and TOC were as follows: quartz (18.6%), kerogen (7.8%), pyrite (3.0%), and plagioclase (1.3%) by volume. Quartz and plagioclase were commonly present as silt-size detrital grains (Figures 3E, F). Mixed-layer illite-smectite, kaolinite, and illite-mica were the predominant clay phases, averaging 14.6%, 11.6%, and 8.0% by volume, respectively. Kaolinite was commonly present as a matrix constituent and as an authigenic mineral in ash beds. Eagle Ford samples in eastern Texas had a much narrower range of TOC (0.64–3.8 vol. %) and a lower maximum TOC (6.8 wt.%) than the deposits in southwestern Texas. Permeability in Eagle Ford samples from the eastern Texas basin averaged 2.36E-07 md, and total porosity averaged 9.0%.

The Maness had the lowest TOC compared to Eagle Ford shales; it also had significantly higher total porosity and lower permeability (average 11.7% and 3.30E-07 md, respectively) than most Eagle

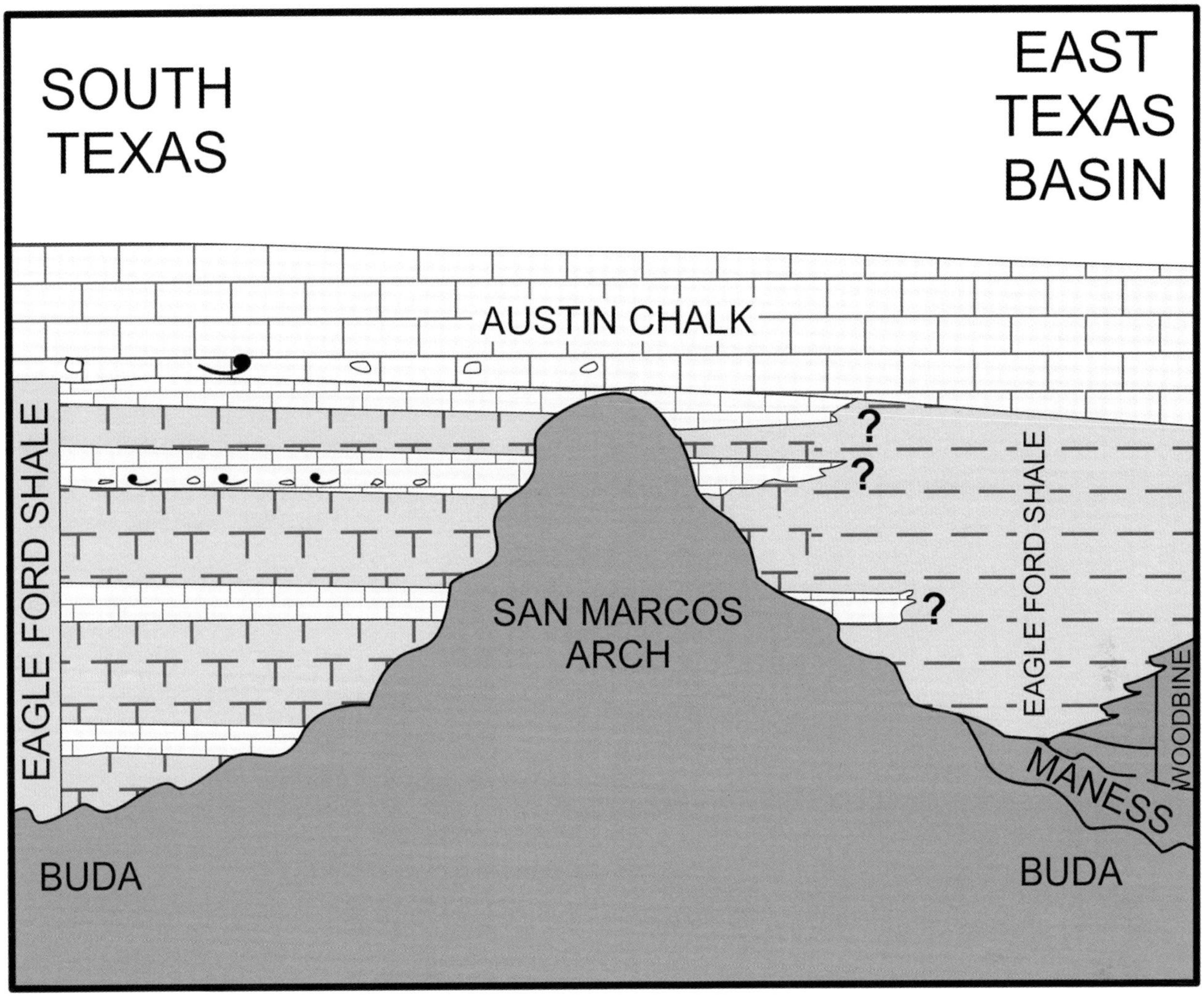

Figure 2. Generalized regional stratigraphic model (west to east) of the study area. Deposits in the Gulf Coast Basin (South Texas) are proximal to distal shelf deposits that are predominantly composed of interbedded marls, wackestones, and packstones. The San Marcos Arch separated the Gulf Coast Basin from the East Texas Basin and restricted the influx of detrital material into the Gulf Coast Basin. Deposits in the East Texas Basin were interbedded, calcareous, laminated, silty shales and scattered marls interpreted as delta to prodelta deposits.

Ford shales, resulting from a higher volume of clay. On average, the Maness consisted of clay (53.9%), quartz (27.8%), calcite (7.1%), and kerogen (4.7%) by volume, with lesser volumes of plagioclase (2.8%) and pyrite (2.5%).

PORE TYPES

Pores are described in this chapter using the classification of Loucks et al. (2010). Pore types in this classification include nonorganic-matter-related interparticle, intracrystalline, and intraparticle pores and organic-matter-related pores or organic nanopores. An additional group, intercrystalline pores, is also described as a subset of the nonorganic-matter-related pores for crystalline matrixes.

Nonorganic-matter-related Pores

Nonorganic-matter-related pores are controlled by mechanical and chemical diagenesis and are divided into interparticle, intraparticle, and intercrystalline pores. Interparticle pores occur between grains and are generally related to the primary pore network. Interparticle pores were the most common pore type in the Eagle Ford and typically occurred between

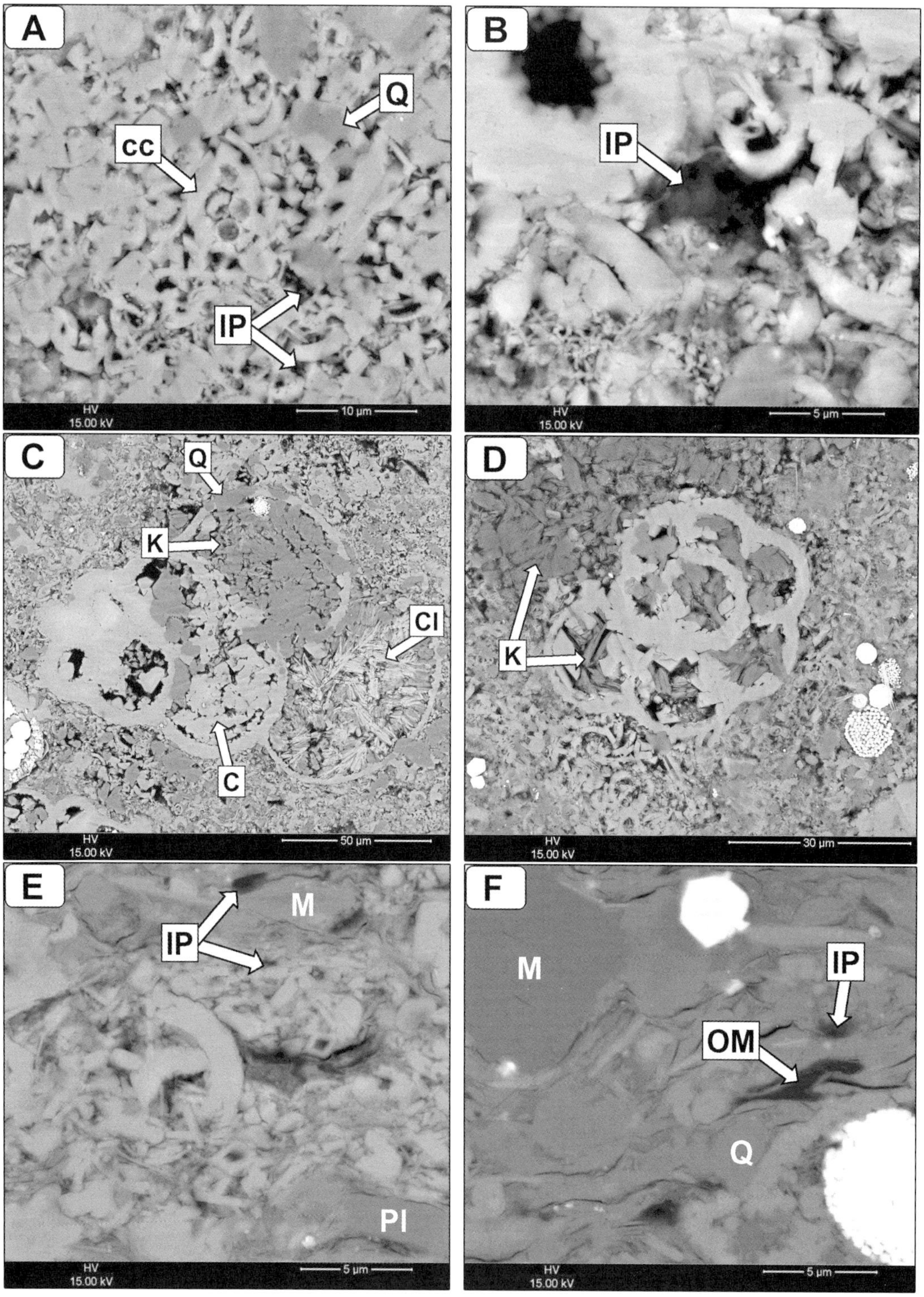

Figure 3. Examples of interparticle pores from argon-ion-milled backscattered electron images. (A) Predominant mineral matrix material in southwestern Texas was predominantly composed of coccolith fragments. Scattered, intact coccolith fragments (cc) were also present. Quartz (Q) was an authigenic mineral that replaces coccolith fragments in the matrix. Abundant interparticle pores (IP) occurred between coccolith fragments. Although most interparticle pores were filled with organic matter, some were open pores. Both organic matter and open pores displayed a low backscatter intensity (black). (B) Most interparticle pores (IP) in southwestern Texas samples ranged from 50 to 300 nm; however, larger, micrometer-scale pores were also common. (C) Foraminifer test walls were commonly replaced by authigenic quartz (Q). Calcite (C), kaolinite (K), and organic matter (black) commonly infilled foraminifer chambers. Chlorite pseudomorphs after kaolinite in foraminifer chambers (Cl) indicated that chlorite in southern Texas Eagle Ford samples likely formed during late diagenesis. (D) Kaolinite (K) was present as foraminifer chamber fill and as an authigenic matrix constituent. (E) Interparticle pores (IP) appeared to be smaller and less well connected in eastern Texas samples. Detrital mica (M) and plagioclase grains (Pl) were abundant. Interparticle pores (IP) were most common in samples that contained scattered fecal pellets. (F) Eastern Texas samples contained significantly more siliciclastic material, such as mica (M), quartz silt (Q), and plagioclase. Interparticle pores appeared to be less well connected and scattered. Organic matter was present (OM), but coccoliths were much less abundant compared with southwestern Texas samples.

coccolith fragments (Figures 3A–D). This pore type generally ranged from less than 100 nm to more than 5 μm in diameter. Smaller interparticle pores tended to appear less well connected than larger-scale pores based on two-dimensional SEM observations.

As defined by Loucks et al. (2010), intraparticle pores occurred within a particle boundary and could be primary (i.e., original pores within grains that are clustered together, such as intercrystalline pores within pyrite framboids) or secondary in origin (i.e., dissolution pores within calcareous fossils and grains or plagioclase grains). Intraparticle pores may also have included primary pores within intact coccospheres, between nannofossil fragments in biogenic aggregates or fecal pellets, within foraminifer test chambers, pores between authigenic clay particles or crystals, and early authigenic phosphate (Figures 4A–E). This pore type tended to be much smaller (<100 nm in diameter) and appeared to be less well connected than interparticle pores. Secondary intraparticle pores that resulted from complete or partial grain dissolution were common within calcareous fossils, carbonate crystals, or feldspar grains (Figure 4F).

Intercrystalline pores generally occurred in recrystallized marls and wackestones. Porosity and effective permeability varied depending on the degree of recrystallization. Predictably, the proportion of preserved, primary interparticle, and intraparticle pores progressively decreased as recrystallization increased (Figures 5A–C). The SEM data also revealed micrometer-scale microfractures not readily visible in thin-section photomicrographs that were infilled with organic matter, calcite, and gypsum (Figures 5D–F), which may have contributed to the total porosity and effective permeability.

Organic-matter-related Pores

Organic-matter-related pores (organic nanopores) in the Eagle Ford were prevalent in organic matter that filled interparticle and intraparticle pores in organic-rich marls. No organic nanopores were documented in samples with a vitrinite reflectance of less than 0.7% or in solid organic particles or particulate carbonaceous material (Figures 6A, B). In samples from the oil window with low vitrinite reflectance (0.6%), it is not likely that a significant amount of gas has been generated that would lead to the development of organic nanopores. Sources of solid particulate organic matter may include terrestrial plant fragments, shallow-water vascular plant detritus, live particulate algaen or waxy insoluble material, soot or ash-derived black carbon, or organic matter rendered unidentifiable by interpolymerization/cross-linking (Iseki, 1981; Volkman and Eiichiro, 2002; Brandes et al., 2004), which may be more resistant to hydrocarbon maturation processes. Given the position of Eagle Ford wells with respect to the paleoshoreline, it is unlikely that vascular plant material (terrestrial or shallow water) was a likely source of particulate carbonaceous matter. In addition, particulate carbon identified in this study did not exhibit any features that would indicate that they were combustion particles. However, differentiating between types of carbonaceous material is especially challenging, requiring a combination of analytical techniques that are outside the scope of FE-SEM analysis (Volkman and Eiichiro, 2002).

Organic matter within micron-scale interparticle pores usually contained well-developed organic nanopores in samples with vitrinite reflectance values greater than 1.0%R_o (Figures 6C, D). Organic nanopores were also observed in organic matter within foraminifer test chambers and may have been combined with intraparticle pores associated with authigenic clays, giving rise to a complex pore network (Figure 6C). Organic nanopores commonly had irregular, ellipsoidal shapes. Diameters typically ranged from less than 10 nm to 1.0 μm and averaged 200 to 300 nm (Figure 6D). Samples that display abundant and well-connected interparticle and intraparticle pores commonly contain the most abundant organic nanopores. Organic matter in recrystallized limestones rarely displays organic nanopores. Overall, organic nanopores were most abundant and displayed the best-developed pore systems in organic matter present between coccolith fragments in distal marls (i.e., Figure 6D) that tended to be less recrystallized and possessed ample maturity (>1.4–1.6%R_o).

Organic maturation in shales occurs throughout the burial process, starting with diagenesis and ending with metagenesis (Dow, 1977; Senftle and Landis, 1991). During diagenesis, biochemical and chemical processes degrade organic matter, breaking it down into smaller and smaller particles. During catagenesis, small molecules are broken off the kerogen matrix, and bitumen (soluble organic matter) is generated. Some of the bitumen is expelled from the source rock, and some is cracked into gas. Finally, metagenesis occurs during late-stage burial and continues through the metamorphic stage. Dry-gas generation occurs during metagenesis. Although these stages of organic matter maturation appear to be discrete phases, in actuality they are gradational. The level of organic nanopore development in the Eagle Ford Shale commonly tracked the level of organic matter maturation based on vitrinite reflectance and FE-SEM data.

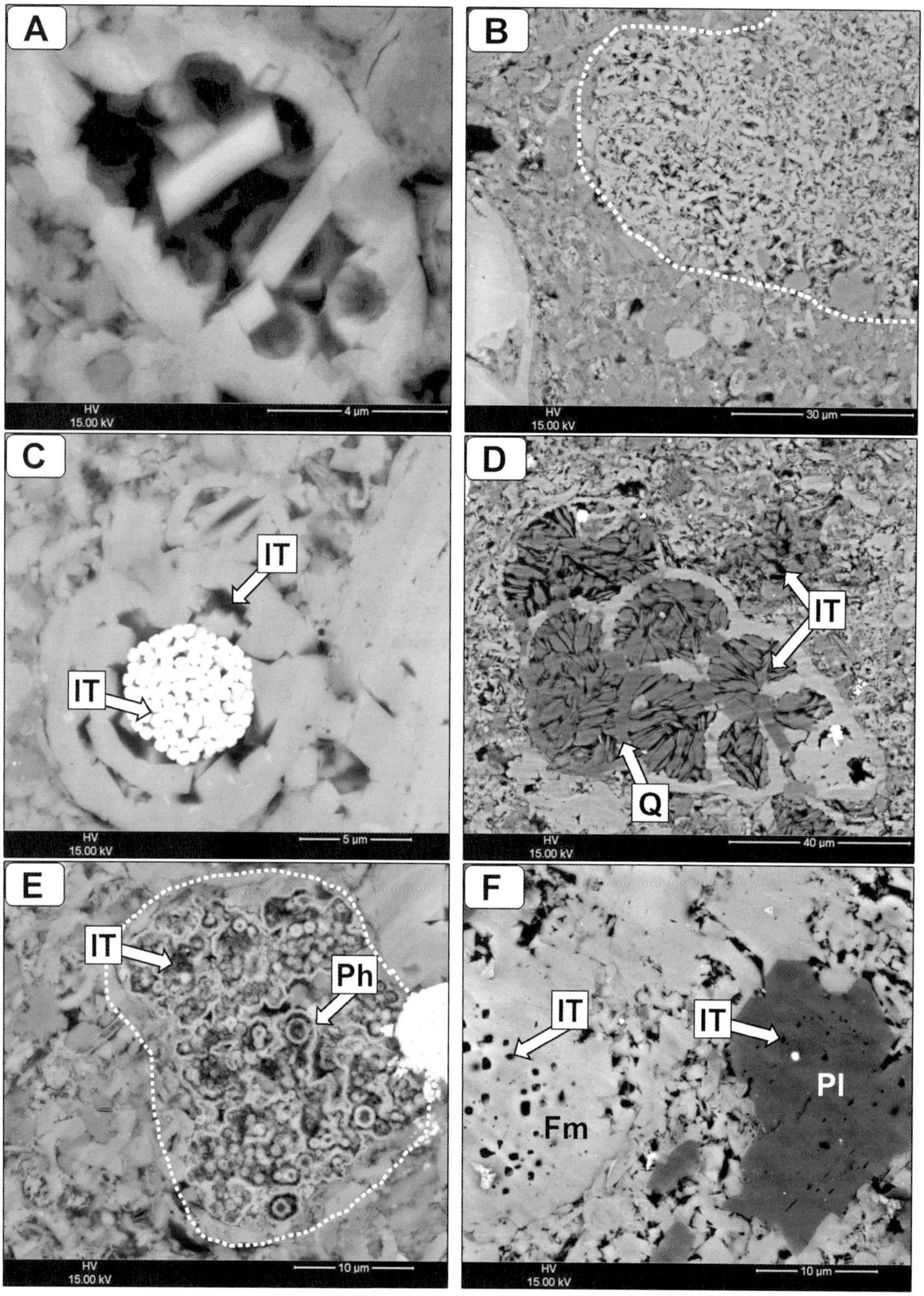

Figure 4. Examples of intraparticle pores (IT) from argon-ion-milled backscattered electron scanning electron microscope (SEM) images, southwestern Gulf Coast Basin. (A) Primary pores preserved within intact coccospheres were several micrometers in diameter. (B) Biogenic aggregates or pellets (encircled by white dashed line) were predominantly composed of coccolith fragments and organic matter. Intraparticle pores within these aggregates ranged from less than 10 to 100 nm, depending on the internal pellet fabric. (C) Intraparticle pores (IT) were present within both foram chambers and pyrite framboids (bright white). (D) Intraparticle pores (IP) present between authigenic clays (IT) within foram chambers and in the matrix. Note calcareous (light gray) foram test walls have been partially replaced by authigenic quartz (dark gray, Q). (E) Patches of early authigenic phosphate (Ph) typically occurred as accretionary masses with abundant intraparticle pores (IT). Intraparticle pores associated with authigenic phosphate commonly contained organic matter (black). (F) Secondary intraparticle pores (IT) typically occurred within recrystallized foraminifera (Fm) and plagioclase grains (Pl).

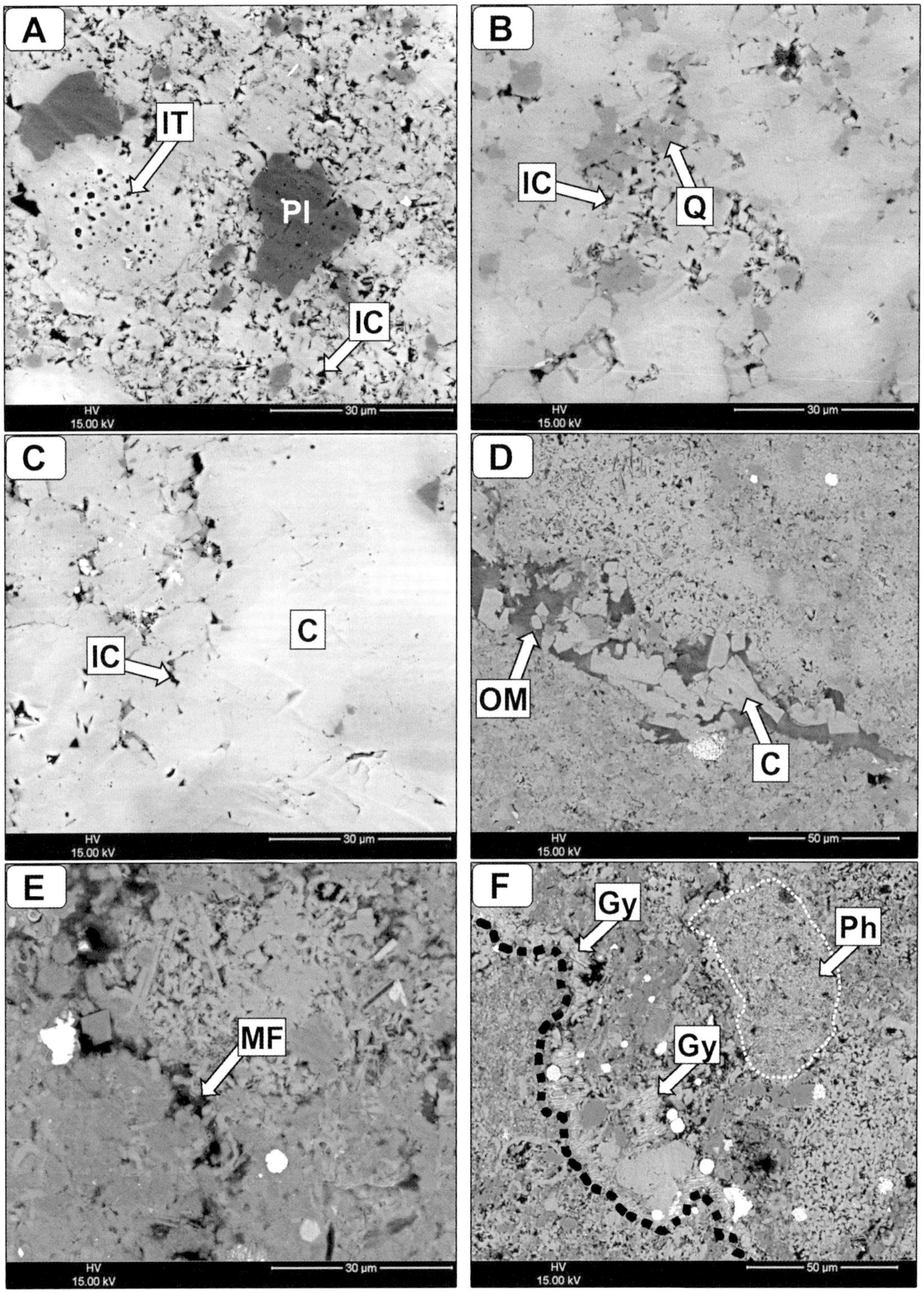

Figure 5. Examples of intercrystalline pores and microfractures from argon-ion-milled backscattered electron scanning electron microscope (SEM) images, southwestern Gulf Coast Basin. (A)–(C) Intercrystalline pores (IC) were the most abundant pore type in recrystallized limestones. Although samples (A)–(C) are compositionally equivalent, calcite (light gray, C) and authigenic quartz (Q) decreased porosity with increasing degree of recrystallization. (D) Calcite spar (light gray) and organic matter (OM) fill a subvertical microfracture. The aperture of the fracture was about 20 μm at the widest point. (E) Smaller near-vertical microfractures (MF) were commonly filled with organic matter (black). (F) A black dashed line highlights the path of a sinuous microfracture that was infilled with gypsum (Gy). A white dashed line indicates a patch of authigenic phosphate (Ph) as indicated by energy dispersive x-ray data.

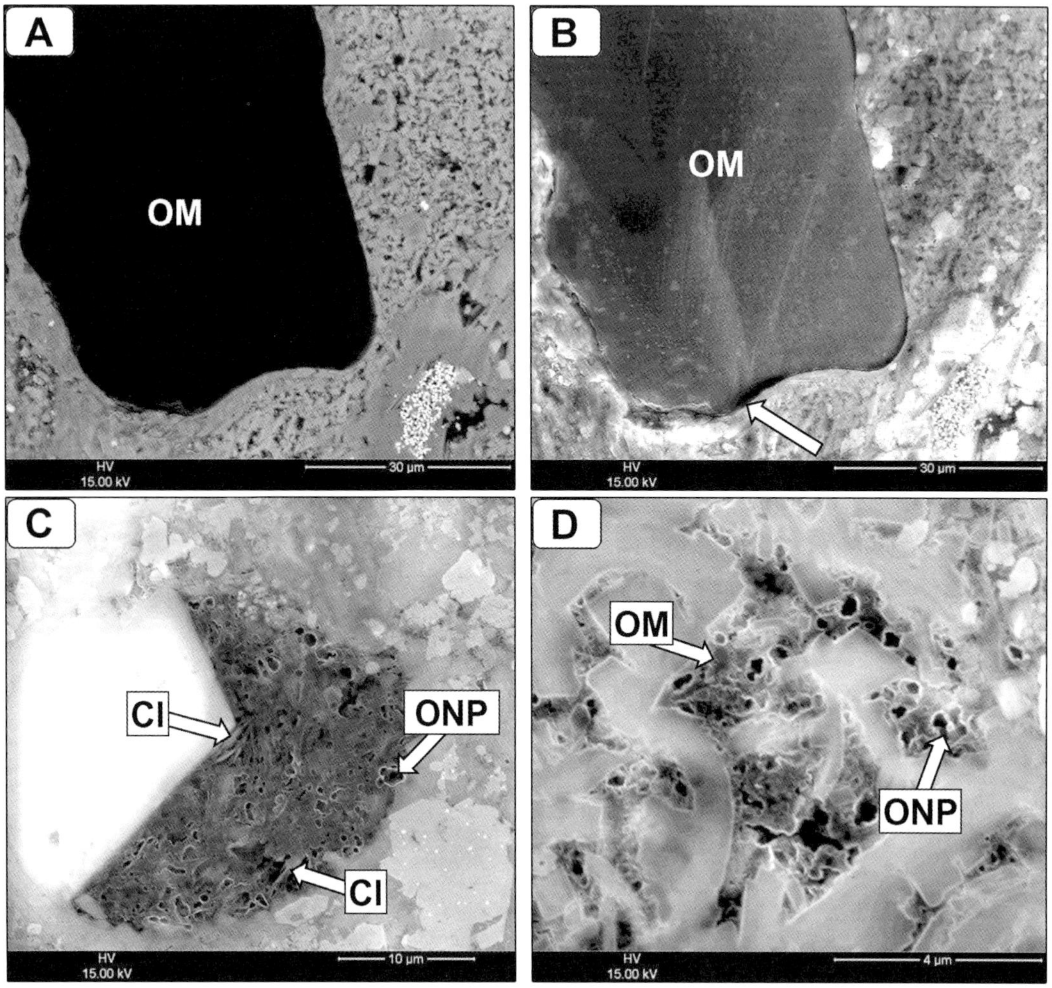

Figure 6. Organic-matter-related pores, southwestern Gulf Coast Basin. (A) Argon-ion-milled backscattered electron image of solid particulate organic matter or carbonaceous material (OM). Coccolith fragments and clay were aligned within a few micrometers around the particle. (B) Secondary electron image, same field of view as Figure 2A. Notice distinct contact with the matrix (white arrow). No organic nanopores were observed within the particulate organic matter, which may be caused by organic matter type or the presence of kerogen. (C) Authigenic clay (Cl) particles within this foram chamber controlled the localized, complex organic nanopore pore system. Organic nanopores (ONP) associated with authigenic clay particles tended to be smaller and well connected. (D) Organic nanopores (ONP) commonly displayed irregular ellipsoidal shapes. Typically, these organic nanopores ranged from less than 10 nm to 1.0 μm in diameter and tended to be most abundant in samples with a vitrinite reflectance between 1.4 and 1.6%R_o. Residual organic matter (OM) is visible between organic nanopores.

With the onset of wet-gas generation and bitumen migration, small organic nanopores began to develop (Figures 7A–C). Maximum organic nanopore abundance and connectivity occurred during late-state catagenesis (Figures 7D–G). As dry gas was generated during late-stage burial, the maximum conversion of bitumen to methane occurred (>1.7%R_o). Only open pores, very small organic nanopores, and residual organic matter were left behind (Figure 7H). In general, there appeared to be a relationship between the abundance of organic nanopores, predominant pore type, maturation (Figure 8), pore connectivity, and permeability similar to that documented in the Barnett Shale (Loucks et al., 2009). However, vertical variability in the presence of organic nanopores (Figure 8), even within similar lithologies that had similar rock properties (independent of stratigraphic position), suggests that maturity, mineralogy, and TOC content were not the only controlling factors in organic nanopore development.

In contrast, forams and coccolith plates were much less abundant in Eagle Ford samples from the eastern Texas basin. Horizontal microburrows were common; fecal pellets composed of coccolith fragments were

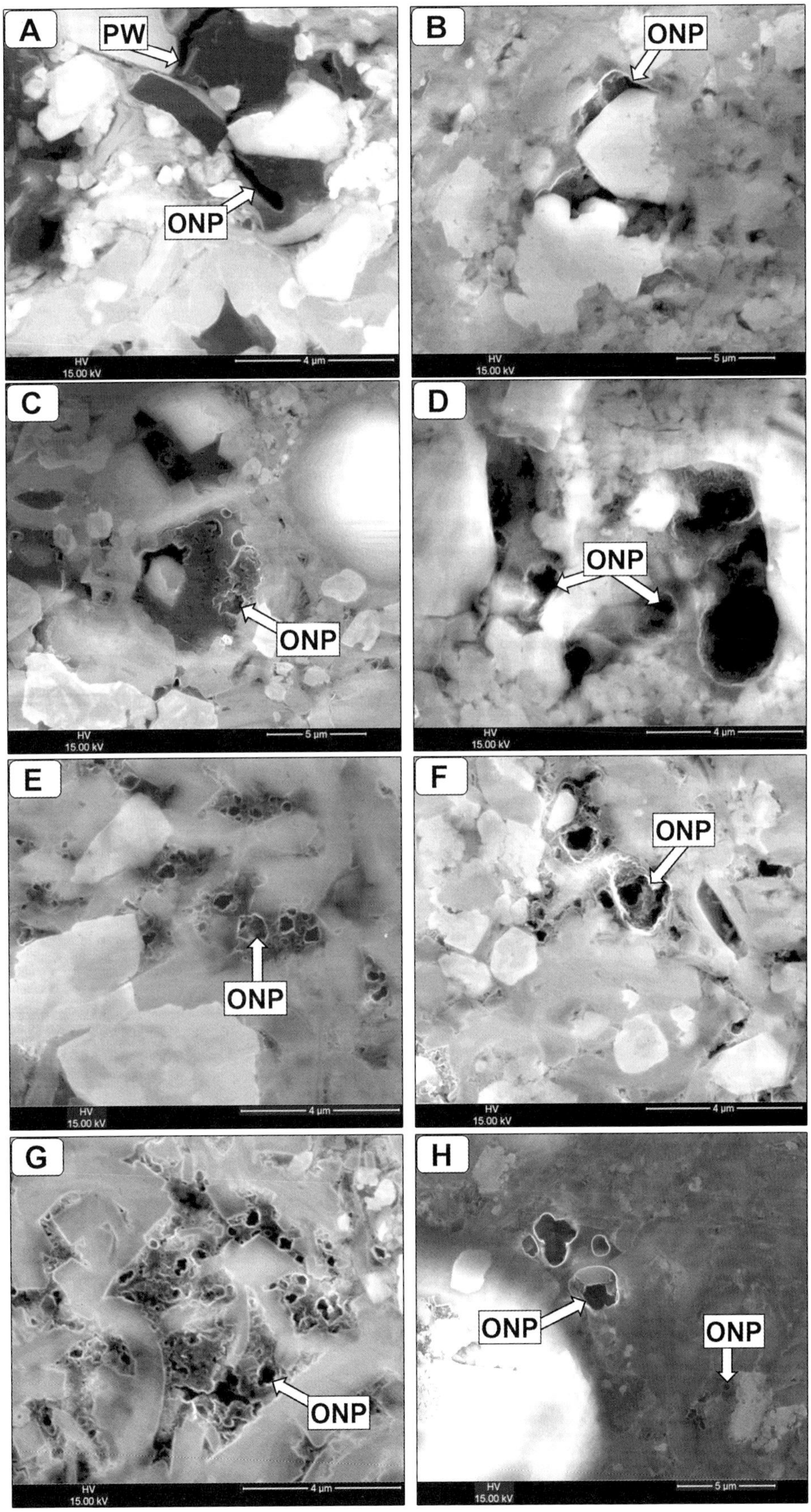

Figure 7. Examples of organic nanopores at various maturity levels. All samples are secondary electron images of foraminifer marls with a TOC of 2.0–4.0 wt.%. Organic nanopore development and apparent connectivity increased with increasing maturity and seemed to be independent of maturity. (A) Samples with vitrinite reflectance values less than 0.7%R_o displayed no organic nanopores. (B) Vitrinite reflectance between 0.7 and 1.0%R_o, small organic nanopores (ONP) typically occurred at the contact between the pore walls and the organic matter and may have extended into the center of the organic matter (black). (C)–(G) As maturity increased, organic nanopores became progressively larger and better connected. The most abundant and best-developed organic nanopores were documented in samples with R_o between 1.4 and 1.6%. (H) At vitrinite reflectance values between 1.6 and 1.8%R_o, organic nanopores were less abundant, and the distribution was patchy. In samples with vitrinite reflectance greater than 1.7%R_o the organic nanopores that remained were much smaller. Only open pores, very small organic nanopores (bottom right), and residual organic matter were left behind. Residual organic matter commonly displayed microfractures (left center).

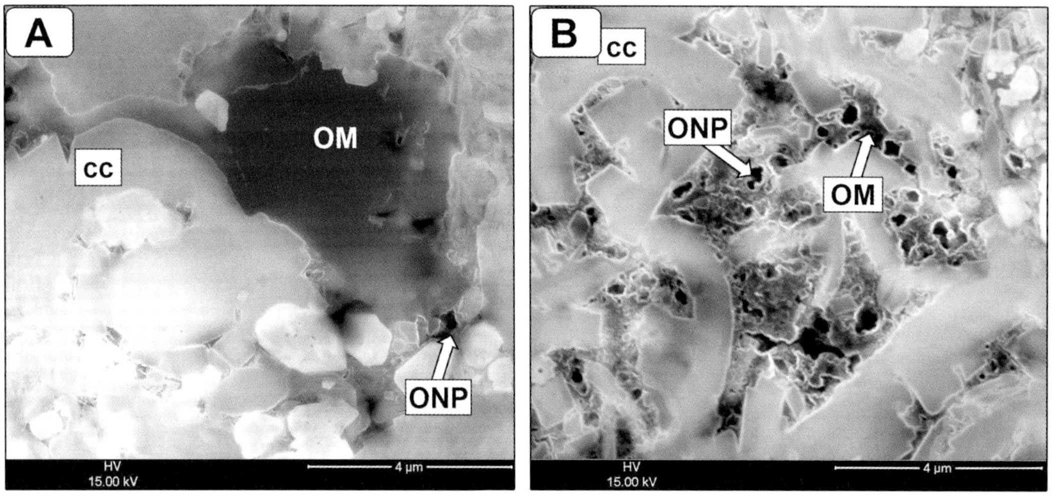

Figure 8. Example of vertical variability in the abundance of organic nanoporosity. (A) Secondary electron image of organic matter (OM) that displays limited organic nanoporosity (ONP) adjacent to the pore wall. Coccolith fragments (cc) were present in the adjacent matrix. (B) Secondary electron image of abundant organic nanopores in a sample 74 ft (22.5 m) below the sample displayed in (A). Organic matter (OM) contained abundant elliptical organic nanopores (ONP) that appeared to be well developed. The matrix of this sample was also composed of coccolith fragments (cc).

scattered throughout burrowed intervals. Detrital clays were more abundant. Poorly connected, interparticle pores were the predominant pore type (Figures 3C, D). No organic nanopores were documented in eastern Texas basin samples. Isolated interparticle pores characterized the Maness Shale, and no organic nanopores were documented. Horizontal microburrows were scattered, and fecal pellets were rare.

CONCLUSIONS

The most abundant pore types in southwestern Gulf Coast Basin samples were interparticle (between coccolith fragments and plates) and intraparticle pores (within coccoliths, locally common fecal pellets, foram chambers, recrystallized foraminifera, plagioclase grains, and pyrite framboids). SEM data revealed that Eagle Ford shales in the southwestern Gulf Coast Basin displayed higher porosity and higher permeability than samples from the Eagle Ford and Maness shales in the eastern Texas basin, which is likely because of the presence of well-connected interparticle and intraparticle pores between abundant coccolith fragments. The relative abundance of interparticle pores in marls varied depending on the abundance and type of fecal pellets present, as well as the presence of authigenic minerals, such as calcite and quartz. Differences in the relative abundance of pore types appeared to be controlled by (1) the ratio of recrystallized limestone to marl (i.e., proximal facies vs. medial or distal facies); (2) the degree of compaction within marls; and (3) the amount of organic matter and clay in the matrix.

Although organic nanopores were typically most abundant in organic-rich marls with abundant coccolith fragments and fecal pellets, vertical variability in the presence and abundance of organic nanopores in all microfacies was common. The variability in the presence and abundance of organic nanopores in similar microfacies in individual wells suggests that the parameters controlling organic nanopores development are not yet fully understood and deserve additional detailed studies.

ACKNOWLEDGMENTS

We would like to express our appreciation to the management of Core Laboratories, in particular Randy Miller, for permission to publish this work. Special thanks to Mark Norini, who compiled the original area map with Roger Barnaby, and to Allen Boren for his assistance with the updated map and illustrative work. We would also like to thank our colleagues Elizabeth Hernandez, Roger Barnaby, and Pat Rush for the discussions and helpful suggestions that greatly enhanced the quality of this chapter. Finally, we are grateful to Randy Miller, Pat Rush, and the reviewers Terrylyn M. Olson and Laura Wray for their careful evaluation and thoughtful comments, which greatly improved the final text.

REFERENCES CITED

Adams, A. E., and J. P. Carr, 2010, Regional depositional systems of the Woodbine, Eagle Ford, and Tuscaloosa of the U.S. Gulf Coast: Gulf Coast Association of Geological Societies Transactions, v. 60, p. 3–27.

Adams, D. D., M. T. Hurtgen, and B. B. Sageman, 2010, Volcanic triggering of a biogeochemical cascade during Oceanic Anoxic Event 2: Nature Geoscience, v. 3, p. 201–204.

Arthur, M. A., and B. B. Sageman, 2004, Sea-level control on source-rock development: Perspectives from the Holocene Black Sea, the Mid-Cretaceous Western Interior Basin of North America, and the Late Devonian Appalachian Basin: SEPM Special Publication, v. 82, p. 35–59.

Brandes, J. A., C. Lee, S. Wakeham, M. Peterson, C. Jacobsen, S. Wirick, and G. Cody, 2004, Examining marine particulate organic matter at sub-micron scales using scanning transmission x-ray microscopy and carbon x-ray absorption near edge structure spectroscopy: Marine Chemistry, v. 92, p. 107–121.

Burton, J. H., D. H. Krinsley, and K. Pye, 1987, Authigenesis of kaolinite and chlorite in Texas Gulf Coast sediments: Clays and Clay Minerals, v. 35, p. 291–296.

Cusack, C., J. Beeson, D. Stoneburner, and G. Robertson, 2010, The discovery, reservoir attributes, and significance of the Hawkville Field and Eagle Ford Shale trend, Texas: Gulf Coast Association of Geological Societies Transactions, v. 60, p. 165–179.

Dawson, W. C., 1997, Limestone microfacies and sequence stratigraphy: Eagle Ford Group (Cenomanian-Turonian) North-Central Texas outcrops: Gulf Coast Association of Geological Societies Transactions, v. 47, p. 99–105.

Dawson, W. C., 2000, Shale microfacies: Eagle Ford Group (Cenomanian-Turonian) North-Central Texas outcrops and subsurface equivalents: Gulf Coast Association of Geological Societies Transactions, v. 50, p. 607–621.

Dawson, W. C., and W. R. Almon, 2010, Eagle Ford Shale variability: Sedimentologic influences on source and reservoir character in an unconventional resources unit: Gulf Coast Association of Geological Societies Transactions, v. 60, p. 181–190.

Donovan, A. D., and T. S. Staerker, 2010, Sequence stratigraphy of the Eagle Ford (Boquillas) Formation in the subsurface of South Texas and outcrops of West Texas: Gulf Coast Association of Geological Societies Transactions, v. 60, p. 861–899.

Dow, W. G., 1977, Kerogen studies and geological interpretations: Journal of Geochemical Exploration, v. 7., p. 79–99.

Iseki, K., 1981, Particulate organic matter transport to the deep sea by salp fecal pellets: Marine Ecology—Progress Series, v. 5, p. 55–60.

Lock, B. E., L. Peschier, and N. Whitcomb, 2010, The Eagle Ford (Boquillas Formation) of Val Verde County, Texas—a window on the South Texas play: Gulf Coast Association of Geological Societies Transactions, v. 60, p. 419–434.

Loucks, R. G., R. M. Reed, S. C. Ruppel, and D. M. Jarvie, 2009, Morphology, genesis, and distribution of nanometer-scale pores in siliceous mudstones of the Mississippian Barnett Shale: Journal of Sedimentary Research, v. 79, p. 848–861.

Loucks, R. G., R. M. Reed, S. C. Ruppel, and U. Hammes, 2010, Preliminary classification of matrix pores in mudrocks: Gulf Coast Association of Geological Societies Transactions, v. 60, p. 435–441.

Luffel, D. L., F. K. Guidry, and J. B. Curtis, 1992, Evaluation of Devonian Shale with new core and log analysis methods: Journal of Petroleum Technology, v. 44, p. 1192–1197.

Mack, G. H., 1992, Paleosols as an indicator of climatic change at the Early-Late Cretaceous boundary, southwestern New Mexico: Journal of Sedimentary Petrology, v. 62, p. 483–494.

Sageman, B. B., and M. A. Arthur, 1994, Early Turonian paleogeographic/paleobathymetric map, Western Interior, U.S., *in* M. V. Caputo, J. A. Peterson, and K. J. Franczyk, eds., Mesozoic systems of the Rocky Mountain region, U.S.A.: Denver, CO, Rocky Mountain Section (Society for Sedimentary Geology), p. 457–469.

Senftle, J. T., and C. R. Landis, 1991, Vitrinite reflectance as a tool to assess thermal maturity, *in* R. K. Merrill, ed., Source and migration processes and evaluation techniques: AAPG Treatise of Petroleum Geology, Handbook of Petroleum Geology, p. 119–125.

Slingerland, R., L. R. Kump, M. A. Arthur, P. J. Fawcett, and B. B. Sageman, 1996, Estuarine circulation in the Turonian Western Interior Seaway of North America: Geological Society of America Bulletin, v. 108, p. 941–952.

Volkman, J. K., and T. Eiichiro, 2002, Chemical and biological studies of particulate organic matter in the ocean: Journal of Oceanography, v. 58, p. 265–279.

Walls, J. D., and S. W. Sinclair, 2011, Eagle Ford shale reservoir properties from digital rock physics: European Association of Geoscientists and Engineers, First Break, v. 29, p. 97–101.

11

Brian Driskill, Joel Walls, Steven W. Sinclair, Juliana DeVito, 2013, Applications of SEM imaging to reservoir characterization in the Eagle Ford Shale, south Texas, U.S.A., *in* W. Camp, E. Diaz, and B. Wawak, eds., Electron microscopy of shale hydrocarbon reservoirs: AAPG Memoir 102, p. 115–136.

Applications of SEM Imaging to Reservoir Characterization in the Eagle Ford Shale, South Texas, U.S.A.

Brian Driskill

Shell Exploration and Production Company, 150 N. Dairy Ashford, Houston, Texas 77079, U.S.A. (e-mail: brian.driskill@shell.com)

Joel Walls and Juliana DeVito

Ingrain, Inc., 3733 Westheimer Rd., Houston, Texas 77027, U.S.A. (e-mails: walls@ingrainrocks.com; devito@ingrainrocks.com)

Steven W. Sinclair

Matador Resources Company, 5400 Lyndon B Johnson Freeway #1500, Dallas, Texas 75240, U.S.A. (e-mail: ssinclair@matadorresources.com)

ABSTRACT

Laboratory measurements of porosity, permeability, and fluid saturation are routinely acquired from core samples to quantify petrophysical data for shale reservoir studies. However, these measurements do not provide information regarding the microstructural character of the mineral matrix, organic matter distribution, and pore network. New imaging techniques using focused ion beam scanning electron microscopy (FIB-SEM) were applied to a subset of prepared core samples from four wells to study the microstructural character of the Cenomanian-Turonian Eagle Ford Formation in the Maverick Basin, south Texas, in the United States. The studied wells span a wide range of organic thermal maturity from the oil window to the dry-gas window. Sixteen digital rock physics (DRP) models were prepared from three-dimensional (3-D) SEM image volumes to quantify volumes of mineral matrix, organic matter, and pore space. The results of this study revealed organic matter pore type, pore size, and permeability vary by stratigraphic interval and the degree of thermal maturity. Pendular organic matter with bubble pores was more common in the oil window. Both pendular and spongy organic matter types with bubble and foam pores were common in the gas/condensate window. Organic matter pores from samples within the dry-gas window had smaller average pore diameters and lower permeability than samples within the oil window. Comparison of core porosity and permeability measurements derived from DRP models with laboratory core measurements showed broad similarities, indicating that despite the small analysis volume involved with FIB-SEM 3-D image volumes, the DRP models prepared in this study were generally representative of the larger core samples.

DOI: 10.1306/13391709M1023587

INTRODUCTION

Historically, shales were thought to perform two key functions: act as a seal for conventional reservoirs and act as a source rock for hydrocarbons. Recently, several shale formations have also proven to be major self-sourcing hydrocarbon reservoirs. Following 20 years of pioneering efforts by Mitchell Energy in establishing commercial gas production from the Barnett Shale in the Fort Worth Basin, Texas, success was finally achieved in about 2000 (Bowker, 2007). Exploration success rapidly expanded in other shale gas plays, including the Fayetteville in Arkansas, the Woodford in Oklahoma, the Haynesville in northwestern Louisiana and eastern Texas, and the Marcellus in Pennsylvania in the United States, and the Horn River and Montney in Canada (Boughal, 2008; Stevens and Kuuskraa, 2009). These results, plus the establishment of liquid production from several shale plays such as the Eagle Ford Formation, confirmed shale as an important North American source of hydrocarbons and spurred a worldwide assessment of shale production potential.

Initially, describing and measuring shale rock properties was a challenge because properties such as porosity and matrix grain composition are below the resolution of optical microscopes. Porosity and permeability measurements were problematical for shales until the Gas Research Institute (GRI) process using crushed samples was developed (Guidry et al., 1995). The GRI process allows relatively quick measurements of both porosity and permeability, and implementations of GRI methods are now standard for shales.

Although the GRI process made it possible to quantify basic reservoir properties, confirmation by visual assessment of shale microstructures was lacking. Scanning electron microscopy (SEM) using broken or mechanically polished samples, in combination with energy-dispersive x-ray (EDX) spectroscopy, provides only limited observations about grain types, size, texture, and mineralogy. Identifying and quantifying natural porosity and permeability were often difficult because of the potential of creating artificial pores during sample preparation and polishing. Also, mechanically polished samples often contain residual topographic features that can influence the interpretability of subtle structures in SEM images.

Recent applications of ion milling have greatly enhanced the ability to interpret shale features with SEM. Ion milling uses an ion beam (usually Ar^+ or Ga^+) to remove small amounts of material to produce an ultraflat surface (around 1×1 mm in area), which can then be imaged with SEM. Systems with focused ion beam (FIB) and SEM facilitate preparation of three-dimensional (3-D) digital image volumes by concatenating a series of SEM images from multiple parallel milled surfaces, similar to the way a closely spaced set of seismic lines can be combined into a 3-D seismic volume.

Scanning electron images from ion-milled samples have revealed details and spatial relationships in shale that previously were difficult to observe. Several published models have described relationships among pore space, grains, and organic matter (Loucks et al., 2009; Wang and Reed, 2009; Loucks et al., 2010; Schieber, 2010; Walls et al., 2011). Although these individual studies relied on small sample sets, a growing consistency exists in observations of individual shales. For instance, in the Jurassic Haynesville Shale, intergranular pores appear to be the dominant pore type (Milner et al., 2010); in the Mississippian Barnett Shale, pores within organic matter are dominant (Loucks et al., 2009; Milner et al., 2010).

Another benefit of ion-milled specimens is that the lack of topography observed in the resulting SEM images makes it possible to more accurately segment them using digital imaging software. Single two-dimensional (2-D) SEM images can be interpreted in terms of grain composition, pore types, and the distribution of organic matter and pores. These segmentations can be easily translated for quantitative or semiquantitative rock component measurements. Using segmented 3-D SEM data sets, digital rock physics (DRP) models can be generated with the goal of understanding not only the volumes of porosity and organic matter in 3-D but also the magnitude of permeability.

This chapter describes the methodology and results of an unconventional reservoir study to identify the distribution of porosity and organic matter, as well as the relationship between porosity and permeability. Sixteen samples from the Eagle Ford were collected from four conventionally cored wells located in Maverick, Dimmit, LaSalle, and Atascosa counties in southern Texas, United States. Well locations are shown in Figure 1 along with the paleogeographic setting. The data set comprises both 2-D SEM images and 3-D FIB-SEM image volumes.

GEOLOGIC SETTING

The Late Cretaceous Eagle Ford marl is stratigraphically positioned between two chalks, the underlying Buda and the overlying Austin (Figure 2). The Eagle Ford occurs in the subsurface and is exposed in outcrops around the Gulf of Mexico coasts of Texas and northeastern Mexico. It was deposited when sea level was rising to the maximum extent during the

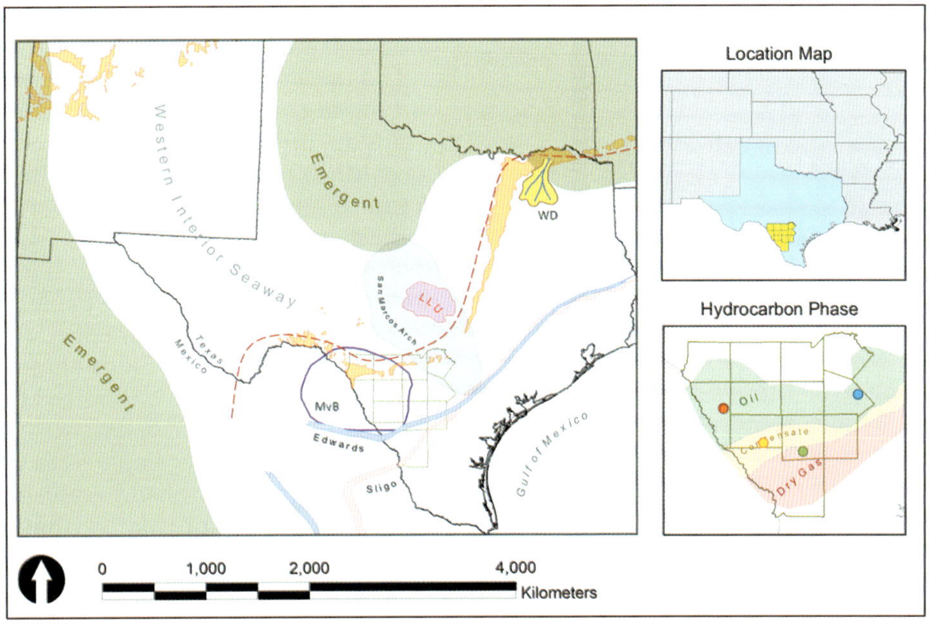

Figure 1. Geographic features along the Texas Gulf coast. Shoreline is for Cenomanian-Turonian (approximated from Blakey, 2011). Prominent features include the Ouachita structural front (red dash); Llano Uplift (LLU; pink polygon); San Marcos Arch (gray oval); Woodbine Delta (WD); Cretaceous shelf edges (Sligo and Edwards; pink and blue ribbons, respectively); Maverick Basin (MvB; purple outline); and Western Interior Seaway. Eagle Ford outcrop is orange (from Stoeser et al., 2007). Llano Uplift outline from Hartman and Scranton (1992); Ouachita Front modified from Morgan (1952); Cretaceous reefs modified from Goldhammer and Johnson (2001). Study area is colored yellow in location map insert and is represented by county outlines on map. Approximate Eagle Ford hydrocarbon phase boundaries in South Texas are shown in lower inset map. Locations of data wells for this report are marked by colored dots on the hydrocarbon phase map: Well NW, red; Well NE, blue; Well SW, yellow; Well SE, green. The Eagle Ford Formation is thicker in the Maverick Basin relative to the remainder of the trend north of the Edwards margin and west of the San Marcos Arch. Modified from Driskill et al., 2012 (1000 km = 621.371 mi).

Phanerozoic, leading to a broad flooded shelf margin along the Texas coast. A dry climate and protection from the Woodbine Delta to the east by the San Marcos Arch limited siliciclastic influx into the Maverick Basin area (Figure 1). The Maverick Basin, a persistent shelf sag basin since at least Pearsall (Aptian) deposition, was located at the southern end of the Western Interior Seaway during Eagle Ford times (Figure 1). Ocean circulation patterns most likely created upwelling zones along the shelf margin that led to a proliferation of microorganisms in the water column. Extensive geochemical and paleontological studies consistently support an interpretation of dysoxic conditions with intermittent anoxia and suboxia, resulting in preservation of high amounts of organic matter in the Eagle Ford (e.g., Bohacs et al., 2011; Driskill et al., 2012; Harbor et al., 2012; Pramudito et al., 2012). Water depths are interpreted to have been less than 100 m (328 ft) in the Maverick Basin and near storm-wave base in the Val Verde County outcrop belt (Driskill et al., 2012).

The Eagle Ford is informally divided into upper and lower members that can be distinguished using log character, microfossils, and texture. In the Maverick Basin, the upper Eagle Ford exhibits lower gamma-ray and higher density log values than the lower Eagle Ford (Figure 3). Outside the Maverick Basin to the east, the upper Eagle Ford progressively thins, and it becomes more difficult to differentiate the upper and lower units. However, the upper and lower Eagle Ford (as well as the underlying Buda and overlying Austin) have distinctive microfossil assemblages that allow for biostratigraphic identification of the upper and lower members even as distinguishing log characteristics are lost (Driskill et al., 2012). Beyond the paleo-Edwards shelf margin to the south, the Eagle Ford increases in thickness; here the upper and lower units are difficult to separate based on log character alone but can be distinguished using microfossils.

LITHOLOGY

Several rock types are commonly observed in Eagle Ford core and outcrop: (1) laterally continuous, dark

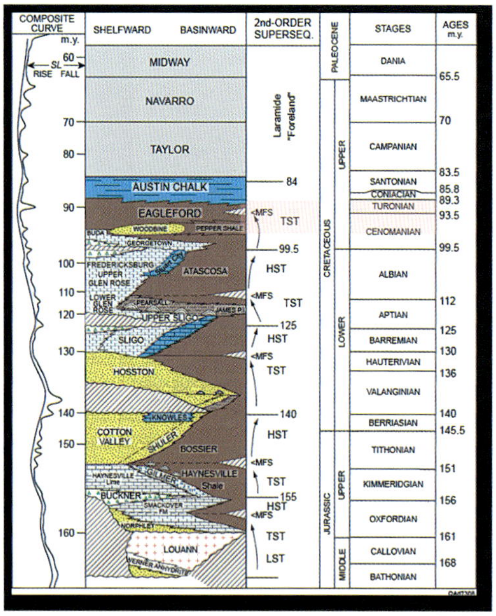

Figure 2. Chronostratigraphic chart for the western Gulf of Mexico margin. The Eagle Ford was deposited during the maximum Phanerozoic sea-level rise (see composite curve). The Eagle Ford is stratigraphically positioned between the Coniacian-Santonian Austin chalk and Cenomanian Buda chalk. The Woodbine and Pepper Shale occur east of the San Marcos Arch, which protected the southern Texas shelf from the Woodbine Delta clastics. Modified from Goldhammer et al. (1991). Sea-level curves modified from Haq et al. (1987): left, long-term curve; right, short-term curve.

gray, carbonaceous, fossiliferous marls interbedded with (2) lighter gray limestones and (3) gray-green altered tuffs. The marls are extremely fine-grain wackestones and packstones with rare foraminiferal grainstones. The marls may be laminated or bioturbated (Figure 4), and locally extensive bioturbation results in massive appearing zones. The limestone units have a replacement texture, are bioturbated (Figure 5), and appear to cap depositional cycles in outcrop and core (Driskill et al., 2012). The limestone units are interpreted as firm grounds that formed during depositional pauses or oxygenation events. Most are under 60 cm (23.6 in.) in thickness, and they are rarely fully resolved on well logs. The tuffs, which are composed mainly of altered clays (principally smectite/illite with minor kaolinite) and feldspar, are common but tend to be thin in core. Most are less than 15 cm (5.9 in.) thick, and many are paper thin and characterized by bright golden mineral fluorescence under ultraviolet light. The thicker tuffs may be bioturbated or show evidence of soft-sediment deformation (Figure 6). Based on limited SEM images, the tuff and limestone beds do not contain sufficient porosity to be considered as potential reservoir quality rock.

The upper and lower Eagle Ford zones have high carbonate contents, on average 60–70%. Carbonate grains are predominantly composed of coccoliths and coccolith fragments, which are present in the mud matrix (Figure 7), with lesser contributions from foraminifera, echinoids, bivalves, and rare calcispheres and ostracods. In general, early calcite cements fill foraminifera tests in the upper Eagle Ford; in the lower Eagle Ford, foraminifera are filled with kaolinite, calcite, or both (Figure 8). Silica observed in the matrix (Figure 7) is thought to be sourced by the altered volcanic ash and replaced radiolaria. Core examination reveals that the upper Eagle Ford contains more foraminiferal packstones and grainstones, and is volumetrically more bioturbated, than the lower Eagle Ford in the Maverick Basin.

Total organic carbon (TOC) content is typically greater in the lower Eagle Ford than the upper Eagle Ford, although an overlap in the TOC ranges exists. Present-day values on the order of 3–10 wt.% for the lower and 2–7 wt.% for the upper Eagle Ford are commonly found. The organic matter is interpreted as amorphous type II kerogen that originated from bacteria, algae, and the soft parts of coccolithophores and foraminifera.

METHODOLOGY

This study reviewed hundreds of 2-D SEM images and created 16 DRP models from four wells in the Eagle Ford play. Wells NW and NE are in the oil window (0.77–0.86%R_o), well SW is in the mixed gas/condensate window (1.25%R_o), and well SE is in the dry-gas window (1.42–1.68%R_o) (Table 1; Figure 1). Figure 9 is a log cross section through the four wells. In this study, SEM data from two sources were used. The larger data set was used to understand the qualitative distribution and types of organic matter, pore space, and matrix minerals. The smaller data set was generated specifically to

Table 1. Thermal Maturity, Depth Range, and FIB-SEM Volume Count for Study Wells

Well Name	FIB-SEM Volumes	Eagle Ford Depth (m)	%R_o Equivalent Range
NW	1	850–975	0.77–0.82
SW	7	2500–2700	1.25
SE	3	3050–3100	1.42–1.68
NE	5	2300–2400	0.79–0.86

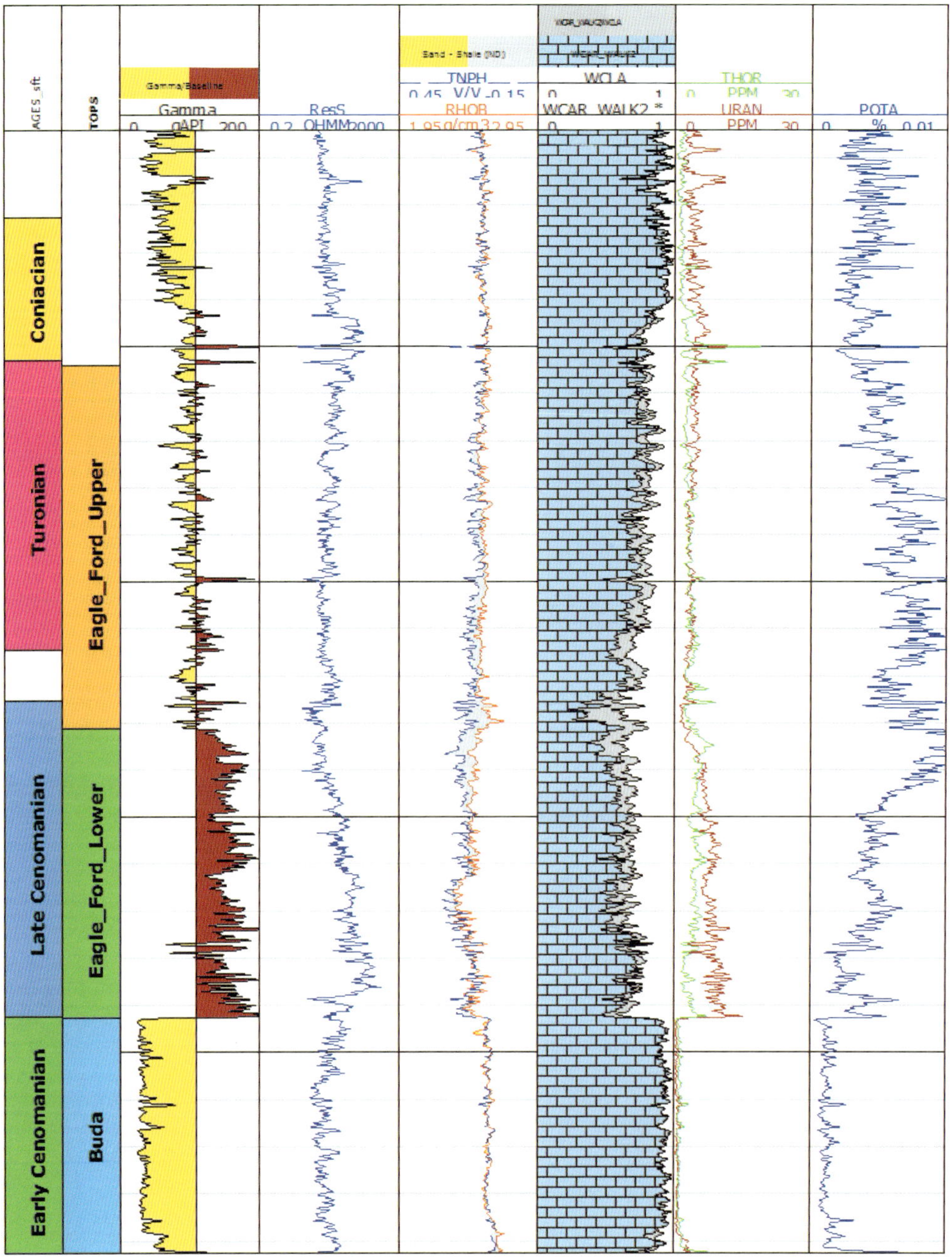

Figure 3. Typical Maverick Basin well log. Lower Eagle Ford is characterized by higher gamma ray (GR), porosity, uranium (U), and thorium (Th). Upper Eagle Ford is characterized by lower GR, lower porosity, higher carbonate content, and higher potassium (K). Austin chalk is characterized by lowest GR, overlying neutron-density (N-D) curves, highest carbonate content, and relative to upper Eagle Ford, lower K. Neutron-density curve separation, corresponding to an increase in clay content, is greatest around the upper/lower Eagle Ford contact and is virtually absent in the Austin chalk. Note that the boundary between the upper and lower Eagle Ford is prominent on the GR, N-D, U, and Th curves. Although not shown, the boundary is also easily recognized on compressional sonic curves. The upper Eagle Ford thins from the Maverick Basin toward the east onto the San Marcos Arch, where the lower Eagle Ford makes up the bulk of the reservoir section. Horizontal lines are 50- and 10-m (164- and 33-ft) interval markers. Track 1: age; track 2: formation; track 3: GR; track 4: deep resistivity; track 5: neutron (blue, on limestone scale) and density (red); track 6: mineralogy from elemental capture spectroscopy (ECS) log (blue blocks = carbonate, gray fill = clay); track 7: uranium (red) and thorium (green) from spectral GR (SGR); track 8: potassium from SGR. All values increase to the right except for neutron porosity. From Driskill et al., 2012.

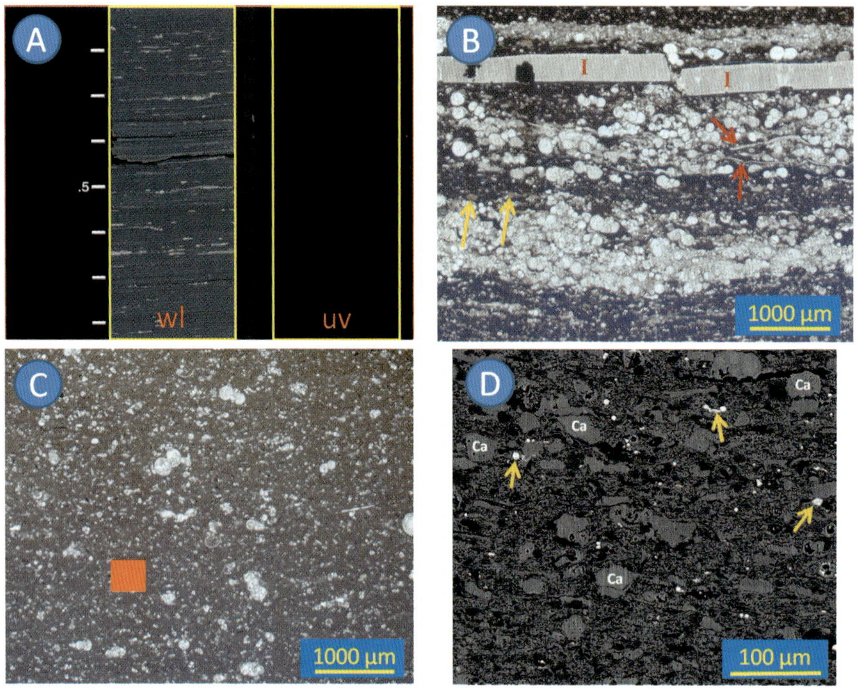

Figure 4. Example Eagle Ford textures. (A) white light (left) and ultraviolet (right) light core photos. Core is dark gray and finely laminated. Laminae are marked by fragments of Inoceramid pelecypods and foraminifera tests (white in photo). No fluorescence is evident. White tick marks are at 2.54-cm (1-in.) intervals. (B) Thin section from lower Eagle Ford showing laminae of microfossil grainstones alternating with dark gray, organic-matter-bearing matrix. Large broken fossil across the top of the image is an Inoceramid (I); some pyrite replacement has occurred. Long, thin fossil fragments in the upper lamina are possibly from a pelagic bivalve (red arrows). Most foraminifera are filled with calcite, although several chambers are filled with kaolinite, giving them a dull brassy color (yellow arrows). (C) Thin section from upper Eagle Ford showing unorganized nature of bioturbated zones. Microfossil tests and fragments are white (calcitic) in gray-brown matrix. Foraminifera are filled with blocky calcite. Red square represents field of view for scanning electron microscopic (SEM) image to the right. (D) Upper Eagle Ford ion-milled surface ×400, 15-kV secondary electron image. Energy-dispersive x-ray spectrum indicate presence of light gray calcite blebs (Ca) in mottled light and dark gray matrix composed of calcite, quartz, sodium plagioclase, illitic clay, and organic matter. Pyrite (yellow arrows) is white. Note that the blebs of calcite have not coalesced as in replacive-textured limestones (see Figure 5). At the SEM scale, laminated and bioturbated zones have similar matrix structures and are not discernible as with thin sections.

create quantitative DRP models. These data were segmented and used for quantitative assessment of porosity, permeability, and organic matter content.

High Electron Beam Energy 2-D SEM Imaging

The 2-D SEM images in Figures 7, 8, 10–12, 17, 19 were captured using a Quanta™ FEG-650 field emission machine in a commercial Houston, Texas, laboratory over a period of several months. These images were used to define overall Eagle Ford matrix composition, pore types, and organic matter location. From these images, sample locations for quantitative analysis were visually chosen using rock texture as a selection criterion. High-resolution (about 500 slices per 30 cm [11.8 in.]), whole-core helical computed tomographic (CT) scans were reviewed to avoid fractures and nonrepresentative areas. Plugs (about 2.5 cm [0.98 in.] diameter) in acceptable areas were taken from preserved core using care to sample perpendicular to any observable depositional fabric. The samples were trimmed and mounted on a stub with carbon paste and carbon coated for one second. A subset of these samples was then imaged with an accelerating voltage of 20 kV, yielding typical SEM images of polished surfaces. The other sample subset was milled with a Jeol SM-09010 cross-section polisher using a 5-kV argon ion beam. The ion milling resulted in a very flat surface over a small area, around 1 × 1 mm. The SEM was set to an accelerating voltage of 15 kV. Fuzzy or imprecise boundaries are observable on some SEM images (Figure 10), which is attributed to focusing issues, residual charging, or choice of voltage.

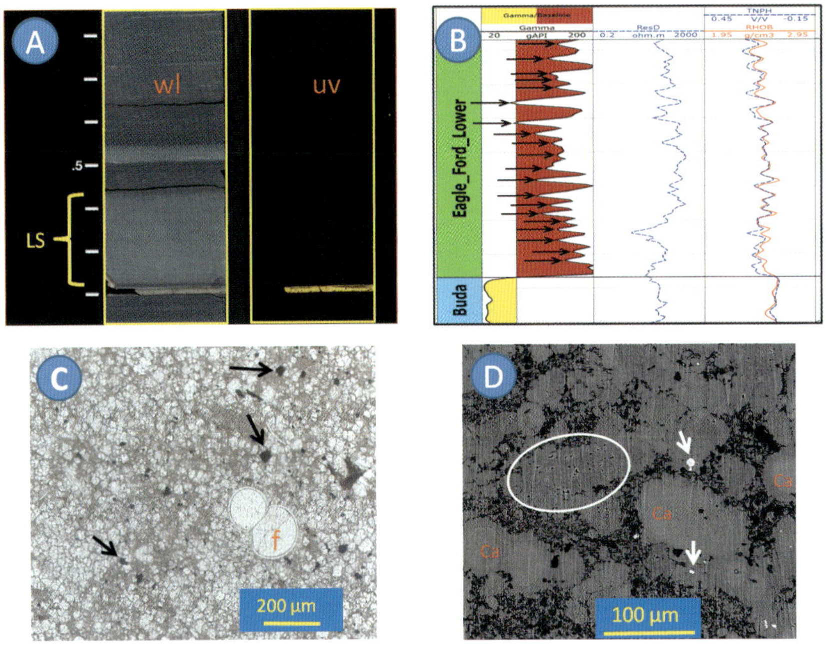

Figure 5. Examples of Eagle Ford limestones with replacive texture. (A) White light (left) and ultraviolet (right) light core photos. Replacive limestone, marked by LS and yellow bracket, is light gray and structureless. Fluorescence in ultraviolet photo is from a thin, altered tuff. (B) Log representation of limestone units (black arrows). Limestones have low gamma-ray (GR) measurements and low porosity (i.e., high density). Not all limestones are resolvable on logs. (C) Upper Eagle Ford thin section showing matrix composed of a light gray-brown mixture of calcite, quartz, sodium plagioclase, illitic clay, and finely divided organic matter with abundant lighter gray interlocking grains of calcite and scattered, irregular grains of black pyrite (black arrows). A foraminifer test (f) filled with blocky calcite is in the right-center of view. (D) Upper Eagle Ford ion-milled surface ×400, 15-kV secondary electron image showing that the calcite grains (Ca) observed in the thin-section image are irregular interlocking blebs, some with apparent intragranular voids (within white ring; voids show white charging around rims). Matrix of calcite, quartz, sodium plagioclase, clay, and organic matter is mottled dark gray-black and light gray. Pyrite framboids and grains (white arrows) occur within matrix and calcite blebs. Other scanning electron microscopic (SEM) images indicated that pore spaces in the matrix were on the order of tens to hundreds of nanometers.

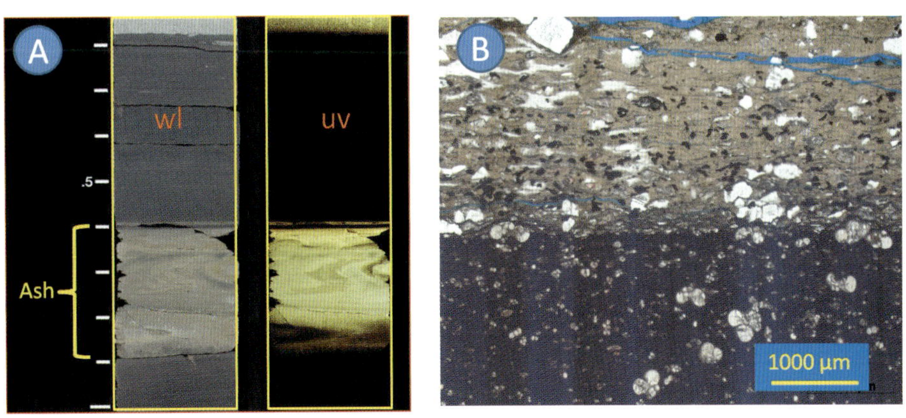

Figure 6. Examples of Eagle Ford altered tuff or ash. (A) White light (left) and ultraviolet light (right) core photos. Ash is principally made of altered clay with plagioclase and pyrite. The ash beds are highly fluorescent. The ashes may be structureless, bioturbated, or laminated. Laminations may, as in this example, be distorted. White tick marks are at 2.54-cm (1-in.) intervals. (B) Thin section with Eagle Ford marl (dark gray) overlain by altered ash (brownish in thin section, but gray-green in core). No scanning electron microscopic (SEM) images of ash layers were obtained. Volumetrically, the ash beds do not represent a significant portion of the Eagle Ford. Pyrolysis of ash samples indicated that they contained virtually no organic matter. Ultraviolet photo from Driskill et al., 2012.

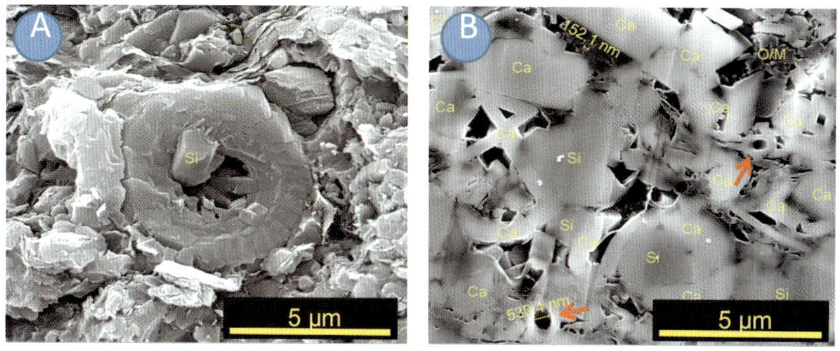

Figure 7. Scanning electron microscopic (SEM) images of Eagle Ford matrix, ×12000. (A) Lower Eagle Ford broken-edge image centered on coccolith with quartz (Si) crystal in central void. (B) Upper Eagle Ford ion-milled surface containing coccolith fragments (e.g., see cross-shape at center left) and many tube-like structures (red arrows), which, in focused ion beam (FIB) SEM volumes, are elongate, hollow cones. Note that pore spaces are on the order of hundreds of nanometers. A small amount of porous organic matter (labeled O/M) is present in upper right. Much of the matrix seemed to be composed of fragments of coccoliths and other calcareous microfossils, along with silica, clay, organic matter, and (although not visible in this image) sodium plagioclase.

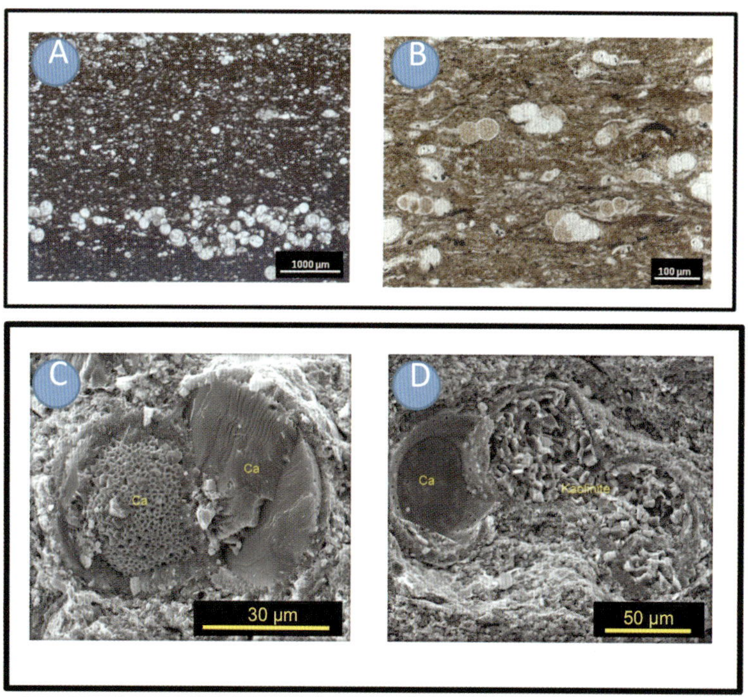

Figure 8. Examples of foraminifer fill in upper and lower Eagle Ford. (A) Upper Eagle Ford thin section showing foraminifera tests filled with calcite (white) in dark gray matrix. (B) Lower Eagle Ford thin section showing foraminifera tests filled with calcite (white) and kaolinite (brown) in brownish-color matrix. (C) Upper Eagle Ford scanning electron microscopic (SEM) image showing foraminifer chambers filled with calcite (Ca). (D) Lower Eagle Ford SEM image showing foraminifer chambers filled with calcite (Ca) and kaolinite. Figures 8A, B from Driskill et al., 2012.

Low Electron Beam Energy SEM Imaging

Scanning electron microscopic images for the DRP analytical program were collected at the Ingrain Inc. laboratory in Houston, Texas (Figures 13–15, 18, 20, 21, 27). High-resolution, dual-energy CT imaging of the whole core, with computed bulk density and photoelectric factor, were used to quality control potential sample locations and to orient plugs taken from the preserved core. Whole-core, dual-energy CT imaging produces image voxel dimensions of about 0.5 mm. The plugs were then micro-CT imaged (20- to 40-μm voxel dimensions) to locate representative areas within the plug for ion milling.

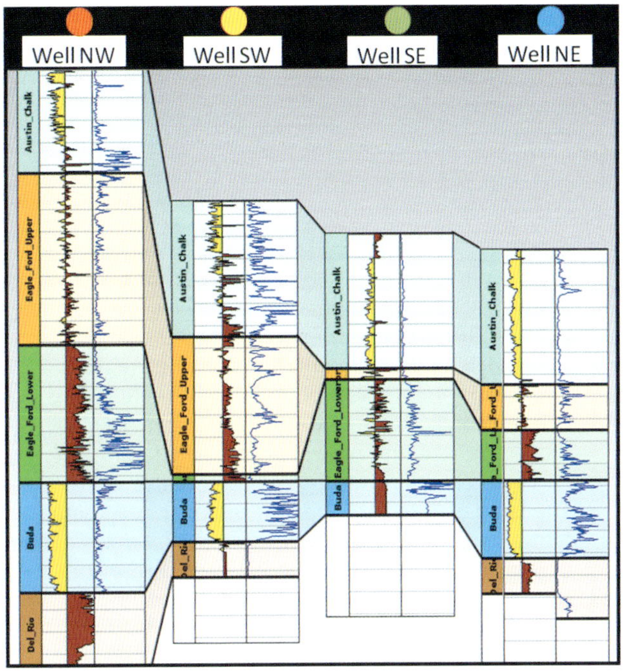

Figure 9. Log section through the four wells in the study. Section is flattened on the base of the lower Eagle Ford. For each well, left track is gamma ray (GR), and right track is deep resistivity on liner scale, both increasing in value toward the right. Depth marks are in increments of 10 m (32.8 ft). Colored dots refer to well locations, Figure 1 inset map.

A working area about 1 mm^2 was milled with an argon-ion mill. A swath of about ten 2-D images was captured at approximately 10 nm per pixel using both secondary electron (SE2) and singly scattered backscattered electron (BSE1) detectors. The vertical orientation is shown for each of the 2-D images. The SE2 images were used primarily to locate porosity and pore boundaries, which are important for segmentation; BSE1 images were helpful in discerning mineralogical variations, organic matter, and grain/crystal boundaries.

No conductive coatings were applied to the milled surfaces, and the 2-D SEM images were obtained at low electron beam energy, about 1 kV. The 2-D SEM images were segmented for matrix minerals, heavy minerals, organic matter, and porosity using a combination of Ingrain proprietary algorithms and commercial 3-D visualization software. These interpretations provided information about mineralogical distribution, porosity types, and organic matter textures.

Although the ion-milled SEM images are presented as 2-D images, they reflect subtle 3-D topographic surfaces. The effects of the residual topography create uncertainty in the segmentation process and the resulting calculated porosity and organic matter content. Careful guidance and review of the segmentation are critical for accurate quantitative results. The 2-D SEM images were used to select areas of interest for 3-D FIB-SEM imaging.

The areas identified for 3-D imaging were mounted and inserted into a FIB-SEM imaging system. This system permits imaging at low energy with no surface coating. The 3-D image data sets were acquired by alternately removing about 10 nm of material with the Ga^+ ion beam from a prepared area and acquiring SE2 and BSE1 images. This process of milling and imaging parallel surfaces was repeated around 600 to 1000 times. Vertical orientation of all images is preserved throughout the process. After capturing all of the individual

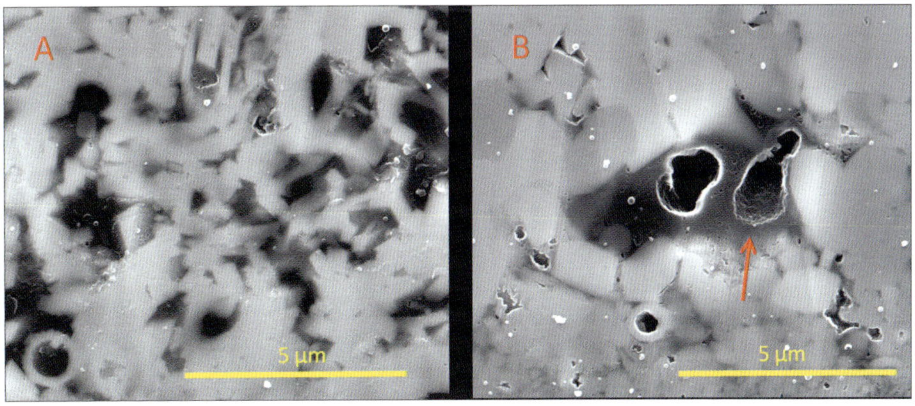

Figure 10. Upper Eagle Ford ion-milled images, ×15000, 15 kV. (A) Note that edges of image elements are fuzzy, which will add uncertainty to segmentation. (B) Image is centered on organic matter containing two large pores. Note that the pore on the right (red arrow) intersects the sample surface at a nonperpendicular angle. The pore wall is almost the same gray-scale color as the organic matter, and the pore rim is not sharp around its circumference, making it difficult to use unsupervised image segmenting for calculating pore and organic matter area. Intragranular pore, yellow arrow.

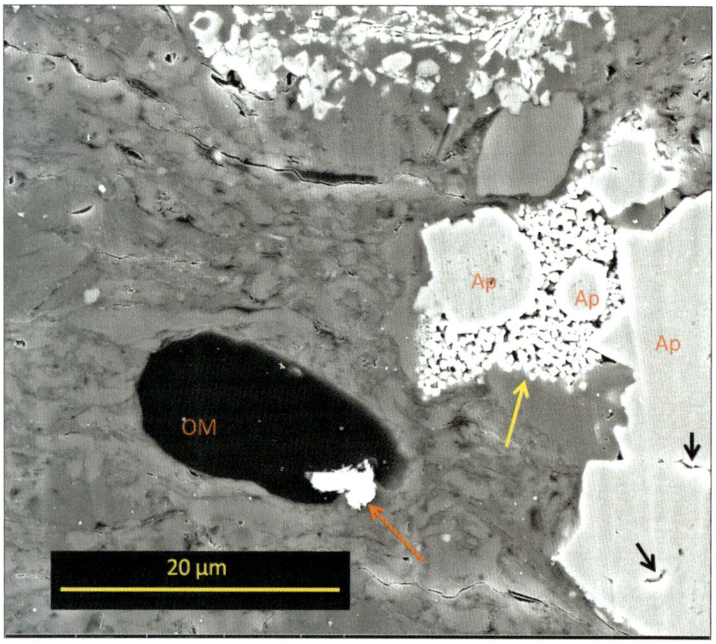

Figure 11. Ion-milled surface in well SW showing loosely packed pyrite framboid (yellow arrow) and small pyrite and/or marcasite crystalline mass that appears to have grown into interior of organic-matter-filled pore (red arrow). Few pyrite masses showed intercrystalline pores, and none that were imaged conclusively contained organic matter. All loosely packed and many tightly packed framboids had intergranular (intercrystalline) organic matter. The large apatite (Ap) mass to the right had some intragranular porosity (black arrows) that may contain organic matter. Matrix was mixture of calcite and quartz. ×3000, 15-kV high voltage, Everhart-Thornley (ET) detector.

FIB-SEM images, they were aligned and merged into separate SE2 and BSE1 3-D objects, with each image voxel having dimensions from 10 to 15 nm. A typical FIB-SEM process was described by Curtis et al. (2010). The FIB-SEM process described in this chapter was described by Huang et al. (this volume).

The 3-D volume was then segmented using Ingrain proprietary 3-D algorithms that separate pore space, porosity associated with organic matter, solid organic matter, and solid matrix framework into separate 3-D volumes. Finally, the connected pore volume object is used in the computation of vertical and horizontal permeability, based on a lattice Boltzmann molecular dynamics algorithm (Tölke et al., 2010). Data reported in this chapter include total porosity, porosity associated with organic matter, and absolute horizontal permeability. It is important to note that the data used to illuminate shale microstructure in the FIB-SEM 3-D volumes represent a very small amount of rock, typically about 2×10^{-9} g as compared to around 300 g of material used for the GRI process in measuring porosity and permeability.

OBSERVATIONS

2-D SEM Images

Figures 10–19 show example 2-D SEM images from the four wells studied. Energy-dispersive x-ray measurements taken across multiple images indicated that calcite and quartz were the predominant constituents, with lesser volumes of clay, sodium plagioclase, organic matter, pyrite, and apatite. The most common recognizable fossils in the matrix were of calcareous coccolithophore plates. It is probable that much of the carbonate matrix material was made of coccolith debris.

Mineral Grains and Cements

Quartz was common, and in some images exceeded calcite in area. Quartz occurred as irregular areas within the

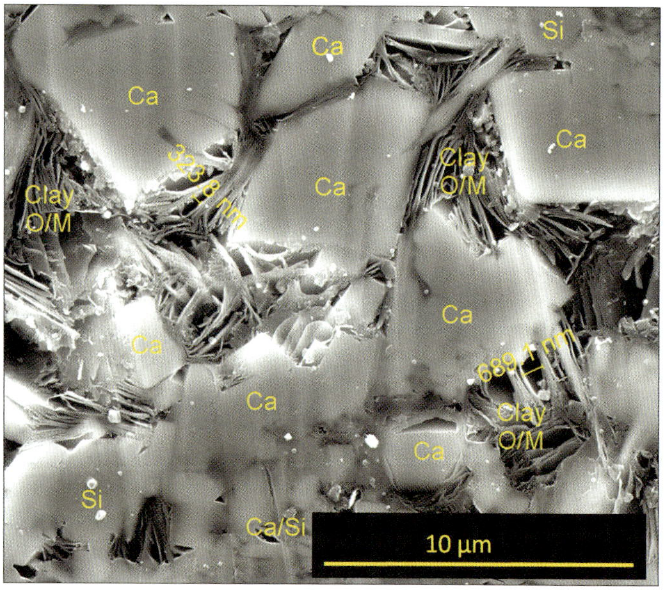

Figure 12. Ion-milled surface in well SW showing mixture of illitic clays and organic matter (Clay O/M) in matrix of calcite (Ca) and quartz (Si). The clays were randomly oriented, and some were bent, characteristic of detrital clays (Schieber, 2010). ×7000, 15-kV high voltage, Everhart-Thornley (ET) detector.

matrix and as authigenic cement identified by euhedral crystals projecting into pore spaces. Similarly, sodium plagioclase occurred as irregular-shape grains in a mosaic of intermixed calcite and quartz and as rare euhedral crystals. Apatite occurred in discrete patches (Figure 11) and in aggregates of very small mineral fragments mixed with organic matter and sometimes clay. At low magnification, the mixed apatite-organic matter-clay aggregates formed bodies interpreted as fecal pellets.

Organic Matter

Organic matter occupied spaces between matrix grains (intergranular), between illite and kaolinite clay particles (interphyllic), and between mineral crystals (intercrystalline), such as within pyrite framboids (Figures 11; 13B; 18A, B) and along crystal edges (Figures 11, 18A).

Clay

Kaolinite was observed within foraminifer chambers only in the lower Eagle Ford Member. In high-magnification images, organic matter could also be identified between kaolinite crystal bundles (Figure 14B). The kaolinite was interpreted as authigenic and appeared to postdate the intragranular organic matter within the foraminifera chambers. Within the matrix, illite was the main clay mineral, which could have both authigenic and detrital aspects (Schieber, 2010). Illite was commonly associated with organic matter (Figures 12; 13A; 14A, C; 15A).

In high thermally mature rocks, empty intergranular (interphyllic or intercrystalline) pores were common (Figures 13A, 15A). In several instances, high magnification of intergranular clay pores showed that thin rims of organic matter coated portions of pore walls. It is possible that intergranular clay pores in high thermally mature rocks were originally filled with organic matter, and that during progressive thermal maturation, transformation of organic matter to hydrocarbons removed much of the pore-filling organic matter.

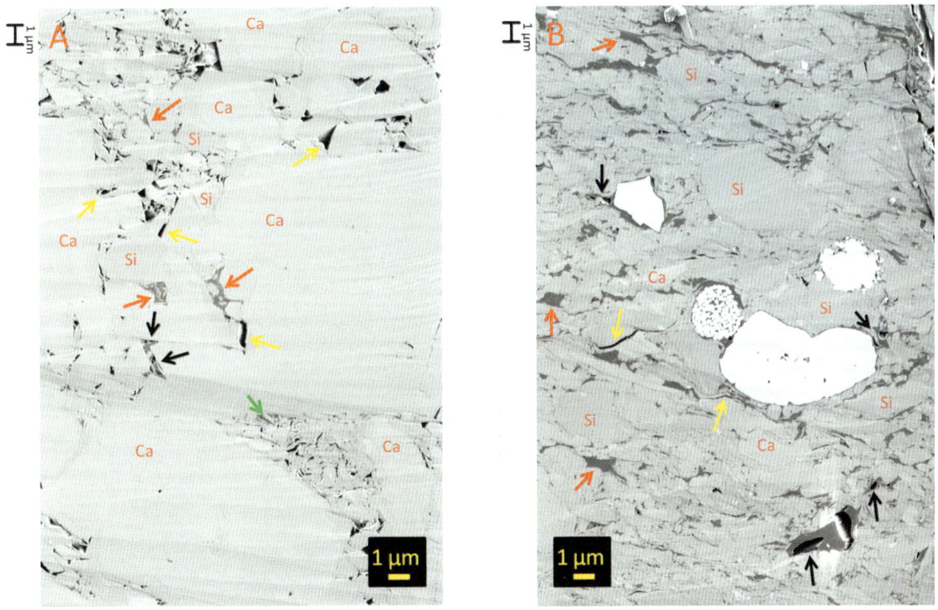

Figure 13. Upper Eagle Ford two-dimensional ion-milled scanning electron microscopic (SEM) images from well NW. (A) Image from low gamma-ray (GR) streak. Matrix was composed of calcite (Ca), quartz (Si), and clay. Organic matter was mainly in spaces between grains (medium gray values, red arrows), although some was associated with clays (green arrow). Pore space (dark gray to black, yellow arrows) was primarily angular and intergranular, occurring between grains of carbonate, quartz, or clay. A few rounded organic matter pores were present (black arrows). There is little organic matter visible in this image, and the organic matter observed was in small isolated pockets. Note that many grain edges along pore boundaries were sharp and angular. (B) Image from high-GR area in lower portion of upper Eagle Ford. In this sample, there was more quartz (Si) than calcite (Ca), and clays were also present. Pyrite (white) was present as isolated intergrown crystals and tightly packed framboids. Pore space was primarily associated with organic matter (either within or at pore boundaries, black arrows) as rounded bubble pores. Much of the organic matter appears solid at this magnification (red arrows). Long pores that look like open fractures may be induced (yellow arrows). Note that the large area of organic matter in the lower right has charging artifacts (white streaks). The pyrite framboid in the image center contained intercrystalline organic matter with extremely small pores (observable at higher magnification).

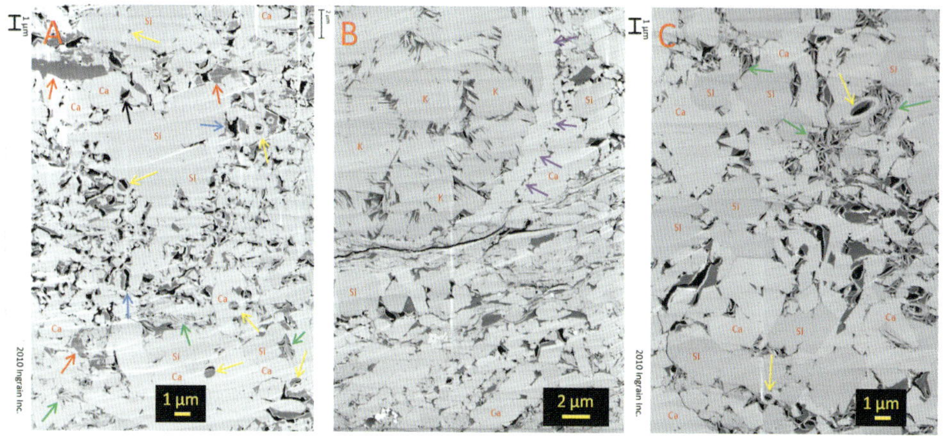

Figure 14. Lower Eagle Ford two-dimensional ion-milled scanning electron microscopic (SEM) images from well NW. (A) Sample was composed of calcite (Ca), quartz (Si), organic matter, and clay (green arrows). Organic matter was mainly intergranular and was solid (red arrows) or contained irregularly shaped, rounded, microsize bubble pores (black arrow). Some of the more irregular pore shapes seem to be a mix of organic matter bubble pores and intergranular pores (blue arrows). Virtually all of the intergranular space had some organic matter in it. It is easy to imagine how, in three dimensions, fluids could flow through this material. Intrapores, which contain organic matter, are indicated by yellow arrows. (B) Sample was composed of calcite (Ca), quartz (Si), and clay (principally kaolinite, K). The upper left quadrant is occupied by a partial foraminfer chamber (purple arrows) filled with kaolinite. Organic matter was intergranular and contained both solid or porous areas. Microscale bubble pores are seen in the organic matter associated with both the kaolinite and matrix grains. Long fracture in center of image was most likely induced. (C) Sample was composed of calcite (Ca), quartz (Si), and clay (green arrows). Pores were predominantly associated with intergranular organic matter and were rounded, irregularly shaped bubble pores. In this sample, many pores are seen within groups of unoriented clay flakes that appeared detrital in origin. Intrapores indicated by yellow arrows. The round or oval-shape intrapores were interpreted as either coccolithophore spines or calcareous sponge spicules. The organic matter within the intrapores was often porous. Figure 14A from Driskill et al., 2012.

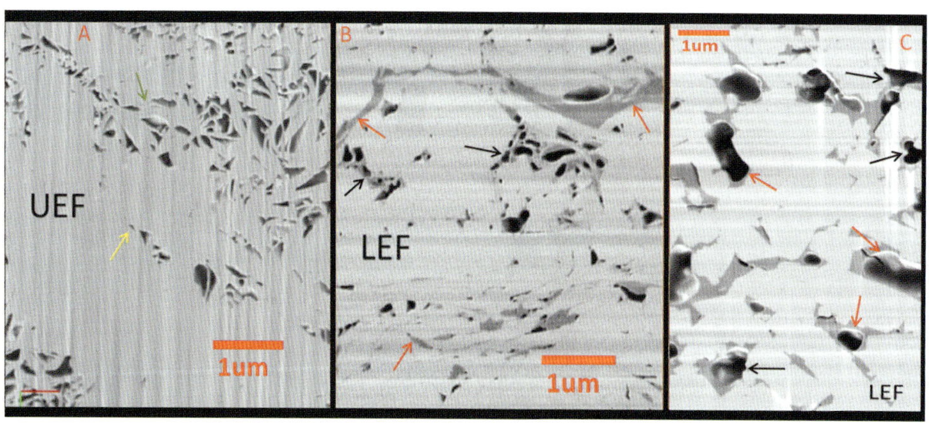

Figure 15. Two-dimensional ion-milled scanning electron microscopic (SEM) images from well NE. (A) Upper Eagle Ford sample composed primarily of calcite. Pores occurred between grains of calcite (yellow arrow) and clay (green arrow). Organic matter was present in some pores, but most pores were empty. Organic matter in this sample was sparse and isolated, similar to Figure 11A. (B) Lower Eagle Ford sample showing solid organic matter (red arrows) and porous organic matter (black arrows). Organic matter occurred in intergranular space. (C) Lower Eagle Ford sample with intergranular organic matter containing microscale rounded bubble pores. Image shows that pore walls in organic matter tended to be more rounded (red arrows); matrix-grain pore walls can add an irregular aspect (black arrows). Linear pores may be induced.

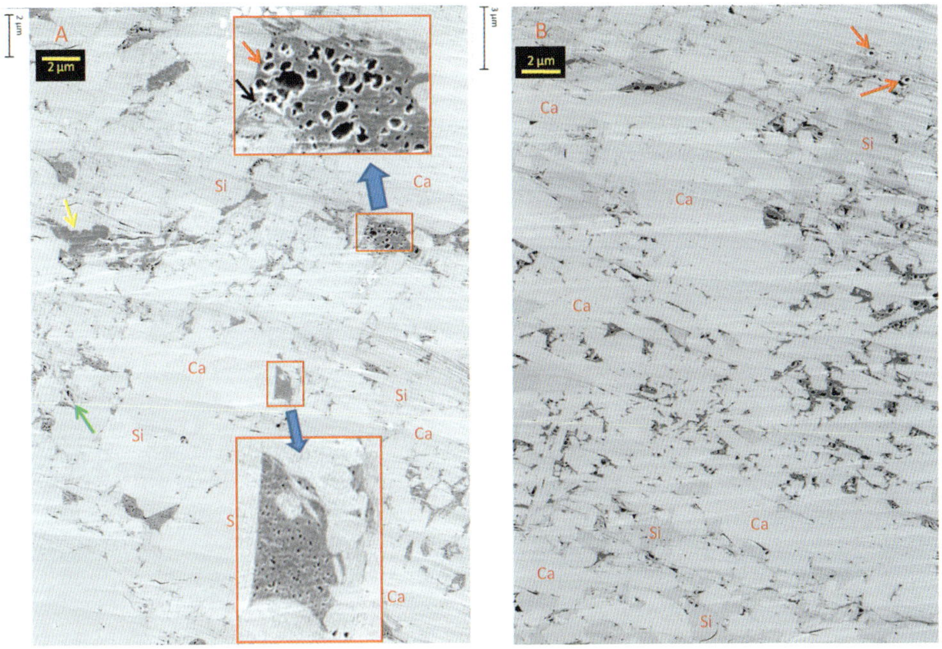

Figure 16. Two-dimensional ion-milled scanning electron microscopic (SEM) images from well SW. (A) Intergranular organic matter contained both large bubble-type pores (red arrow) and much smaller foam pores (black arrow). See insets for details; upper inset contains both bubble and foam pores; lower inset contains only foam pores. Unlike the organic matter pores observed in wells NE and NW, these pores had rough-looking walls. Even areas with solid-appearing organic matter contained nanopores (yellow arrow). Organic matter occurred in isolated patches. Clays are rare in this image (green arrow). (B) Sample with porous intergranular organic matter. Both bubble and foam pores had rough walls as in (A). Note two small ovals in upper right (red arrows); focused ion beam (FIB) SEM volumes show that in three dimensions these structures took the form of elongate, hollow cones.

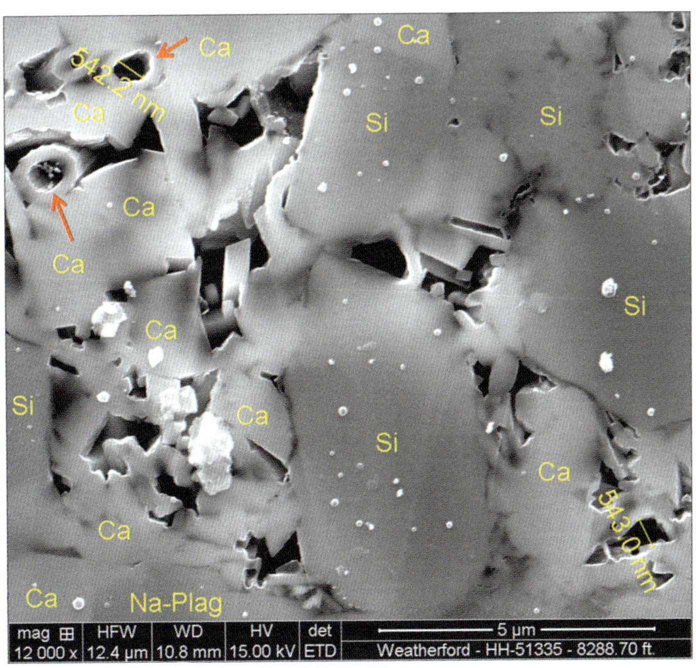

Figure 17. Upper Eagle Ford two-dimensional ion-milled scanning electron microscopic (SEM) images from well SW. No organic matter is observed in pores. Intrapores, interpreted as either coccolithophore spines or sponge spicules, are indicated in red (see also red arrows in Figure 16B). Pore diameters are measured in nanometers. Coccolith marked with black arrow.

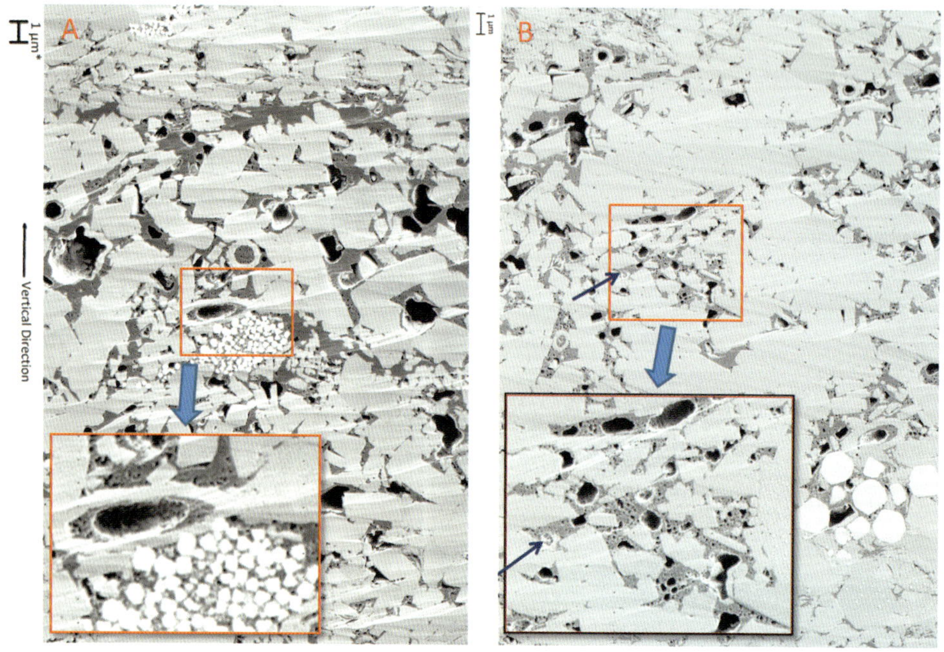

Figure 18. Lower Eagle Ford two-dimensional ion-milled scanning electron microscopic (SEM) images from well SE. (A) Intergranular organic matter contained both bubble and foam pores. Note that bubble pore walls were perforated by smaller foam pores (inset). Loosely packed pyrite framboids (white) with porous organic matter were present. Red arrows mark intrapores. (B) Image similar to (A) showing porous intergranular organic matter. Loosely packed pyrite crystals (white) were associated with porous organic matter. Many of the matrix grains had flat and angular edges, which could imply matrix recrystallization. Even organic matter that appeared solid often contained nanometer-scale pores (inset, black arrows). Intrapores marked by red arrows.

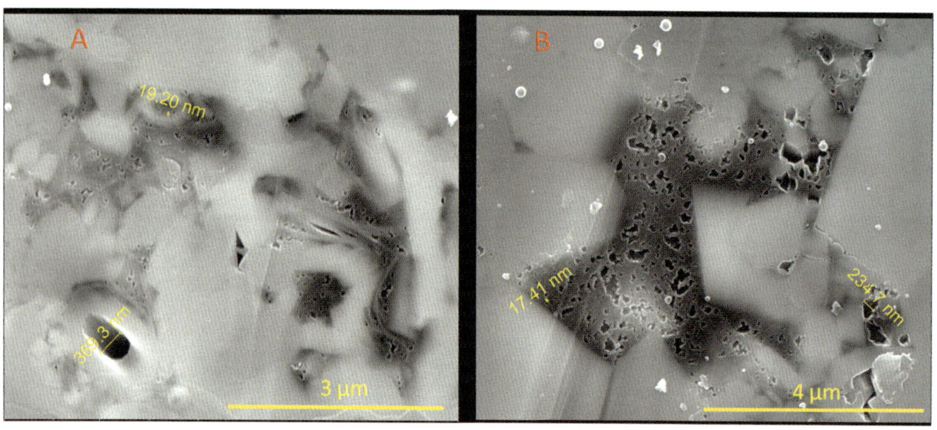

Figure 19. Two-dimensional ion-milled scanning electron microscopic (SEM) images showing irregular walls of nanopores in organic matter. (A) ×24000, high-voltage, 15-kV, Everhart-Thornley (ET) detector. (B) ×18000, high-voltage 15-kV, ET detector.

Pores

Pores in the Eagle Ford were intergranular (Figures 7B, 12, 13A, 15A, 17, 21); intragranular (Figures 7A, B; 11; 13A; 14A, C; 17; 18A, B; 19A; 21B); or within organic matter (Figures 10B; 13A, B; 14A, B, C; 15B, C; 16A, B; 18A, B; 19A, B; 21A, B, C). Intergranular porosity was more common in the upper Eagle Ford and occurred as either phyllosilicate framework pores or grain framework pores as described by Schieber (2011). Intergranular pores were irregularly shaped (interpreted as a result of compaction) or geometrically shaped, with pore geometry defined by bounding crystal faces or situated between clay flakes or grains (Figures 10A; 12; 14C; 15A, B, C; 16B; 17; 18A, B; 19). It was not uncommon to see crystal faces encroaching into intergranular pore space; this was a result of a pervasive recrystallization in the Eagle Ford that is difficult to detect in thin section. Intragranular pores were the least commonly observed. Intragranular pores included primary pores (Figures 14A, C; 15B;

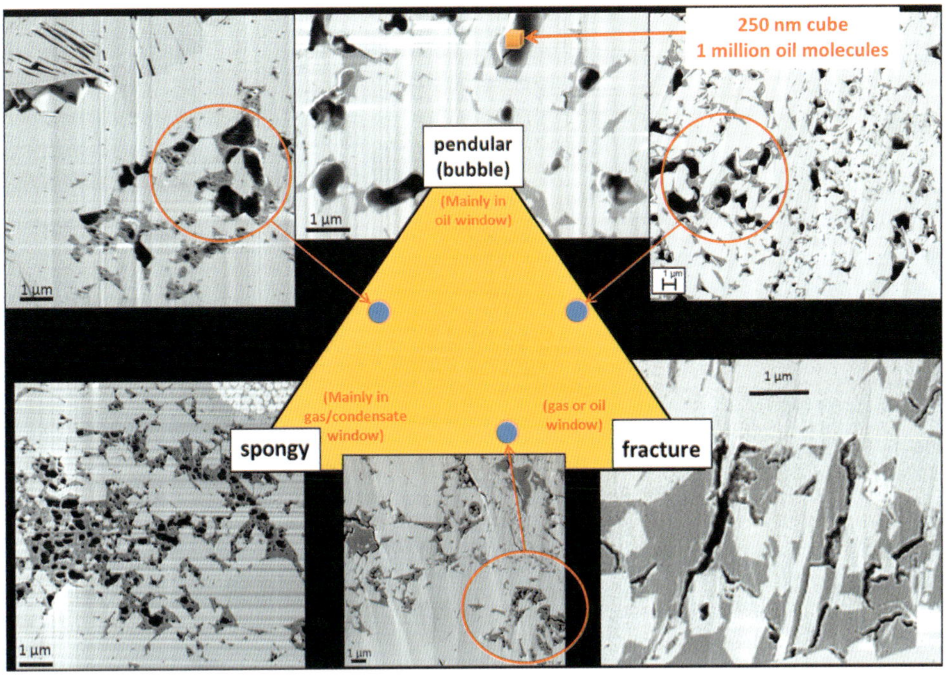

Figure 20. Three classes of organic matter textures were observed in organic rich shales. Pendular organic matter porosity was more common in oil-window samples, and spongy organic matter pores were more common in the gas/condensate window in the Eagle Ford samples. Fracture-type organic porosity was not common in Eagle Ford but has been observed in a wide range of thermal maturation conditions in other formations.

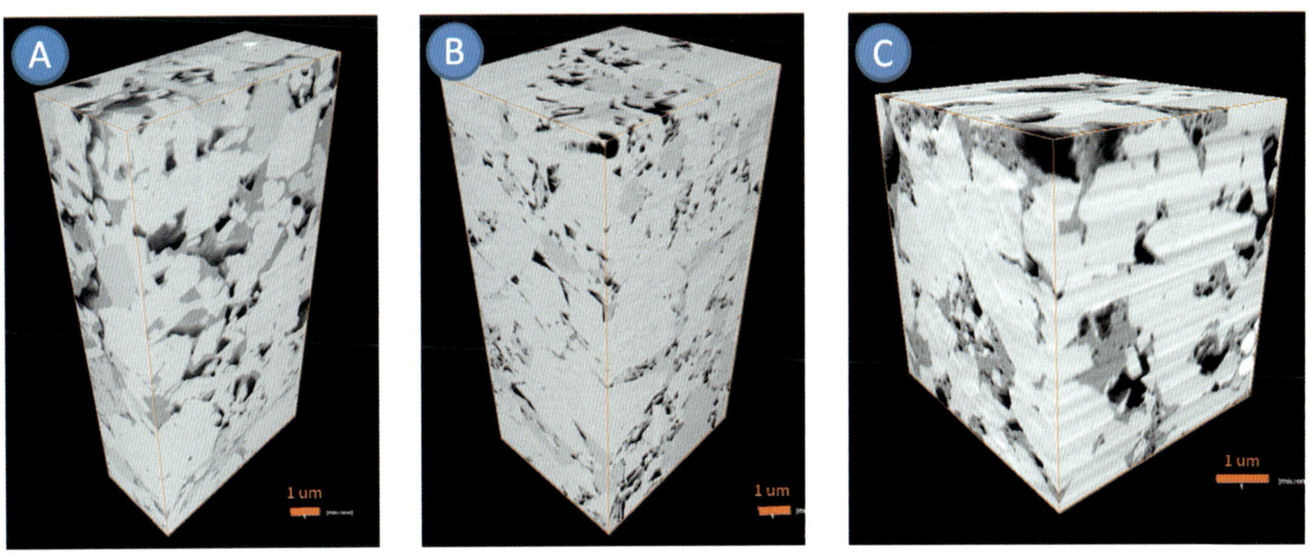

Well NW, LEF, Oil Window
Porosity – 6.2%
Kerogen – 11.5%
Perm = 2640 nD
High calcite
Pendular organics

Well SW, UEF, Gas/condensate window
Porosity – 6.7 %
Kerogen – 1.4 %
Perm = 551 nD
Variable minerals,
Abundant inter-granular pores,
spongy organics

Well SE, LEF, Gas/condensate window
Porosity – 11.5 %
Kerogen – 7.7 %
Perm = 1034 nD
High calcite,
Large pendular OM pores plus smaller
spongy OM pores

Figure 21. 3D FIB-SEM volumes constructed from grey scale SEM images. Organic matter is medium grey and pore space is dark grey/black. Lighter mineral matrix is predominantly calcite and quartz). Pyrite is white. A. Lower Eagle Ford sample from Well NW in the oil window. B. Upper Eagle Ford sample from well SW in the gas window. C. Lower Eagle Ford sample from well SE in the gas/condensate window.

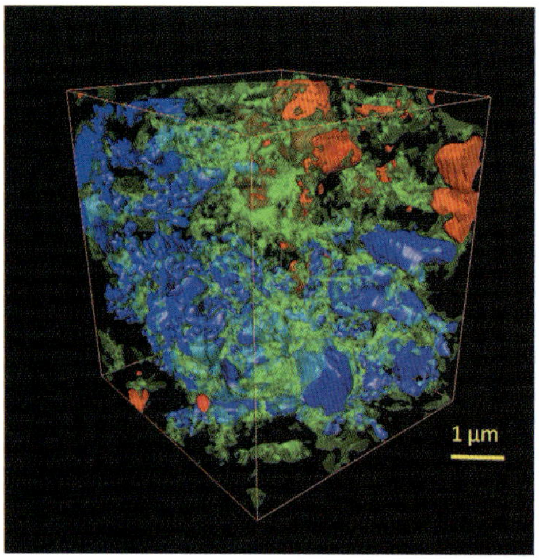

Figure 22. Segmented three-dimensional focused ion beam-scanning electron microscopic (FIB-SEM) volume from lower Eagle Ford sample, well SE. Connected porosity is blue, disconnected porosity is red, organic material is green, and framework minerals are transparent. Total porosity was 11.5%, and absolute permeability was 1034 nD.

16; 17), which Schieber (2011) referred to as intrapores, isolated pores within coalesced crystal masses (Figure 14A), and pores caused by dissolution, referred to as solvopores (Schieber, 2011).

Pores in organic matter tended to be micrometer scale and rounded in the oil window unless they impinged on grain boundaries (Figures 13; 14A, B, C). In more thermally mature rocks, micrometer-scale organic matter pores were more irregular in shape and tended to be perforated with nanometer-scale pores. Pores within organic matter in samples within the condensate and dry-gas windows were dominated by nanometer-scale pores (Figures 10B; 16A; 18A, B; 19).

3-D SEM Volumes

The 3-D FIB-SEM data supported observations from the 2-D data sets. Examples of unsegmented 3-D FIB-SEM volumes for upper and lower Eagle Ford are shown in Figure 21. After segmentation, a 3-D volume could be displayed as in Figure 22; minerals are rendered transparent, and the internal connected porosity, disconnected porosity, and organic matter are shown by different colors. The lattice Boltzmann computational algorithm (Tölke et al., 2010) was applied to determine the absolute permeability created by the connected porosity. Total porosity is the sum of connected and disconnected porosity. Organic matter volume is the bulk volume of the solid organic matter excluding internal pores.

A cross plot of total porosity versus organic matter content shows a positive correlation between traditional TOC data (demarked with "Lab2" suffix) and FIB-SEM data from the upper and lower Eagle Ford (Figure 23). A dashed line indicates the overall trend in the data, but numerous outliers exist in both the physical TOC data and the SEM sample

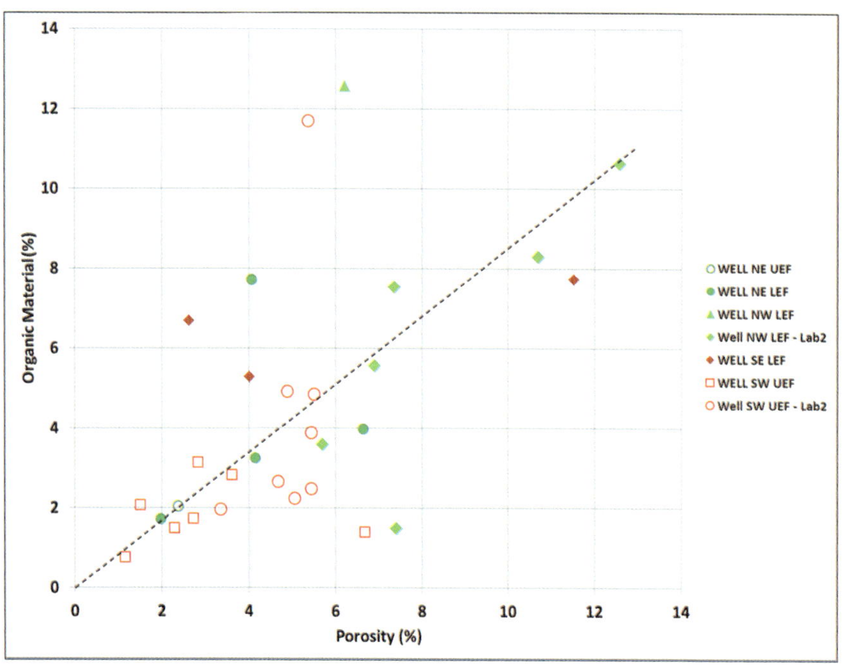

Figure 23. Cross plot of total porosity versus volume of solid organic material from three-dimensional focused ion beam-scanning electron microscopic (FIB-SEM) volumes.

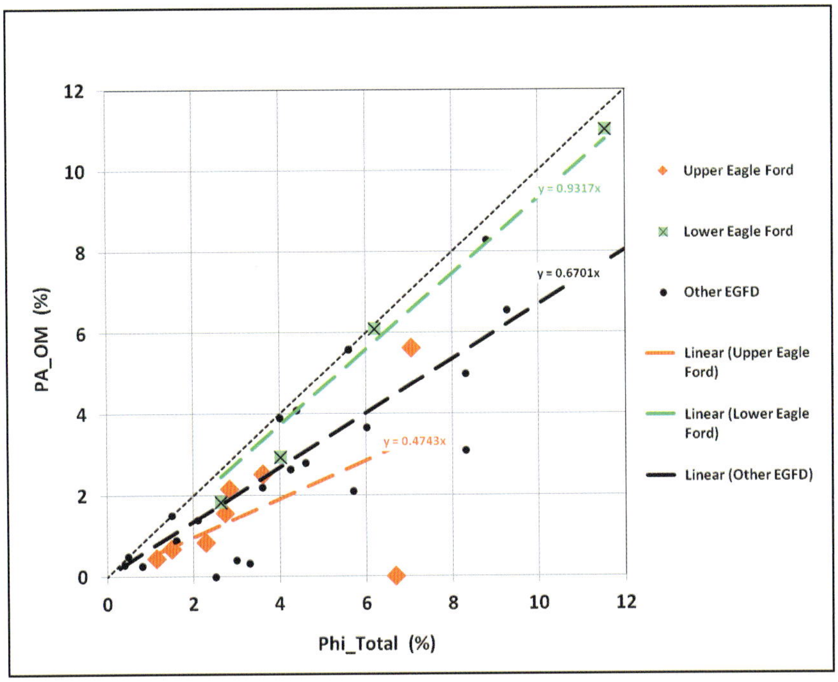

Figure 24. Cross plot of total porosity (Phi_Total) versus porosity associated with organic matter (PA_OM) as derived from three-dimensional digital rock physics (DRP) models. Regression for the Eagle Ford as a whole (black dashed line) implied that 67% of the total porosity was associated with organic matter: 93% in the lower Eagle Ford and 47% in the upper Eagle Ford. The upper Eagle Ford had more intergranular pore space devoid of organic matter than the lower member. Large symbols are from this study, small black dots are from other Eagle Ford wells.

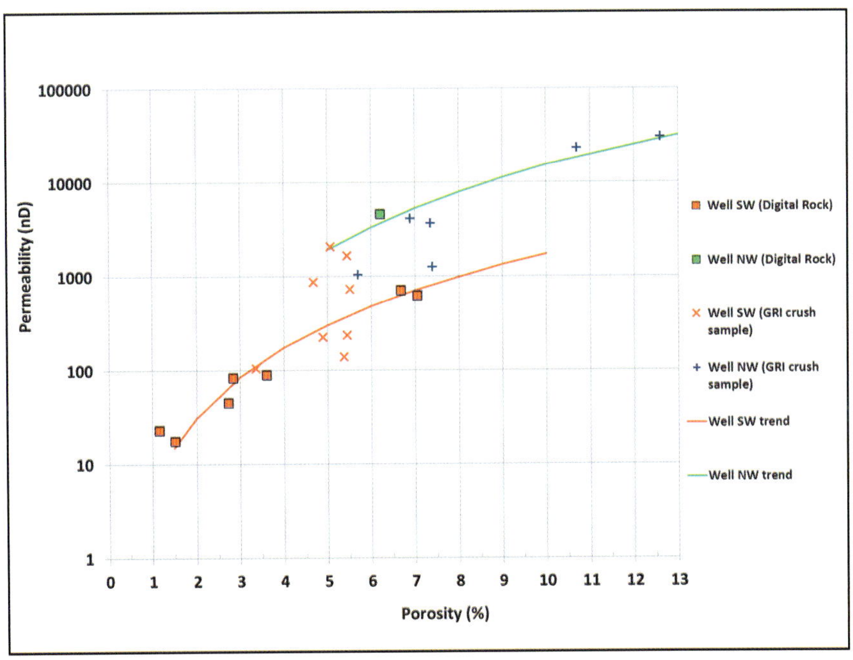

Figure 25. Cross plot of total porosity versus absolute permeability. Data derived from digital rock physics (DRP) models (boxes) and Gas Research Institute (GRI) crushed rock analyses (X's and +'s). Data from the two methods fall within the same space over three orders of magnitude in permeability.

data that are attributable to both natural variability in the Eagle Ford Shale and differences in the volume of the analyzed samples. Figure 24 shows a plot of porosity associated with organic matter versus total porosity. The four lower Eagle Ford samples had about 93% porosity associated with organic matter, compared to about 47% for the upper Eagle Ford. Combining all samples, the Eagle Ford had about 67% porosity associated with organic matter. Having computed porosity associated with organic matter, the fraction of original solid organic matter that had been converted to porosity (and

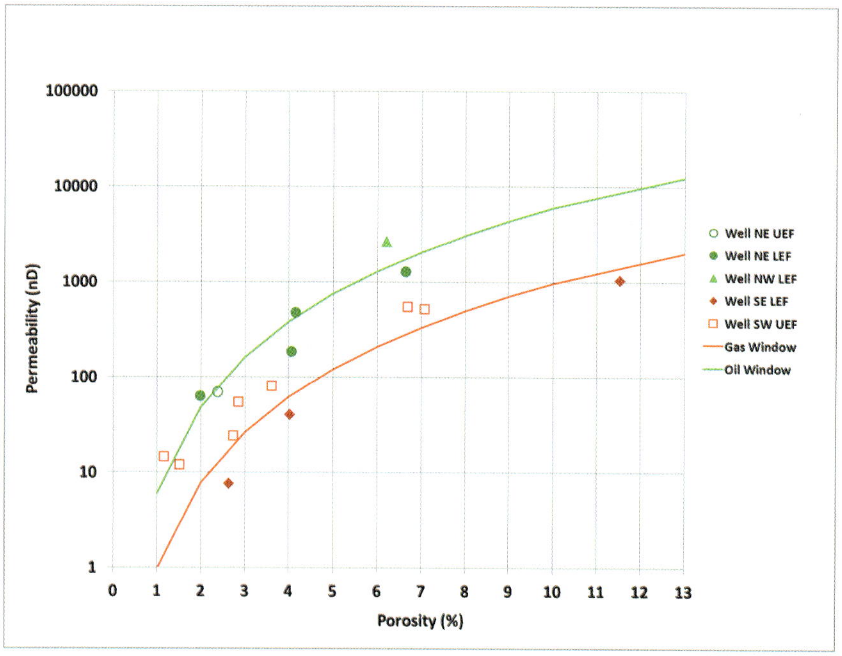

Figure 26. Cross plot of total porosity versus absolute permeability derived from the digital rock physics (DRP) models. Models indicated that the Eagle Ford in the oil window (green symbols) tended to have a more favorable porosity-permeability relationship than the Eagle Ford in the condensate/dry-gas window (red symbols).

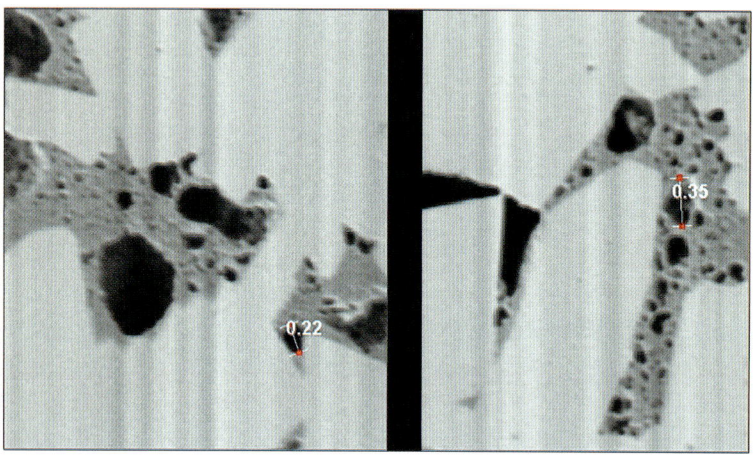

Figure 27. Figure illustrating typical maximum sizes of spongy organic matter pores. Both samples A and B are from well SW.

preserved) can also be calculated. This fraction is referred to as the *conversion ratio* (CR). Conversion ratio is defined by the equation:

$$CR = \frac{PA\_OM}{(PA\_OM + solid\ OM)}$$

where PA_OM is porosity associated with organic matter. For example, the SE well sample shown in Figure 22 had about 11% porosity associated with organic matter and 7.7% solid organic matter. From the equation, the conversion ratio is 59%. The average conversion ratio for all of the 16 FIB-SEM samples was about 39%. Direct quantification of porosity associated with organic matter separately from intergranular or intragranular porosity is important for understanding fluid saturation and fluid distribution. It is interpreted that only hydrocarbons occupy pores associated with organic matter in thermally mature rocks, whereas both hydrocarbons and water may occupy intergranular pores.

Physical laboratory-based porosity and permeability data derived from the GRI process were available for comparison to the 3D FIB-SEM computed data sets for two wells. On a plot of porosity versus permeability, the GRI and FIB-SEM data generally follow a similar trend, but details differed for each data set (Figure 25). For instance, the gas-mature upper Eagle Ford FIB-SEM data plotted along a trend (1.2–7 porosity units [pu] and 27–700 nD); the GRI data clustered in a narrower porosity range (3.4–5.5 pu) but exhibited a wider permeability range (100–2000 nD). These differences were interpreted to be related to the scales of each measurement type (hundreds of grams versus nanograms of material for the GRI and FIB-SEM process, respectively) and depth differences for each type of sample. Although an attempt was made to colocate the SEM image and GRI samples, that was not always possible. Consequently, for some SEM data sets, the nearest GRI samples were up to 15 cm (5.9 in.) distant.

ORGANIC MATTER PORES

The character of the observed organic matter pores was classified as pendular, spongy, and fracture based on a modification of the porosity classification scheme of Walls and Sinclair (2011). Pendular organic matter appeared to fill small intergranular and grain contact regions, leaving open pore space in the larger voids (Figure 20). Pores in this material had smooth, curved, concave surfaces facing pore interiors and often appeared bubble-like. "Spongy" organic matter was commonly observed in thermally mature gas shales. Both were interpreted to be the result of conversion of solid (or semisolid) organic matter to hydrocarbons. In-situ fracture porosity in organic matter was characterized by widely spaced irregular fracture faces and was often found in combination with one of the other two organic matter pore types. Very narrow cracks with matching opposing fracture faces were interpreted as "induced" fractures formed by cooling or degassing of the organic matter during core retrieval.

The SEM image data indicate that the oil-window samples were primarily dominated by pendular organic matter pores and had a higher permeability at any level of porosity than the gas-/condensate-window samples, which were dominated by spongy organic matter pores (Figure 26). In addition, the lower Eagle Ford samples had higher average porosity and permeability than the upper Eagle Ford samples. The two curved lines on Figure 26 indicate approximate permeability versus total porosity trends for the oil-window and gas-window samples.

Some samples had both extensive sponge-like and pendular organic matter pores. An example was used to estimate the approximate upper limit of pore diameter for sponge-like pores. This upper limit was about 0.3 μm (Figure 27). Next, pore size distribution was computed from the segmented 3D FIB-SEM volumes (Figure 28). A dashed vertical line in this figure shows the approximate upper limit of spongy pore diameter in these Eagle Ford samples. This figure also shows that, in general, the oil-window samples (green lines) had larger pore sizes than the gas-window samples (red lines). This is because oil-window samples typically had more pendular organic matter and less spongy organic matter or intergranular porosity than gas-window samples. It also shows that upper Eagle Ford samples (thin lines) had smaller pores, mainly because they had more intergranular pores than lower Eagle Ford. Permeability values for the samples in Figure 28 are shown next to the sample labels.

Schieber (2011) referred to pendular organic matter porosity as "bubble" pores and spongy organic matter porosity as "foam" pores. The pendular (bubble) pores were seen in both upper and lower Eagle Ford from samples within both the oil and gas/condensate windows. Within the oil-mature lower Eagle Ford, blebs and threads of organic matter with closely spaced bubble pores were common. The 3D FIB-SEM volumes showed that these bubbles often connected outside the plane of the image and acted as a conduit for hydrocarbons. It was also noted that porosity outside the organic matter was rare in the numerous lower Eagle Ford samples reviewed. Spongy pores were commonly seen in more mature gas- or condensate-window Eagle Ford samples. Images from the Eagle Ford within the gas/condensate window often exhibited bubble pores in organic matter that was also profuse with minute spongy pores, providing a dual-pore system (Figures 16A, B; 18; 19).

DISCUSSION

The current assumption is that originally solid (or semisolid) organic matter develops porosity during thermal maturation (Loucks et al., 2009; Passey et al., 2010). The degree and nature of organic matter pores may be related to the type of organic matter. Prior to the onset of thermal maturation, organic matter likely filled all of the available nonmineral space in the compacting rock, leaving virtually no porosity or fluid flow path. Pendular or bubble pores probably develop first in the oil window. With increasing thermal maturation, gas evolves, creating spongy organic matter with foam porosity. However, some samples within the gas window exhibit only spongy organic matter pores. This may be caused by several factors: the initial organic matter composition (with varying reaction kinetics), a more rapid rate of burial and maturation, or more likely, the reconfiguration of kerogen with in-situ gas expansion that obscures preceding (e.g., pendular) textures. The shapes and sizes of the three main pore types (pendular, spongy, and fracture) seem to be influenced by organic matter rigidity, with the pendular pores associated with the least-rigid solid organic matter and the in-situ fracture pores associated with the most rigid or brittle organic

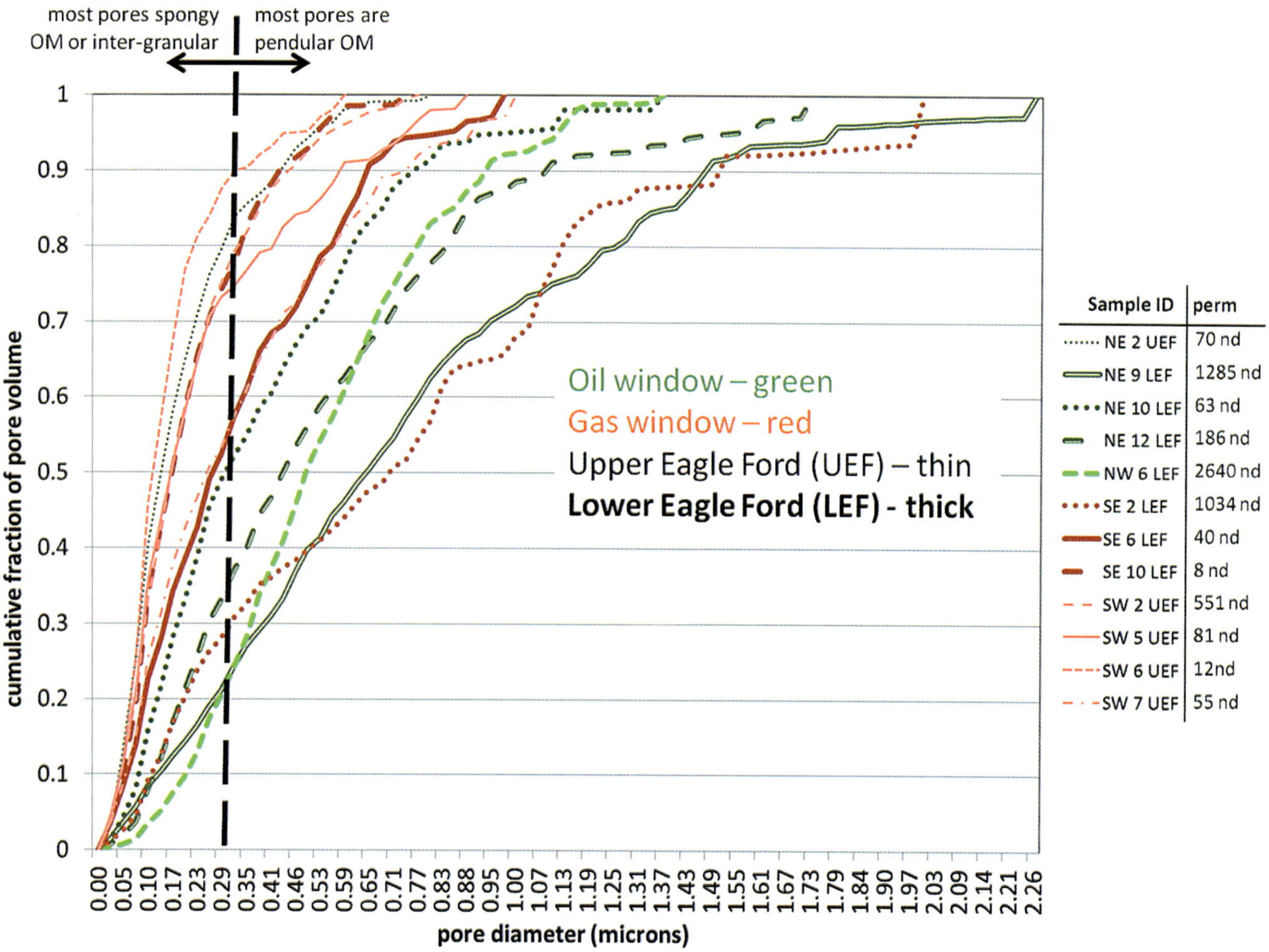

Figure 28. Pore diameter versus cumulative fraction of pore volume. Vertical dashed line shows approximate upper boundary of spongy-type organic pores and intergranular pores.

matter. Some samples had little porosity in the organic matter even though the samples were thermally mature for hydrocarbon generation. This may be because of initial organic matter composition, or the porosity may have been created then lost because of compaction and expulsion of hydrocarbons. Considerable uncertainty remains, especially about how the organic matter composition varies on the nanoscale and how organic matter porosity is affected by local variations in mineralogy.

CONCLUSIONS

The data and images analyzed revealed important insights into the micro- and nanoscale attributes of the Eagle Ford unconventional reservoir. Some of these observations are as follows:

- Matrix was composed primarily of calcite and quartz with lesser amounts of clay, sodium plagioclase, organic matter, pyrite, and apatite based on EDX measurements.
- Matrix grains often exhibited a recrystallized appearance.
- Organic matter was primarily in intergranular spaces; the volume of intragranular organic matter was very small.
- Predominant pore type in Eagle Ford was dependent on zone (upper or lower) and thermal maturity.
- Pendular organic matter with bubble pores was more common in the oil window (wells NW and NE).
- Both pendular and spongy organic matter types with bubble and foam pores were common in gas/condensate window (wells SW and SE).
- Wells SW and SE (more mature, dry gas, and gas/condensate) had smaller average pore sizes and

lower permeability than wells NW and NE (less mature, oil window).

- Different porosity-permeability trends reflected physical differences between upper and lower Eagle Ford and between the oil window and gas window.
- The average porosity associated with organic matter as a fraction of total porosity was about 93% in the lower Eagle Ford and about 47% in the upper Eagle Ford. The average for all samples was about 67%.
- The average conversion ratio (percentage of solid organic matter that had converted to organic matter porosity) for all samples was about 39%.

ACKNOWLEDGMENTS

We wish to thank Shell Exploration and Production Company and Matador Resources Company for permission to publish this chapter. Also, we appreciate clarifying technical discussions with Robert Campbell, Katrin Ruckweid, Aysen Ozkan, and Andrew Govert. We also wish to acknowledge the constructive comments by Laura Wray and an anonymous reviewer. Editorial assistance from Wayne Camp, Elizabeth Diaz, Paul Smith, and Sarah Rilling-Hall was very helpful in completing this chapter.

REFERENCES CITED

Blakey, R., 2011, Regional paleogeography, paleogeography and geologic evolution of North America, North American paleogeographic maps, Late Cretaceous (85Ma), http://cpgeosystems.com/nam.html (accessed October, 2011).

Bohacs, K., J. D. Ottmann, R. Lazar, M. Dumitrescu, R. Klimentides, J. Schieber, R. Montelli, C. Liu, and J. Shamrock, 2011, Genetic controls on the occurrence, distribution, and character of reservoir-prone strata of the Eagle Ford Group and related rocks: Houston Geological Society, Applied Geoscience Mudrocks Conference, Houston, Texas, February 7–8, 37 p.

Boughal, K., 2008, Unconventional plays grow in number after Barnett Shale blazed the way: World Oil, August, p. 77–81.

Bowker, K. A., 2007, Barnett Shale gas production, Fort Worth Basin: Issues and discussion: AAPG Bulletin, v. 91, p. 523–533, doi: 10.1306/06190606018.

Curtis, M. E., R. J. Ambrose, C. H. Sondergeld, and C. S. Rai, 2010, Structural characterization of gas shales on the micro- and nano-scales: SPE, Canadian Unconventional Resources and International Petroleum Conference, October 19–21, Calgary, Alberta, Canada, SPE Paper 137693, 15 p., doi: 10.2118/137693-MS.

Driskill, B., N. Suurmeyer, S. Rilling-Hall, A. Govert, and A. Garbowicz, 2012, Reservoir description of the subsurface Eagle Ford Formation, Maverick Basin Area, South Texas, USA: SPE, EAGE Annual Conference and Exhibition, June 4–7, Copenhagen, Denmark, SPE Paper 154528, 23 p., doi: 10.2118/154528-MS.

Goldhammer, R. K., and C. A. Johnson, 2001, Middle Jurassic-Upper Cretaceous paleogeographic evolution and sequence-stratigraphic framework of the Northwest Gulf of Mexico rim, *in* C. Bartolini, R. T. Buffler, and A. Cantú-Chapa, eds., The western Gulf of Mexico Basin: Tectonics, sedimentary basins, and petroleum systems: AAPG Memoir, v. 75, p. 45–81.

Goldhammer, R. K., P. J. Lehmann, R. G. Todd, J. L. Wilson, W. C. Ward, and R. C. Johnson, 1991, Sequence stratigraphy and cyclostratigraphy of the Sierra Madre Oriental, northeast Mexico, a field guidebook: Gulf Coast Section, Society for Sedimentary Geology (SEPM), 85 p.

Guidry, K., D. Luffel, and J. Curtis, 1995, Development of laboratory and petrophysical techniques for evaluating shale reservoirs final technical report: Gas Research Institute, Chicago, Report GRI-95/0496, 286 p.

Harbor, R., S. C. Ruppel, and T. Hentz, 2012, Facies and lithostratigraphy of the Upper Cretaceous (Turonian-Cenomanian) Eagle Ford Formation, Central and South Texas (abs.): AAPG Annual Convention and Exhibition, April 22–25, Long Beach, California.

Hartman, B. M., and D. F. Scranton, 1992, Geologic map of Texas: Bureau of Economic Geology, University of Texas at Austin, scale 1:500,000, 4 sheets.

Haq, B. U., J. Hardenbol, and P. R. Vail, 1987, Chronology of fluctuating sea levels since Triassic: Science, v. 235, p. 1156–1167.

Loucks, R. G., R. M. Reed, S. C. Ruppel, and D. M. Jarvie, 2009, Morphology, genesis, and distribution of nanometer-scale pores in siliceous mudstones of the Mississippian Barnett Shale: Journal of Sedimentary Research, v. 79, p. 848–861, doi: 10.2110/jsr.2009.092.

Loucks, R. G., R. M. Reed, S. C. Ruppel, and U. Hammes, 2010, Preliminary classification of matrix pores in mudrocks: Gulf Coast Association of Geological Societies, Transactions, v. 60, p. 435–441.

Milner, M., R. McLin, and J. Petriello, 2010, Imaging and porosity in mudstones and shales: Comparison of secondary and ion-milled backscattered SEM methods: SPE, Canadian Unconventional Resources and International Petroleum Conference, October 19–21, Calgary, Alberta, Canada, SPE Paper 138975, 10 p.

Morgan, H. J., 1952, Paleozoic beds south and east of Ouachita folded belt: AAPG Bulletin, v. 36, p. 2266–2274.

Passey, Q. R., K. M. Bohacs, W. L. Esch, R. Klimentidis, and S. Sinha, 2010, From oil-prone source rock to gas-producing shale reservoir: Geologic and petrophysical characterization of unconventional shale-gas reservoirs: SPE, Chinese Petroleum Society and Society of Petroleum Engineers International Oil and Gas Conference and Exhibition in China, June 8–10, Beijing, China, SPE Paper 131350, 29 p., doi: 10.2118/131350-MS.

Pramudito, A., A. A. Miceli Romero, A., Donovan, W. Li, S. Staerker, and T. McClain, 2012, Geochemical fingerprints of the Eagle Ford sequences: Learning and implications from the outcrops in Terrell and Val Verde Counties, Texas (abs.): AAPG Annual Convention and Exhibition, April 22–25, Long Beach, California.

Schieber, J., 2010, Common themes in the formation and preservation of intrinsic porosity in shales and mudstones—illustrated with examples across the Phanerozoic: SPE, Unconventional Gas Conference, February 23–27, Pittsburgh, Pennsylvania, SPE Paper 132370, 9 p., doi: 10.2118/132370-MS.

Schieber, J., 2011, Shale microfabrics and pore development—an overview with emphasis on the importance of depositional processes: Canadian Society of Petroleum Geologists, Canadian Society of Exploration Geophysicists, and Canadian Well Logging Society Joint Annual Convention, May 9–12, Calgary, Alberta, Canada, 4 p.

Stevens, S., and V. Kuuskraa, 2009, Seven plays dominate North American activity: Oil and Gas Journal, September 28, p. 39–49.

Stoeser, D. B., G. N. Green, L. C. Morath, W. D. Heran, A. B. Wilson, D. W. Moore, and B. S. Van Gosen, 2007, Preliminary integrated geologic map databases for the United States: Central states: Montana, Wyoming, Colorado, New Mexico, North Dakota, South Dakota, Nebraska, Kansas, Oklahoma, Texas, Iowa, Missouri, Arkansas, and Louisiana: USGS, Open File Report 2005-1351, vers. 1.2, http://Pubs.usgs.gov/of/2005/1351/ (accessed May 30, 2012).

Tölke, J., C. Baldwin, Y. Mu, N. Derzhi, Q. Fang, A. Grader, and J. Dvorkin, 2010, Computer simulations of fluid flow in sediment: From images to permeability: Society of Exploration Geophysicists, The Leading Edge, v. 29, p. 68–74, http://dx.doi.org/10.1190/1.3284055.

Walls, J. D., and S. W. Sinclair, 2011, Eagle Ford shale reservoir properties from digital rock physics: European Association of Geoscientists and Engineers, First Break, v. 29, p. 97–101.

Walls, J., E. Diaz, S. W. Sinclair, and B. Driskill, 2011, Pore scale imaging and reservoir properties in Eagle Ford Shale (abs.): Society of Petrophysicists and Well Log Analysts, Fall Workshop, Computational Petrophysics, October 9–12, Ashville, North Carolina.

Wang, F. P., and R. M. Reed, 2009, Pore networks and fluid flow in gas shales: SPE, Annual Technical Conference and Exhibition, October 4–7, New Orleans, SPE Paper 124253, 8 p., doi: 10.2118/124253-MS.

12

James M. Rine, Erin Smart, William Dorsey, Kultaransingh Hooghan, Michael Dixon, 2013, Comparison of porosity distribution within selected North American shale units by SEM examination of Argon-ion-milled samples, *in* W. Camp, E. Diaz, and B. Wawak, eds., Electron microscopy of shale hydrocarbon reservoirs: AAPG Memoir 102, p. 137–152.

Comparison of Porosity Distribution within Selected North American Shale Units by SEM Examination of Argon-ion-milled Samples

James M. Rine, Erin Smart, William Dorsey, Kultaransingh Hooghan, and Michael Dixon

Weatherford Laboratories, 5200 N. Sam Houston Parkway West, Suite 500, Houston, Texas, 77086, U.S.A. (e-mails: james.rine@weatherfordlabs.com, erin.smart@weatherfordlabs.com, will.dorsey@weatherfordlabs.com, bobby.hooghan@weatherfordlabs.com, mike.dixon@weatherfordlabs.com)

ABSTRACT

The distribution of nanometer-size pores in ten selected Eagle Ford Group, Haynesville, Marcellus, and Barnett shale samples was similar when comparing relative numerical abundances of maximum pore diameters but not when comparing relative abundances of pore areas (pore sizes). Differences also existed between units in the association of pores with organic material. Pores were measured on argon-ion-milled (AIM) samples and examined with a field emission environmental scanning electron microscope (SEM). One Haynesville sample was also evaluated using a focused ion beam (FIB) SEM to compare to the AIM results. With the AIM samples, pore types were subdivided into three categories—organic pores, mixed matrix/organic pores, and matrix pores—based on the amount and type of material (organic or inorganic) surrounding the pores. Organic pores are pores generally associated with kerogen macerals, whereas mixed matrix/organic pores are pores that are probably associated with bitumen or pyrobitumen. Matrix pores are not associated with any organic matter. Within the sample set studied, only the Barnett samples contained pores almost exclusively within organic particles. The majority of the maximum pore diameters were less than 100 nm within all the samples examined. Only the Barnett samples, however, had a majority of their pore areas (or porosity) comprised of pores less than 10,000 nm^2 (which is the area of an equidimensional pore with the maximum pore diameter of 100 nm).

INTRODUCTON

Electron microscopic examination of pores within mudrocks (shales and mudstones) has become much more sophisticated over the last few years, driven not only by the intense economic interests in shale gas and oil but also by the technological improvements that allow characterization of nanometer-scale features. Since 2007, electron microscopy techniques have evolved sufficiently to allow researchers to

DOI: 10.1306/13391710M1023588

identify and quantify nanometer-scale pores. Previous studies characterized mudstone porosity by documenting the frequency distribution of pore sizes. This was done by comparing maximum pore diameters (Reed and Loucks, 2007; Loucks et al., 2009) and by analyzing pore volumes (Sondergeld et al., 2010). The objective of this chapter is to demonstrate other approaches to analyzing porosity by categorizing different pore type distributions and by measuring relative abundances of pore area and volume in four mudrock units: the Mississippian Barnett, the Upper Jurassic Haynesville, the Middle Devonian Marcellus formations, and the Upper Cretaceous Eagle Ford Group. Although we believe the data sets presented in this chapter are insufficient to make broad conclusions concerning the pore systems in these rock units, it is hoped that the approaches demonstrated here will assist successive researchers with more comprehensive data sets.

This chapter compares pore type character and pore size distribution between the shales based on the results of two-dimensional (2-D) field emission environmental scanning electron microscopic (FE-ESEM) examinations of ten argon-ion-milled (AIM) processed samples (Table 1). This chapter also compares results of FE-SEM examination of one AIM Haynesville sample to the three-dimensional (3-D) focused ion beam (FIB) SEM examination of the same sample.

PREVIOUS STUDIES

Groundbreaking work at the Texas Bureau of Economic Geology in the use of argon-ion-beam milling (Reed and Loucks, 2007; Loucks et al., 2009) led to the positive identification and quantification of nanometer-scale pores observed in SEM images. Prior to this work, the amount of microporosity could be measured by core analysis, but the distribution and types of pores could only be inferred by thin-section petrography (generally with spiked resins responsive to epifluorescence) or by examination of broken sample surfaces using a standard SEM. Based on our observations of Barnett and Haynesville thin sections, however, epifluorescence is commonly not observed even in samples with significant measured porosities (5 to 10%). This is probably caused by the limited ability of the spiked resins to fully penetrate these low-permeability samples. Furthermore, SEM examination of broken mudrock sample surfaces generally cannot differentiate true pores from sampling-artifact holes caused by plucking.

The Loucks et al. (2009; p. 851) study identified nanometer-scale pores and quantified pore size distribution by first using computer software (JMicrovision) to "outline and measure all individual pores." Pore diameters were then point counted from secondary electron (SE) SEM images. Rine et al. (2010), using SE FE-SEM images, hand-digitized each pore outline and colored each pore area, which in turn were analyzed using the software ImageJ (Abramoff et al., 2004) to provide pore dimensions and pore areas. Rine et al. (2010) used SE and backscattered electron (BSE) images for pore type classifications (i.e., organic vs. matrix pores). Image analysis software, such as Avizo Fire 6.3, is used to determine porosity and pore types from 3-D data sets obtained with FIB-SEM (Ambrose et al., 2010; Sondergeld et al., 2010).

METHODOLOGY

This comparison study examined four samples from two wells in the Barnett Shale (Fort Worth Basin of central Texas), four samples from two wells in the Haynesville Formation (in the East Texas Basin), two samples from one well in the Marcellus (Appalachian Basin), and two samples from one well in the Eagle Ford Group (Maverick Basin) (Figure 1; Table 1). Samples were selected based on range of total organic carbon (TOC), crushed rock (matrix) porosity values (Luffel and Guidry, 1992), and sample availability.

Sample Preparation and Microscopy

Samples were mounted on stubs and hand-polished (using 600- or 1000-grit sandpaper) prior to argon-ion milling with a Jeol SM-09910 mill. Fluids were not removed from the samples prior to processing. The advantage of argon-ion milling over mechanical polishing is that it can create a polished surface with minimal artifacts, such as surface abrasion marks and grinding debris (Loucks et al., 2009). After a light coating with gold, samples were examined with an FEI Quanta FEG-650 field emission environmental SEM (FE-ESEM) at magnifications ranging from ×500 to ×200,000 to determine the range of pore sizes present in each mudrock unit. Although representative areas of the milled surfaces (approximately 1 mm^2) in each sample were targeted for examination, undoubtedly some sample bias exists toward areas of higher pore density in order to acquire a sufficient number of pores to compare. The FE-ESEM is used to obtain high-resolution BSE and SE electron images 2-D of the milled surface. The BSE images show compositional variations, and the SE images highlight topography, such as depressions

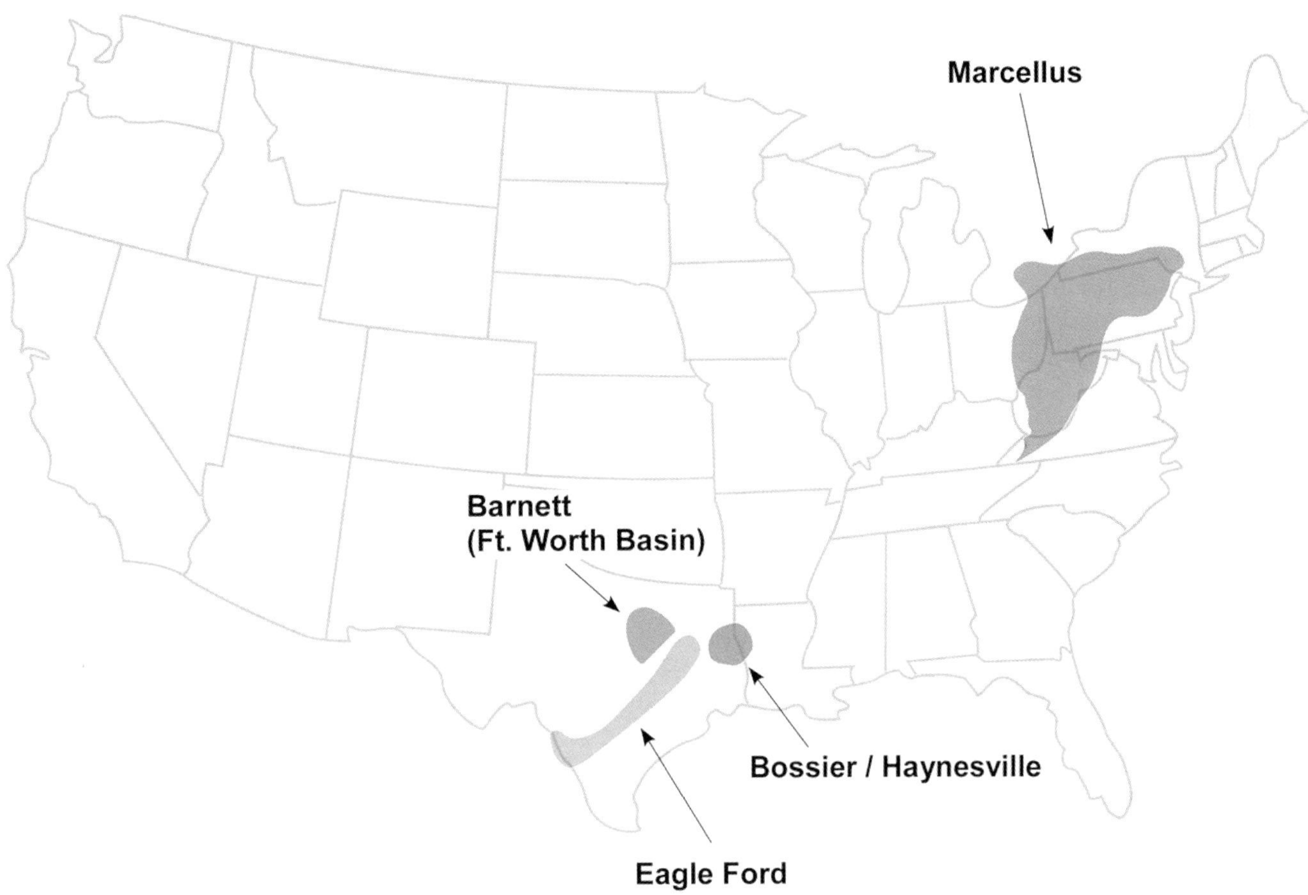

Figure 1. Map depicts general boundaries of basins where samples from this survey originated. Table 1 lists samples used for this study, including counties where wells are located.

Table 1

Formation	County, State	Sample ID	Sample Depth Top (ft)	Sample Depth Btm (ft)	Crushed-Rock Porosity (He)	LTOC (wt %)
Barnett	Wise County, TX	B1	7131.00	7131.10	6.4	7.76
Barnett	Wise County, TX	B2	6796.00	6796.25	1.1	1.15
Barnett	Wise County, TX	B3	6685.30	6685.65	4.2	7.39
Haynesville	Shelby County, TX	H2	12079.35	12079.60	6.1	5.46
Haynesville	Red River Parish, LA	H3	12457.10	12457.20	7.0	1.49
Haynesville	Red River Parish, LA	H4	12668.30	12668.40	8.2	2.46
Marcellus	Somerset County, PA	M1	5491.00	5491.05	3.0	0.19
Marcellus	Somerset County, PA	M2	5631.00	5631.05	4.5	3.59
Austin Chalk - Eagle Ford	DeWitt County, TX	EF1	14335.80	14336.00	10.3	1.83
Austin Chalk - Eagle Ford	DeWitt County, TX	EF2	14325.10	14325.30	9.7	4.17

caused by pores. An advantage of FE-ESEM is twofold. First, native-state samples (including samples that are wet or oily) can be examined, generally at low vacuum. Second, high-resolution 2-D images (down to nanometer size range) can document the characteristics of very-fine-grain unconventional reservoir rock samples, generally at high vacuum. For this study, all samples were examined using high vacuum pressure (to improve image resolution) and with an electron beam energy of 15 kV.

Pore Identification (FE-ESEM)

By FE-ESEM, pores were identified with SE images based on their degree of darkness (relatively deep portions of pores are generally black) and their generally bright edges (Figure 2). The edges of the pores are brighter because of the higher SE signal along the juncture of the inside surface of the pore and outer milled surface (Reimer, 1993). The pore areas bounded by these bright edges are often larger than the pore areas that are black because they include relatively shallow portions of a pore that are visible in the SE image (Figure 2). Because image-processing software available to us, such as Avizo Fire 6.3 software and Adobe Photoshop 6.0, only delineate the black (relatively deep) portions of the pores, this study hand-digitized each pore outline, then filled each pore area with a designated color.

Pore types were subdivided into three categories: organic pores, mixed matrix/organic pores, and matrix pores. Organic pores were defined based on the presence of organic material along three or more sides of the pore or greater than 75% of the diameter in the case of a circular pore. Mixed matrix/organic pores were defined as pores with one or two sides abutting organic material or less than 75% of the diameter in the case of a circular pore. The rationale for distinguishing a mixed matrix/organic pore was that these pores may be more likely associated with bitumen or pyrobitumen (secondary expelled and migrated hydrocarbons) versus an organic pore, which was more likely present within kerogen macerals deposited contemporaneous with the surrounding matrix material. A matrix pore was defined as a pore not contiguous with organic material.

The presence of organic matter was identified using BSE images because of the ability of BSE to record variations in composition (mean atomic number; Figure 3). Materials with low mean atomic number, such as carbon, exhibit low BSE intensity and are scaled as darker regions in SEM images. We consider this criterion of delineating pore types based on the relationship of the pore to organic material to be less interpretive than the pore classification schemes used by Loucks et al. (2010), which defined pores as "interparticle" or "intraparticle" and requires delineating particle boundaries.

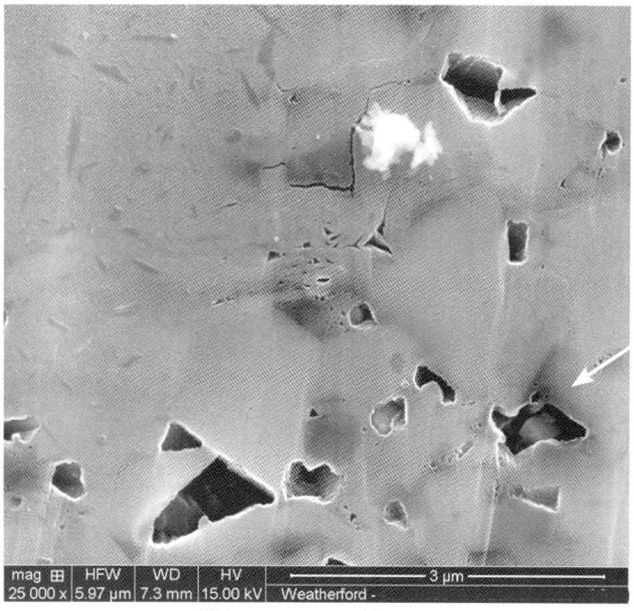

SE Image 25000X

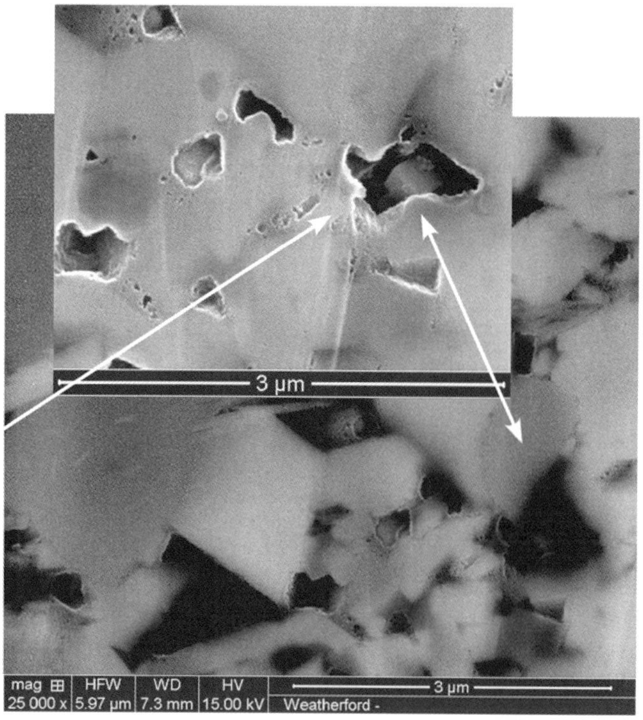

BSE Image 25000X

Figure 2. Pores were identified with secondary electron (SE) images based on their darkness (central portions of pores are generally black) and their generally bright edges. Note that some pore areas within the SE images bounded by bright edges are commonly larger than the black area within the pore. Also, some pores in the SE images do not include black areas. Black areas within the backscattered electron (BSE) image depict the presence of pore space or organic material.

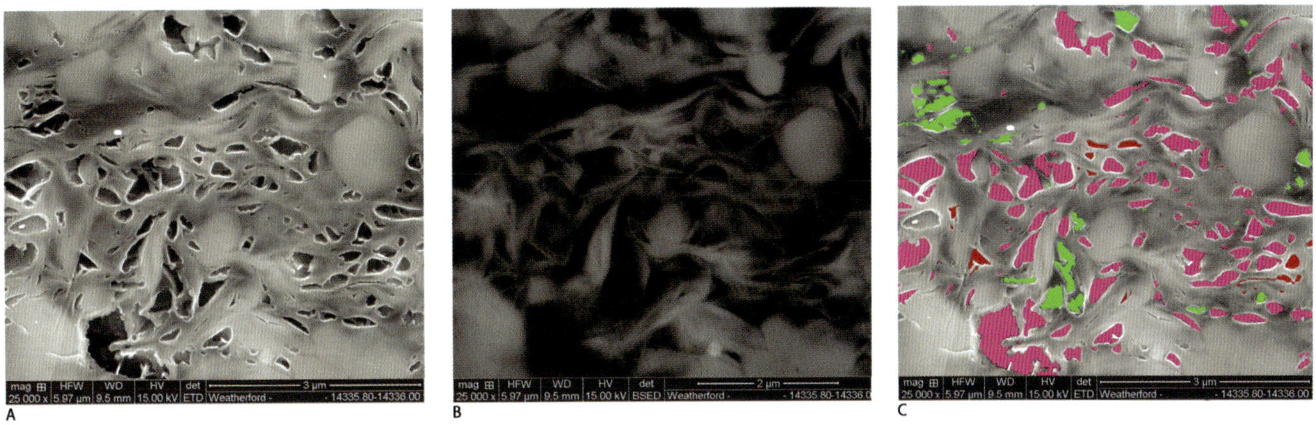

A B C

Figure 3. Images are of an Eagle Ford Group sample (EF2; Table 1). The images, which are of the same field of view, include a secondary electron (SE) image (A), a backscattered electron (BSE) image (B), and an analyzed image (C) in which pores have been colored and delineated as organic (green), mixed matrix/organic (pink), and matrix (red) pores. Pores were identified with SE images based on their degree of darkness (central portions of pores are generally black) and their generally bright edges. Black areas within the BSE image depict the presence of pore space or organic material. Note that the majority of the matrix/organic pores had rounded borders and may be pores within skeletal particles such as foraminifera tests.

"Pore digitizing" used Adobe Photoshop 6.0 software over a background SE image. Pores were outlined and areas colored on separate layers for each pore type, then saved as JPEG files against a white background. ImageJ software analyzed these separate layers to determine number of pores, pore area, and other dimensional data.

FE-ESEM Image Analysis

Pore images derived from the FE-ESEM images, were analyzed using ImageJ (IJ 1.45m), which is a public domain digital image-processing and analysis software available from the United States National Institute of Health (http://rsbweb.nih.gov/ij/index.html). This software processes individual JPEG pore type images in two steps. First, it converts them to binary images. Second, the ImageJ software analyses the binary images, providing both a summary table, which includes the number of pores and percentage pore area, and a complete size analysis of the individual pores within the image analyzed. The complete analysis of the individual pores includes maximum diameter (termed *feret* by Image J), which is the longest distance between two points on a region of interest (in this case, an individual image of a digitized pore), and minimum diameter (*minimum feret*). Also included are the areas of the binary image of each individual pore. Although pore area is 2-D, it is commonly used (such as with thin-section point-counting methods) as an estimate for porosity, which is a volume parameter. The ImageJ analyses are unitless; consequently, the scale bars were analyzed from numerous JPEG SE images to determine that the ImageJ unit for images used in this study equals 5.795 $\pm$0.02 nm. To convert ImageJ pore areas to square nanometers, area values were multiplied by the square of 5.795 nm ($x \times 33.582$ nm^2).

FIB-SEM Image Analysis

FIB-SEM analysis was done on one Haynesville Formation sample (H4). The sample was scanned for potential locations for carrying out an Auto Slice & View™ (AS&V) data collection regime (Sondergeld et al., 2010). One location was selected, and initial scanning and subsequent slicing were carried out using an FEI Company (Hillsboro, OR) Helios 650 small dual-beam (SDB) system (Auto Slice and View G3 software package). The AS&V was carried out on the location covered by the Platinum pad. Gallium-ion-beam milling was carried out using 430-pA beam current, tilted to 53°. Under normal conditions, in an SDB, milling is carried out at 52° to the imaging beam. Process optimization dictated an additional 1° tilt for milling. Imaging was carried out at 45° incidence, 3-mm working distance. Final horizontal field width was 5.30 μm. The system was set up to mill and image up to 350 slices, each slice being 10 nm (0.010 μm thick. Imaging resolution was 2048 × 1768 pixels. The individual voxel size was x = 0.0013 μm; y = 0.0018 μm; z = 0.0100 μm. As each individual slice was milled, the AS&V S/W took a reference image followed by a zoomed-in location selected while the run was set up. Subsequently, individual slices were aligned using the

Avizo Fire 6.3 software package. Data were then interpreted and assimilated using the various filtering and segmentation routines available in the software package.

FE-ESEM PORE TYPES AND DISTRIBUTION

Pore Types

The Barnett stands out when comparing analyzed FE-ESEM images of all the mudrock samples examined in this study by containing almost exclusively organic pores (Figure 4). Loucks et al. (2009) and Ambrose et al. (2010) also determined that porosity within the Barnett was almost exclusively within organic material. Only a fraction of 1% of the pores in the Barnett samples examined in this study were interpreted as matrix/organic pores, and none were interpreted to be strictly matrix pores (Table 2).

Pores interpreted to be combined mixed matrix/organic pores were the most abundant pore type within the Eagle Ford, Marcellus, and Haynesville samples examined in this study (Figures 3, 5, 6, respectively). In comparing the geometry of these pores, the pores of

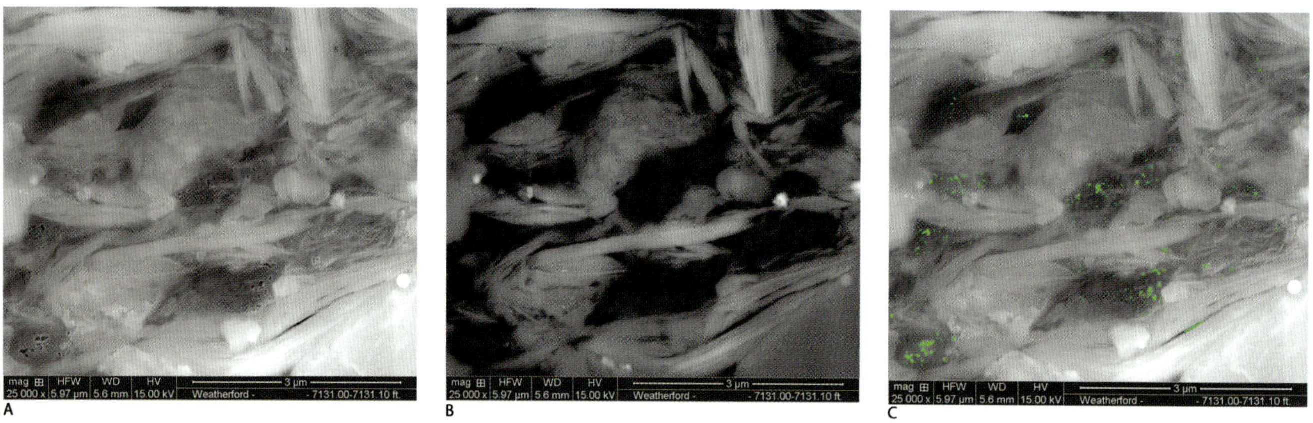

Figure 4. Images are of a Barnett sample (B1; Table 1). The images, which are of the same field of view, include a secondary electron (SE) image (A), a backscattered electron (BSE) image (B), and an analyzed image (C) in which pores have been colored and delineated as organic (green), mixed matrix/organic (pink), and matrix (red) pores. Note that no pore types other than organic pores were identified in this field of view. Pores were identified with SE images based on their degree of darkness (central portions of pores are generally black) and their generally bright edges. Black areas within the BSE image depict the presence of pore space or organic material.

Table 2

FORMATION	SAMPLES	PORE TYPE	COUNT	TOTAL AREA (ImageJ)	TOTAL AREA (nm^2)	AREA FRACTION (POROSITY %)	MAX AREA (nm^2)	MIN AREA (nm^2)	MAX DIAMETER (nm)
Barnett	SUM (B1+B2+B3)	Organic	798	32197	1.08E+06	0.75	1.07E+05	34	769
Barnett	SUM (B1+B2+B3)	Mat/Org	10	385	1.29E+04	0.02	2.46E+04	772	234
Barnett	SUM (B1+B2+B3)	Matrix	0	0	0.00E+00	0.00	0.00E+00	0	0
Barnett	SUM (B1+B2+B3)	Totals	1616	65164	2.19E+06	0.77			
Haynesville	SUM (H2+H3+H4)	Organic	527	37339	1.25E+06	1.15	6.00E+04	34	654
Haynesville	SUM (H2+H3+H4)	Mat/Org	216	82574	2.77E+06	2.77	1.40E+05	34	1166
Haynesville	SUM (H2+H3+H4)	Matrix	343	28602	9.61E+05	0.62	3.30E+04	34	481
Haynesville	SUM (H2+H3+H4)	Totals	2172	297030	9.97E+06	4.53			
Marcellus	SUM (M1+M2)	Organic	264	75973	2.55E+06	1.05	6.43E+05	34	1550
Marcellus	SUM (M1+M2)	Mat/Org	95	77526	2.60E+06	1.11	3.82E+05	34	1903
Marcellus	SUM (M1+M2)	Matrix	77	40942	1.37E+06	0.67	3.56E+05	201	1643
Marcellus	SUM (M1+M2)	Totals	436	194441	6.53E+06	2.83			
Eagle Ford	SUM (EF1+EF2)	Organic	335	84021	2.82E+06	2.21	5.36E+05	34	1282
Eagle Ford	SUM (EF1+EF2)	Mat/Org	291	324066	1.09E+07	7.58	6.85E+05	34	2553
Eagle Ford	SUM (EF1+EF2)	Matrix	178	53422	1.79E+06	1.07	2.68E+05	34	1387
Eagle Ford	SUM (EF1+EF2)	Totals	804	461509	1.55E+07	10.86			

the Eagle Ford (Figure 3) appeared to be more rounded than the matrix/organic pores of the Marcellus and Haynesville, which appeared to have more planar-edge boundaries (Figures 5, 6, respectively). The more rounded pore shapes were interpreted as pores within skeletal particles such as foraminifera tests, which are common within the Eagle Ford. The more angular pore shapes could be interpreted as being between particles or the interparticle pores of Loucks et al. (2010).

Abundance and Size Distribution

The frequency distribution of maximum pore diameters was similar for all the mudrock samples studied, with the most numerically abundant diameters less than 100 nm (Figures 7A, 8A, 9A, 10A). This frequency distribution was consistent with the results reported in the work of Loucks et al. (2009) for the Barnett Shale. The distribution was also consistent for organic pores

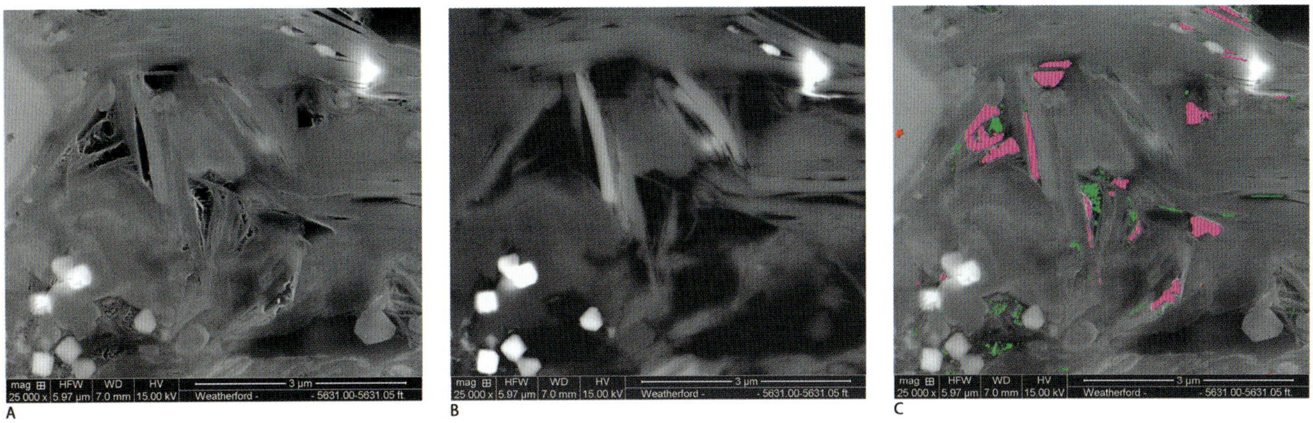

Figure 5. Images are of a Marcellus sample (M2; Table 1). The images, which are of the same field of view, include a secondary electron (SE) image (A), a backscattered electron (BSE) image (B), and an analyzed image (C) in which pores have been colored and delineated as organic (green), mixed matrix/organic (pink), and matrix (red) pores. Pores were identified with SE images based on their degree of darkness (central portions of pores are generally black) and their generally bright edges. Black areas within the BSE image depict the presence of pore space or organic material. Note that the majority of the matrix/organic pores have one or more straight borders and may abut inorganic particles.

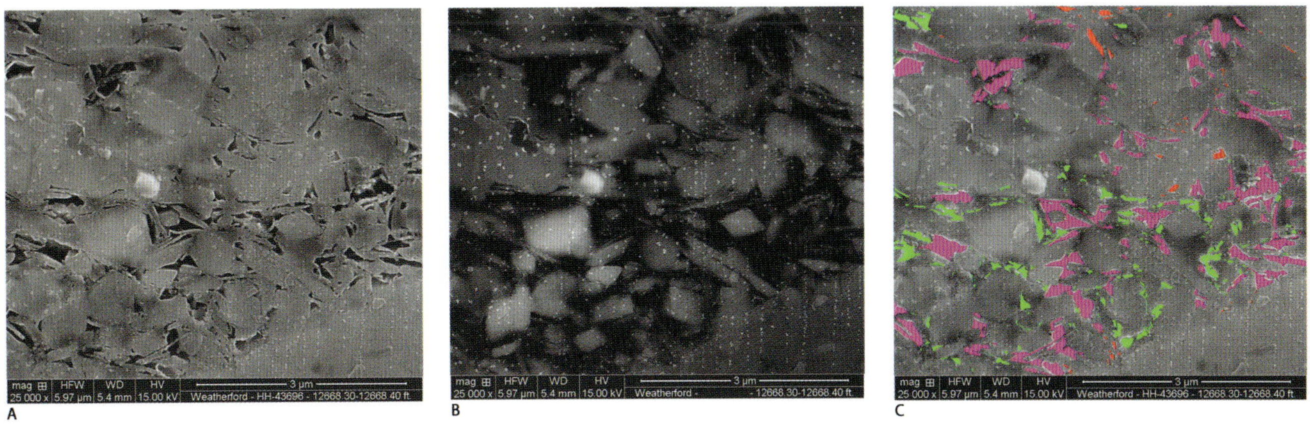

Figure 6. Images are of a Haynesville sample (H4; Table 1). The images, which are of the same field of view, include a secondary electron (SE) image (A), a backscattered electron (BSE) image (B), and an analyzed image (C) in which pores have been colored and delineated as organic (green), mixed matrix/organic (pink), and matrix (red) pores. Pores were identified with SE images based on their degree of darkness (central portions of pores are generally black) and their generally bright edges. Black areas within the BSE image depict the presence of pore space or organic material. Note that the majority of the matrix/organic pores have one or more straight borders and may abut inorganic particles.

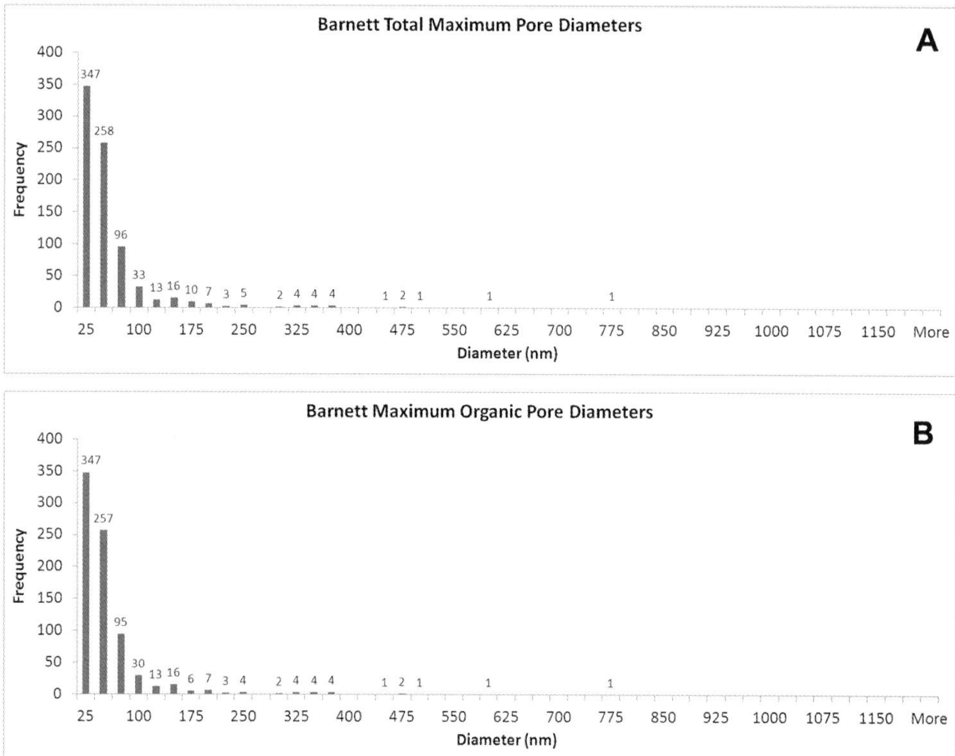

Figure 7. Histograms show the frequency distribution of pore sizes, based on maximum pore diameter, present within the Barnett samples examined in this study. The number of pores present within each diameter range is plotted at the top of each diameter range bar. (A) Total pore distribution; (B) organic pore distribution. The size distribution of matrix/organic pores present within the Barnett samples was not charted since only 10 pores of 808 total pores were so designated.

within all four units studied (Figures 7B, 8B, 9B, 10B). The size distributions for the mixed matrix/organic and matrix pores within the studied Eagle Ford, Marcellus, and Haynesville samples, however, were not as skewed to the smaller pore diameters (Figures 7C, D; 8C, D; 9C, D; 10C, D). (The Barnett samples examined only had a few matrix/organic pores, and their distribution was not charted.)

In addition to comparing pore size frequency distribution, the relative abundance of total pore areas was analyzed to understand and visualize the importance of less-numerous but larger pores. The relative abundance of pore areas was determined by subdividing the pores within size units. A total of 29 units were delineated, with the smallest pores in the 0 to 625 nm^2 (the square of 25 nm) subdivision and the largest subdivision being pore areas more than 490,000 nm^2 (the square of 700 nm). The sum of pore area within each subdivision was then divided by the total pore area of the sample to determine the percentage of the total pore area of that subdivision (Figure 11; Table 3). Within the Barnett Shale, over 55% of the porosity (pore area) resides within pores less than 10,000 nm^2 (a pore diameter of 100 nm). This pore distribution was not the same, however, within the other units. For the Haynesville samples, that fraction of the total porosity in pores smaller than 10,000 nm^2 was less than 35%. It was 7% for the Eagle Ford and 10% for the Marcellus (Table 3; Figure 11). (It should be noted that these pore area distributions may only apply to the few samples examined in this study and not to the formations generally.)

COMPARISON OF FIB-SEM AND FE-ESEM ANALYSES

Focused ion beam-SEM analysis was done on one Haynesville Shale (sample H4; Figure 12) to compare with pore distribution analysis of an FE-ESEM sample. The objective of this single comparison was to determine if the mostly computer-automated analysis of the FIB-SEM images produced similar results to the more interpretive analysis of the hand-digitized FE-ESEM images. Total porosities for the two analyses were similar (9.0% for the FIB-SEM and 10.8% for the FE-ESEM; Table 4) with differences probably caused by scale and location on the sample analyzed. These results were roughly equal to the crushed rock porosity results (8.2%; Table 1). In comparing the pore type distributions in the FIB-SEM sample with the pore type distributions of the Haynesville samples (Table 4), both results showed that "organic" pores were the least plentiful. It is important to note that the pore type definitions used for the FIB-SEM and the FE-ESEM samples were not the same. The Avizo Fire 6.3 software package interprets organic pores in

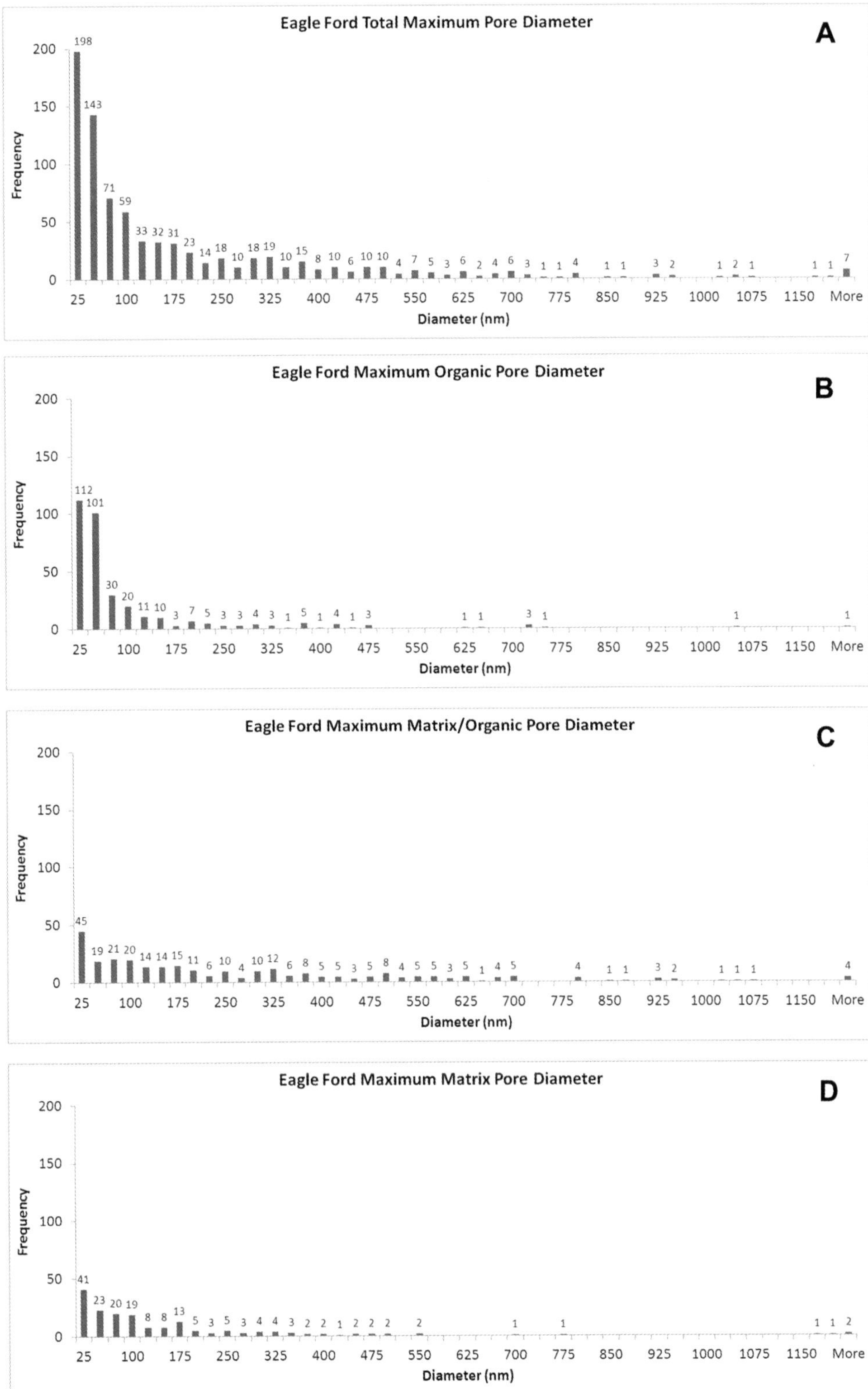

Figure 8. Histograms show the frequency distribution of pore sizes, based on maximum pore diameter, present within the Eagle Ford samples examined in this study. The number of pores present within each diameter range is plotted at the top of each diameter range bar. (A) Total pore distribution; (B) organic pore distribution; (C) matrix/organic pore distribution; (D) matrix pore distribution.

the FIB-SEM images as only those pores surrounded on six sides by organic matter, whereas the FE-ESEM analyzed samples delineated as organic pores based on the presence of organic material along three or four sides of the pore (or greater than 75% of the diameter in the case of a circular pore). Consequently, a closer comparison of the FE-ESEM results would exclude organic pores with only three sides adjoining

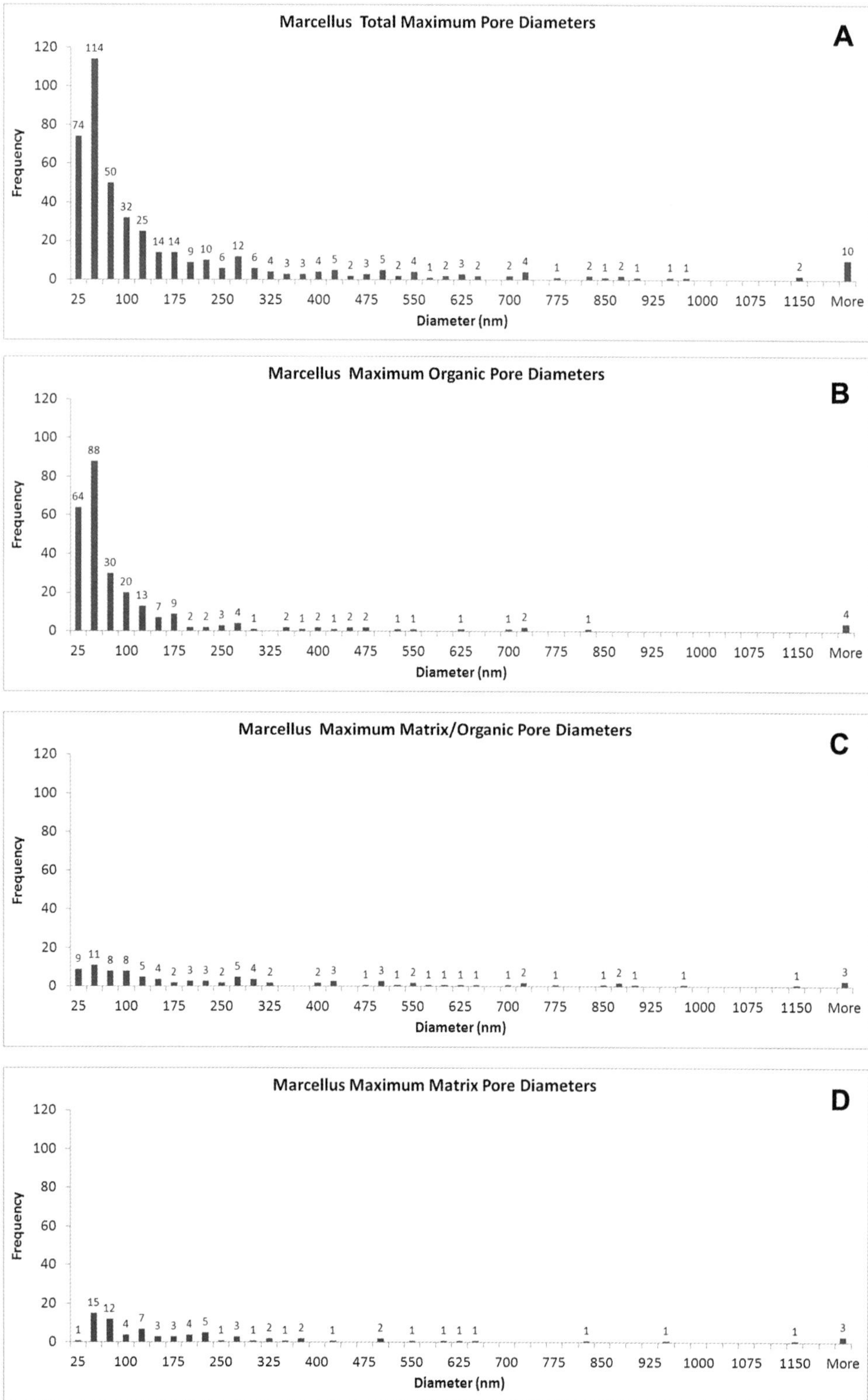

Figure 9. Histograms show the frequency distribution of pore sizes, based on maximum pore diameter, present within the Marcellus samples examined in this study. The number of pores present within each diameter range is plotted at the top of each diameter range bar. (A) Total pore distribution; (B) organic pore distribution; (C) matrix/organic pore distribution; (D) matrix pore distribution.

organics and consolidate mixed matrix/organic pores with matrix pores (FIB-SEM nonorganic porosity). Only the distribution of the total pore volumes was graphed in Figure 13 because of the different pore type designations between the FIB-SEM and FE-ESEM samples.

The frequency distribution of total pore volumes within the one Haynesville FIB-SEM sample was similar

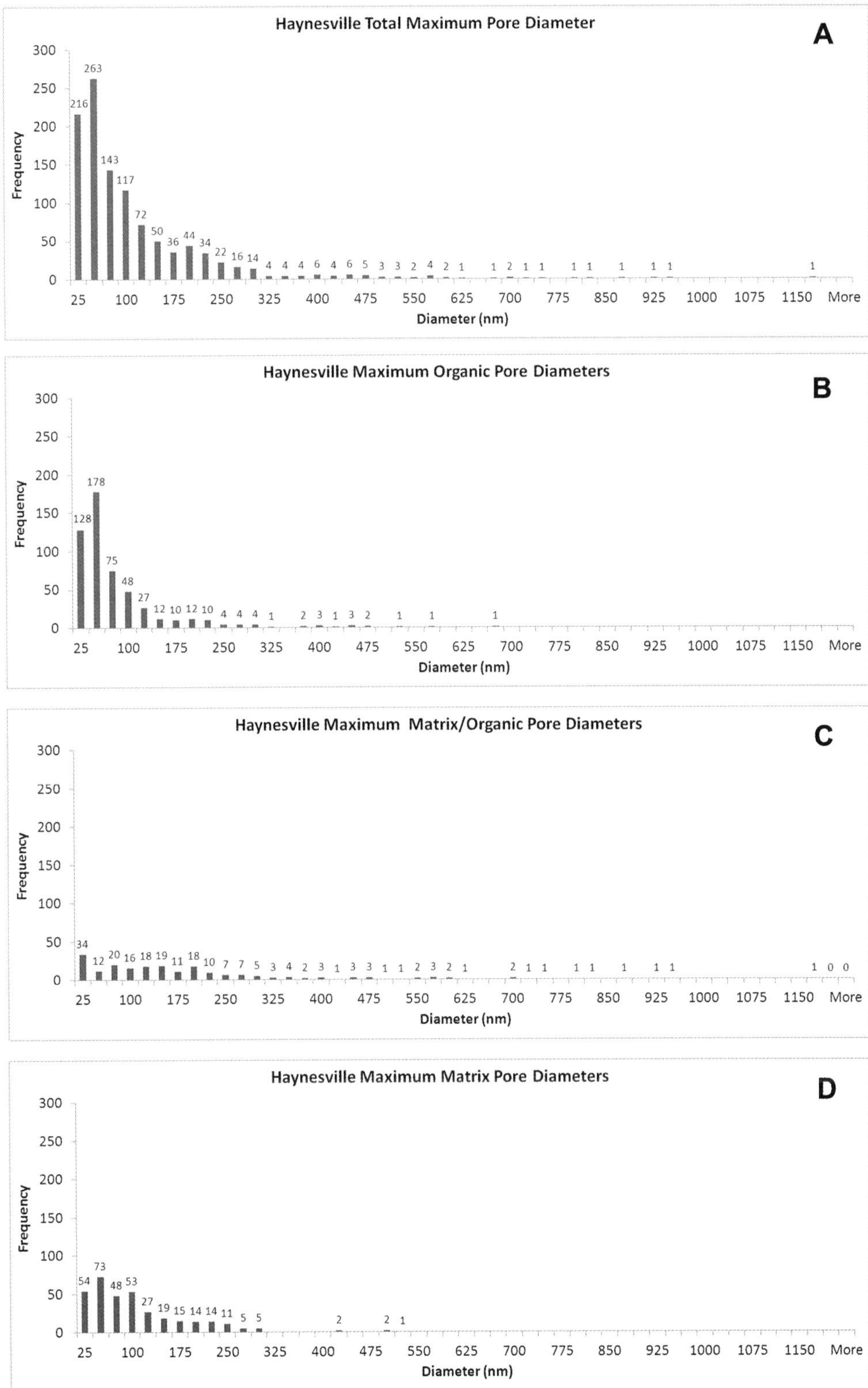

Figure 10. Histograms show the frequency distribution of pore sizes, based on maximum pore diameter, present within the Haynesville samples examined in this study. The number of pores present within each diameter range is plotted at the top of each diameter range bar. (A) Total pore distribution; (B) organic pore distribution; (C) matrix/organic pore distribution; (D) matrix pore distribution.

to the histogram of Haynesville total pore sizes based on maximum pore diameter (Figures 10A, 13A). The most numerically abundant pores had volumes equal to or less than 1.00E+06 nm^3 (the cube of 100 nm; Figure 13A).

Examination of how total porosity (pore volume) was distributed between the pore sizes present within this one Haynesville sample showed a distribution not as skewed to the smallest pores (Figure 13B). In fact,

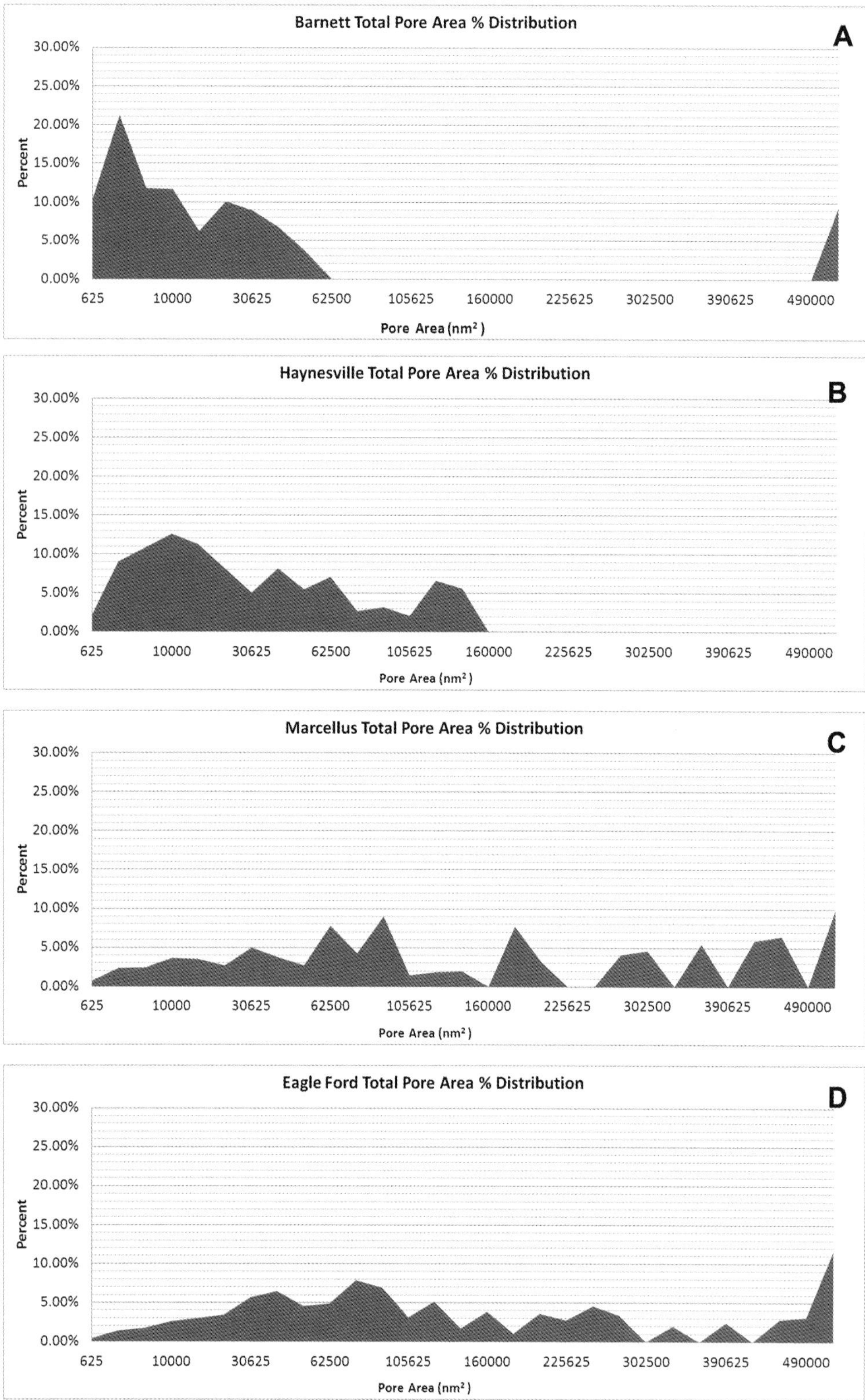

Figure 11. Charts show the differences in the relative abundance of total pore areas (two-dimensional pore sizes) between the samples from the Barnett (A), Haynesville (B), Marcellus (C), and Eagle Ford (D). The relative abundance of pore areas was determined by first subdividing the pores within size units. A total of 29 units were delineated, with the smallest pores in the 0- to 625-nm^2 (the square of 25 nm) subdivision and the largest subdivision being all pore areas more than 490,000 nm^2 (the square of 700 nm). The sum of pore area within each subdivision was then divided by the total pore area of the sample to determine the percentage of the total pore area of that subdivision. Whereas all the unit samples examined in this study had median maximum pore diameters of less than 100 nm, only the Barnett samples had a majority of their pore area (or porosity) comprised of pores less than 10,000 nm^2 (the square of 100 nm).

the larger pores appeared more volumetrically abundant in this FIB-SEM analysis than they appeared in the FE-ESEM examination (Figure 11B). The chart in Figure 13B was constructed similar to the total pore area charts in Figure 11 such that the pore volume delineations were similar, with the largest subdivision being all pore areas more than 3.43E+08 nm^3 (the cube of 700 nm) but with the smallest pore volume

Table 3

Formations	Barnett	Haynesville	Eagle Ford	Marcellus
Unit Divisions (nm^2)	Volume %	Volume %	Volume %	Volume %
625	10.5%	2.1%	0.4%	0.7%
2500	21.3%	9.0%	1.4%	2.3%
5625	11.8%	10.8%	1.7%	2.4%
10000	11.7%	12.6%	2.6%	3.6%
15625	6.2%	11.3%	3.0%	3.5%
22500	10.1%	8.2%	3.4%	2.7%
30625	8.9%	5.0%	5.7%	5.0%
40000	6.7%	8.2%	6.5%	3.7%
50625	3.6%	5.5%	4.6%	2.7%
62500	0.0%	7.1%	4.9%	7.8%
75625	0.0%	2.7%	7.9%	4.2%
90000	0.0%	3.2%	6.9%	9.0%
105625	0.0%	2.1%	3.1%	1.4%
122500	0.0%	6.6%	5.1%	1.8%
140625	0.0%	5.6%	1.7%	2.0%
160000	0.0%	0.0%	3.8%	0.0%
180625	0.0%	0.0%	1.0%	7.7%
202500	0.0%	0.0%	3.6%	3.2%
225625	0.0%	0.0%	2.8%	0.0%
250000	0.0%	0.0%	4.6%	0.0%
275625	0.0%	0.0%	3.4%	4.1%
302500	0.0%	0.0%	0.0%	4.6%
330625	0.0%	0.0%	2.0%	0.0%
360000	0.0%	0.0%	0.0%	5.5%
390625	0.0%	0.0%	2.4%	0.0%
422500	0.0%	0.0%	0.0%	5.9%
455625	0.0%	0.0%	2.8%	6.4%
490000	0.0%	0.0%	3.1%	0.0%
>490000	9.4%	0.0%	11.6%	9.8%
	100.0%	100.0%	100.0%	100.0%

subdivision being 0 to 1.56E+07 nm^3 (the cube of 250 nm). The sums of these pore areas within each subdivision are also shown in Table 5.

DISCUSSION

Although the data sets were limited, and thus likely not representative of any of the formations studied in general, it is hoped that the approaches described in this chapter will help other researchers more fully use their electron microscopy data sets. Some specific observations regarding the approaches presented in this chapter are discussed in the following.

A major benefit of electron microscopy, in addition to resolving nanometer-scale pores, is the ability to quantify the pores present using digital image analysis software. Quantification of pores with FE-ESEM samples, however, is presently labor intensive, whether the pore diameters are individually measured as with the Loucks et al. (2009) study or hand-digitized and processed with ImageJ software. The ability to delineate pore and pore types automatically from 2-D FE-ESEM images would greatly enhance the value of such examinations. Focused ion beam-SEM quantitative analyses are already automated, but the disadvantage of the FIB-SEM is that sample size is an order of magnitude smaller than the already small FE-ESEM image size. An optimum approach may be to combine the two methodologies.

Delineation of pore types with regard to their association with organic material is a critical component of electron microscopic analysis of pores within shale hydrocarbon reservoirs. With FE-ESEM samples, the presence of organic matter is based on identification of organic matter from BSE images. With FIB-SEM

Table 4

A. FIB-SEM	
FIB-SEM SAMPLE H4	**Percentage (volume %)**
Total porosity*	9.00
Porosity in organics	0.15
Nonorganic porosity	8.52
Organics	16.32
High density	0.19
Carbonate	Cannot distinguish carbonate materials from high-density materials; gray scales are too similar.

* Voxel sizes: $0.0013 \times 0.0018 \times 0.010$ μm.

B. FE-ESEM	
FE-ESEM SAMPLE H4	**Percentage (area %)**
Total porosity	10.80
Organic pores	2.80
Mixed matrix/organic pores	8.52
Matrix pores	0.50

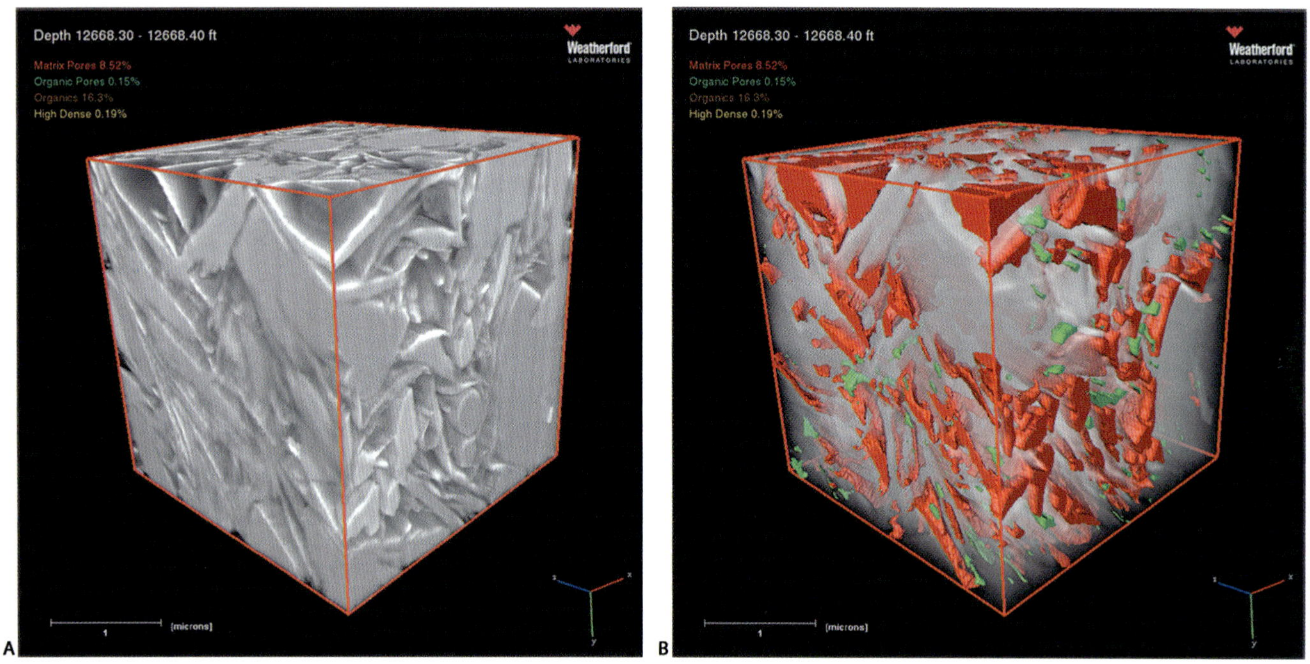

Figure 12. The three-dimensional (3-D) blocks depict one focused ion beam scanning electron microscopy (FIB-SEM) analysis done on a portion of the Haynesville sample H4. (A) Gray scale image of the FIB-SEM-analyzed sample. (B) The sample with interpreted distribution of matrix or nonorganic pores (red) and organic pores (green). The dimensions of the block were 5.30 μm wide by 3.5 μm thick. The total porosity of the FIB-SEM sample was calculated to be 9.0%, whereas the porosity by the argon-ion-milled (AIM) SEM survey was 10.8%. The crushed-rock porosity determination for this sample was 8.2% (Table 1).

examinations combined with Avizo analysis, a majority of pores interpreted as organic pores are probably within kerogen because these pores are completely surrounded by organic material. This study introduced a mixed matrix/organic pore designation to address those pores that may be associated with bitumen or pyrobitumen (secondary expelled and migrated hydrocarbons).

A better understanding of the distribution of porosity relative to pore size is important and should be considered in addition to the numerical frequency distribution of pore sizes. This is especially true for study units other than the Barnett that are not so dominated by a single pore type. Whereas the Haynesville, Marcellus, Eagle Ford, and Barnett samples examined in this study all had median maximum pore diameters

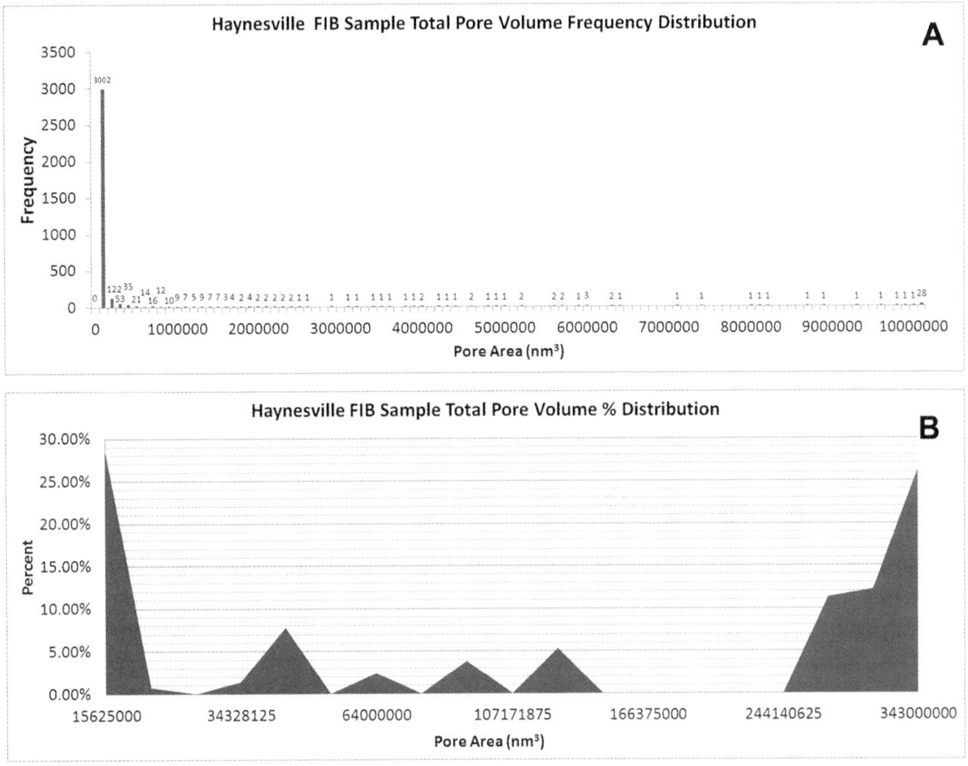

Figure 13. Frequency distribution of total pore volumes within the one Haynesville sample examined by focused ion beam scanning electron microscopy (FIB-SEM) (A) was similar to the histogram of Haynesville total pore sizes, based on maximum pore diameter (Figure 10A), with the most numerically abundant pores having volumes equal to or less than 1.00E+06 nm^3 (the cube of 100 nm). (B) Shows the distribution of total porosity (or total pore volume) between the pore sizes and reveals a distribution not as skewed to the smaller pore sizes. The chart in (B) was constructed using pore volume unit delineations that were the cubed pore diameter units in the Figure 11 charts, except the smallest unit in (B) (1.56E+07 nm^3) is the cube of 250 nm.

Table 5

Unit Divisions (nm^3)	Volume %
1.56E+07	28.93%
2.08E+07	0.73%
2.70E+07	0.00%
3.43E+07	1.36%
4.29E+07	7.77%
5.27E+07	0.00%
6.40E+07	2.39%
7.68E+07	0.00%
9.11E+07	3.79%
1.07E+08	0.00%
1.25E+08	5.30%
1.45E+08	0.00%
1.66E+08	0.00%
1.90E+08	0.00%
2.16E+08	0.00%
2.44E+08	0.00%
2.75E+08	11.29%
3.08E+08	12.19%
3.43E+08	26.25%

of less than 100 nm, only the Barnett samples had a majority of their pore area (or porosity) comprised of pores less than 10,000 nm^2 (a pore diameter of 100 nm).

CONCLUSIONS

The major conclusions derived from this study are as follows:

1. The distribution of nanometer-size pores in ten selected Eagle Ford Group, Haynesville, Marcellus, and Barnett formation samples were similar when comparing relative numerical abundances of maximum pore diameters but not when comparing relative abundances of pore areas (pore sizes). Consequently, relative distributions of pore areas should be considered when comparing mudrock units.
2. Pore types, which were subdivided into three categories (organic pores, mixed matrix/organic pores, and matrix pores), based on analysis of both SE and BSE images, showed significant distribution differences between mudrock units. Within the sample set studied, only the Barnett samples contained pores almost exclusively within organic particles.
3. For the characterization of nanometer-scale pores, SEM examination of AIM samples was a far superior methodology to thin-section petrography and standard (broken-sample) SEM examinations.

ACKNOWLEDGMENTS

We thank BP America Production Company, Samson Resources, and two anonymous companies for permitting us to examine their samples and to publish the results of those analyses. We also thank Weatherford Laboratories for the time and resources to conduct this study and, in particular, Kellen Guzman, Natalia Gonzalez, and Jeanette Kennedy for their help in processing and displaying the analytical results. Finally, we are grateful to Wayne Camp, William Demis, Keith E. Goggins, and an anonymous reviewer for their help in improving this chapter.

APPENDIX: AIM FE-ESEM IMAGE AND FIB-SEM DATA SETS

The tabular pore size data sets from the FE-ESEM and FIB-SEM examinations used in this chapter are provided as an appendix to this chapter in the CD-ROM accompanying this volume.

REFERENCES CITED

Abramoff, M. D., P. J. Magalhaes, S. J. Ram, 2004, Image processing with ImageJ: Biophotonics International, v. 11, Issue 7, pp. 36–42.

Ambrose, R. J., R. C. Hartman, M. Diaz-Campos, I. Y. Akkutlu, and C. H. Sondergeld, 2010, New pore-size considerations for shale gas in place calculations: Society of Petroleum Engineers Unconventional Gas Conference, February 23–25, Pittsburgh, Pennsylvania, SPE Paper 131772, 17 p., doi: 10.2118/131772-MS.

Loucks, R. G., R. M. Reed, S. C. Ruppel, and D. M. Jarvie, 2009, Morphology, genesis, and distribution of nanometer-scale pores in siliceous mudstones of the Mississippian Barnett Shale: Journal of Sedimentary Research, v. 79, p. 848–861.

Loucks, R. G., R. M. Reed, S. C. Ruppel, and U. Hammes, 2010, Primary classification of matrix pores in mudrocks: Gulf Coast Association of Geological Societies Transactions, v. 60, p. 435–441.

Luffel, D. L., and F. K. Guidry, 1992, New core analysis methods for measuring rock properties of Devonian Shale: Journal of Petroleum Technology, v. 4, no. 11, p. 1184–1190.

Reed, R. M., and R. G. Loucks, 2007, Imaging nanoscale pores in the Mississippian Barnett Shale of the northern Fort Worth Basin (abs.): AAPG, Annual Convention, Abstracts Volume, v. 16, 115 p.

Reimer, L., 1993, Image formation in low voltage scanning electron microscopy: Bellingham, Washington, SPIE Optical Engineering, 143 p.

Rine, J. M, W. Dorsey, M. Floyd, and P. Lasswell, 2010, A comparative SEM study of pore types and porosity distribution in high to low porosity samples from selected gas-shale formations: Gulf Coast Association of Geological Societies Transactions, v. 60, p. 825.

Sondergeld, C. H., R. J. Ambrose, C. S. Rai, and J. Moncrieff, 2010, Micro-structural studies of gas shales: Proceedings of the Society of Petroleum Engineers Unconventional Gas Conference, February 23–25, Pittsburgh, Pennsylvania, SPE Paper 131771, 17 p., doi: 10.2118/131771-MS.

U.S. National Institute of Health, ImageJ (IJ 1.45m), http://rsbweb.nih.gov/ij/index.html (accessed May 4, 2012).

13

Juergen Schieber, 2013, SEM observations on ion-milled samples of Devonian Black Shales from Indiana and New York: The petrographic context of multiple pore types, *in* W. Camp, E. Diaz, and B. Wawak, eds., Electron microscopy of shale hydrocarbon reservoirs: AAPG Memoir 102, p. 153–171.

SEM Observations on Ion-milled Samples of Devonian Black Shales from Indiana and New York: The Petrographic Context of Multiple Pore Types

Juergen Schieber
Indiana University, Department of Geological Sciences, 1001 E. 10th St., Bloomington, Indiana, 47405, U.S.A. (e-mail: jschiebe@indiana.edu)

ABSTRACT

Middle Devonian Marcellus Shale and late Devonian New Albany Shale samples were argon-ion milled and studied by scanning electron microscope for petrographic features and pore development. Petrographic observations revealed that a finite number of pore types exist in spite of considerable variability in composition, depositional setting, and compaction history. The four major pore types are framework pores, framework shelter pores, dissolution pores within inorganic grains, and organic matter pores. Framework pores occur in open spaces of the grain fabric. The most common framework pores are defined by phyllosilicate (clays, micas) and carbonate grains (biogenic, diagenetic) and also occur in areas with abundant diagenetic silica. Framework pores associated with carbonate grains can be well developed where abundant skeletal debris provides shelter porosity. Phyllosilicate framework pores increase in abundance with increasing clay content, but more critically, their abundance in compacted shales hinges on the presence of pressure shadows generated adjacent to mechanically competent grains (quartz, feldspar, dolomite, calcite, pyrite) that resist compaction. Dissolution pores in inorganic grains were predominantly encountered in association with calcite and dolomite grains. In places, other minerals, such as pyrite, also can show dissolution effects, but do not contribute significantly to overall porosity. Dissolution pores probably reflect decreased pH associated with the formation of carboxylic and phenolic acids at elevated temperatures (about 80 to 120° C). At low carbonate contents (a few percent), dissolution pores constitute only isolated porosity, but in shale intervals that contain abundant carbonate, or where carbonate grains were concentrated into laminae, this pore type may be an important facilitator of gas storage and transmission. Pores within organic matter are maturity dependent and restricted to amorphous organic matter (bituminite/amorphinite) in thermally mature samples ($>0.6\%R_o$). Grain and fabric shrinkage during core storage and sample processing produced pore artifacts as well (false pores), and their correct identification is critical for accurate petrographic pore assessment.

DOI: 10.1306/13391711M1023589

INTRODUCTION

Typical shales are sedimentary rocks with a dominant grain size below 63 μm and constitute approximately two-thirds of the sedimentary rock volume (Schieber, 1998). Shales function as the seals that prevent hydrocarbons from escaping from their reservoir, and a subset of carbonaceous (organic-rich) shales serves as source rocks. The first natural gas well in the United States tapped Devonian black shales beneath Fredonia, New York. This well was drilled in 1825, almost four decades before the Drake Oil Well near Titusville, Pennsylvania, which is considered to mark the birth of the oil industry. Though shale gas has been produced in the eastern United States for more than a century, the contribution of shale gas to overall production has been marginal. In the past decade, however, shale gas has become a growing source of natural gas in the United States, spurred by technological advances, such as horizontal drilling and hydraulic fracturing techniques. Increased shale gas production has been essential to offset declining production from conventional gas reservoirs.

Early studies of gas-producing shales focused on Devonian black shales of the eastern United States, strata that have long been known to produce natural gas from fractured as well as nonfractured shale intervals (Broadhead et al., 1982). Petrographic examination of porosity in shales requires electron microscopy. Attempts to produce sufficiently thin and undisturbed samples of shale for transmission electron microscopy (TEM) were long hampered by the difficulties of avoiding mechanical damage inflicted by microtome knives (Davies et al., 1990). The application of argon-ion milling, a sample preparation technique developed by material scientists (Bollinger and Fink, 1980), has helped to overcome this limitation and has enabled imaging of pore spaces and mineral interrelationships at the magnifications needed for shale studies (Jiang et al., 1990; Hover et al., 1996; Rask et al., 1997; Schieber, 1998, 2002). Although shale pores are readily observed in TEM images of argon-ion-milled shales (e.g., Hover et al., 1996), the preferred technique for the study of pores in shales is the scanning electron microscopic (SEM) observation of argon-ion-milled surfaces (Tomutsa et al., 2007; Loucks et al., 2009; Schieber, 2010, 2011). In this study, the focus was on pores that are intrinsic to the shale matrix, although it is acknowledged that fractures can be a significant contributor to the overall porosity of shale intervals.

METHODS

Samples of the Devonian New Albany Shale were collected from a core drilled in Daviess County, Indiana, stored at the core library of the Indiana Geological Survey. Samples of the Devonian Marcellus Shale were provided by Taury Smith of the New York State Museum from a core drilled in northern Chenango County, New York. For location of samples within these wells relative to stratigraphic position and members, see Table 1. Thin slices of these samples were mounted on a custom-designed sample holder and mechanically polished on the top surface. The sample holder was then placed into a GATAN 600 Duomill, and the polished surface was argon-ion milled at a low incident angle for 1 to 3 hours. Milled surfaces were then examined by SEM without conductive coating

Table 1

Well: 1-3 Kavanaugh, Daviess County, Indiana				Well: Beaver Meadow #1, Chenango County, New York			
	Sample	Depth in Meters	Member		Sample	Depth in Meters	Member
Top New Albany Shale		610.11		Top of core		575.5	
	x	612.65	Clegg Creek		x	566.9	Oatka Creek
	x	616.53	Clegg Creek		x	573.9	Oatka Creek
	x	619.88	Clegg Creek		x	582.8	Cherry Valley
	x	622.04	Camp Run		x	588.9	Union Springs
black/gray split	x	623.18	Camp Run	Base Marcellus Shale	x	591.3	Union Springs
	x	634.48	Morgan Trail			592	
	x	637.08	Selmier				
	x	641.02	Selmier				
	x	641.62	Blocher				
	x	644.34	Blocher				
	x	648.85	Blocher				
Base New Albany Shale		649.11					

(FEI Quanta 400 FEG, low-vacuum mode). Our SEM was tested to design specifications just prior to our examination of specimens for this study, and objects of 5 nm size or larger were resolvable. For comparison purposes, a subset of these samples was also polished via edge milling with a GATAN Ilion ion mill. All samples were viewed on surfaces oriented perpendicular to bedding.

Although the ability to make distinctions between organic matter types is critical to the interpretation of the presented research, SEM imaging by itself can only identify organic matter as such based on low backscatter intensities. In this study, the differentiation of amorphous organic matter versus structured organic matter (such as *Tasmanites*) benefited from prior observation of these components in the immature New Albany Shale with fluorescent light microscopy. *Tasmanites*, for example, has an intense yellow fluorescence under UV (ultraviolet) light. In addition, it is a collapsed feature (presence of medial inclusions) of uniform thickness with rounded terminations that shows deformation between harder grains without significant change of thickness. Whereas the UV fluorescence is a matter of chemical composition and fades with maturation, the latter features are morphological in nature and are still readily observed when the sediment has been heated beyond maturity. Given that the New Albany and the Marcellus are both Devonian in age, it can thus safely be assumed that what has the morphology of *Tasmanites* in the mature Marcellus is indeed a collapsed *Tasmanites* cyst.

Amorphous organic matter (bituminite/amorphinite) in the New Albany Shale characteristically is intimately intermingled with mineral matter (clays, silt), shows a dull brownish fluorescence under UV light, is squeezed into interstitial spaces between mineral grains, and lacks any characteristic morphology. The assumption was made that in the mature Marcellus Shale, irregularly shaped and squeezed organic matter with mineral inclusions was the direct equivalent to amorphous organic matter (bituminite/amorphinite) as observed in the New Albany Shale.

Identification of the third organic component considered here, bitumen, is more problematic. In thermally immature rocks, bitumen has not yet evolved, and thus a morphological comparison cannot be made for this component between the New Albany and the Marcellus. The rationale for suggesting a bituminous nature for part of the amorphous organic matter in the Marcellus is the observation of large bubble-like structures of some hundred nanometers in diameter. The bubbles suggest formation in some sort of fluid phase, and the fact that the enclosing organic matter is still in the rock points to a fluid of high viscosity. Thus, one can envision that thermal cracking of bitumen within pore spaces produces gas bubbles within the bitumen. Supporting observations for this view have been reported in a study of maturation effects in the Jurassic Posidonia Shale (Bernard et al., 2012).

NEW ALBANY SHALE

Overview

A 40 m (131 ft) thick core (well 1-3 Kavanaugh, Daviess County) from southern Indiana was examined and sampled (Table 1). The core contains all the stratigraphic subdivisions of the New Albany Shale as

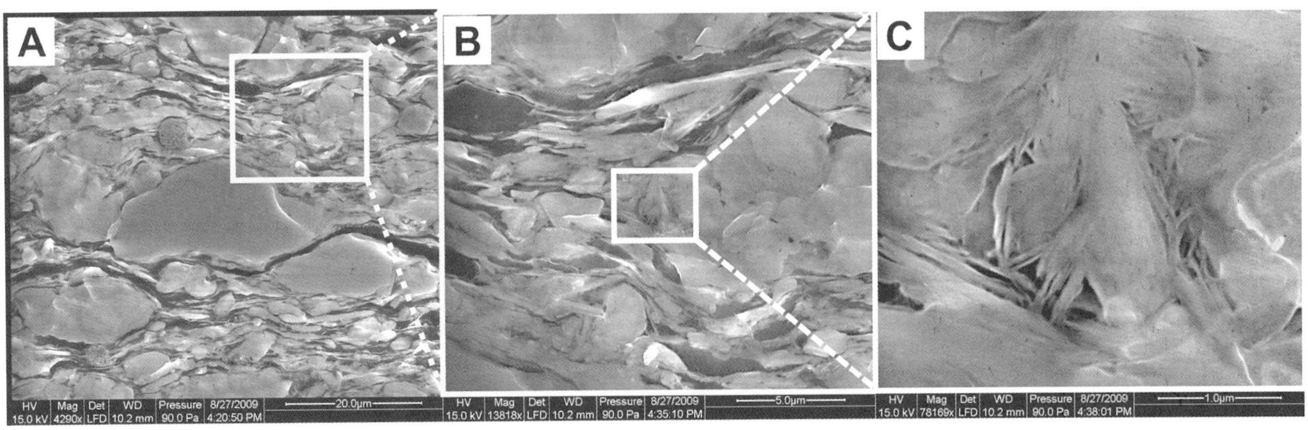

Figure 1. Fabric and phyllosilicate framework pores from the New Albany Shale. (A) Low-magnification secondary electron scanning electron microscopic (SEM) image showing planar fabric caused by compaction and compression around resistant larger grains. Dark streaks consist of kerogen. (B) Enlargement of gray-framed area in (A). Planar-compressed fabric still dominant. (C) Enlargement of white-framed area in (B). Shows largely randomly oriented clays that define triangular openings, here termed phyllosilicate framework pores.

recognized by Lineback (1970). In ascending order, these are the Blocher, Selmier, Morgan Trail, Camp Run, and Clegg Creek members. This core has been studied more recently by Lazar (2007), and a summary of its lithology, sedimentary structures, macrofossils, trace fossils, and total organic carbon content (TOC) was provided by Lazar and Schieber (2004). The 13 samples that were examined include all of the stratigraphic members of the New Albany Shale. The samples include three samples with measured TOC less than 5 wt.%, two samples with TOC values slightly above 5 wt.%, and eight samples with TOC values in excess of 8 wt.%. Measured vitrinite reflectance values indicate that the New Albany Shale in Daviess County is immature (0.57%R_o, Comer et al., 1994).

Observations

Visual estimates from numerous SEM images suggested that framework pores defined by phyllosilicate grains are the most ubiquitous pore type in all New Albany Shale samples. Visual estimates from numerous SEM images suggested that framework pores defined by phyllosilicate grains are the most ubiquitous pore type in all New Albany Shale samples (Figure 1). These pores consist of triangular openings defined by a latticework of randomly oriented clay mineral platelets (Figure 2), ranging in size from 10 nm to more than a micrometer, and occur within a largely compaction-dominated, planar fabric (Figure 2). Phyllosilicate framework pores are best developed in pressure

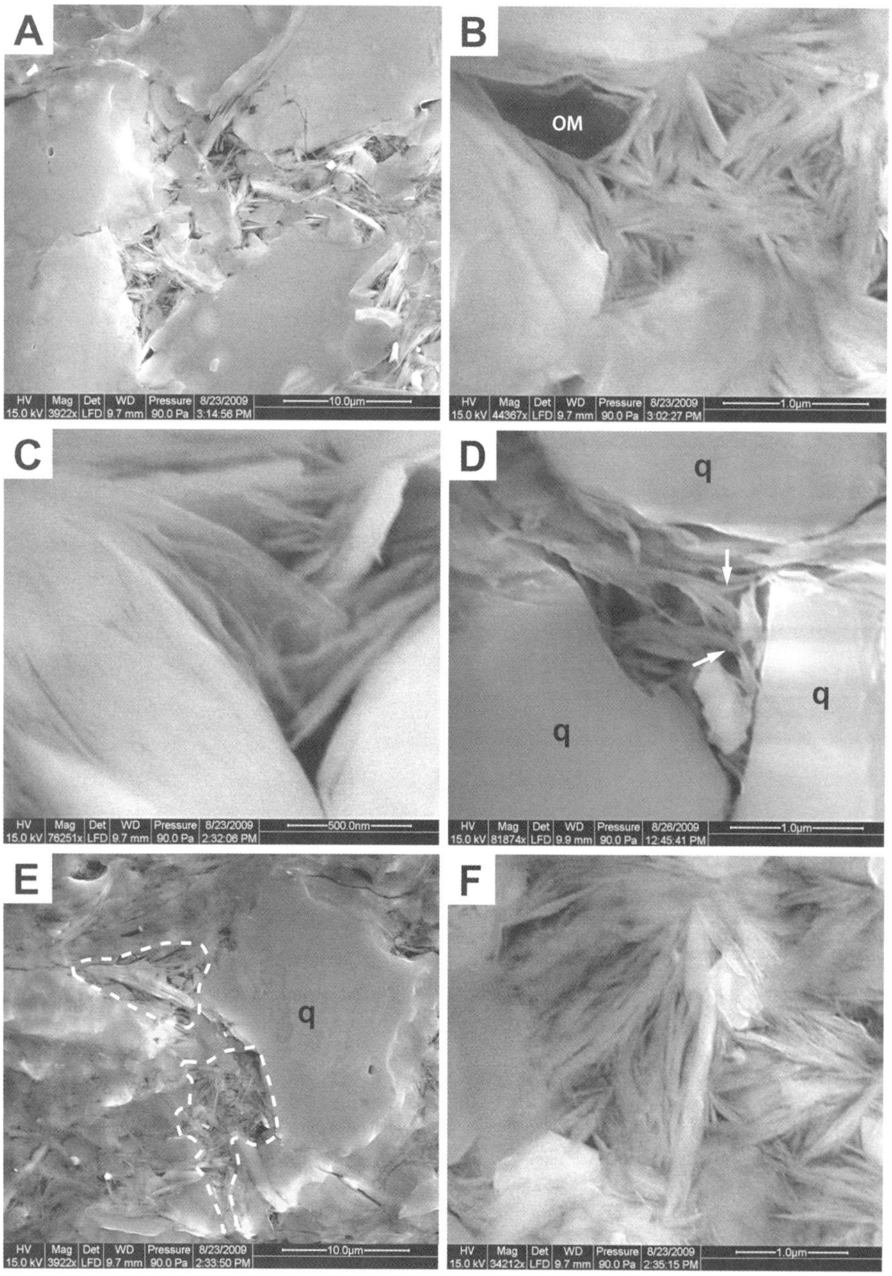

Figure 2. (A) Secondary electron scanning electron microscopic (SEM) images illustrating domains of randomly oriented clay minerals in a silt-rich shale interval of the New Albany Shale. The silt grains provide protection from compaction pressures and allow triangular pores, defined by phyllosilicate platelets, to remain open. (B) Close-up of randomly oriented framework of phyllosilicates with triangular pores. OM marks a piece of probable detrital organic matter. (C) High-magnification image of phyllosilicate framework with triangular pores. (D) Space between small quartz silt grains (q) protects phyllosilicate framework from collapse. Bending of clay flakes (arrows) suggests that at least in part the interstitial clays are of detrital origin. (E) Domains of randomly oriented clay minerals (dashed outlines) with triangular framework pores in pressure shadow of quartz silt grain (q). (F) Close-up of framework texture of randomly oriented phyllosilicates with triangular pores. The central, vertical-oriented clay flake seems to have pierced over and underlying clay flakes because of compaction forces. Images (A) through (F) are from samples that contained approximately 5% total organic carbon (TOC), were rich in quartz silt and clay, and were of medium-to-dark-gray color.

shadows adjacent to larger compaction-resistant grains (rigid grains including fossil debris, silt, and other mineral grains) and in spaces between such grains (Figure 3). In shales, these observed framework pores are analogous to intergranular pores in sandstones. Shales differ from sandstones, however, in that they are generally subject to severe mechanical compaction associated with dewatering during burial. The term framework pore does highlight pore preservation caused by the resistance to compaction by a variety of mineral frameworks.

The pore-defining clay mineral platelets of phyllosilicate framework pores are typically at most a few micrometers in size and appear to have multiple origins. They show textures suggestive of detrital origin, such as bending and splintering caused by compaction, as well as piercing and differential compaction when in near-vertical orientation (Figures 3, 4). Clay minerals that are likely of diagenetic origin tend to be finer grained and more idiomorphic appearing (Hover et al., 1996). They are present in the New Albany Shale but do not appear to be of volumetric importance.

In addition to framework formation by phyllosilicates, early diagenetic growth of carbonate and silicate grains can provide compaction-sheltered spaces that preserve pore space during burial. Figure 5A shows an example of such a pore defined by dolomite grains. Even though the space is now occupied by clay minerals, compaction sheltering by dolomite grains allows open framework pores between clays. In Figure 4B, pore preservation is seen among early diagenetic quartz grains.

In clay-rich shales with low amorphous kerogen content (<5 wt. % TOC) (Figure 3), framework pores appear to remain open through burial history. In contrast, in the samples with more than 5 wt.% TOC, the dispersed amorphous organic matter appears to fill most of the framework pores (Figure 9).

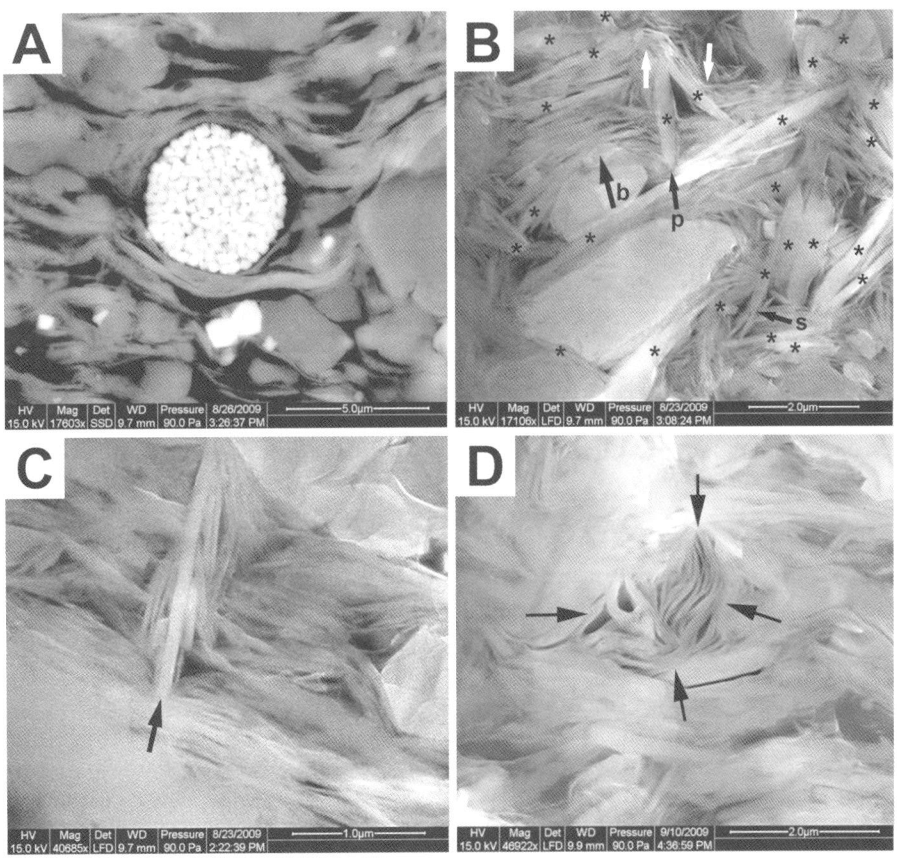

Figure 3. (A) Scanning electron microscopic (SEM) backscattered image that shows bending of detrital clays, mica flakes, and kerogen blebs around pyrite framboid (center) and quartz grains (total organic carbon [TOC] approximately 9%). (B) Random clay fabric with phyllosilicate framework pores. Clay flakes marked with * are considered clearly detrital and define a lattice framework of compaction-protected space. Black arrows: b = compactional bending of clays around quartz grains; s = splintering of clay flake due to compression; and p = piercing by clay flake. White arrows mark wrapping of clays because of compaction. (C) Subvertical clay flake pierces underlying clays because of compaction (black arrow). (D) Black arrows define an area where compression of a vertically oriented phyllosilicate grain produced pores because of splits between phyllosilicate sheets. All images from the New Albany Shale. Images (B)–(D) are secondary electron images; samples contained approximately 5% TOC.

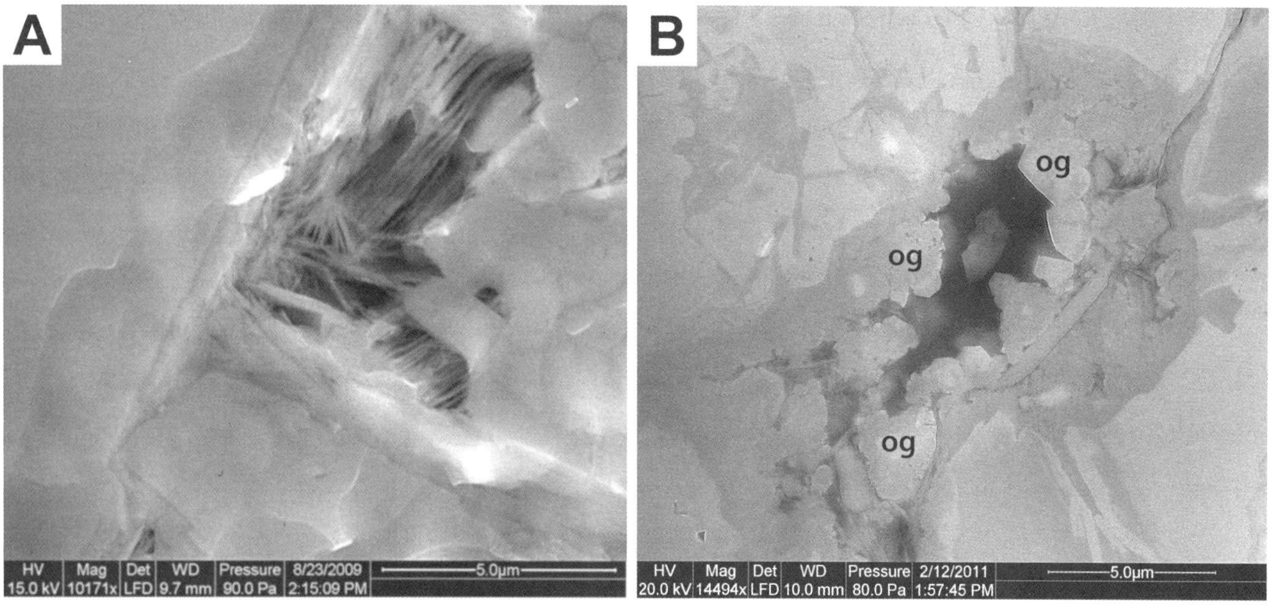

Figure 4. (A) A framework pore between dolomite grains that is partially filled with clay minerals. The clays form frameworks and define pores (total organic carbon [TOC] approximately 5%). (B) Amorphous organic matter (OM) (black) fills intergranular pore partially lined by quartz overgrowths (og) between quartz silt grains (q). Note lack of OM pores in this low-maturity ($<0.6\%R_o$) sample (TOC approximately 10%). Scanning electron microscopic (SEM) secondary electron images from the New Albany Shale.

Whereas pressure shadows are most helpful for preserving phyllosilicate framework pores, clay mineral clamping and pore propping by diagenetic minerals can also provide protection from compactional closure. Clamping of clay platelets in the margins of expanding diagenetic minerals effectively cements phyllosilicate frameworks in place and protects them from compactional collapse (Figures 6B, C, D). Likewise, diagenetic mineral grains can prop open triangular spaces between clay platelets and thus prevent their collapse during compaction (Figures 7A, B, C).

Dissolution pores are commonly observed when carbonate grains (typically dolomite and calcite) are scattered through the shale matrix, or where carbonate grains have been reworked into thin lag deposits. Dissolution rarely removes entire grains (Figure 7A). More commonly, partial dissolution is observed along the carbonate grain margins, resulting in corroded seams and generally vuggy (honeycombed) carbonate minerals (Figures 6B, C). These corroded seams range in width from a few tens to several hundreds of nanometers. Dissolution pores are either irregular in shape or retain rectangular outlines because of the structure of the carbonate substrate (Figure 6D). In charge contrast imaging and under cathodoluminescence, many carbonate grains show zonation that reflects variable contents of magnesium and iron in these grains, as verified by energy-dispersive spectrometry (EDS). No signs of collapse into aforementioned dissolution cavities have been observed, and they do not appear to be filled with secondary minerals. The described style of carbonate dissolution is a common type of secondary porosity in the New Albany Shale. Dissolution features were also observed in association with pyrite grains (Figure 7), but appear much rarer than carbonate dissolution.

The organic matter observed in the New Albany core was largely composed of two types of kerogen: structured organic matter and amorphous organic matter. Structured organic matter retains the morphologic features of the originating organism (e.g., alginites such as *Tasmanites* cysts) (Figures 8A, B). *Tasmanites*, the cyst stage of fossil marine algae, is abundant in eastern U.S. black shale beds and is the most readily identified variety of structured organic matter throughout the Devonian. When examined with epifluorescence microscopy (UV light), *Tasmanites* will fluoresce bright yellow in immature samples, whereas the interstitial amorphous (unstructured) organic matter will produce a dull brownish fluorescence color, a typical UV response for this type of maceral (Taylor et al., 1998). Amorphous organic matter occurs as discontinuous laminae and irregular patches that fill spaces between mineral grains (Figures 7A; 8C, D). The latter form of organic matter is commonly described as amorphinite/bituminite by organic

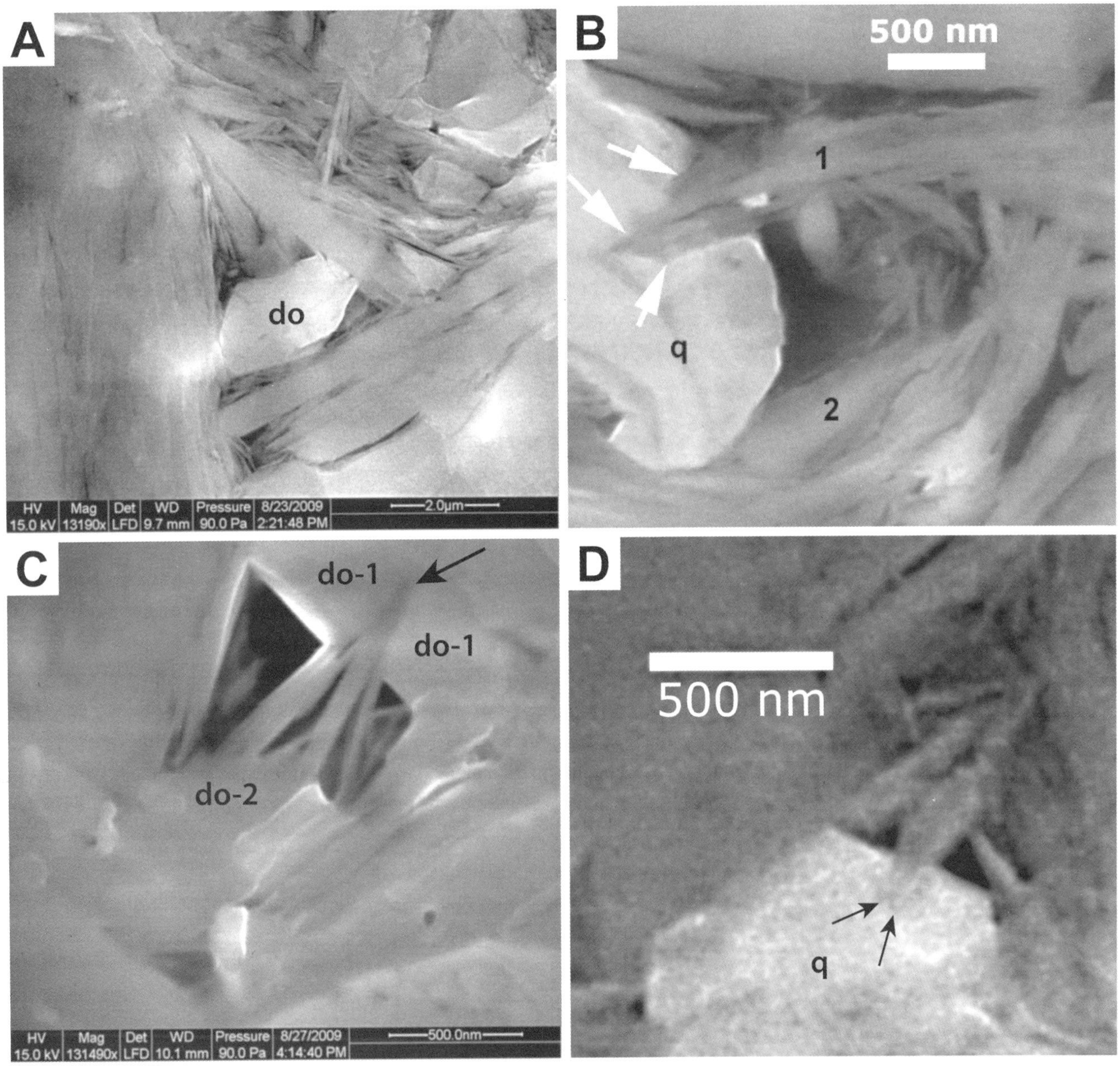

Figure 5. (A) Triangular phyllosilicate framework pores. Dolomite grain (do) partially fills and props open a space between clay platelets. (B) Close-up from Figure 2F shows phyllosilicate framework and an authigenic quartz grain (q). Clay platelet marked 1 has been clamped in place by the growing quartz grain (arrows). The authigenic quartz clamps clay platelets and props open pore spaces. (C) Clamping and propping by authigenic dolomite. Dolomite grain marked do-1 clamps clay platelets (arrow) in place, and dolomite grain marked do-2 props open pore space. (D) Clamping of clay platelet (arrow) by authigenic quartz grain (q). Scanning electron microscopic (SEM) secondary electron images from the New Albany Shale. Total organic carbon (TOC) values for these samples were in the 3 to 5% range.

petrographers (Taylor et al., 1998) and is considered to represent bacterially degraded marine organic matter (phytoplankton, zooplankton). In the thermally immature New Albany Shale, both organic matter categories (structured vs. amorphous) develop smooth ion-milled surfaces and lack pores that are visible at the image resolution of a field emission SEM.

MARCELLUS SHALE

Overview

Samples from a 90 m (295 ft) thick Marcellus core (Beaver Meadow 1 well, Chenango County, New York) were provided by Taury Smith of the New York

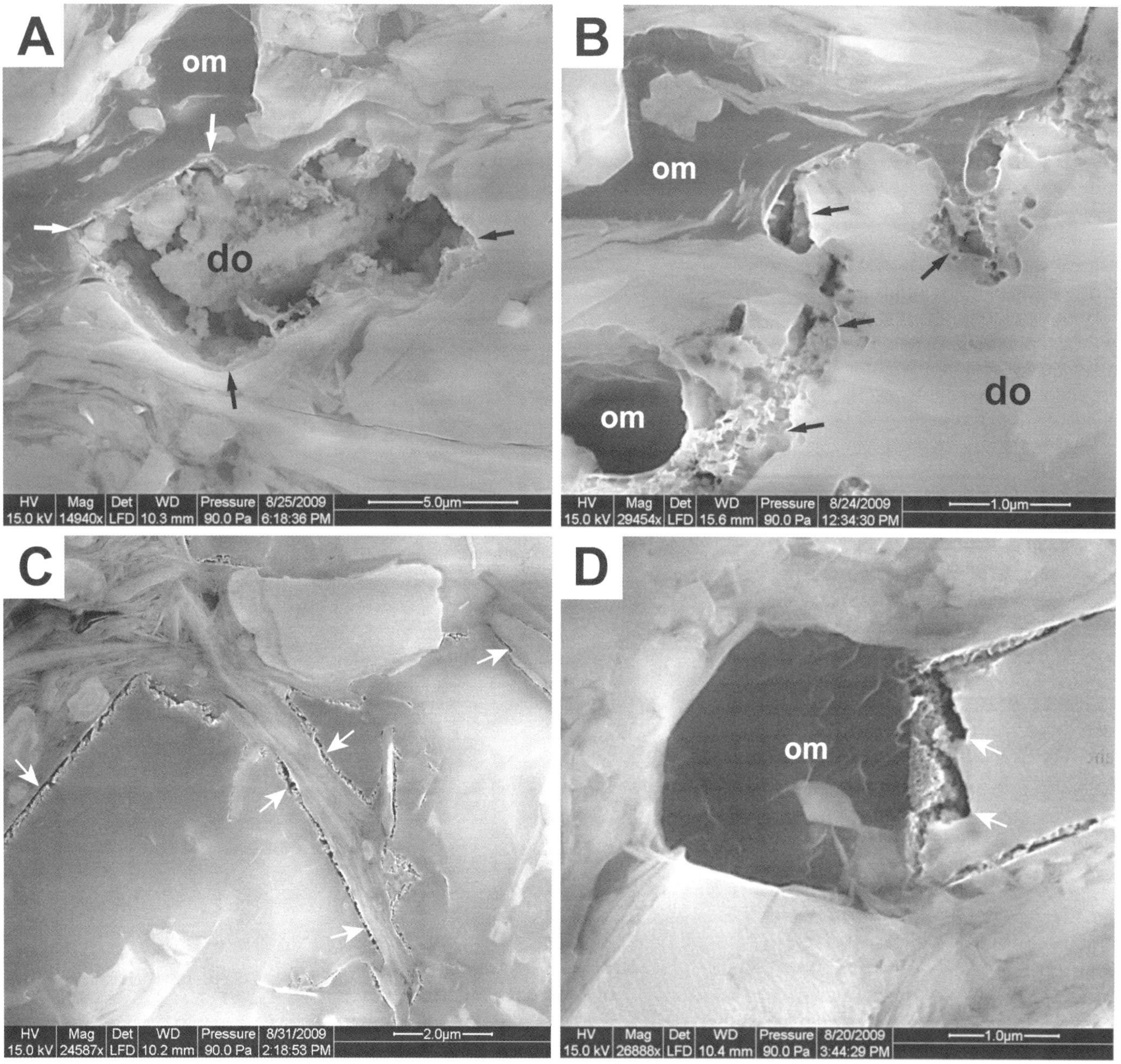

Figure 6. Manifestations of carbonate dissolution in the New Albany Shale. (A) A secondary pore caused by partial dissolution of a dolomite (do) grain. The dolomite nature of the residual material in the center was verified with energy-dispersive spectrometry (EDS) analysis. Pore margins marked by white and black arrows; nonporous organic matter marked om. (B) Corrosion and porosity formation (arrows) along the margins of a dolomite grain. Nonporous organic matter marked om. (C) Dissolution along margins of dolomite grains. Arrows point to dissolution porosity. (D) Dolomite grain margin etched along crystallographic boundaries (arrows), as well as honeycombed dolomite. Nonporous organic matter fills pore. Scanning electron microscopic (SEM) secondary electron images from the New Albany Shale. Total organic carbon (TOC) values for these samples were in the 8 to 9% range.

State Museum (see Table 1 for stratigraphic position and member of samples). Stratigraphic subdivisions present in the core include (in ascending order) the Union Springs Member, the Cherry Valley Limestone, and the Oatka Creek Member (T. Smith, 2010, personal communication). A total of five samples, spanning the 566.9 to 591.3 m (1860 to 1940 ft) depth interval, were examined. Measured vitrinite reflectance values show the Marcellus to be gas mature (2.1%R_o, Weary et al., 200).

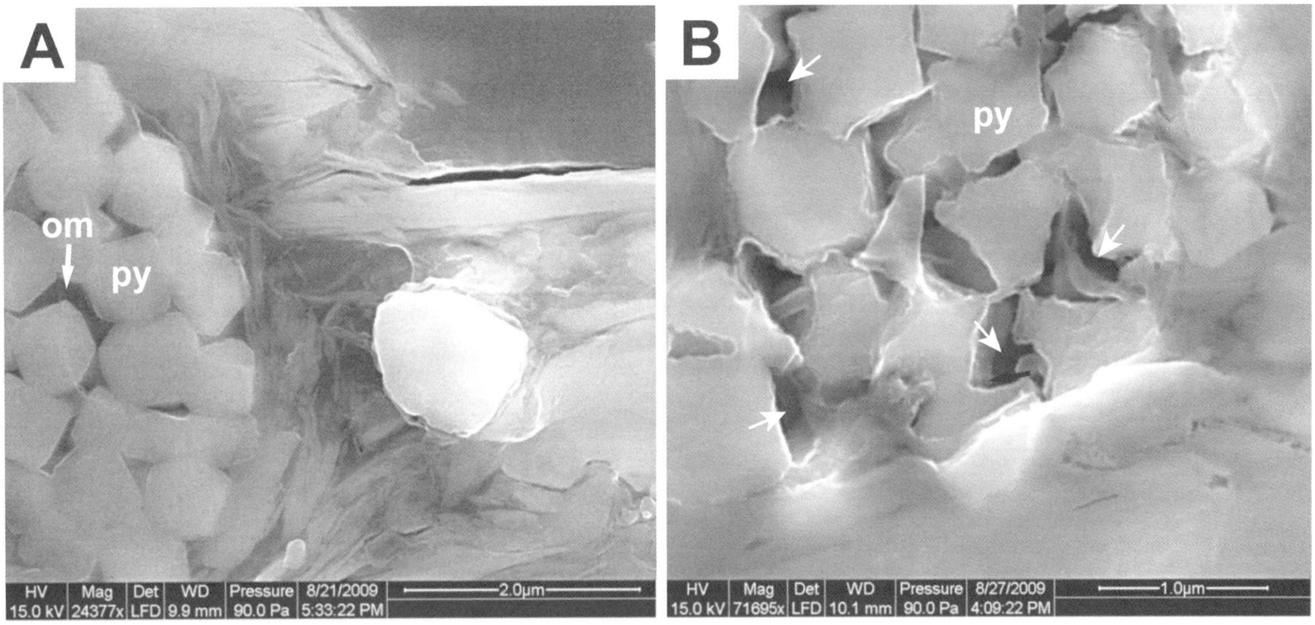

Figure 7. Secondary electron scanning electron microscopic (SEM) images showing manifestations of pyrite dissolution in the New Albany Shale. (A) The left third of the image consists of a pyrite framboid that shows an infill of organic matter (om) between pyrite (py) crystallites. This is very typical for pyrite framboids in the New Albany and many other Devonian black shales. Note also that the framework pores of clays in the central portion of the image are filled with nonporous amorphous kerogen, a typical feature of immature ($<0.6\%R_o$) New Albany samples. The horizontal pore in the upper-right quadrant of the image, at the boundary between dark organic matter and clay mineral flake, is an artifact caused by sample expansion. (B) Close-up of pyrite framboid where pyrite crystallites have undergone corrosion along margins and secondary pores (arrows) are visible between pyrite crystallites. Total organic carbon (TOC) values for samples were approximately 8% (A) and 3% (B).

The shale matrix in the Marcellus Shale consists of a mixture of clay minerals (dominantly illite, Figure 9), quartz grains, carbonate grains (dolomite and calcite), and organic matter. The latter consists of structured kerogen (retaining original structures and outlines) and unstructured kerogen that fills interstitial spaces between mineral grains. Some of the intervals sampled include common to abundant calcareous shell debris derived from ostracods, tentaculites (styliolinids), and small brachiopods (Figure 10).

Observations

Framework pores defined by phyllosilicate grains are ubiquitous in the Marcellus Shale (Figure 9) and consist of triangular openings defined by a latticework of randomly oriented clay mineral platelets (Figure 9). These triangular pores range in size from 10 nm to more than a micrometer and tend to be best developed in areas that were protected from the full force of compactional pressures. Protected areas occur in pressure shadows adjacent to larger compaction-resistant, rigid grains.

In beds and laminae with calcareous shell material, interior cavities of fossils provided shelter from compaction (Figure 10B). The shale matrix within such cavities shows abundant uncollapsed phyllosilicate frameworks with preserved pores (Figures 10C, D). That fossil cavities stayed open for some time during shale burial is indicated by calcite infills (Figure 10A), and whereas true shelter porosity was not seen in the studied samples, it is likely to occur elsewhere in the Marcellus Shale.

In the examined Marcellus samples, carbonate grains and authigenic cements are common features. Preservation of framework pores between carbonate grains is widespread, and these pores range from several hundred nanometers to more than a micrometer in size (Figures 11A, B, C). The carbonate grains appear to have facilitated preservation of phyllosilicate frameworks within carbonate-rich bedsets and laminae. Clay mineral clamping and pore propping by carbonate minerals (Figure 9C, D, E, F) is commonly observed

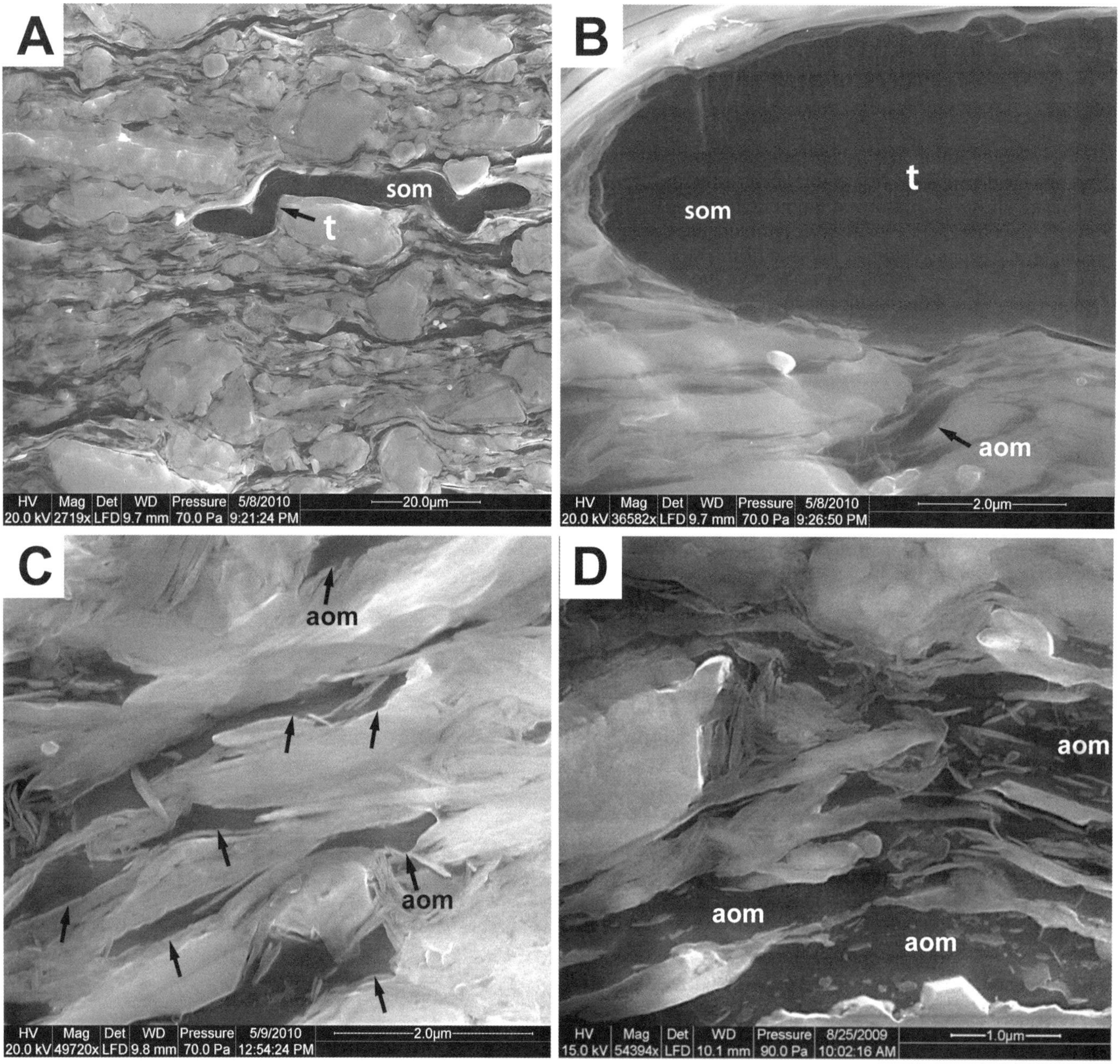

Figure 8. Secondary electron scanning electron microscopic (SEM) images of immature ($<0.6\%R_o$) New Albany Shale. (A), (B) The object t is a collapsed *Tasmanites* cyst. The ion-milled surface of this structured kerogen maceral (som) is smooth and shows no visible pores at maximum SEM magnification. Under ultraviolet (UV) light, the *Tasmanites* macerals exhibit bright yellow fluorescence, consistent with the low thermal maturity. Nonstructured or amorphous organic matter (aom) is also present. (C), (D) Amorphous organic matter is intimately mingled with mineral matter. This amorphous kerogen also shows a smooth ion-milled surface, lacks visible pores, and has a dull brownish fluorescence under UV light. Total organic carbon (TOC) values for these samples were in the 8 to 10% range.

in the Marcellus Shale. In samples where authigenic silica cement is abundant (Figure 11D), it parallels the effects of carbonate cementation, and helps to preserve phyllosilicate frameworks from compactional collapse, as well as defining interstitial pores.

Dissolved margins (Figures 11E, F) were observed on carbonate grains but were uncommon. Their presence suggests late diagenetic dissolution effects associated with organic acids (phenolic and carboxylic acids). Because the lower members of the Marcellus

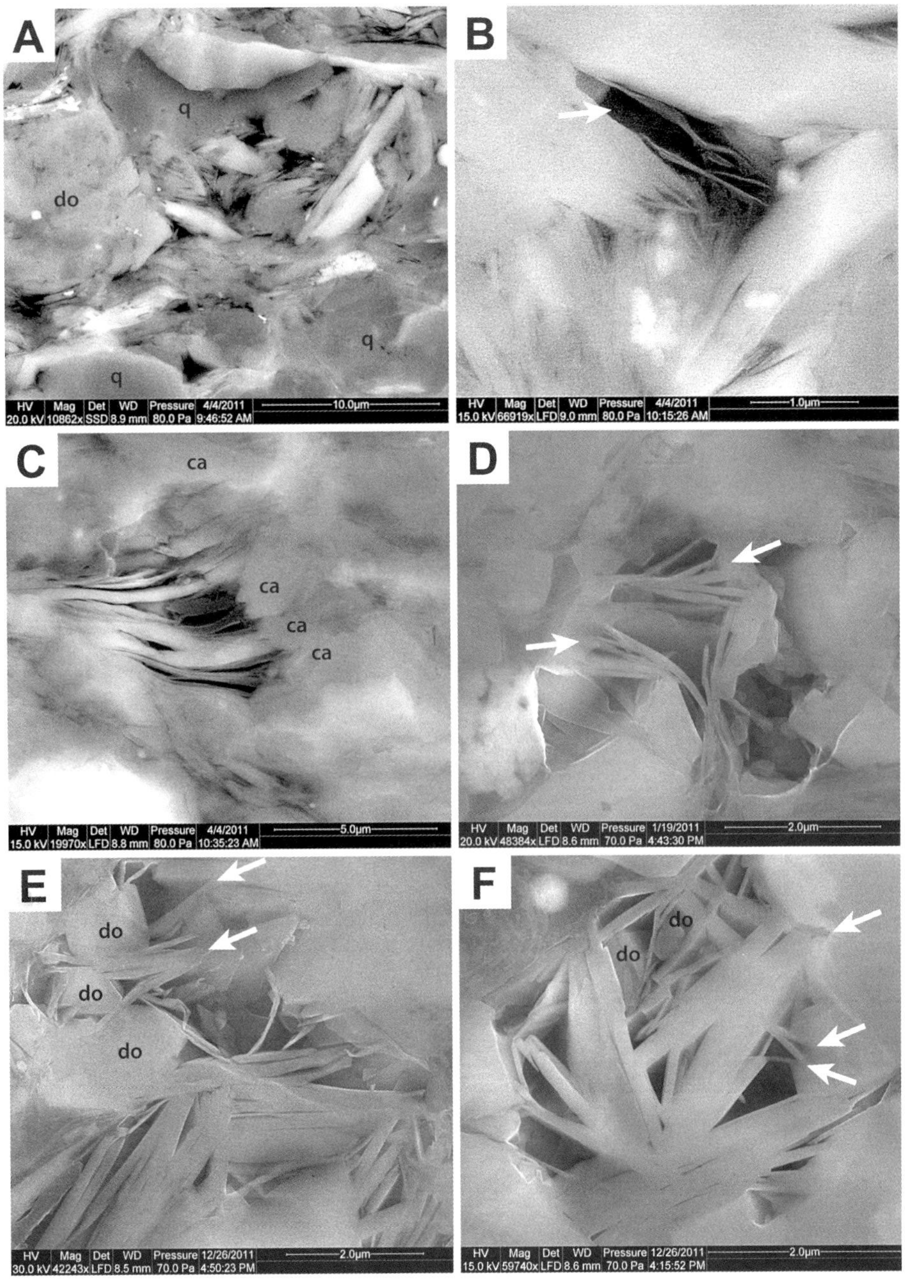

Figure 9. Fabric and phyllosilicate framework pores in samples from the Marcellus Shale. (A) Backscattered electron scanning electron microscopic (SEM) image showing shale matrix consisting of quartz silt grains (q), dolomite grains (do), and clay mineral flakes. Note framework texture of clay flakes in central portion of image. (B) Close-up of large pore (arrow) in framework texture defined by clay minerals (illite). (C) In center of image, clay mineral flakes with intervening pores. To the right (arrows), the pores have been prevented from collapsing by authigenic calcite grains (ca) that propped open the pore space and clamped the clay flakes in place. (D) Phyllosilicate framework and clamping of clay mineral flakes (arrows). (E) Phyllosilicate framework with propping of pores by diagenetic dolomite (do) and clamping of clay mineral in diagenetic carbonate (arrows). (F) Phyllosilicate framework with triangular pores. Some pores propped by diagenetic dolomite. Also, some clay flakes clamped by diagenetic carbonate. (B)–(F) Secondary electron SEM images.

Shale, as well as the lower portion of the Oatka Creek Member, are carbonate rich (unpublished data, L.B. Smith), the system was well buffered against strong pH excursions (unlike the generally carbonate-poor New Albany Shale). Thus, this type of dissolution feature may be a localized phenomenon. Carbonate dissolution pores may be more prevalent in other areas of the basin.

Pores in organic matter (organic matter pores) occur in all samples and are restricted to the unstructured, amorphous organic matter. Amorphous organic matter may include material that was partially degraded by microbes (Pacton et al., 2011), is disseminated throughout the rock matrix, and also fills open spaces in fossil cavities. Structured kerogen (e.g., remains of marine algae such as *Tasmanites*) is devoid of organic matter pores (Figures 12A, B). Amorphous kerogen contains abundant pores in the thermally mature Marcellus Shale samples (Figure 12C). Two types of pores have been observed within amorphous kerogen: (1) abundant tiny pores in the 10 to 50 nm range (foam pores) and (2) a superimposed small population

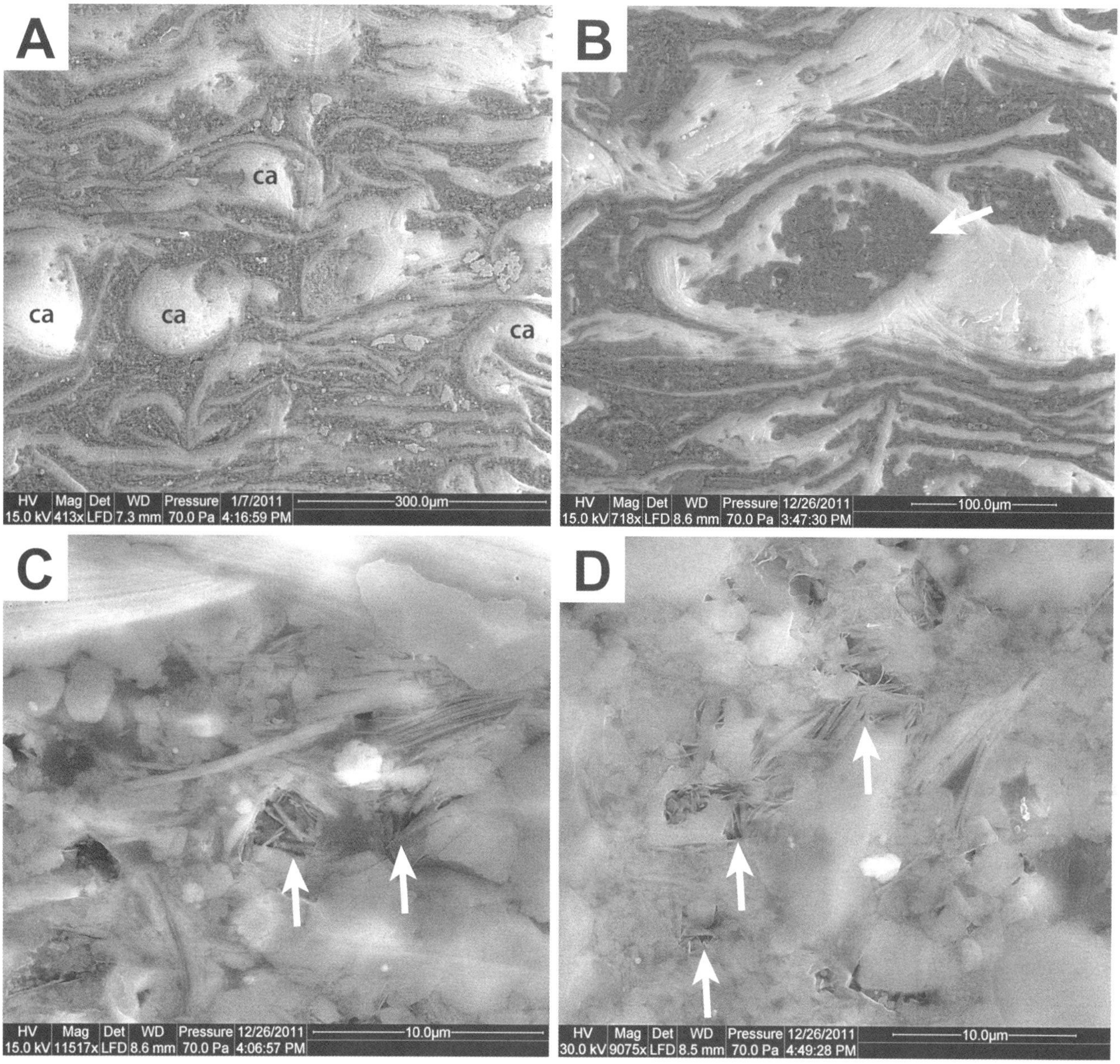

Figure 10. Fossil debris (ostracod shells) shelters framework textures and associated pores from compaction in the Marcellus Shale. (A) Overview secondary electron image with abundant ostracod shells in shale matrix. A number of sheltered spaces are filled with calcite cement (ca). (B) Closer view of a sheltered space (arrow) that is filled with shale matrix. (C), (D) Close-ups of sheltered space from image (C). Arrows point to areas with preserved phyllosilicate framework textures and preserved porosity.

of larger pores (bubble pores) in the 100 to 1000 nm size range (Figures 12D, E, F). Texturally this results in the overall appearance of a sponge. The rounded bubble pores in particular suggest formation in some sort of fluid phase, quite possibly bitumen (Bernard et al., 2012). The close spatial association with foam pores suggests a like origin for these smaller organic matter pores as well. A definite correlation of organic matter pores in the Marcellus with bitumen, however, will require direct confirmation of porous bitumen by other analytical methods (e.g., organic petrography). Nonetheless, the fact that the suspected bitumen areas are typically only a few micrometers in size (Figure 12) makes organic petrographic observations a challenging task. Based on observations of the five Marcellus Shale samples in this study, it appears that organic matter pores may well be the dominant pore type. However, in certain horizons, contributions by phyllosilicate and carbonate-defined frameworks may add to the overall porosity.

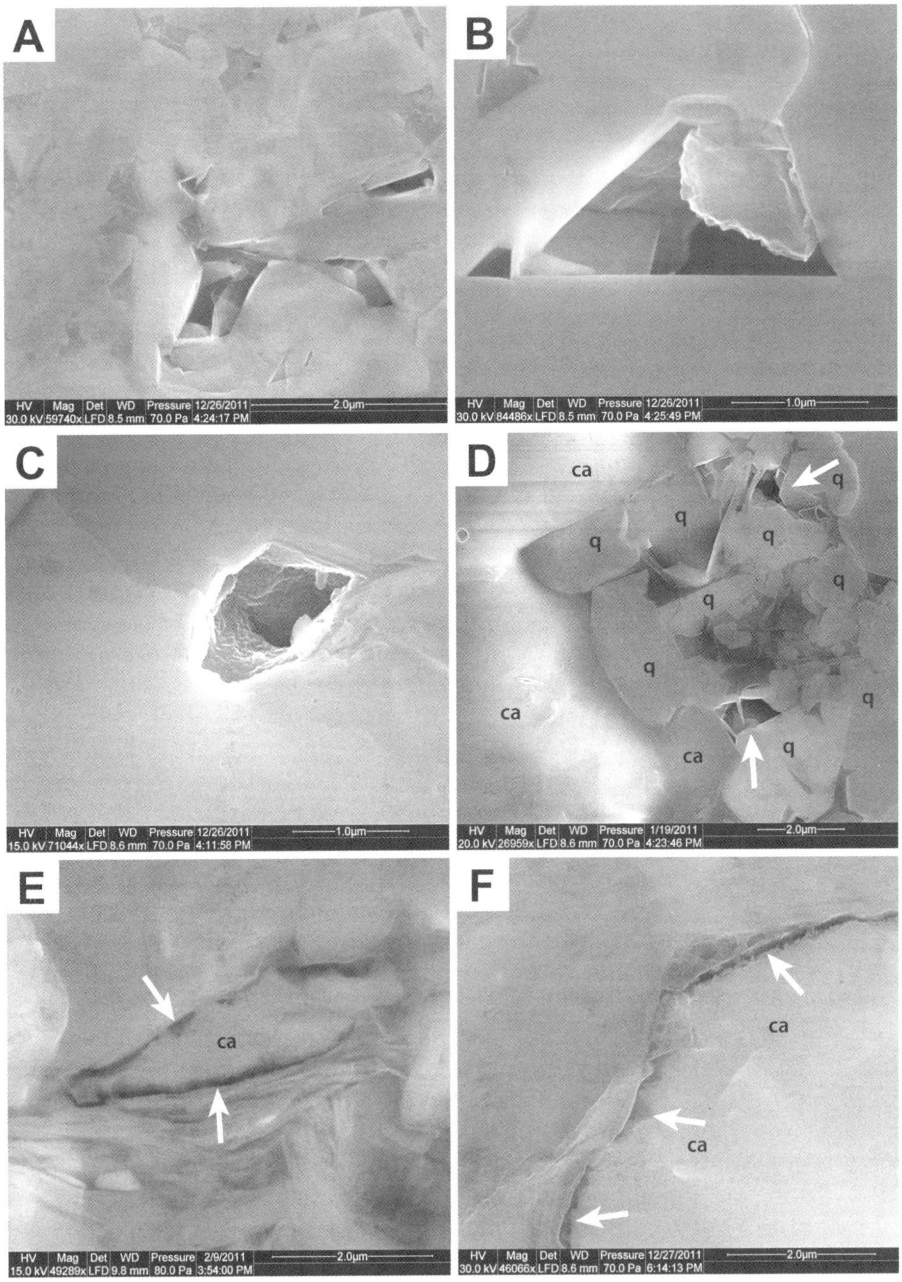

Figure 11. Secondary electron images illustrating pore types in carbonate-rich portions of the Marcellus Shale. (A) Interlocking carbonate grains defining carbonate framework pores. (B) Framework pores bounded by crystal faces of calcite. (C) Framework pore between calcite grains. The dimpled surface in the interior of the pore probably consists of kerogen residues. (D) Calcite grains (ca) surround an area with authigenic quartz (q). Arrows point to pores within quartz grain framework. (E), (F) Carbonate dissolution pores. The rims of calcite grains are partially dissolved (arrows), giving rise to secondary porosity.

DISCUSSION OF PORE OBSERVATIONS

Phyllosilicate framework pores, as illustrated for the New Albany and Marcellus shales, are ubiquitous in many shale successions with an appreciable clay component (Schieber, 2010). These pores have been documented in high-resolution TEM studies of other shale units (e.g., Yau et al., 1987, their figure 2a; Hover et al., 1996, their figure 5). In these earlier studies (Yau et al., 1987; Li et al., 1995; Hover et al., 1996; Rask et al., 1997; Masuada, 2001) the samples were ion milled to a thickness of about 100 to 200 nm, and although the focus was on clay minerals and the smectite-to-illite transformation during burial diagenesis, the pores are clearly visible in published images. Because of the very thin and fragile nature of these TEM samples, however, whether a given triangular opening between clay minerals was a bona fide pore or simply an ion-milling artifact was not easy to assess. Thus, the potential significance of these triangular features, here described as phyllosilicate framework pores, went unrecognized for decades.

Phyllosilicate transformation in the "Zone of Intermediate Burial Diagenesis" (Surdam et al., 1991) has been documented from many sedimentary basins

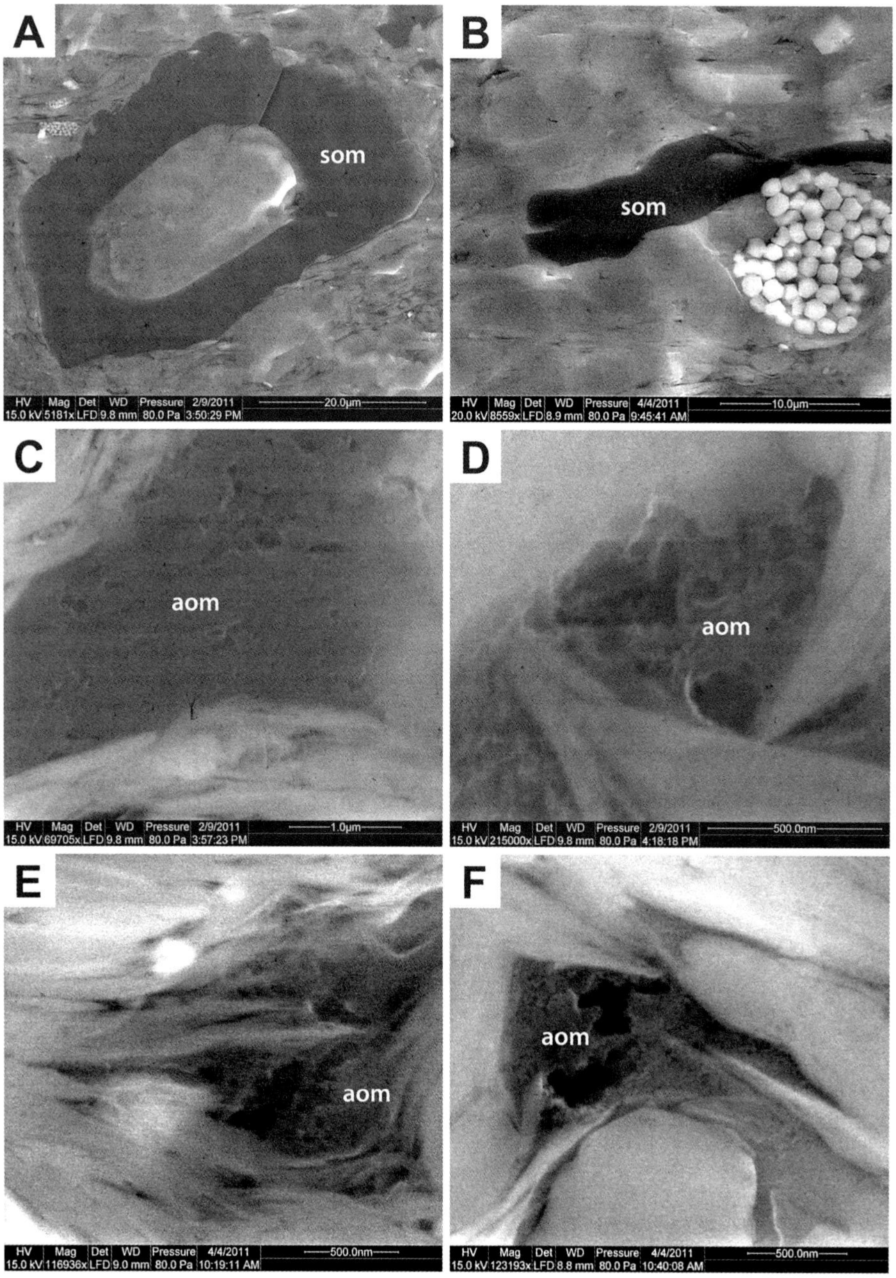

Figure 12. Secondary electron scanning electron microscopic (SEM) images illustrating various forms of organic matter in gas-mature (>2.0%R_o) Marcellus Shale. (A), (B) Structured organic matter (som) in the form of filled (A) and collapsed *Tasmanites* algal cysts (B). The ion-milled surface of this structured kerogen is smooth and shows no visible pores at highest SEM magnification. (C) Interstitial amorphous organic matter (aom). This kerogen material could potentially be bitumen or pyrobitumen and shows pores in the 100 to 500 nm size range. (D) The kerogen/amorphous organic matter (aom) occupies interstitial space between clay minerals. The most abundant pores are in the 20 to 50 nm size range, but a few larger pores in the 100 to 200 nm size range are also present. (E), (F) Kerogen (bitumen?) in interstitial spaces of clay mineral frameworks. In these images, a bimodal pore size distribution is clearly visible. The kerogen groundmass is riddled with small "foam" pores in the 10 to 50 nm size range, and within this foam, larger "bubble" pores that measure 100 to 500 nm in size occur.

and involves the change from smectite to interlayered illite/smectite or to chlorite and/or interlayered chlorite/smectite (e.g., Li et al., 1995; Hover et al., 1996). Contingent on local factors, these phyllosilicate fabrics develop at a burial depth of several kilometers, and the implicit idea has been that they are not predicated by original depositional fabrics. It could thus be argued that phyllosilicate framework pores are to be expected in any shale succession with a sufficient clay content that has been buried deep enough for the onset of the smectite-illite transformation.

In contrast, the observation that phyllosilicate framework pores in the New Albany and Marcellus seem to be associated with lattice frameworks of detrital clay flakes (Figure 3) that imparted partial protection from compaction, and with pressure shadows of compaction-resistant larger grains (Figure 2), suggests that initial depositional fabric leaves a significant imprint on the developing diagenetic fabric parameters. The edge-to-face particle arrangements seen in detrital clay lattice frameworks (Figure 2B) could be a depositional fabric component that reflects deposition of clays from flocculation. This type of fabric has been observed in other ancient shales and has been attributed to a flocculation origin (e.g., O'Brien et al., 1994). Floccule formation occurs in clay suspensions regardless of salinity (Schieber et al., 2007) and is generally considered a fabric characteristic for clay deposition in a wide range of environments (Bennett et al., 1991). Whereas these initial card-house structures are prone to flattening via compaction (Bennett et al., 1991), partial preservation in pressure shadows, such as illustrated in Figures 2E, F, and 3B, C, D, is to be expected. In fact, next to clays, silt grains of variable composition are the most common component of many shales (Potter et al., 2005), and it is probably not a coincidence that a negative correlation between silt content and seal quality has been observed in other shales (Dawson and Almon, 2002). Observations presented here predict that with increasing silt content there should be an increase in the proportion of "pressure-shadowed" rock volume containing phyllosilicate framework pores. This in turn should translate into an increase in porosity/permeability and result in a decrease of seal quality. Thus, the empirical relationship observed by Dawson and Almon (2002) appears to have a quite rational textural explanation.

The diagenetic dimension of phyllosilicate framework porosity is illustrated in Figure 13. Transmission electron microscopic images from the Cretaceous Mancos Shale (Figures 13A, B, C) show sharply defined triangular pores as small as 10 nm, and mineral growth within these clay mineral defined voids (Figure 13A). The fill is apatite in Figure 13A, but quartz is probably most common, followed by carbonates. The clay platelets in Figures 13A, B, are quite uniform and idiomorphic, and Figure 13C shows thin clay packages with layer terminations in the upper half of the image and clays with a fine-grain felted look in the lower half of the image. All of these features suggest diagenetic growth related to the smectite-illite transformation (e.g., Li et al., 1995; Hover et al., 1996). Although diagenetic cementation can potentially occlude phyllosilicate framework pores, it more typically serves to keep them open either because diagenetic cements act as a proppant or by partially engulfing clay platelets during growth and clamping them in place. Whereas fundamentally any diagenetic mineral can function in this capacity, authigenic carbonates and quartz are most commonly observed.

Because the smectite-to-illite conversion overlaps with the onset of chemical reactions involving kerogens, the latter have a tendency to fill phyllosilicate framework pores. In the New Albany Shale, for example, samples with high TOC values typically show kerogen filling phyllosilicate framework pores (Figure 8). In New Albany samples with comparatively low organic matter contents (e.g., Figure 2), phyllosilicate framework pores have remained open even after deep burial and a long diagenetic history.

Secondary porosity development through corrosion of carbonate grains (Figures 6, 11) implies acidic pore waters and a drop of pH. Research on oil field brines and pore water evolution during burial (e.g., Hayes, 1991; Surdam et al., 1991) suggests that the initial stages of smectite transformation are followed by (and overlap with) decarboxylation of kerogen and a buildup of carboxylic and phenolic acids in

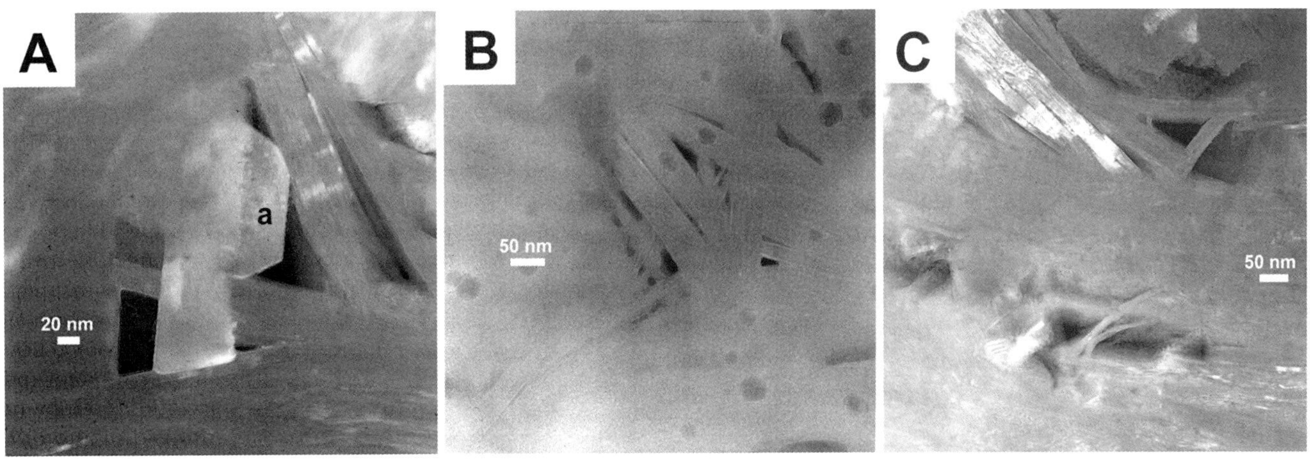

Figure 13. Examples of phyllosilicate framework pores of likely diagenetic origin. All surfaces perpendicular to bedding; images oriented stratigraphic up. (A)–(C) Phyllosilicate framework pores in Mancos Shale (Cretaceous) from the Book Cliffs of Utah. Transmission electron microscopic (TEM) images of samples that were ion milled from both sides. Randomly oriented matrix clays define triangular pores that range in size from 10 to 100 nm. Central pore in (A) is partially filled by diagenetic apatite (marked a).

the temperature range from 80 to 120°C (MacGowan and Surdam, 1990). In sandstones, this stage is associated with formation of secondary porosity (e.g., Hayes, 1991; Surdam et al., 1991) and destruction of carbonates. It seems reasonable to conjecture that the observed carbonate dissolution in the New Albany and Marcellus Shales was likewise linked to organic acid formation. Because the initial amount of organic matter in a shale will place limits on the amount of organic acids that can be generated this way, it seems likely that dissolution porosity of this type is best developed in organic-rich shales. In the overall context, and given that no diagenetic clays were observed in any dissolution cavities, one may conclude that in the studied examples the smectite conversion was largely concluded by the time organic acid production peaked and induced carbonate dissolution.

Because the main phase of hydrocarbon generation and migration follows the formation of diagenetic phyllosilicate frameworks and carbonate dissolution pores (e.g., Hayes, 1991; Surdam et al., 1991), this provides an appealing and logical explanation for the occurrence of amorphous kerogen as well as bitumen, a solid hydrocarbon residue, within phyllosilicate framework pores of thermally mature samples, such as the Marcellus Shale. The amorphous kerogen that occupied pores in the immature New Albany samples may represent an analogous precursor material that would have been present in the immature Marcellus Shale. In the latter, with increasing thermal maturation, the amorphous kerogen may first have converted to bitumen that subsequently developed pores as gaseous hydrocarbons exsolved from the void-filling bitumen. The organic matter pores observed within the gas-mature ($>2.0\%R_o$) Marcellus Shale (Figure 12) are consistent with such a scenario. Because of the higher temperatures required for this step, abundant organic matter pores should only be expected in rocks that have reached sufficient thermal maturity ($>0.6\%R_o$). The observation that well-developed organic matter pores only occurred in the Marcellus Shale samples, a rock unit that has been heated to the dry-gas window ($>2.0\%R_o$; Weary et al., 2000), whereas the similarly organic-rich but immature New Albany Shale ($<0.6\%R_o$; Comer et al., 1994) showed no organic matter pores (Figure 7), supports this assumption. This conclusion is also supported by the work of Loucks and others (2009), who suggested a thermal maturity threshold of greater than $0.7\%R_o$ for the formation of organic matter pores in the Mississippian Barnett Shale of the Fort Worth Basin, Texas. If this reasoning applies broadly, it may also imply that well-developed organic matter porosity (as observed in the Marcellus examples) should occur only within organic-rich (and oil-prone) shales that have reached the gas window or approximately greater than $1.1\%R_o$ (Dow, 1977; Tissot et al., 1987).

In thermally immature rocks, ion-milled amorphous organic material (kerogen) has a smooth appearance, as exhibited by samples from the Devonian New Albany Shale (Figure 8). In rocks that have reached thermal maturity, voids of variable size (a few nanometers to more than a micrometer) are observed in interstitial areas that were once filled by nonporous amorphous kerogen and later converted to bitumen. These were exemplified by samples from the Marcellus Shale (Figure 12). It appears that thermal maturation and migration of hydrocarbons left behind an organic residue that texturally resembles a sponge, consisting of a matrix of organic material with abundant tiny foam pores and superimposed larger bubble pores (Figure 12). The observation that the structured *Tasmanites* kerogen macerals do not exhibit organic matter porosity under the SEM in spite of clearly different levels of thermal maturation between the New Albany (immature) and Marcellus (gas mature) supports the interpretation that organic matter pores are largely restricted to bitumen retained within pore spaces that were filled with amorphous kerogen prior to gas generation.

CONCLUSIONS

Although only two shale units were compared in this study, the pore types illustrated are similar in style to pores that I observed in several other shale units and shale gas reservoirs. Of course, any given shale unit is likely to have unique qualities that affect pore preservation and formation and as such will be different in detail.

Framework, dissolution, and organic matter pore development appear to be an integral part of the depositional, diagenetic, and catagenic history of a given shale succession, particularly the formation of new clay minerals and the maturation of organic matter. It is a well-known fact that freshly deposited modern muds have high water contents, on the order of 80–90% (e.g., Soutar and Crill, 1977; Schimmelmann et al., 1990; Parthenaides, 1991). As water is expelled during burial, the original flocculated fabric collapses and has its best chance of preservation in areas where compaction is minimized by rigid grains (pressure shadows) or within fossil interiors. In addition to this inherited depositional component, phyllosilicate framework pores have a diagenetic component that relates to clay mineral growth during the smectite-to-illite conversion. Phyllosilicate framework pores as described here may be the reason for the observed correlation between silt content and permeability of other shale units (Dawson and Almon, 2002). To a point, higher clay contents imply more abundant phyllosilicate pores, but high clay contents (50% or more) are probably detrimental because they imply lower silt contents and fewer pressure shadows.

Early diagenetic cements (carbonate, quartz), especially when abundant, can prevent fabric collapse through bridge formation between rigid grains and enable preservation of phyllosilicate frameworks with pores. The presence of these cements may also lead to preservation of open pores between adjacent diagenetic grains.

Secondary carbonate dissolution pores rimming carbonate grains appear to form later in diagenetic history and are probably related to formation of carboxylic and phenolic acids in the course of kerogen maturation. Isolated pores are generated at low carbonate contents, and connected pores may result when carbonate grains are abundant (>10 wt.%) or have been winnowed into laminae.

In shales with high organic matter content (>5 wt.% TOC), framework pores may be initially filled by nonporous organic matter (amorphous kerogen) that may be converted to porous bitumen (pyrobitumen?) at elevated thermal maturity. Shales with lower organic matter content (about 1–5 wt.%) may preserve many of the framework pores in spite of considerable burial and thermal maturation. In the latter instance, framework pores are likely connected and thus potentially able to transmit gas. Void-filling bitumen appears to contain abundant organic matter pores once a shale reaches thermal maturity (>0.6%R_o). The exact relationship between the level of maturity and organic matter pore development is a topic that requires further investigation. Structured kerogen macerals, such as alginite (e.g., *Tasmanites*), do not appear to develop organic matter pores, irrespective of thermal maturity. Whether thermal maturation of structured kerogen indeed does not produce pores or whether the pores are simply too small to be visible at SEM magnifications (<5 nm) is another question that deserves some attention.

Although fracture systems are an important element for shale gas production, at a fracture spacing on the order of decimeters to several meters, gas still has to migrate out of blocks of fracture-bounded shale and into these fractures. Thus, the intrinsic porosity described here is a crucial variable that needs to be understood for a realistic evaluation of long-term production from a given shale unit. The totality of pores in a shale are survivors as well as products of diagenetic and catagenic history, and the described pore types should be a common element in all potential shale gas successions.

Matrix pores as described here represent a key control for gas transfer from the shale matrix to the fractures (natural or artificial) and strongly affect the potential for commercial gas production. Porosity and permeability of shales are a multifaceted topic of considerable economic significance, and this contribution gives a shale petrographer's perspective on pore formation and pore preservation in shales. Petroleum engineers use different methodologies to quantify porosity, such as mercury porosimetry (Thompson et al., 1987) and nitrogen low-pressure isotherm analysis (Sing, 2001). Porosity and permeability data derived by these methods, however, need not necessarily show a close match to independently derived porosity and permeability data because multiple simplifications and assumptions exist (Sing et al., 1985) that fail to do justice to the inherent complexity of fine-grain sediments (Soeder, 1988). Petrographic evaluation of porosity provides an invaluable reality check for these method-driven measurements of porosity and permeability.

ACKNOWLEDGMENTS

I would not have come this far in my work on shale fabrics had it not been for a few lucky coincidences along the way. A bit over two decades ago, David Krinsley provided the initial impetus to incorporate various SEM-based techniques into my research. His enthusiasm was essential at a time when sample preparation techniques still imposed limits on what could be seen via SEM. A decade later, Michael Coviello, a material sciences TEM specialist at UT Arlington, taught me a lot about ion milling samples for electron microscopy. At the same time, Howard Arnott, a UT Arlington biologist, provided access to his well-equipped electron microscopy lab and introduced me to the bag of tricks that biologists use to work with soft and fragile samples. What I learned in these labs has been extremely helpful in developing successful protocols (not all shales are created equal) for ion milling shale samples. More recently, a National Science Foundation (NSF) equipment grant (EAR-0318769) provided funds for the purchase of the analytical SEM (FEI Quanta 400 FEG) that was used for acquiring most of the images used in this contribution. John Rupp of the Indiana Geological Survey and Taury Smith of the New York State Museum were most helpful with providing access to core material and background information about the studied samples. Ion milling and SEM observation of Marcellus Shale samples were generously supported by Malcolm Pirnie Inc. GATAN Inc. loaned a beta model of their Ilion ion mill to the IU Shale Research Lab.

REFERENCES CITED

Bennett, R. H., N. R. O'Brien, and M. H. Hulbert, 1991, Determinants of clay microfabric signatures—Processes and mechanisms, *in* R. H. Bennett, W. R. Bryant, and M. H. Hulbert, eds., Microstructure of fine-grained sediments: New York, Springer-Verlag, p. 5–32.

Bernard, S., B. Horsfield, H. M. Schulz, R. Wirth, A. Schreiber, and N. Sherwood, 2012, Geochemical evolution of organic-rich shales with increasing maturity: A STXM and TEM study of the Posidonia Shale (Lower Toarcian, northern Germany): Marine and Petroleum Geology, v. 31, p. 70–89.

Bollinger, D., and R. Fink, 1980, A new production technique: Ion milling: Solid-State Technology, p. 79–84.

Broadhead, R. F., R. C. Kepferle, and P. E. Potter, 1982, Stratigraphic and sedimentologic controls of gas in shale—Example from Upper Devonian of northern Ohio: AAPG Bulletin, v. 66, p. 10–27.

Coleman, J. L., R. C. Milici, T. A. Cook, R. R. Charpentier, M. Kirschbaum, T. R. Klett, R. M. Pollastro, and C. J. Schenk, 2011, Assessment of undiscovered oil and gas resources of the Devonian Marcellus Shale of the Appalachian Basin province, 2011: USGS Fact Sheet 2011–3092, 2 p.

Comer, J. B., T. Hamilton-Smith, and W. T. Frankie, 1994, Source rock potential, *in* N. R. Hasenmueller and J. B. Comer, eds., Gas potential of the New Albany Shale (Devonian and Mississippian) in the Illinois Basin: Illinois Basin Consortium, Gas Research Institute 92-0391/ Illinois Basin Studies 2, p. 47–54.

Davies, D. K., W. R. Bryant, R. K. Vessell, and P. J. Burkett, 1990, Porosities, permeabilities and microfabrics of Devonian shales, *in* R. H. Bennett, W. R. Bryant, and M. H. Hulbert, eds., Microstructure of fine-grained sediments: From mud to shale: New York, Springer-Verlag, p. 109–119.

Dawson, W., and W. R. Almon, 2002, Top seal potential of Tertiary deep-water shales, Gulf of Mexico: Gulf Coast Association of Geological Societies Transactions, v. 52, p. 167–176.

Dow, W. G., 1977, Kerogen studies and geological interpretations: Journal of Geochemical Exploration, v. 7, p. 79–99.

Hayes, J. B., 1991, Porosity evolution of sandstones related to vitrinite reflectance: Organic Geochemistry, v. 17, p. 117–129.

Hover, V. C., D. R. Peacor, and L. M. Walter, 1996, STEM/ AEM evidence for preservation of burial diagenetic fabrics in Devonian shales: Journal of Sedimentary Research, v. 66, p. 519–530.

Jiang, W.T., D. R. Peacor, and E. J. Essene, 1990, Transmission electron microscopic study of coexisting pyrophyllite and muscovite—Direct evidence for the metastability of illite: Clays and Clay Minerals, v. 38, p. 225–240.

Lazar, O. R., 2007, Redefinition of the New Albany Shale of the Illinois Basin: An integrated stratigraphic, sedimentologic, and geochemical study: unpublished Ph.D. thesis, Indiana University, Bloomington, Indiana, 336 p.

Lazar, O.R., and J. Schieber, 2004, Summary of observations from a 40-meter cored interval of the New Albany Shale in well 1-3 Kavanaugh, Daviess County, Indiana, *in* J. Schieber and O. R. Lazar, eds., Devonian black shales of the eastern U.S: New insights into sedimentology and stratigraphy from the subsurface and outcrops in the Illinois and Appalachian basins: Field Guide for the 2004 Great Lakes Section SEPM Annual Field Conference, Indiana Geological Survey Open File Study 04-05, p. 1–6.

Lineback, J. A., 1970, Stratigraphy of the New Albany Shale in Indiana: State of Indiana, Department of Natural Resources, Geological Survey Bulletin, v. 44, 73 p.

Loucks, R. G., R. M. Reed, S. C. Ruppel, and D. M. Jarvie, 2009, Morphology, genesis, and distribution of nanometer scale pores in siliceous mudstones of the Mississippian Barnett Shale: Journal of Sedimentary Research, v. 79, p. 848–861.

MacGowan, D. B., and R. C. Surdam, 1990, Carboxylic acid anions in formation waters, San Joaquin Basin and Louisiana Gulf Coast, U.S.A—Implications for clastic diagenesis: Applied Geochemistry, v. 5, p. 687–701.

O'Brien, N. R., C. E. Brett, and W. L. Taylor, 1994, Microfabric and taphonomic analysis in determining sedimentary processes in marine mudstones: Example from Silurian of New York: Journal of Sedimentary Research, v. A64, p. 847–852.

Pacton, M., G. E. Gorin, and C. Vasconcelos, 2011, Amorphous organic matter—Experimental data on formation and the role of microbes: Review of Paleobotany and Palynology, v. 166, p. 253–267.

Parthenaides, E., 1991, Effect of bed shear stresses on the deposition and strength of deposited cohesive muds, *in* R. H. Bennett, W. R. Bryant, and M. H. Hulbert, eds., Microstructure of fine-grained sediments: New York, Springer-Verlag, p. 175–183.

Potter, P. E., J. B. Maynard, and P. J. Depetris, 2005, Mud and mudstones: Berlin, Springer-Verlag, 297 p.

Rask, J. H., L. T. Bryndzia, N. R. Braunsdorf, and T. E. Murray, 1997, Smectite illitization in Pliocene-age Gulf of Mexico mudrocks: Clays and Clay Minerals, v. 45, p. 99–109.

Schieber, J., 1998, Deposition of mudstones and shales: Overview, problems, and challenges, *in* J. Schieber, W. Zimmerle, and P. Sethi, eds., Shales and mudstones (vol. 1): Basin studies, sedimentology and paleontology: Stuttgart, Schweizerbart'sche Verlagsbuchhandlung, p. 131–146.

Schieber, J., 2002, Sedimentary pyrite: A window into the microbial past: Geology, v. 30, p. 531–534.

Schieber, J., 2004, SEM and TEM study of silica diagenesis in shales from the Cretaceous Blackhawk Formation, Book Cliffs, Utah: Implications for sealing capacity: AAPG Annual Meeting in Dallas, Abstract Volume, p. A124.

Schieber, J., 2010, Common themes in the formation and preservation of intrinsic porosity in shales and mudstones—Illustrated with examples across the Phanerozoic: Society of Petroleum Engineers Unconventional Gas Conference, February 23–25, 2010, Pittsburgh, Pennsylvania, SPE Paper 132370, 10 p., doi: 10.2118/132370-MS.

Schieber, J., 2011, Shale microfabrics and pore development—An overview with emphasis on the importance of depositional processes, *in* D. A. Leckie and J. E. Barclay, eds.,

Gas shale of the Horn River Basin: Calgary, Canadian Society of Petroleum Geologists, p. 115–119.

Schieber, J., J. B. Southard, and K. G. Thaisen, 2007, Accretion of mudstone beds from migrating floccule ripples: Science, v. 318, p. 1760–1763.

Schimmelmann, A., Lange, C. B., and Berger, W. H., 1990, Climatically controlled marker layers in Santa Barbara basin sediments and fine-scale core-to-core correlation: Limnology and Oceanography, v. 35, p. 165–173.

Shirley, K., 2001, Shale gas exciting again: AAPG Explorer, March, p. 1–4.

Sing, K. S. W., 2001, The use of nitrogen adsorption for the characterization of porous materials: Colloids and Surfaces A: Physicochemical and Engineering Aspects, v. 187–188, p. 3–9.

Sing, K. S. W., D. H. Everett, R. A. W. Haul., L. Moscou, R. A. Pierotti, J. Rouquerol, and T. Siemieniewska, 1985, Reporting physisorption data for gas/solid systems with special reference to the determination of surface area and porosity: Pure and Applied Chemistry, v. 57, p. 603–619.

Soeder, D. J., 1988, Porosity and permeability of eastern Devonian gas shale: SPE Formation Evaluation, vol. 3, no. 2, p. 116–124.

Soutar A., and P. A. Crill, 1977, Sedimentation and climatic patterns in the Santa Barbara Basin during the 19th and 20th centuries: GSA Bulletin, v. 88, p. 1161–1172.

Surdam, R. C., D. B. MacGowan, and T. L. Dunn, 1991, Predictive models for sandstone diagenesis: Organic Geochemistry, v. 17, p. 243–253.

Taylor, G. H., M. Teichmüller, and C. Davis, 1998, Organic petrology: Stuttgart, Gebrüder Borntreager, 704 p.

Thompson, A. H., A. J. Katz, and R. A. Raschke, 1987, Mercury injection in porous media: A resistance devil's staircase with percolation geometry: Physical Review Letters, v. 58, p. 29–32.

Tissot, B. P., R. Pelet, and P. Ungerer, 1987, Thermal history of sedimentary basins, maturation indices, and kinetics of oil and gas generation: AAPG Bulletin, v. 71, p. 1445–1466.

Tomutsa, L., D. Silin, and V. Radmilovic, 2007, Analysis of chalk petrophysical properties by means of submicron-scale pore imaging and modeling: SPE Reservoir Evaluation and Engineering, v. 10, p. 285–293.

Yau, Y. C., D. R. Peacor, and S. D. McDowell, 1987, Smectite-to-illite reactions in Salton Sea shales: A transmission and analytical electron microscopy study: Journal of Sedimentary Petrology, v. 57, p. 335–342.

Weary, D. J., R. T. Ryder, and R. Nyahay, 2000, Thermal maturity patterns (CAI and $\%R_o$) in the Ordovician and Devonian rocks of the Appalachian basin in New York: USGS Open-File Report 00-496, USGS, Reston, Virginia.

Glossary

Argon (Ar)-Ion Milling (AIM): Process of creating ultrasmoothed surface using an accelerated beam of ionized argon gas to remove (sputter) small amounts of material (hundreds of microns deep) from the specimen (a single cut is typically 1.5 mm wide). Rock samples are usually milled perpendicular to bedding. Some argon-ion mills include the capability to cool the specimen using liquid nitrogen to eliminate any potential thermal alterations.

Artifact: Artificial product or formation of a microscopic specimen created during sample preparation or by electron beam reactions that is not indicative of the true specimen.

Atomic Number (Z): Number of protons in the nucleus of an atom.

Backscattered Electrons (BSEs): Electrons produced by elastic collisions of primary beam electrons with atomic nuclei of the specimen. The number of backscattered electrons (backscatter intensity) increases with the mean atomic number (Z) of the area sampled by the electron probe. Backscattered electron images from flat (polished or ion-milled) surfaces reflect compositional variations of the specimen.

Bitumen: Chemically defined as the fraction of naturally occurring solid organic matter in sedimentary rocks that is soluble in organic solvents and fusible.

Charging Effect: A negative charge can accumulate on nonconductive materials and deflect secondary electrons, producing image artifacts. Charging can be reduced by applying a conductive coating (e.g., gold or carbon) or by operating at low accelerating voltage or low vacuum pressure with uncoated specimens.

Clay: May refer to either a sedimentary particle size (<4 μm in diameter) or a type of phyllosilicate mineral classified into several groups, including chlorite, illite, kaolinite, mixed layer, and smectite.

Coccolith: Calcareous nannofossil (<63 μm in diameter) formed by coccolithophores (calcareous single-cell phytoplankton). Tests of coccolithophores are made up of numerous coccoliths and are known as coccospheres (although not necessarily spherical). Common component of chalk and Mesozoic pelagic calcareous sedimentary rocks.

Computed Tomography (CT): Computer-generated, three-dimensional image created from x-ray images taken around an axis of rotation.

Cross Section Polisher (CP): *See* argon-ion milling.

Digital Rock Physics (DRP): Analytical technique combining high-resolution digital image analysis and numerical simulation to characterize physical rock properties using a variety of image sources, such as x-ray computed tomography (CT) and scanning electron microscopy (SEM).

Edge Effect: Secondary electron yield is highest at thin edges of three-dimensional objects that result in bright rims surrounding sharp edges in secondary electron images.

Element Map: Spatial distribution of a specific element by recording x-ray (EDS) measurements during a raster scan of the specimen. Element maps are commonly displayed in false color as montages of multiple elements or as overlays to produce a composite element map.

Energy Dispersive X-ray Spectrometry (EDS or EDXS): Analytical technique used to identify the elemental composition of a specimen by measuring the energy spectrum of x-rays emitted from a specimen resulting from inelastic collisions of electron bombardment of atomic nuclei.

Environmental Scanning Electron Microscope (ESEM): Type of electron microscope operated under low-vacuum pressure enabling examination of uncoated conductive samples in their natural state, which may contain volatiles such as oil or water, without producing charging effects.

Field Emission Scanning Electron Microscope (FE-SEM): Type of scanning electron microscope equipped with a field emission source that produces an electron beam with a narrower radius than a conventional tungsten filament electron gun, resulting in higher-resolution SEM images.

Fissility: Ability to easily split a rock into thin planes nearly parallel to bedding.

Focused Ion Beam Scanning Electron Microscope (FIB-SEM): Type of scanning electron microscope that combines both electron and ion beam columns within

DOI: 10.1306/13391716M1021681

one instrument, providing dual-source imaging and gallium (Ga)-ion milling capabilities. Thin slices of material are removed during ion milling, and specimens can be alternately milled and imaged to produce a three-dimensional digital image.

Foraminiferan: Marine unicellular organisms that form tests (shells) composed of calcium carbonate, agglutinated sediment particles, or rarely silica. Both planktonic (floating) and benthic (bottom-dwelling) forms exits. Range from Cambrian to present.

Framboid: Microscopic spherical mineral crystal aggregates (commonly pyrite). Named after the French word *framboise* (raspberry) reflecting the similar morphological appearance.

Kerogen: Chemically defined as the fraction of solid organic matter in sedimentary rocks that is insoluble in organic solvents. Classified by atomic H/C and O/C ratios as type I, II, III, and IV.

Maceral: Preserved remains of predominantly plant material observed by reflected or transmitted light microscopy. Classified into three main groups: liptinite, vitrinite, and inertinite.

Micrometer or Micron (μm): Metric unit of measure equal to one-millionth of a meter (1×10^{-6} m) or one-thousandth of a millimeter (0.001 mm).

Mudrock: Colloquial term generally used to include both shale and mudstone. Not equivalent to the formal term *mudstone*, which is restricted from shale (*see* Mudstone).

Mudstone: Fine-grain clastic sedimentary rock composed of lithified mud (mixture of silt- and clay-size particles) that does not exhibit fissility.

Nanometer (nm): Metric unit of measure equal to one-billionth of a meter (1×10^{-9} m) or one-thousandth of a micrometer (0.001 μm).

Pyrobitumen: Insoluble and nonvolatile solid residue derived from the thermal degradation of hydrocarbons. May yield bitumen on heating.

QEMSCAN: Acronym for quantitative evaluation of minerals by scanning electron microscopy. Computer-automated microanalysis device combining EDS and SEM to produce mineral maps. Digital image analysis provides statistical evaluation of mineral, grain and pore frequency and size distributions.

Rock-Eval Pyrolysis: Method to determine source rock quality by programmed heating of sample in a pyrolysis oven in an inert atmosphere. Measurements include amount of free hydrocarbons (S_1), hydrocarbons generated by thermal cracking of kerogen (S_2), and amount of CO_2 produced (S_3).

S2 Yield: Amount of hydrocarbons in milligrams derived from labile organic matter on pyrolysis per gram of sample.

Scanning Electron Microscope (SEM): Type of electron microscope that produces images by scanning an electron beam in a rectangular pattern (raster scan) over the specimen surface.

Scanning Transmission X-ray Microscope (STXM): Type of microscope that focuses monoenergetic x-rays through a thin section of a specimen. Images are produced by monitoring the x-ray signal transmitted through the sample as it is scanned across the specimen.

SE2: Notation is used to differentiate secondary electrons produced by backscattered electrons exiting the specimen's surface instead of being produced from the primary electron beam. Some compositional contrast is contained within the SE2 signal because of the greater penetration level of backscattered electrons.

Secondary Electron (SE): Low-energy (<50 eV) electrons originating in the specimen that are ejected from the surface because of ionization by an incident primary beam electron. The number of secondary electrons ejected (secondary electron yield) decreases with increasing incident angle between the electron beam and specimen surface. Secondary electron images reveal high-resolution topographic features of the specimen's surface.

Shale: Fine-grain clastic sedimentary rock composed of lithified mud (mixture of silt- and clay-size particles) that exhibits fissility.

Source Rock Analyzer (SRA): Portable pyrolysis instrument. *See* Rock-Eval.

Tasmanites: Organic microfossils interpreted as unicellular prasinophyte algal cysts named from occurrences in Permian oil shales in Tasmania. Type of kerogen maceral (alginate).

T_{max}: Thermal maturity indicator corresponding to temperature (in celsius) at maximum S_2 production from pyrolysis analysis. Three main categories: immature (400–430°C), mature oil (435–450°C), and overmature (>450°C).

TOC: Fraction of organic carbon content of sample in weight percent (wt. %).

Transmission Electron Microscope (TEM): Type of electron microscope that produces images by transmitting a beam of high-energy electrons through an ultrathin specimen.

Vitrinite Reflectance (VR_O): Thermal maturity indicator. Optical microscope measurement of the mean random reflectance measured in oil (%R_o) of polished surfaces of vitrinite macerals dispersed in sedimentary rocks.

Vitrinite reflectance increases with thermal maturity. There are three main categories: immature (<0.5%R_o), mature (0.5–1.35%R_o), and overmature (>1.35%R_o) with respect to generation of liquid hydrocarbons.

Vitrinite Reflectance Equivalent (VR_E): Thermal maturity indicator. Optical microscope measurement of the mean random reflectance measured in oil (%R_o) of polished surfaces of nonvitrinite organic matter (typically bitumen) converted to equivalent vitrinite reflectance from empirical data.

X-ray Absorption Near-Edge Spectroscopy (XANES): Analytical technique used to identify materials by measuring the intensity of photoelectrons created by absorbed x-rays from a specimen under a scanning transmission x-ray microscope.

X-ray Powder Diffraction (XRD): Analytical method to determine mineral composition based on identification of unique patterns of x-rays diffracted by crystalline materials (diffraction spectra). Typically measured from pulverized (powder) samples.

Image Catalog

Table 1. List of Formations Contained in the Image Catalog (Part 2)

Map Number	Formation	Age (Epoch)
1	Antelope	Late Miocene
2	Bakken	Late Devonian
3	Barnett	Mississippian
4	Boquillas	Late Cretaceous
5	Bossier	Late Jurassic
6	Chattanooga	Late Devonian
7	Devonian Shale	Late Devonian
8	Devonian Limestone	Devonian
9	Eagle Ford	Late Cretaceous
10	Fayetteville	Mississippian
11	Floyd	Mississippian
12	Green River	Eocene
13	Haynesville	Late Jurassic
14	HRZ	Late Cretaceous
15	Manning Canyon	Mississippian
16	Marcellus	Middle Devonian
17	Monterey	Late Miocene
18	Mowry	Late Cretaceous
19	Mulky	Pennsylvanian
20	New Albany	Middle Devonian
21	Niobrara	Late Cretaceous
22	Ohio	Late Devonian
23	Paradox	Pennsylvanian
24	Pearsall	Early Cretaceous
25	Pierre	Late Cretaceous
26	Spraberry	Early Permian
27	Tuscaloosa	Late Cretaceous
28	Utica	Middle Ordovician
29	Wolfcamp	Early Permian
30	Woodford	Late Devonian

Map number corresponds with numbers shown on Formation Distribution Map.

DOI: 10.1306/13391717M1021681

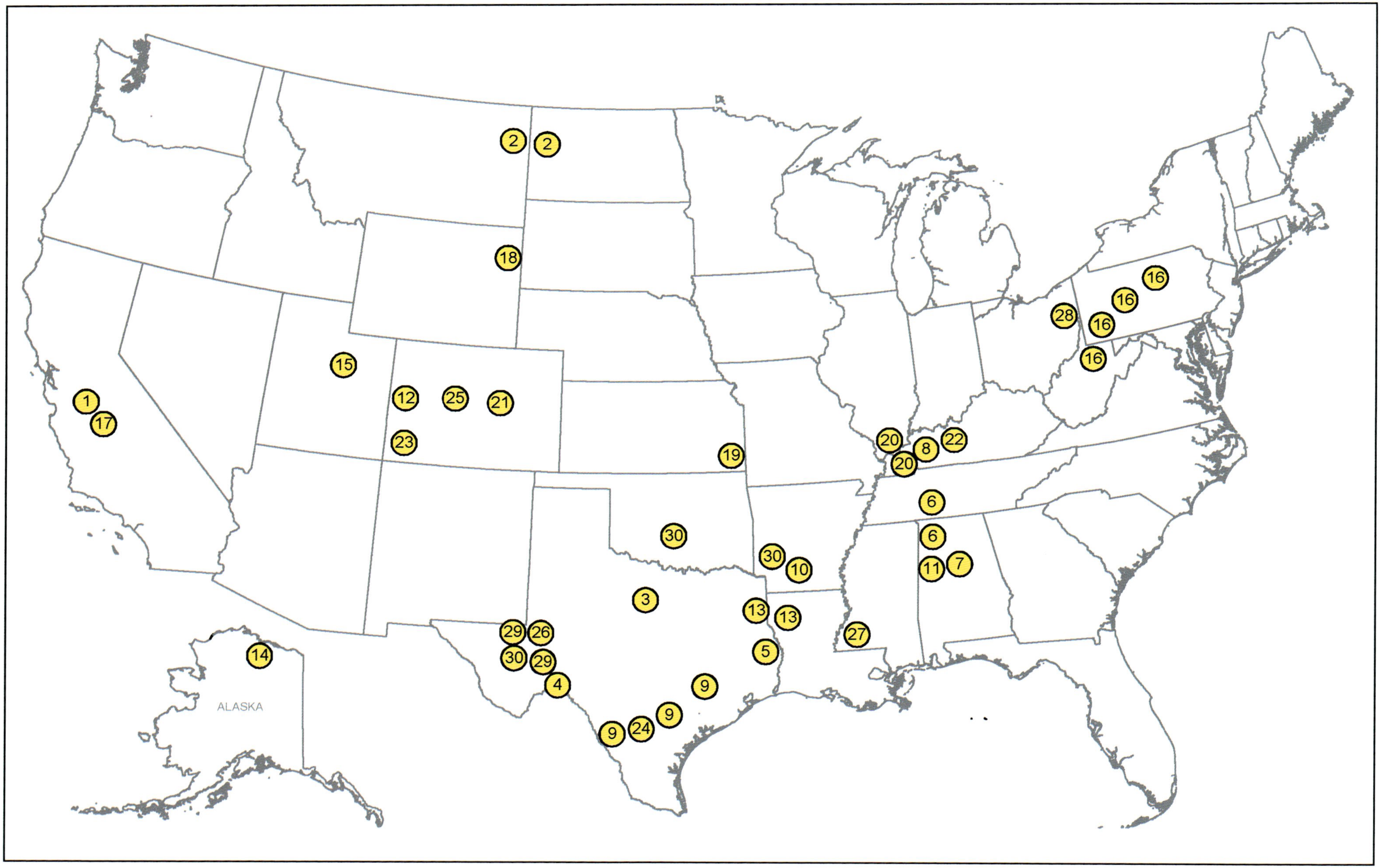
ALASKA

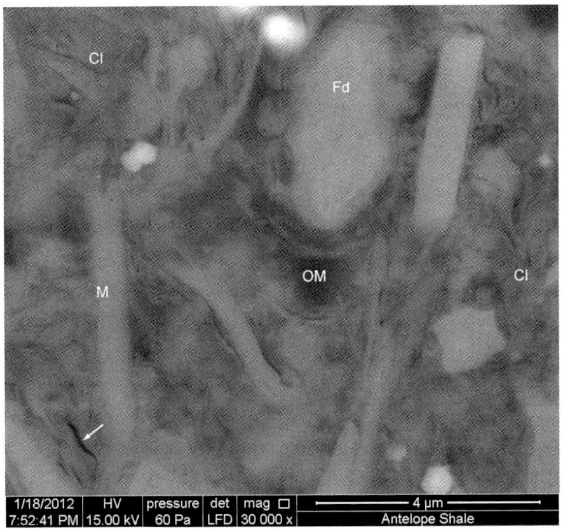

F.1.Antelope
Antelope Shale (Late Miocene)
San Joaquin Basin, California
Ar-ion milled, SE image
Thermal maturity: 0.41%R_o
Submitted by B. Wawak and T. Gentzis, Core Laboratories
Various clay-size particles seen in this image include mica (M), other phyllosilicate minerals (Cl, clays), organic matter (OM), and feldspar (Fd). White particles are pyrite. Interparticle pores associated with clays (arrow) are also present.

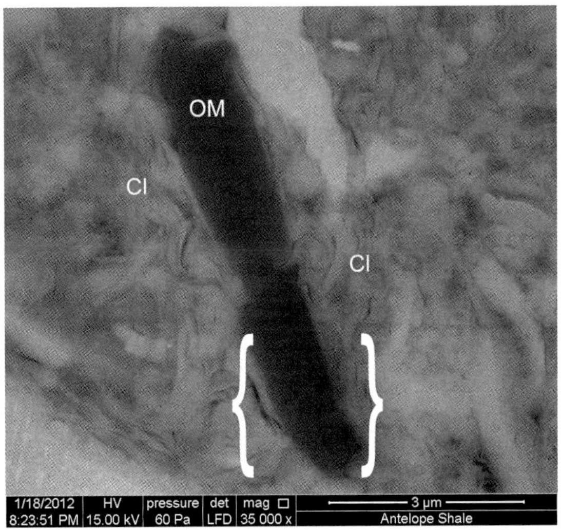

F.2.Antelope
Antelope Shale (Late Miocene)
San Joaquin Basin, California
Ar-ion milled, SE image
Thermal maturity: 0.41%R_o
Submitted by B. Wawak and T. Gentzis, Core Laboratories
Silt-size particle of smooth-textured organic matter (OM) in a matrix of clays and quartz. Most of particles in adjacent matrix are clays (Cl) associated with some interparticle pores. Area indicated by brackets is enlarged in Figure 3.

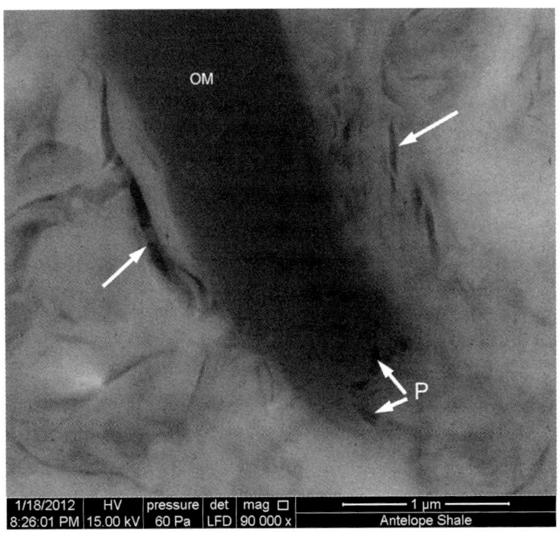

F.3.Antelope
Antelope Shale (Late Miocene)
San Joaquin Basin, California
Ar-ion milled, SE image
Thermal maturity: 0.41%R_o
Submitted by B. Wawak and T. Gentzis, Core Laboratories
A higher-magnification view of part of Figure 2 showing small, irregularly shaped nanopores (P) at edge of the particle of organic matter (OM) and interparticle pores associated with adjacent clay minerals (arrows).

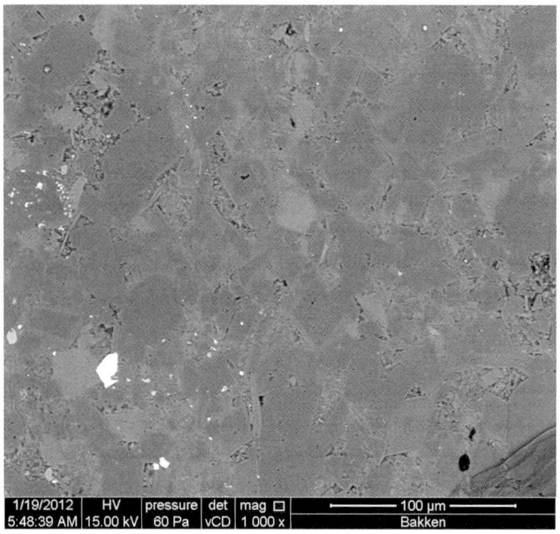

F.4.Bakken

Bakken Formation (Late Devonian)
Williston Basin, North Dakota
Ar-ion milled, BSE image
Thermal maturity: 0.87%R_o
Submitted by B. Wawak and T. Gentzis, Core Laboratories
Relatively low-magnification view showing the crystalline texture of this dolomitic mudstone. Most of the dolomite crystals have light-color ferroan rims around darker cores of nonferroan dolomite. Pinpoint pores are relatively abundant in the cores of the dolomite crystals. White particles are pyrite.

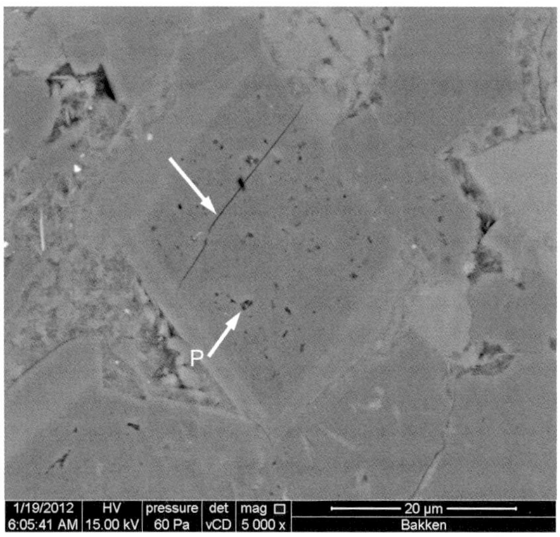

F.5.Bakken

Bakken Formation (Late Devonian)
Williston Basin, North Dakota
Ar-ion milled, BSE image
Thermal maturity: 0.87%R_o
Submitted by B. Wawak and T. Gentzis, Core Laboratories
Dolomite crystal containing a microfracture that extends across the width of the crystal (arrow). Fracture appears to be confined to the nonferroan core of the crystal. Note also abundance of intracrystalline pores (P). The rim is lighter color in this backscattered electron image because of the occurrence of elevated amounts of iron.

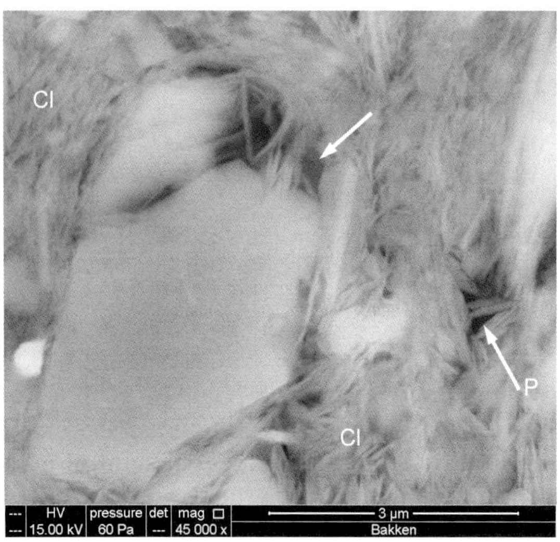

F.6.Bakken

Bakken Formation (Late Devonian)
Williston Basin, North Dakota
Ar-ion milled, BSE/SE image
Thermal maturity: 0.87%R_o
Submitted by B. Wawak and T. Gentzis, Core Laboratories
Intercrystalline pores filled with phyllosilicate minerals, predominantly clays (Cl) with small amount of organic matter (arrow). Note occurrence of interparticle pore (P) associated with clays. Large crystal in lower left side of image is dolomite.

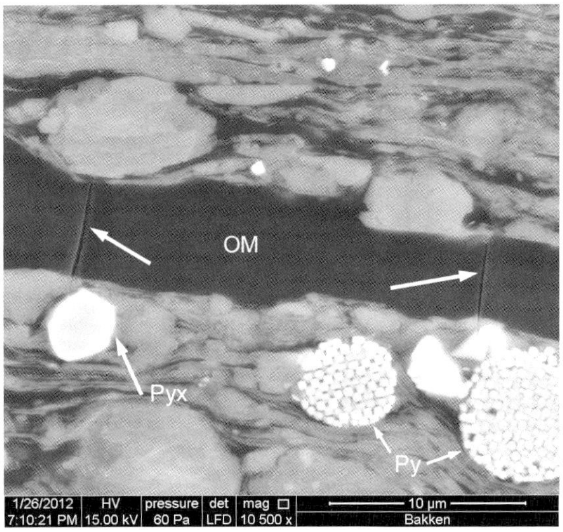

F.7.Bakken
Bakken Formation (Late Devonian)
Williston Basin, Montana
Ar-ion milled, SE image
Thermal maturity: 2.27%R_o
Submitted by B. Wawak and T. Gentzis, Core Laboratories
Large discrete particle of organic matter (OM) with fractures extending across the body of the particle (arrows). Note also large crystal of pyrite (Pyx) and pyrite framboids (Py). Near-parallel orientation of phyllosilicate minerals with large elongated particle of organic matter shows oriented fabric of this sample.

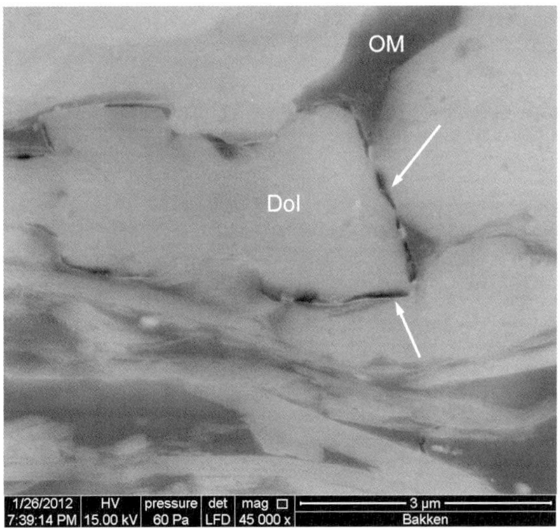

F.8.Bakken
Bakken Formation (Late Devonian)
Williston Basin, North Dakota
Ar-ion milled, BSE/SE image
Thermal maturity: 2.27%R_o
Submitted by B. Wawak and T. Gentzis, Core Laboratories
Elongated pores (arrows) occurring along the edges of a dolomite (Dol) grain. These pores likely are a result of partial dissolution of the dolomite grain by organic acids associated with the organic matter (OM).

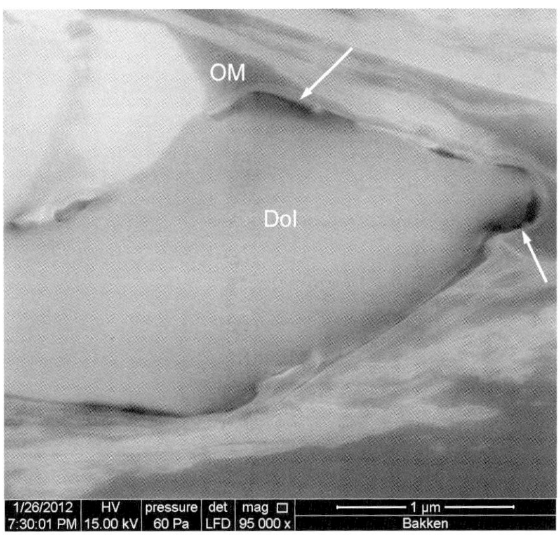

F.9.Bakken
Bakken Formation (Late Devonian)
Williston Basin, Montana
Ar-ion milled, SE image
Thermal maturity: 2.27%R_o
Submitted by B. Wawak and T. Gentzis, Core Laboratories
Pores (arrows) associated with edges of dolomite (Dol) grains. These pores likely are a result of partial dissolution of the dolomite by organic acids associated with the organic matter (OM). Except for these open pores, organic matter (dark gray) fills all of the interparticle pores in this part of the sample.

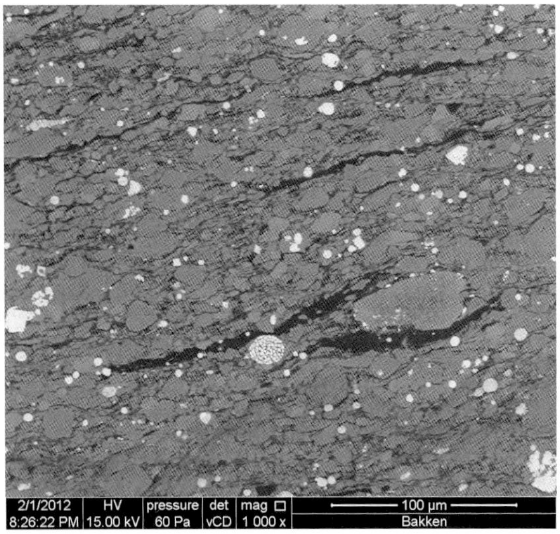

F.10.Bakken

Bakken Formation (Late Devonian)
Williston Basin, North Dakota
Ar-ion milled, BSE image
Thermal maturity: 0.70%R_o
Submitted by B. Wawak and T. Gentzis, Core Laboratories
Relatively low-magnification overview showing oriented fabric of sample. The fabric is accentuated by subparallel alignment of elongated organic matter (black). Note also an abundance of pyrite (white) that occurs as individual crystals and framboids.

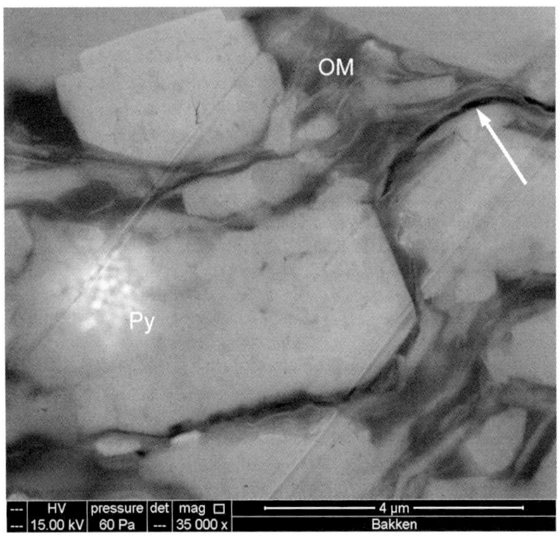

F.11.Bakken

Bakken Formation (Late Devonian)
Williston Basin, North Dakota
Ar-ion milled, BSE image
Thermal maturity: 0.70%R_o
Submitted by B. Wawak, Core Laboratories
Organic matter (OM) fills interparticle pore in this part of the sample. The only open pores are elongated pores associated with the edges of grains (arrow), which may be stress-relief fractures caused by decompression of the core as it is retrieved from the subsurface. Note pyrite (Py) within silt-size particle.

F.12.Bakken

Bakken Formation (Late Devonian)
Williston Basin, North Dakota
Ar-ion milled, BSE image
Thermal maturity: 0.70%R_o
Submitted by B. Wawak and T. Gentzis, Core Laboratories
Nonporous organic matter (OM) filling intraparticle pores within a pyrite framboid (Py).

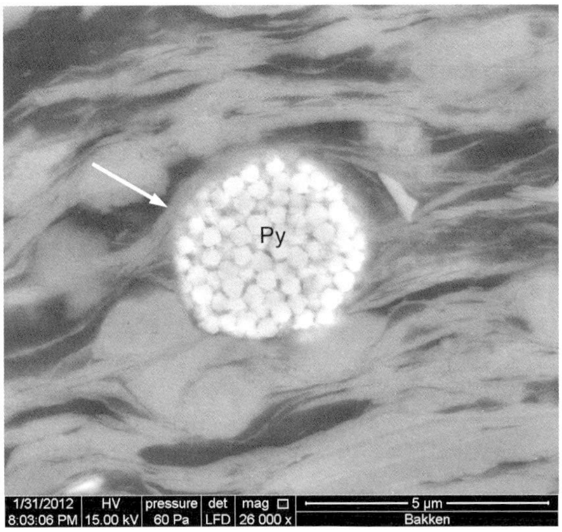

F.13.Bakken
Bakken Formation (Late Devonian)
Williston Basin, North Dakota
Ar-ion milled, BSE image
Thermal maturity: 0.70%R_o
Submitted by B. Wawak and T. Gentzis, Core Laboratories
Organic matter (black) filling interparticle pores associated with clays and other detrital grains. Note clays draped over (arrow) pyrite framboid (Py). Subparallel orientation of elongated phyllosilicate minerals defines organized fabric in this part of the sample.

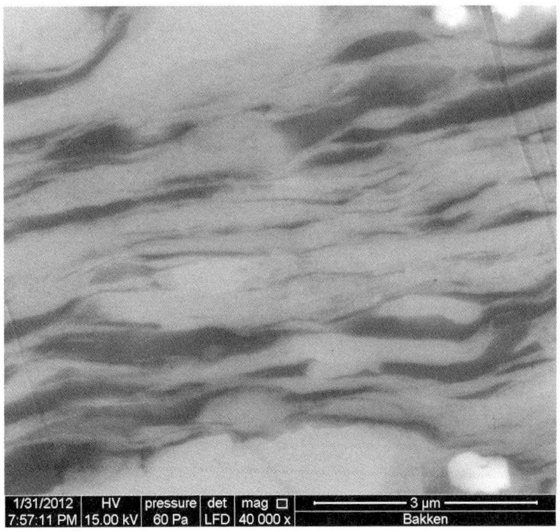

F.14.Bakken
Bakken Formation (Late Devonian)
Williston Basin, North Dakota
Ar-ion milled, BSE image
Thermal maturity: 0.70%R_o
Submitted by B. Wawak and T. Gentzis, Core Laboratories
Organic matter (black) filling interparticle pores associated with clays and other detrital grains. Organized fabric created by parallel orientation of phyllosilicate minerals. Grooves in upper right corner are artifacts caused by argon-ion milling.

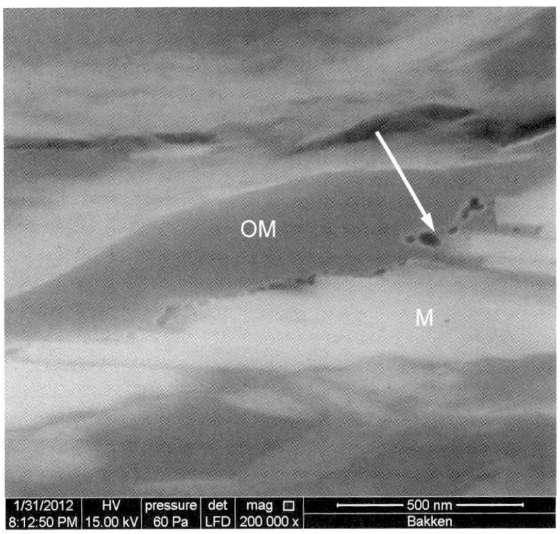

F.15.Bakken
Bakken Formation (Late Devonian)
Williston Basin, North Dakota
Ar-ion milled, BSE image
Thermal maturity: 0.70%R_o
Submitted by B. Wawak and T. Gentzis, Core Laboratories
Nanopores developed at the interface between phyllosilicate minerals (M) and organic matter (OM). The long dimension of the indicated pore (arrow) is 61 nm.

F.16.Barnett
Barnett Formation (Mississippian)
Fort Worth Basin, Texas
Depth: 4200–4300 ft (1280–1311 m)
Ar-ion milled, BSE image
Thermal maturity: 1.2–1.3%R_o
Submitted by D. Jennings, Core Laboratories
Authigenic phosphate dispersed throughout this silty shale is preferentially concentrated in interparticle pores. Box indicates high-magnification area (Figure 15).

F.17.Barnett
Barnett Formation (Mississippian)
Fort Worth Basin, Texas
Depth: 4200–4300 ft (1280–1311 m)
Ar-ion milled, BSE image
Thermal maturity: 1.2–1.3%R_o
Submitted by D. Jennings, Core Laboratories
High-magnification reveals that authigenic phosphate displays hollow tubes that are randomly oriented, suggesting that it precipitated during early burial as a result of microbial processes.

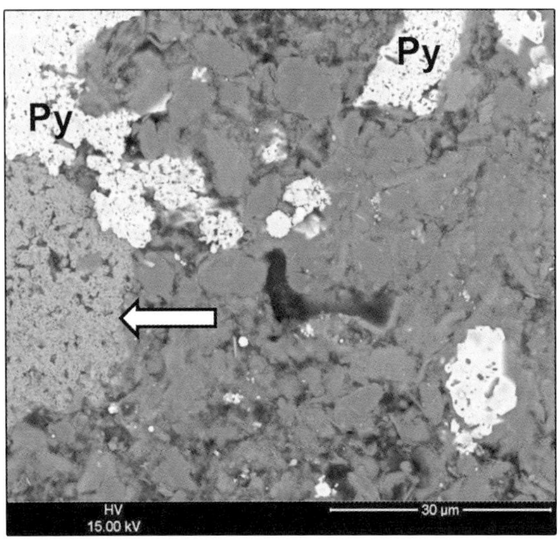

F.18.Barnett
Barnett Formation (Mississippian)
Fort Worth Basin, Texas
Depth: 4100–4200 ft (1220–1280 m)
Ar-ion milled, BSE image
Thermal maturity: 0.9–1.0%R_o
Submitted by D. Jennings, Core Laboratories
Authigenic phosphate with a microbial fabric (arrow) is partially replaced by pyrite (Py), indicating that phosphate in this sample precipitated early in the burial process and was subsequently replaced by pyrite.

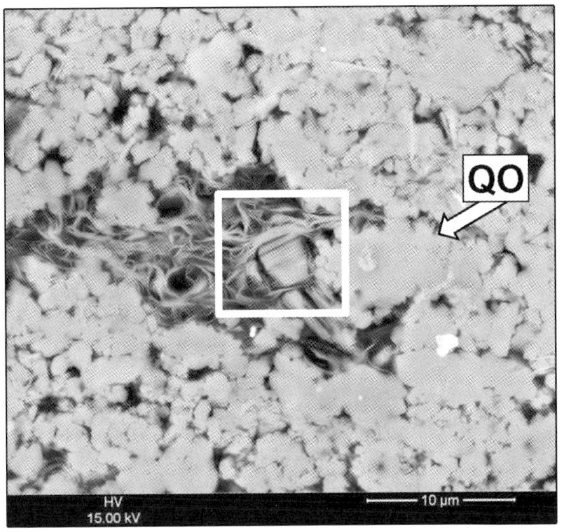

F.19.Barnett
Barnett Formation (Mississippian)
Fort Worth Basin, Texas
Depth: 6400–6500 ft (1951–1981 m)
Ar-ion milled, BSE image
Thermal maturity: N/A
Submitted by D. Jennings, Core Laboratories
Micrometer-scale, vuggy intercrystalline pores are abundant (black). Quartz silt grains commonly display quartz overgrowths (QO) that reduce porosity and restrict pore connectivity. Authigenic clay forming intracrystalline and intraparticle pore networks within micron-scale interparticle pores. White rectangle shows the area imaged at higher magnification.

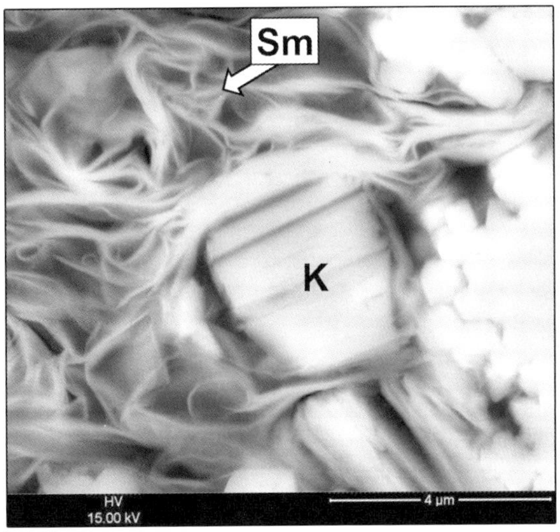

F.20.Barnett
Barnett Formation (Mississippian)
Fort Worth Basin, Texas
Depth: 6400–6500 ft (1951–1981 m)
Ar-ion milled, BSE image
Thermal maturity: N/A
Submitted by D. Jennings, Core Laboratories
Smectite (Sm) and kaolinite (K) form complex pore networks that significantly decrease permeability within intrinsic pores.

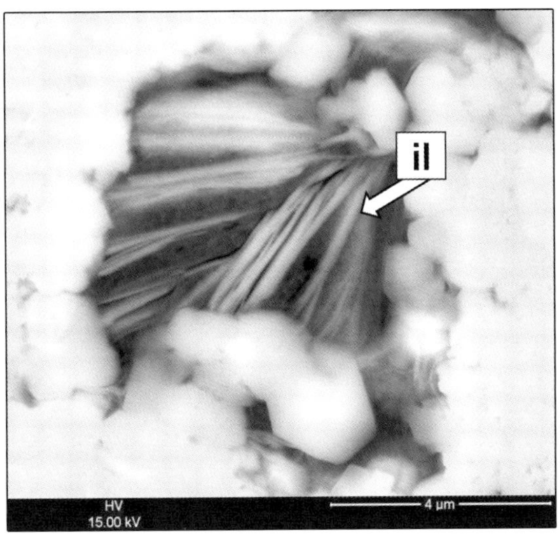

F.21.Barnett
Barnett Formation (Mississippian)
Fort Worth Basin, Texas
Depth: 6400–6500 ft (1951–1981 m)
Ar-ion milled, BSE image
Thermal maturity: N/A
Submitted by D. Jennings, Core Laboratories
Authigenic illite (il) also commonly forms bridged pore networks within micrometer-scale, intrinsic intercrystalline pores. Black organic matter is present between illite particles.

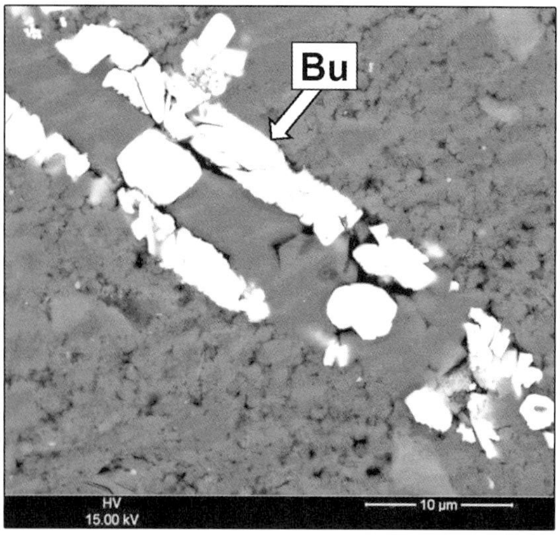

F.22.Barnett
Barnett Formation (Mississippian)
Fort Worth Basin, Texas
Depth: 12400–12500 ft (3780–3810 m)
Ar-ion milled, BSE image
Thermal maturity: N/A
Submitted by D. Jennings, Core Laboratories
Walls of this quartz-filled microburrow (Bu) have been replaced by marcasite (white).

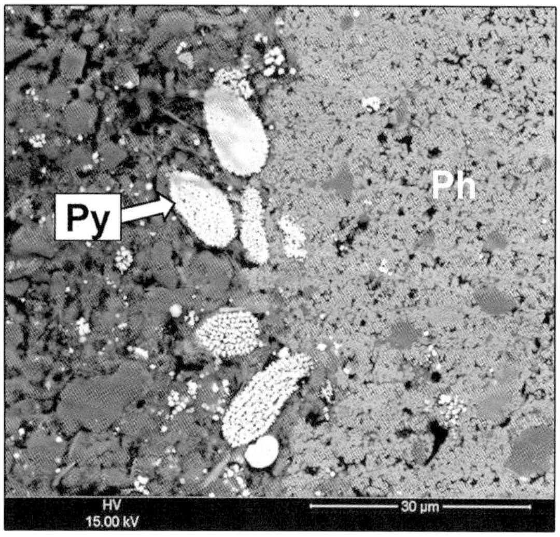

F.23.Barnett
Barnett Formation (Mississippian)
Fort Worth Basin, Texas
Depth: 6100–6200 ft (1859–1890 m)
Ar-ion milled, BSE image
Thermal maturity: 0.8–0.9%R_o
Submitted by D. Jennings, Core Laboratories
Ovoid pyrite (Py) framboids organized around authigenic phosphate (Ph) are a similar size and display an organized structure, suggesting that they are pyrite-filled microburrows or mineralized fecal pellets.

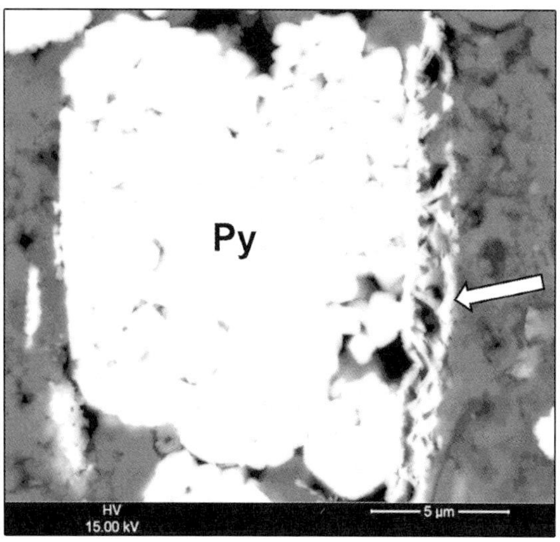

F.24.Barnett
Barnett Formation (Mississippian)
Fort Worth Basin, Texas
Depth: 12400–12500 ft (3780–3810 m)
Ar-ion milled, BSE image
Thermal maturity: N/A
Submitted by D. Jennings, Core Laboratories
Remnant, microburrow walls (arrow) occur with oval, pyrite-replaced fecal pellets (Py). The morphology of burrow walls suggests that the original pyrite burrow has been replaced by marcasite.

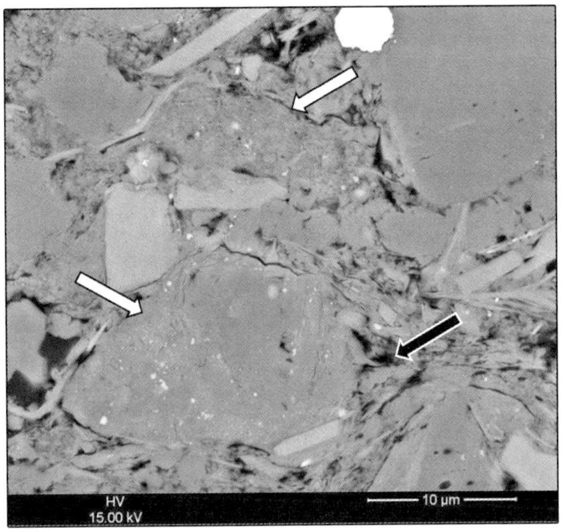

F.25.Barnett
Barnett Formation (Mississippian)
Fort Worth Basin, Texas
Depth: 4200–4300 ft (1280–1311 m)
Ar-ion milled, BSE image
Thermal maturity: 1.3–1.4%R_o
Submitted by D. Jennings, Core Laboratories
Description: Differential compaction around the clay aggregates (white arrows) in this silty shale suggest that the clay was deposited as individual clay floccules. Notice the pressure shadow pores (or phyllosilicate framework pores) that developed during compaction at either side of the large floccule (i.e., black arrow), suggesting that the floccule was a consolidated grain prior to significant compaction.

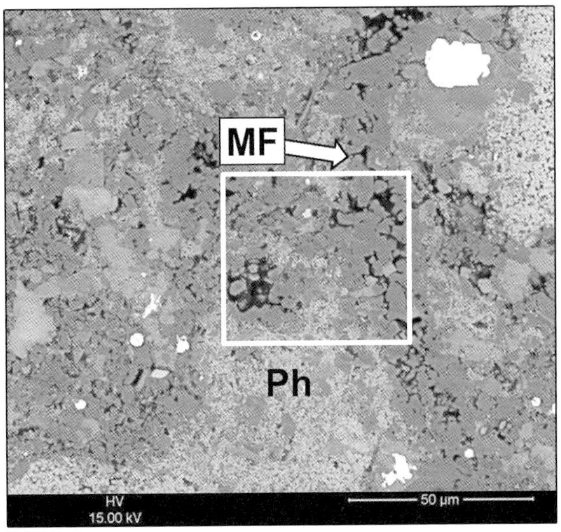

F.26.Barnett
Barnett Formation (Mississippian)
Fort Worth Basin, Texas
Depth: 6200–6300 ft (1890–1920 m)
Ar-ion milled, BSE image
Thermal maturity: N/A
Submitted by D. Jennings, Core Laboratories
Description: Quartz crystals (dark gray) partially fill subvertical microfractures (MF) in a phosphate-rich (Ph), silty shale. White rectangle shows enlarged area (Figure 27).

F.27.Barnett
Barnett Formation (Mississippian)
Fort Worth Basin, Texas
Depth: 6200–6300 ft (1890–1920 m)
Ar-ion milled, BSE image
Thermal maturity: N/A
Submitted by D. Jennings, Core Laboratories
Description: Intercrystalline pores occur along the mineralized fracture (white arrow) and enhance microporosity.

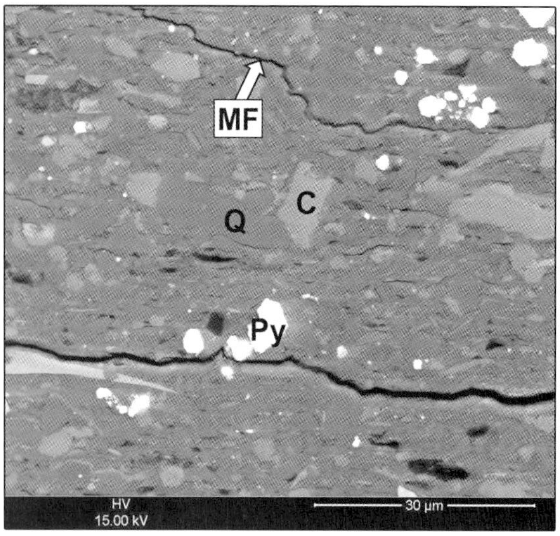

F.28.Barnett
Barnett Formation (Mississippian)
Delaware Basin, Texas
Depth: 8000–8100 ft (2438–2469 m)
Ar-ion milled, BSE image
Thermal maturity: 1.2–1.3%R_o
Submitted by D. Jennings, Core Laboratories
This wavy microlaminated, calcareous, silty shale contains quartz silt (Q), finely crystalline pyrite (Py), and calcite grains (C). Nanometer-scale, interparticle pores are the most abundant pore type present. Bedding plane parallel, artificial fractures and desiccation-enhanced microfractures (MF) are common.

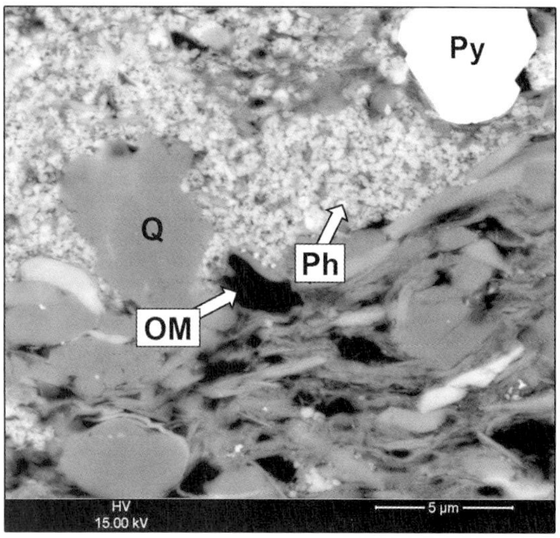

F.29.Barnett
Barnett Formation (Mississippian)
Delaware Basin, Texas
Depth: 8100–8200 ft (2469–2499 m)
Ar-ion milled, BSE image
Thermal maturity: N/A
Submitted by D. Jennings, Core Laboratories
The predominant pore types in this wavy microlaminated, silty shale include elongate, micrometer-scale, interparticle pores and subrounded nanometer-scale micropores. Pyrite (Py), quartz silt (Q), and clay are the most abundant constituents. Many of the interparticle pores contain organic matter (OM). Patches of microbially mediated phosphate (Ph) are locally present.

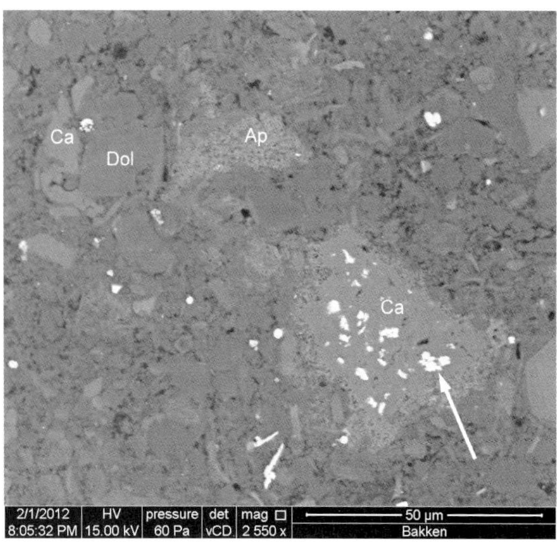

F.30.Barnett
Barnett Formation (Mississippian)
Fort Worth Basin, Texas
Ar-ion milled, BSE image
Thermal maturity: 0.74%R_o
Submitted by B. Wawak and T. Gentzis, Core Laboratories
There is a variety of minerals identified in this sample including pyrite (arrow) within a calcite grain (Ca), apatite (Ap), and dolomite (Dol). The intergranular pores are mostly filled with organic matter (black).

F.31.Barnett
Barnett Formation (Mississippian)
Fort Worth Basin, Texas
Ar-ion milled, BSE image
Thermal maturity: 0.74%R_o
Submitted by B. Wawak and T. Gentzis, Core Laboratories
Image shows a silt-size titanium mineral (Ti). EDX analysis identified titanium and oxygen in this grain, suggesting it is a titanium oxide paramorph (possibly rutile, brookite, or anatase). Light color of the titanium mineral is an indication of its greater density compared to the nearby grains, which are predominantly quartz and phyllosilicate minerals.

F.32.Barnett
Barnett Formation (Mississippian)
Fort Worth Basin, Texas
Ar-ion milled, BSE image
Thermal maturity: 0.74%R_o
Submitted by B. Wawak and T. Gentzis, Core Laboratories
EDX data indicate the primary elements in the lighter-color material are calcium, phosphorous, fluorine, and oxygen, suggesting this mineral is apatite. This apatite (Ap) has a botryoidal texture that suggests its origin is associated with microbial processes. Organic matter (black) fills most of the interparticle and intraparticle pores.

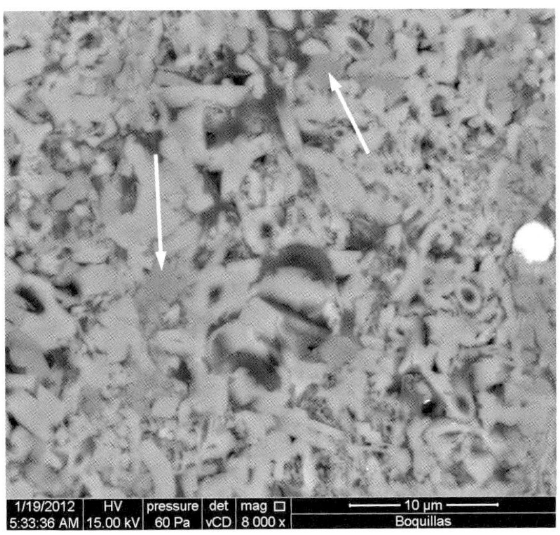

F.33.Boquillas
Boquillas Formation (Late Cretaceous)
Val Verde Basin, Texas
Ar-ion milled, BSE image
Thermal maturity: 0.45%R_o
Submitted by B. Wawak and T. Gentzis, Core Laboratories
This part of the sample consists predominantly of broken pieces of coccoliths and scattered clay particles (arrows) that are primarily kaolinite. Interparticle and intraparticle pores are partially filled with organic matter (dark color).

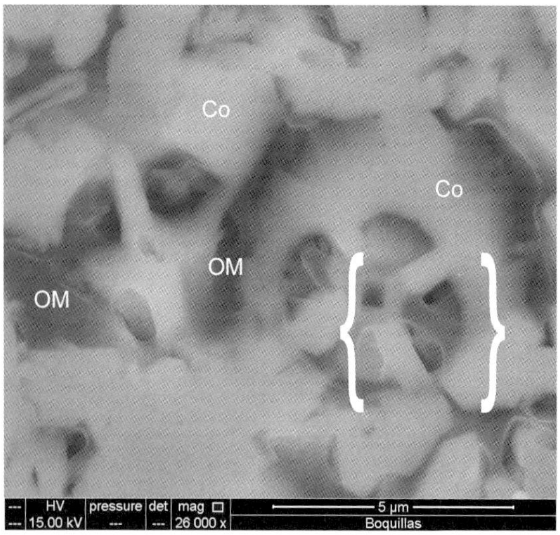

F.34.Boquillas
Boquillas Formation (Late Cretaceous)
Val Verde Basin, Texas
Ar-ion milled, BSE/SE image
Thermal maturity: 0.45%R_o
Submitted by B. Wawak and T. Gentzis, Core Laboratories
Organic matter (OM) partially filling the inter- and intraparticle pore space associated with coccoliths (Co). Brackets indicate area enlarged in Figure 35.

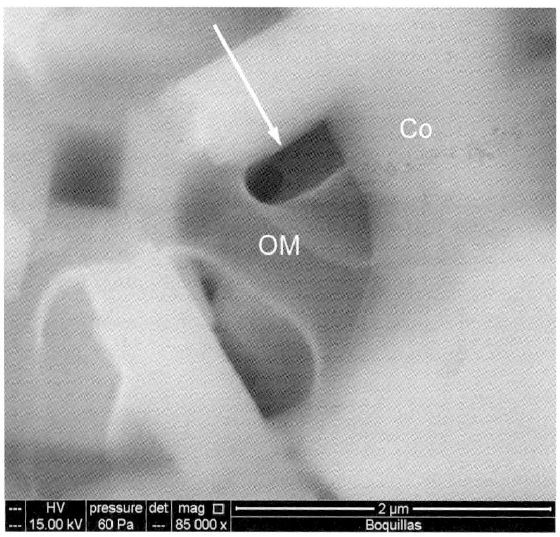

F.35.Boquillas
Boquillas Formation (Late Cretaceous)
Val Verde Basin, Texas
Ar-ion milled, BSE/SE image
Thermal maturity: 0.45%R_o
Submitted by B. Wawak and T. Gentzis, Core Laboratories
High-magnification view of area indicated by brackets in Figure 34 showing a "bridge" of organic matter (OM) within an intraparticle pore associated with a coccolith (Co). The red arrow indicates the remaining unfilled pore space.

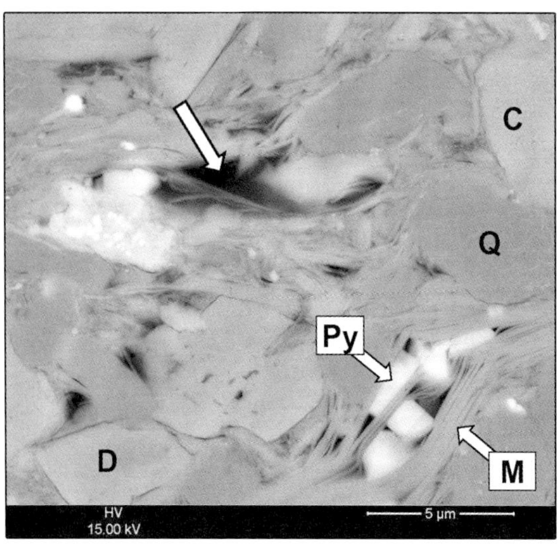

F.36.Bossier
Upper Bossier Formation (Jurassic)
East Texas Basin, Texas
Depth: 13400–13500 ft (4084–4115 m)
Ar-ion milled, BSE image
Thermal maturity: N/A
Submitted by D. Jennings, Core Laboratories
This microlaminated, silty shale contains quartz silt (Q), abundant tabular flakes of mica (M), carbonate grains (C), and scattered dolomite rhombs (D). Pyrite (Py) is present between degraded mica particles (M). Interparticle pores are the most abundant pore type. Organic matter is commonly present in pressure shadow pores (white arrow).

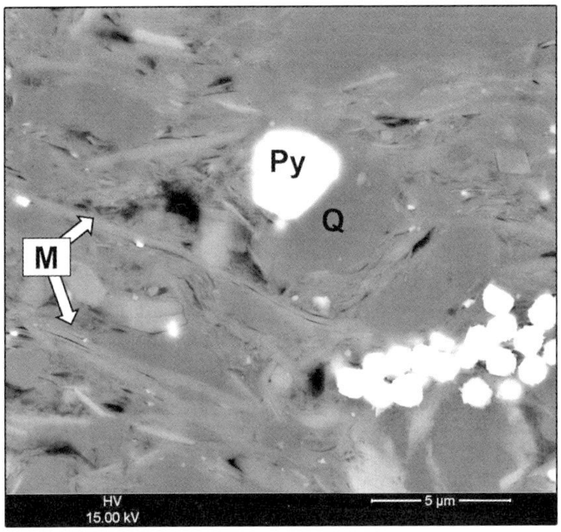

F.37.Bossier
Upper Bossier Formation (Jurassic)
East Texas Basin, Texas
Depth: 13400–13500 ft (4084–4115 m)
Ar-ion milled, BSE image
Thermal maturity: 1.3–1.4%R_o
Submitted by D. Jennings, Core Laboratories
This microlaminated, silty shale contains quartz silt (Q), abundant tabular flakes of mica (M), and finely crystalline pyrite (Py). Interparticle pores (black) are the most abundant pore type. Detrital and authigenic clays form bridged pores and pressure shadow pores at the edge of framework grains.

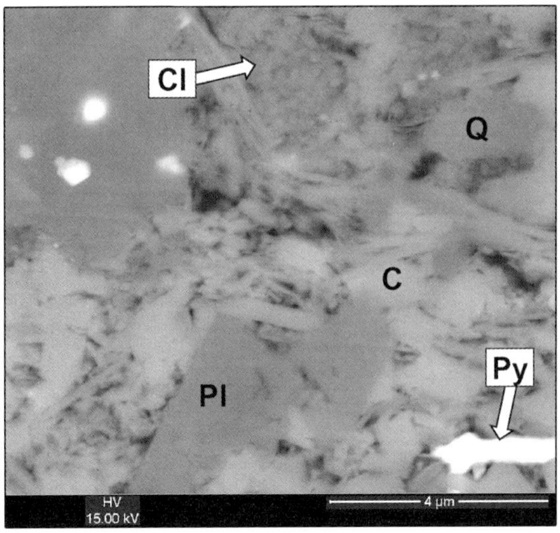

F.38.Bossier
Lower Bossier Formation (Jurassic)
East Texas Basin, Texas
Depth: 13900–14000 ft (4237–4267 m)
Ar-ion milled, BSE image
Thermal maturity: 1.2–1.3%R_o
Submitted by D. Jennings, Core Laboratories
Quartz (Q) and plagioclase (Pl) in this calcareous, silty shale are predominantly present as detrital, silt-size, grains. Calcite crystals (C) and coccolith fragments make up the largest portion of the calcite fraction. Authigenic clay particles (Cl) occlude interparticle pores. Euhedral pyrite (Py) is also present.

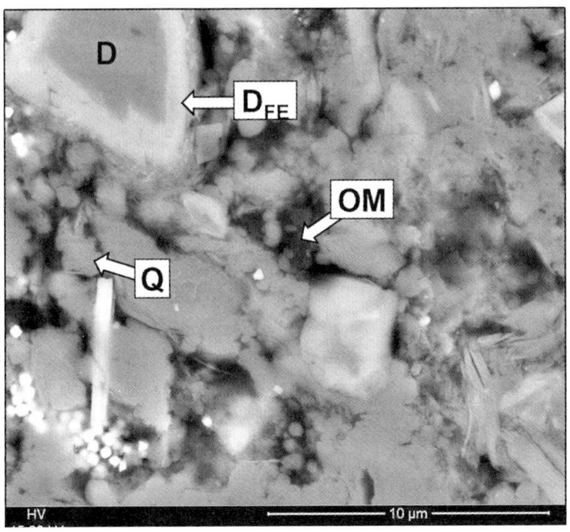

F.39.Chattanooga
Lower Chattanooga Formation (Late Devonian)
Black Warrior Basin, Alabama
Depth: 8300–8400 ft (2530–2560 m)
Ar-ion milled, BSE image
Thermal maturity: N/A
Submitted by D. Jennings, Core Laboratories
Zoned dolomite rhombs (D) possess ferroan dolomite rims (D_{FE}). Authigenic quartz (Q) is a common matrix constituent. Intercrystalline and interparticle pores appear to be well connected and filled with organic matter (OM).

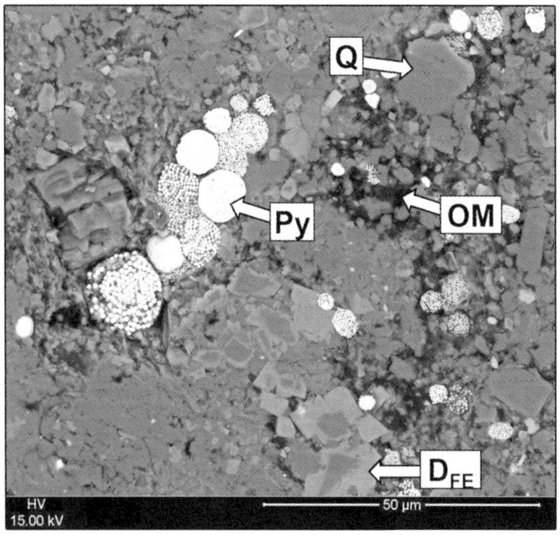

F.40.Chattanooga
Lower Chattanooga Formation (Late Devonian)
Black Warrior Basin, Alabama
Depth: 8300–8400 ft (2530–2560 m)
Ar-ion milled, BSE image
Thermal maturity: N/A
Submitted by D. Jennings, Core Laboratories
Pyrite framboids (Py) that are disseminated throughout the matrix in this silty shale commonly occur as clusters. Many pyrite framboids are degraded. Quartz is present as detrital silt (medium-grain grains) and as an authigenic matrix mineral. Ferroan dolomite cement (D_{FE}) and zoned dolomite rhombs are also scattered throughout the matrix. Abundant, well-connected, interparticle pores and intercrystalline pores are the most abundant pore types. Organic matter (OM) is abundant.

F.41.Chattanooga
Lower Chattanooga Formation (Late Devonian)
Black Warrior Basin, Alabama
Depth: 4500–4600 ft (1372–1402 m)
Ar-ion milled, BSE image
Thermal maturity: N/A
Submitted by D. Jennings, Core Laboratories
Nanometer-scale phosphate coccoids (Ph) and organic matter (OM) commonly fill interparticle and intracrystalline pores in this silty shale.

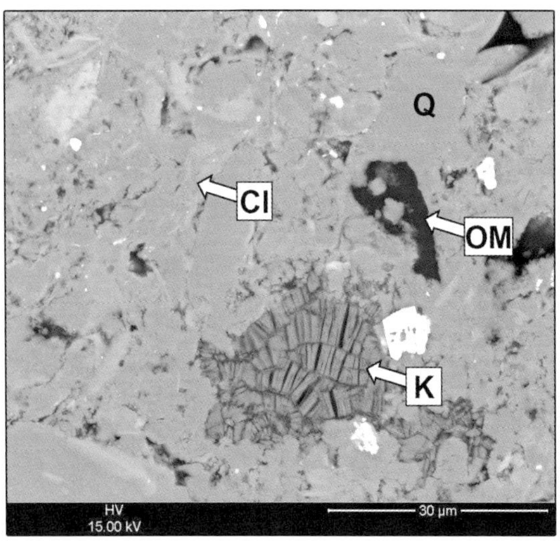

F.42.Chattanooga
Chattanooga Formation (Late Devonian)
Appalachian Basin, Tennessee
Depth: 42400–4300 ft (1280–1310 m)
Ar-ion milled, BSE image
Thermal maturity: N/A
Submitted by D. Jennings, Core Laboratories
This massive-appearing, silty shale contains quartz silt (Q), finely crystalline pyrite (bright white), and patches of kaolinite (K).
Intercrystalline pores occur between kaolinite particles. Organic material (OM) is also scattered throughout the matrix. Nanometer-scale interparticle pores are the most abundant pore type and appear to be poorly connected because of the presence of authigenic clay (Cl) that occludes interstitial space.

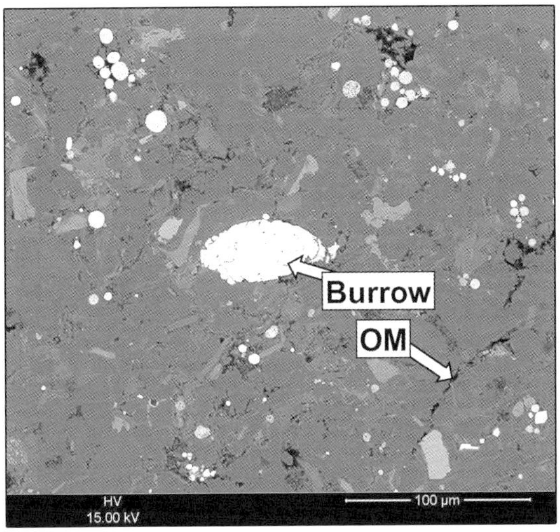

F.43.Chattanooga
Chattanooga Formation (Late Devonian)
Appalachian Basin, Tennessee
Depth: 4600–4700 ft (1402–1372 m)
Ar-ion milled, BSE image
Thermal maturity: N/A
Submitted by D. Jennings, Core Laboratories
Matrix material is compacted around oval pyrite features, suggesting that they are pyrite-filled, horizontal microburrows. Microfractures filled with organic matter (OM) typically have micron-scale apertures.

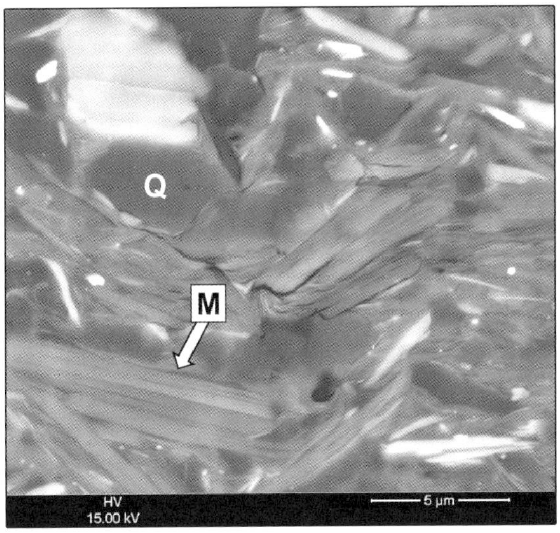

F.44.Chattanooga
Chattanooga Formation (Late Devonian)
Appalachian Basin, Tennessee
Depth: 4200–4300 ft (1280–1310 m)
Ar-ion milled, BSE image
Thermal maturity: N/A
Submitted by D. Jennings, Core Laboratories
Interparticle pores are rare in this wavy, laminated, micaceous, silty shale. Quartz silt (Q), pyrite grains (bright white), and abundant mica (M) are the predominant constituents. Desiccation-enhanced pores and circumgranular cracks are common.

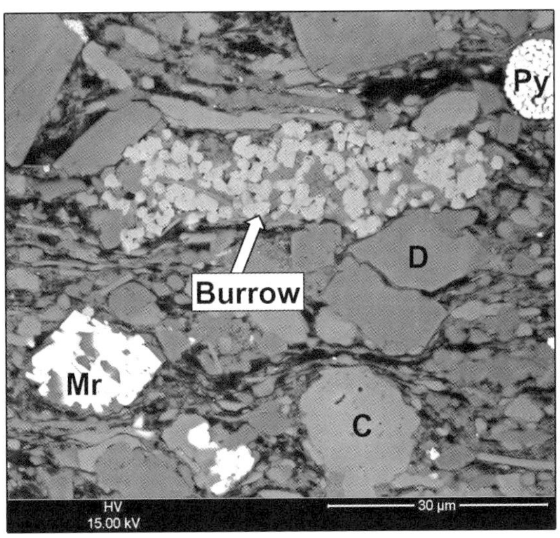

F.45.Devonian
Devonian Limestone (Devonian)
Illinois Basin, Kentucky
Depth: 3800–3900 ft (1158–1189 m)
Ar-ion milled, BSE image
Thermal maturity: N/A
Submitted by D. Jennings, Core Laboratories
Dolomite rhombs (D) and calcite grains (C) are the most abundant nonmatrix constituents in this microlaminated, dolomitic siltstone. Framboidal pyrite (Py) is typically degraded. Marcasite (Mr) is common. Horizontal, slightly compacted microburrows are filled with phosphate and authigenic clay. Interparticle nanopores are present within microburrows.

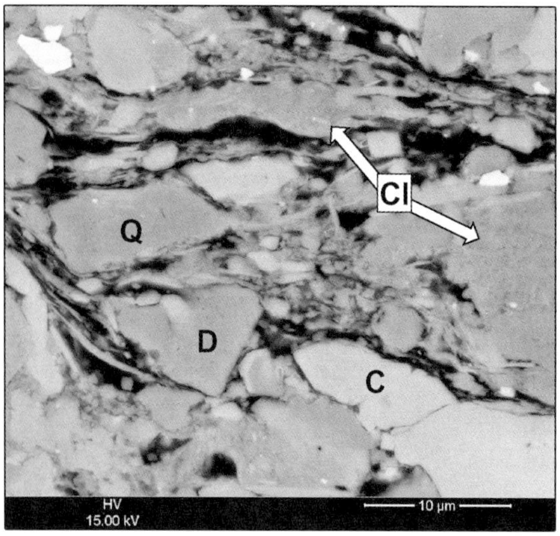

F.46.Devonian
Devonian Limestone (Devonian)
Illinois Basin, Kentucky
Depth: 3800–3900 ft (1158–1189 m)
Ar-ion milled, BSE image
Thermal maturity: N/A
Submitted by D. Jennings, Core Laboratories
Clay aggregates (Cl) occur with dolomite rhombs (D), quartz silt (Q), and calcite grains (C) in this microlaminated, dolomitic siltstone. Organic matter (black) is present along bedding plane partings.

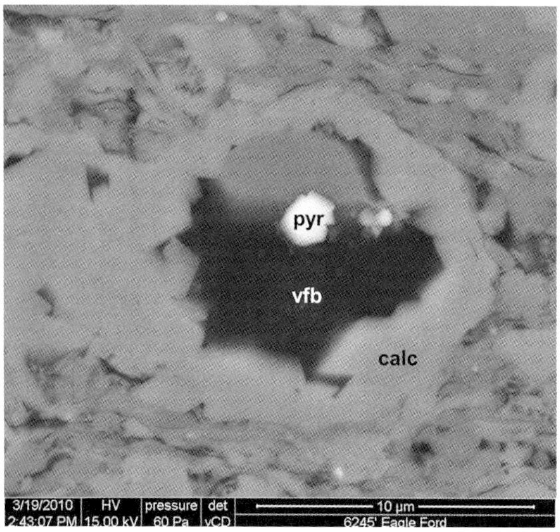

F.47.Eagle Ford
Eagle Ford Formation (Late Cretaceous)
Maverick Basin, Dimmitt County, Texas
Depth: 6245 ft (1903 m)
Ar-ion milled, BSE image
Thermal maturity: 1.60%R_o
Submitted by W. Camp, Anadarko Petroleum
Interior void of calcareous foraminifer test partially filled with bitumen (vfb). Pyrite (pyr) and other minerals occupy upper one-third of void space and pre-date the bitumen fill.

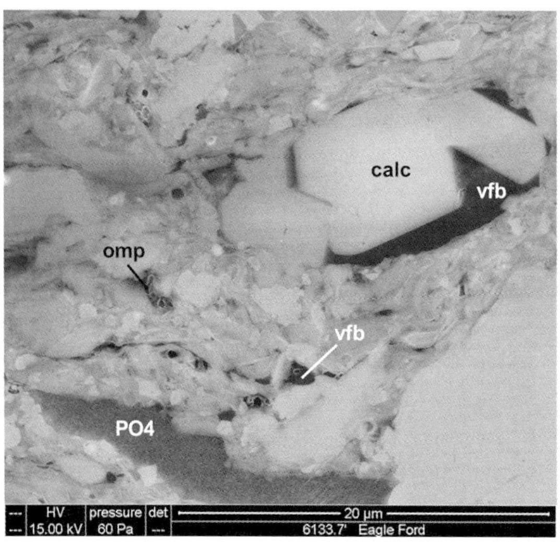

F.48.Eagle Ford
Eagle Ford Formation (Late Cretaceous)
Maverick Basin, Dimmitt County, Texas
Depth: 6133.7 ft (1869 m)
Ar-ion milled, superimposed SE and BSE image
Thermal maturity: 1.60%R_o
Submitted by W. Camp, Anadarko Petroleum
Superimposed SE and BSE image combines textural and compositional information within a single image. Euhedral calcite cement (calc) and bitumen (vfb) fill elliptical void. Bitumen also fills most of surrounding intergranular and intragranular pore space. Organic matter pores (omp) are locally present within some of the void-filling bitumen, as revealed by the SE image. A phosphate grain (PO4) is also present.

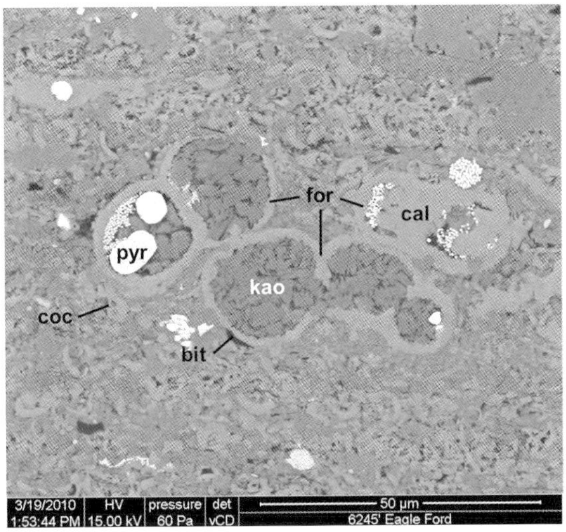

F.49.Eagle Ford
Eagle Ford Formation (Late Cretaceous)
Maverick Basin, Dimmitt County, Texas
Depth: 6245 ft (1903 m)
Ar-ion milled, BSE image
Thermal maturity: 1.6%R_o
Submitted by W. Camp, Anadarko Petroleum
Calcareous planktonic foraminifer test chambers (for) filled with authigenic kaolinite (kao), calcite (cal), pyrite (pyr), and bitumen (bit). Pyrite occurs as individual crystals and framboidal aggregates. Bitumen is also observed filling pores and as disseminated organic matter within laminae composed of coccolith (coc) fragments.

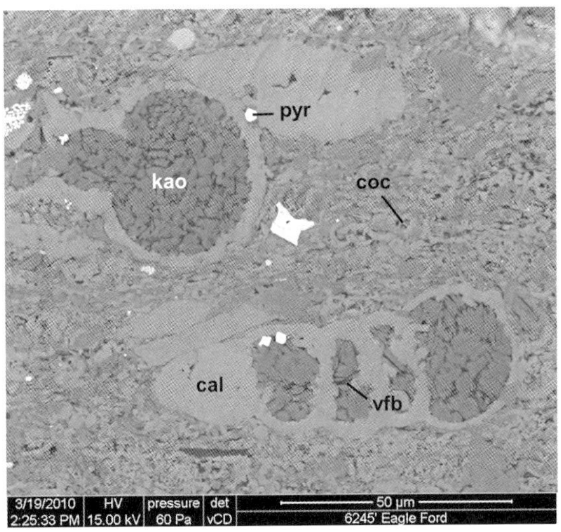

F.50.Eagle Ford
Eagle Ford Formation (Late Cretaceous)
Maverick Basin, Dimmitt County, Texas
Depth: 6245 ft (1903 m)
Ar-ion milled, BSE image
Thermal maturity: 1.6%R_o
Submitted by W. Camp, Anadarko Petroleum
Kaolinite booklets (kao) and void-filling bitumen (vfb) fill internal chambers of calcareous planktonic foraminifer tests (cal). Bitumen is also observed filling pores within laminae composed of coccolith (coc) fragments. Pyrite (pyr) is also present as a minor constituent.

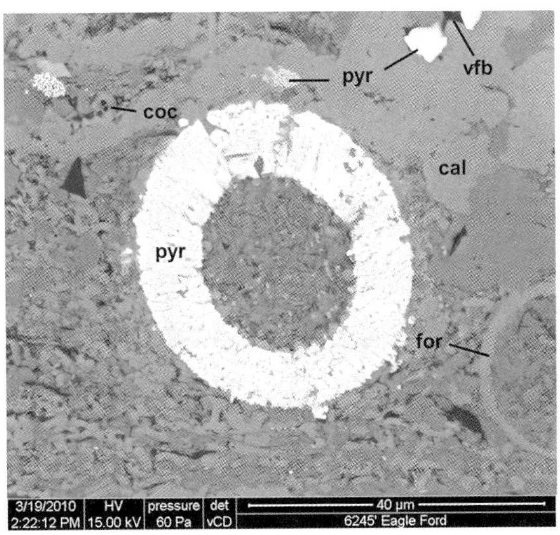

F.51.Eagle Ford
Eagle Ford Formation (Late Cretaceous)
Maverick Basin, Dimmitt County, Texas
Depth: 6245 ft (1903 m)
Ar-ion milled, BSE image
Thermal maturity: 1.6%R_o
Submitted by W. Camp, Anadarko Petroleum
Pyrite (pyr) replacing unidentified fossil (possibly echinoderm spine). Pyrite also occurs as single crystals and elliptical aggregates (framboids) in upper portion of image. Most of the calcite (cal) present is derived from calcareous fossil fragments, including foraminifera (for) and coccoliths (coc). Bitumen (vfb) fills larger pore spaces.

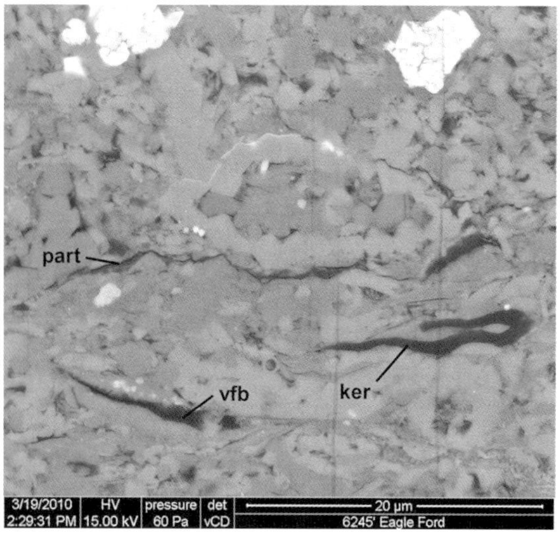

F.52.Eagle Ford
Eagle Ford Formation (Late Cretaceous)
Maverick Basin, Dimmitt County, Texas
Depth: 6245 ft (1903 m)
Ar-ion milled, BSE image
Thermal maturity: 1.6%R_o
Submitted by W. Camp, Anadarko Petroleum
Irregular shape of organic matter suggests this is a remnant of a structured kerogen maceral (ker). Void-filling bitumen (vfb) fills larger voids and smaller interparticle and intraparticle pores. Bedding plane partings (part) are also present.

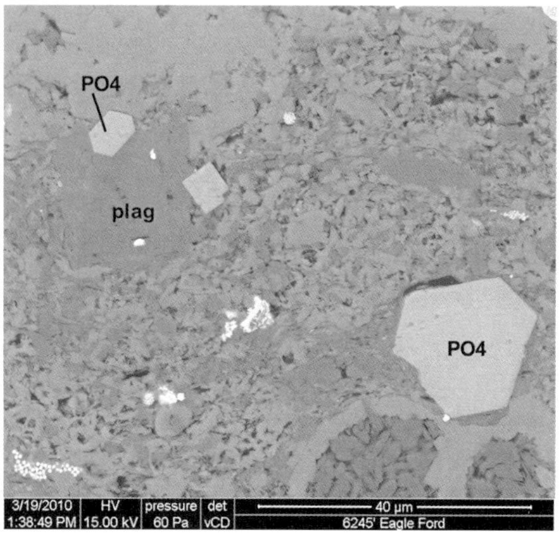

F.53.Eagle Ford
Eagle Ford Formation (Late Cretaceous)
Maverick Basin, Dimmitt County, Texas
Depth: 6245 ft (1903 m)
Ar-ion milled, BSE image
Thermal maturity: 1.6%R_o
Submitted by W. Camp, Anadarko Petroleum
Authigenic euhedral apatite crystals (PO4) appear to partially replace euhedral plagioclase (plag).

F.54.Eagle Ford
Eagle Ford Formation (Late Cretaceous)
Maverick Basin, Dimmitt County, Texas
Depth: 6245 ft (1903 m)
Ar-ion milled, SE image
Thermal maturity: 1.6%R_o
Submitted by W. Camp, Anadarko Petroleum
Sponge-like texture of porous pyrobitumen (bit) fills both interparticle pores and intraparticle pores within calcareous coccolith fragments (coc) and minor clay grains. Pores within pyrobitumen presumably formed during thermal cracking of oil to gas during burial.

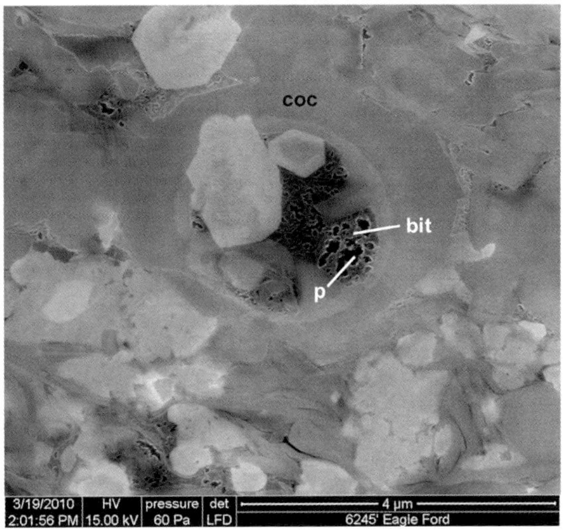

F.55.Eagle Ford
Eagle Ford Formation (Late Cretaceous)
Maverick Basin, Dimmitt County, Texas
Depth: 6245 ft (1903 m)
Ar-ion milled, SE image
Thermal maturity: 1.6%R_o
Submitted by W. Camp, Anadarko Petroleum
Sponge-like texture of porous pyrobitumen (bit) fills intraparticle pores within calcareous coccolith fragments (coc) and minor interparticle pores between clay grains. Pores (p) within pyrobitumen presumably formed during thermal cracking of oil to gas during burial.

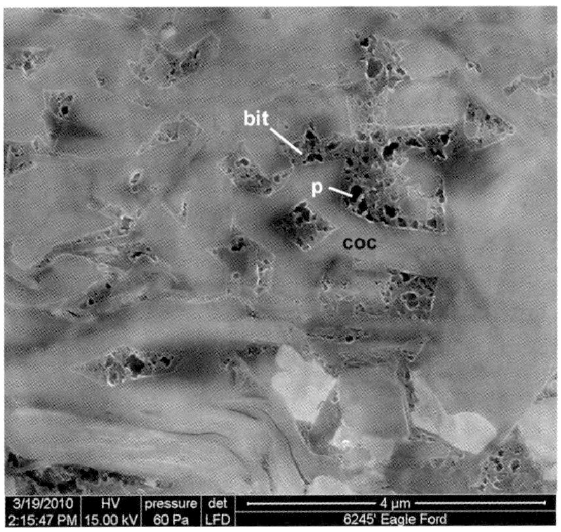

F.56.Eagle Ford
Eagle Ford Formation (Late Cretaceous)
Maverick Basin, Dimmitt County, Texas
Depth: 6245 ft (1903 m)
Ar-ion milled, SE image
Thermal maturity: 1.6%R_o
Submitted by W. Camp, Anadarko Petroleum
Sponge-like texture of porous pyrobitumen (bit) filling intraparticle and interparticle pores between calcareous coccolith fragments (coc). Pores (p) within pyrobitumen presumably formed during thermal cracking of oil to gas during burial.

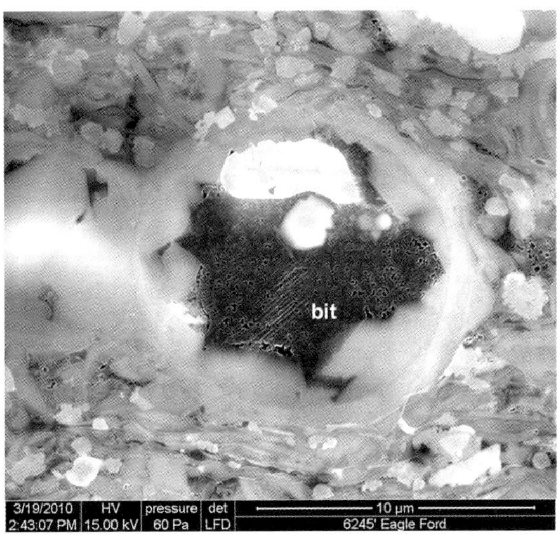

F.57.Eagle Ford
Eagle Ford Formation (Late Cretaceous)
Maverick Basin, Dimmitt County, Texas
Depth: 6245 ft (1903 m)
Ar-ion milled, SE image
Thermal maturity: 1.6%R_o
Submitted by W. Camp, Anadarko Petroleum
Sponge-like texture of porous pyrobitumen (bit) filling intraparticle pores within calcareous fossil fragments. Pores within pyrobitumen presumably formed during thermal cracking of oil to gas during burial.

F.58.Eagle Ford
Eagle Ford Formation (Late Cretaceous)
Maverick Basin, Dimmitt County, Texas
Depth: 6245 ft (1903 m)
Ar-ion milled, SE image
Thermal maturity: 1.6%R_o
Submitted by W. Camp, Anadarko Petroleum
Bedding plane parting (part) cross cuts mineral grains and void-filling bitumen (bit). Organic matter pores (omp) are evident from the sponge-like texture within some of the bitumen. Partings may create mechanical weakness planes that may respond to hydraulic fracture stimulation and facilitate hydrocarbon flow. Origin of fracture is uncertain and may be a result of core extraction, desiccation, or sample preparation.

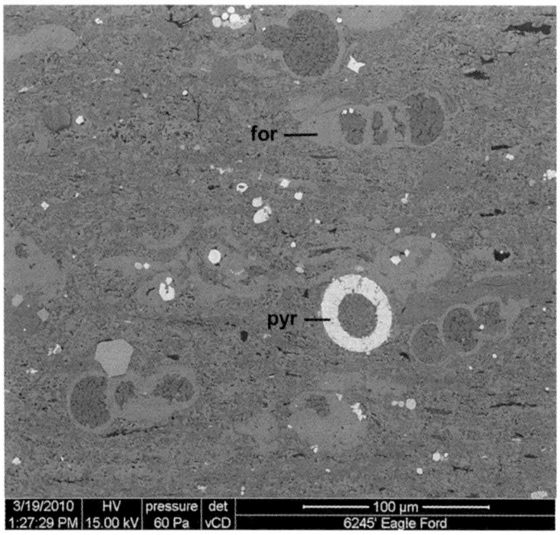

F.59.Eagle Ford
Eagle Ford Formation (Late Cretaceous)
Maverick Basin, Dimmitt County, Texas
Depth: 6245 ft (1903 m)
Ar-ion milled, BSE image
Thermal maturity: 1.6%R_o
Submitted by W. Camp, Anadarko Petroleum
Calcareous planktonic foraminifer (for) tests (up to very fine sand size) with long axes oriented parallel to bedding. Clay laminae are locally deformed because of differential compaction. Interior test chambers are filled with kaolinite. Pyrite (pyr) encircles kaolinite and appears to have replaced an unidentified fossil (possible echinoderm spine).

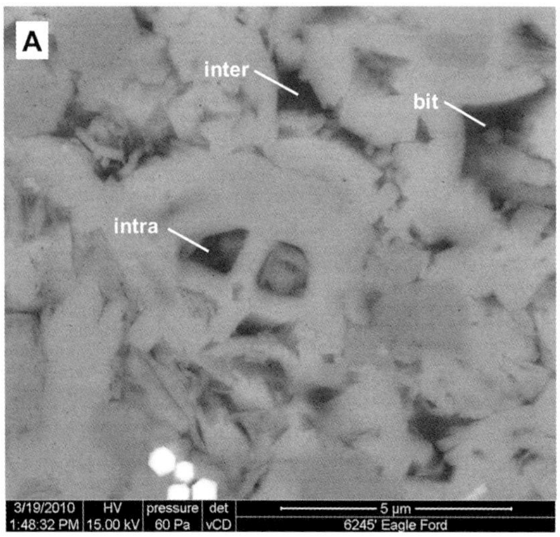

F.60.Eagle Ford
Eagle Ford Formation (Late Cretaceous)
Maverick Basin, Dimmitt County, Texas
Depth: 6245 ft (1903 m)
Ar-ion milled, SE and BSE images
Thermal maturity: 1.6%R_o
Submitted by W. Camp, Anadarko Petroleum
Calcareous coccolith fragments make up a large portion of the calcite present in the Eagle Ford marlstones. (F.60) Primary intergranular (inter) and intragranular (intra) pores within the coccolith-rich laminae remained open during early burial and were later filled with oil (now bitumen, bit) during catagenesis. Secondary electron image (F.61) reveals irregular-shape organic matter pores (omp) within bitumen that are an order of magnitude smaller than the granular pores. The sponge-like texture of the void-filling bitumen is presumably a product of advanced thermal maturation and conversion of bitumen to pyrobitumen.

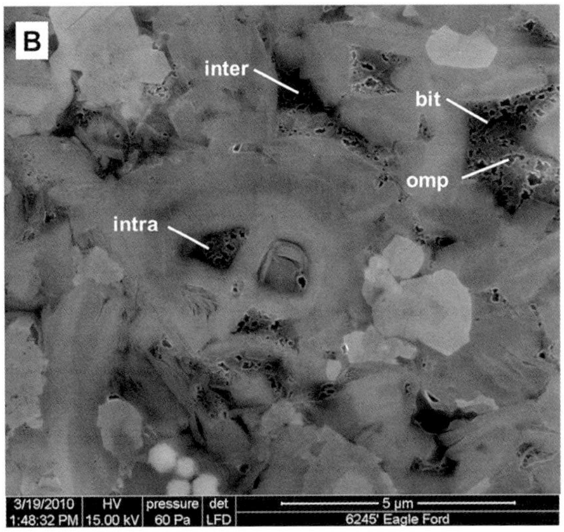

F.61.Eagle Ford
Eagle Ford Formation (Late Cretaceous)
Maverick Basin, Dimmitt County, Texas
Depth: 6245 ft (1903 m)
Ar-ion milled, SE and BSE images
Thermal maturity: 1.6%R_o
Submitted by W. Camp, Anadarko Petroleum
Calcareous coccolith fragments make up a large portion of the calcite present in the Eagle Ford marlstones. (F.60) Primary intergranular (inter) and intragranular (intra) pores within the coccolith-rich laminae remained open during early burial and were later filled with oil (now bitumen, bit) during catagenesis. Secondary electron image (F.61) reveals irregular-shape organic matter pores (omp) within bitumen that are an order of magnitude smaller than the granular pores. The sponge-like texture of the void-filling bitumen is presumably a product of advanced thermal maturation and conversion of bitumen to pyrobitumen.

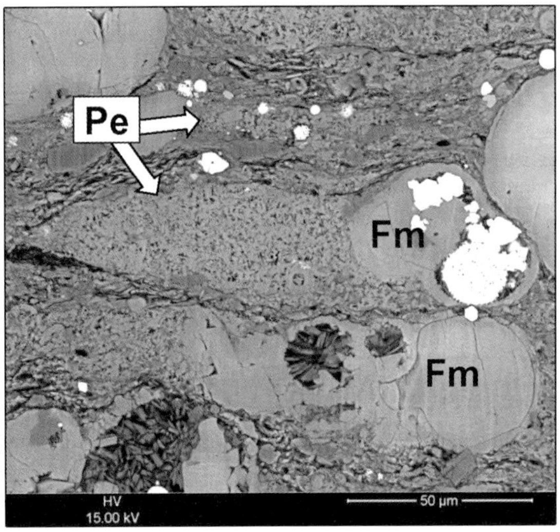

F.62.Eagle Ford
Eagle Ford Formation (Late Cretaceous)
East Texas Basin, Texas
Depth: 7700–7800 ft (2347–2377 m)
Ar-ion milled, BSE image
Thermal maturity: 0.8–0.9%R_o
Submitted by D. Jennings, Core Laboratories
Biogenic aggregates or fecal pellets (Pe) with an open internal fabric are compacted parallel to bedding and usually occur in laminated, distal, inner-shelf deposits. Differential compaction at the edge of the aggregates indicates that they were delivered to the seafloor as discrete particles. Globose foraminifera (Fm) are commonly associated with these aggregates.

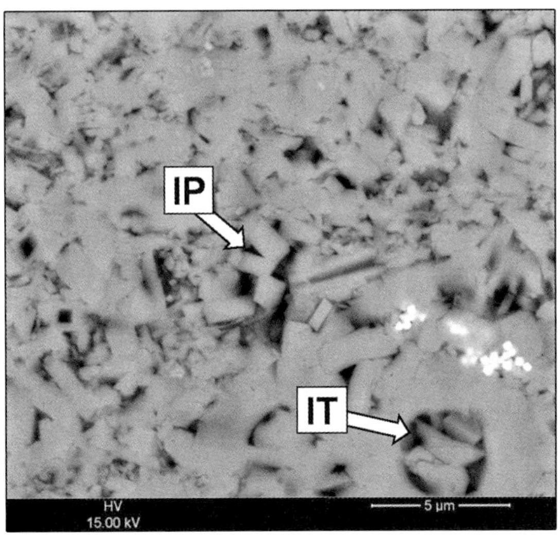

F.63.Eagle Ford
Eagle Ford Formation (Late Cretaceous)
Southwest Gulf Coast Basin, Texas
Depth: 7700–7800 ft (2347–2377 m)
Ar-ion milled, BSE image
Thermal maturity: 0.8–0.9%R_o
Submitted by D. Jennings, Core Laboratories
Biogenic aggregates or pellets are predominantly composed of calcareous coccolith and foraminifer fragments. Well-connected interparticle pores (IP) enhance the porosity of pellet-rich deposits. Intraparticle pores (IT) are also preserved within foraminifer chambers and intact coccolith plates.

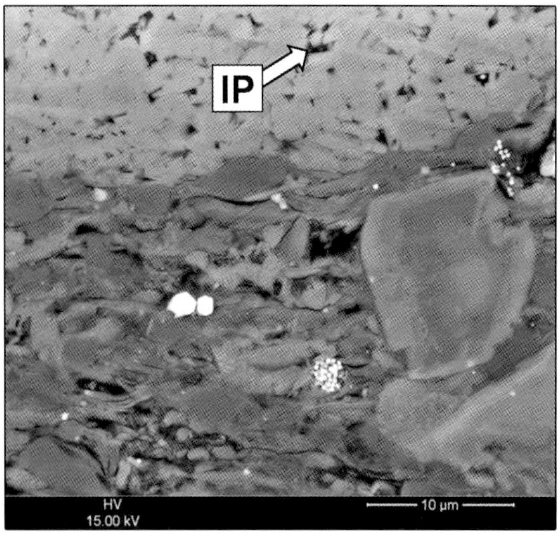

F.64.Eagle Ford
Eagle Ford Formation (Late Cretaceous)
Southwest Gulf Coast Basin, Texas
Depth: 11700–11800 ft (3566–3597 m)
Ar-ion milled, BSE image
Thermal maturity: 1.1–1.2%R_o
Submitted by D. Jennings, Core Laboratories
Coccolith fragments in larger biogenic aggregates are tightly packed and display evidence of recrystallization. Interparticle pores (IP) appear to be poorly connected and much smaller than in smaller pellets with a more open internal fabric.

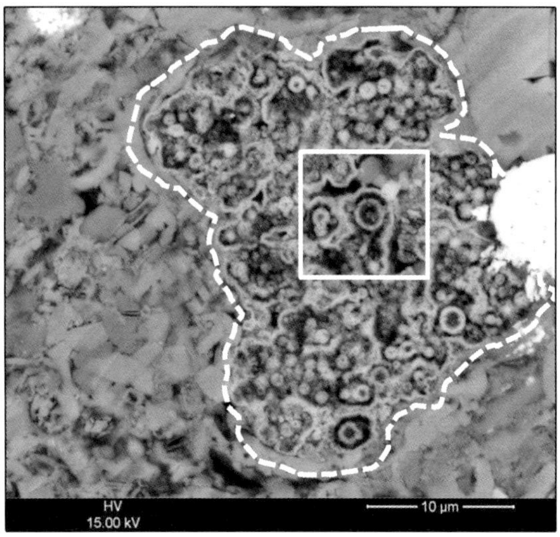

F.65.Eagle Ford
Eagle Ford Formation (Late Cretaceous)
Southwest Gulf Coast Basin, Texas
Depth: 9600–9700 ft (2926–2957 m)
Ar-ion milled, BSE image
Thermal maturity: N/A
Submitted by D. Jennings, Core Laboratories
Patches of microbially mediated, phosphatic, accretionary masses (outlined by a white dashed line) display irregular boundaries with the surrounding matrix. Degraded pyrite (bright white grains) associated with authigenic phosphate provides evidence for an incursion of more oxygenated surface to midcolumn waters and possibly long-lasting high rates of sulfate reduction. Area of high-magnification image indicated by white box.

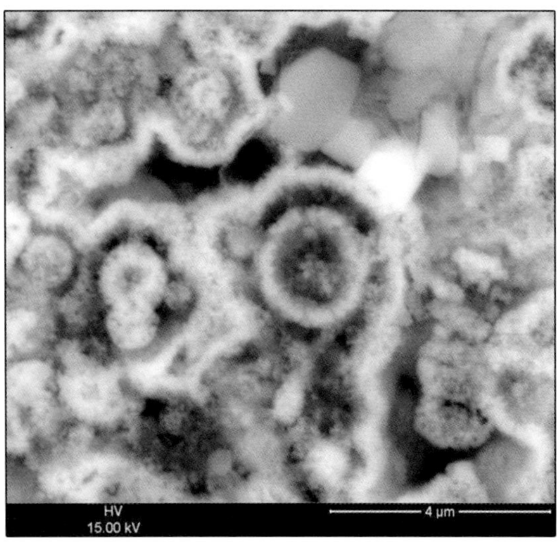

F.66.Eagle Ford
Eagle Ford Formation (Late Cretaceous)
Southwest Gulf Coast Basin, Texas
Depth: 9600–9700 ft (2926–2957 m)
Ar-ion milled, BSE image
Thermal maturity: N/A
Submitted by D. Jennings, Core Laboratories
Fine-grain, phosphatic, accretionary masses are interpreted as a product of the activity of sulfur-oxidizing and sulfate-reducing bacteria, organic matter mineralization, and Fe(III) reduction. The highest phosphate concentrations in modern sediments occur immediately underneath a layer of phytodetritus (biogenic aggregates), suggesting that phosphate is remobilized from the mineralization of newly settled organic matter.

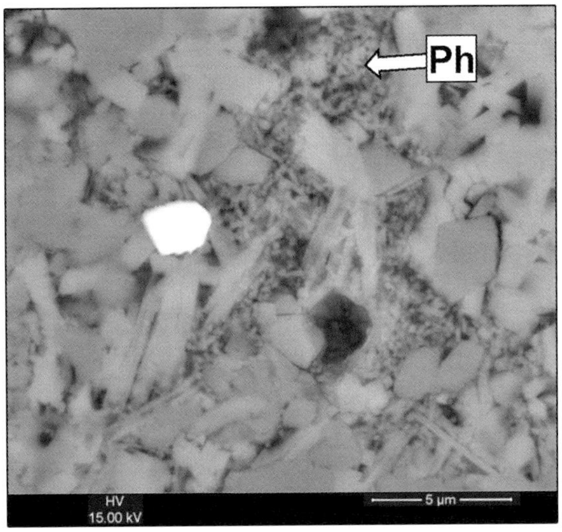

F.67.Eagle Ford
Eagle Ford Formation (Late Cretaceous)
Southwest Gulf Coast Basin, Texas
Depth: 12700–12800 ft (3871–3901 m)
Ar-ion milled, BSE image
Thermal maturity: N/A
Submitted by D. Jennings, Core Laboratories
Patches of fine-grain, microbially mediated phosphate tubes (Ph) are common in organic-rich marls.

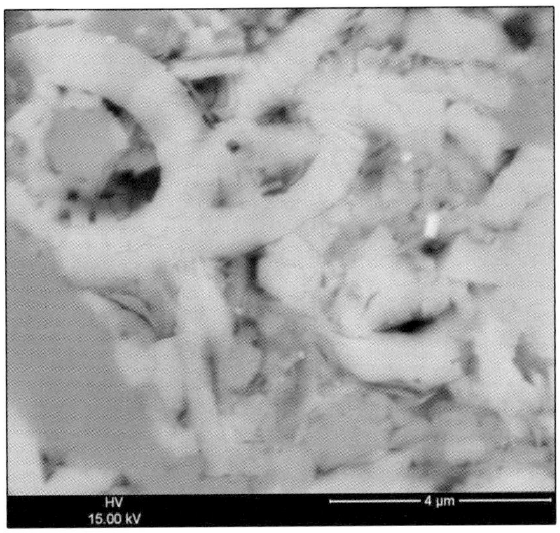

F.68.Eagle Ford
Eagle Ford Formation (Late Cretaceous)
Southwest Gulf Coast Basin, Texas
Depth: 13800–13900 ft (4206–4237 m)
Ar-ion milled, BSE image
Thermal maturity: N/A
Submitted by D. Jennings, Core Laboratories
Coccolith plates and plate fragments are abundant and constitute most of the matrix in many southwestern Texas Eagle Ford samples. Despite their abundance, they are not visible in thin-section photomicrographs because of their small size. Several species are preserved. The abundance of coccoliths as the main matrix constituent results in the presence of abundant, well-connected interparticle pores (black).

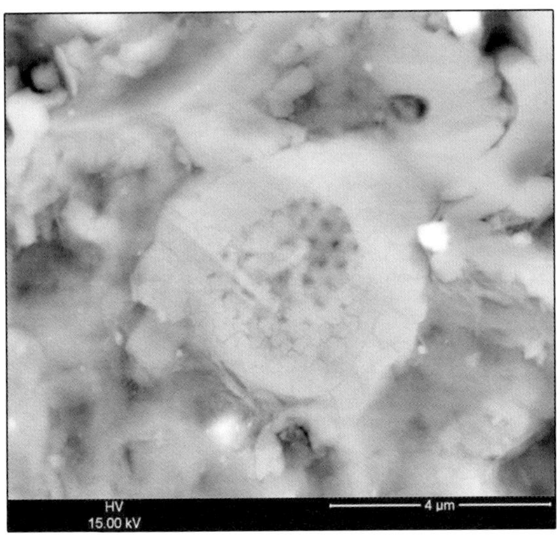

F.69.Eagle Ford
Eagle Ford Formation (Late Cretaceous)
Southwest Gulf Coast Basin, Texas
Depth: 14100–14200 ft (4298–4328 m)
Ar-ion milled, BSE image
Thermal maturity: N/A
Submitted by D. Jennings, Core Laboratories
The small, circular coccolith plate in this image displays several nanometer-scale intraparticle pores and cycles of elements surrounding the central area.

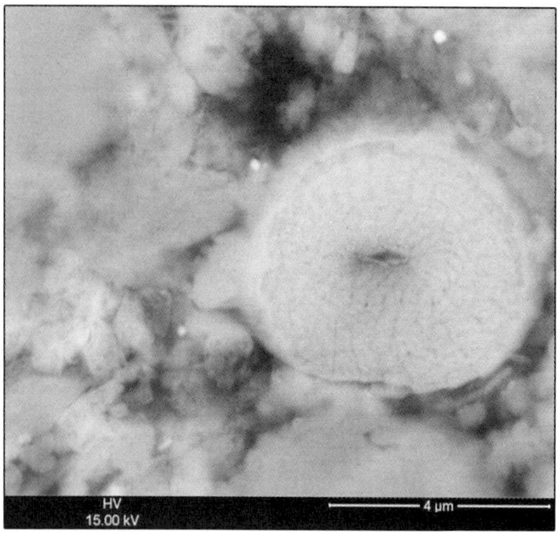

F.70.Eagle Ford
Eagle Ford Formation (Late Cretaceous)
Southwest Gulf Coast Basin, Texas
Depth: 14100–14200 ft (4298–4328 m)
Ar-ion milled, BSE image
Thermal maturity: N/A
Submitted by D. Jennings, Core Laboratories
The medium-size coccolith plate in this image displays a nanometer-scale intraparticle pore in the central area.

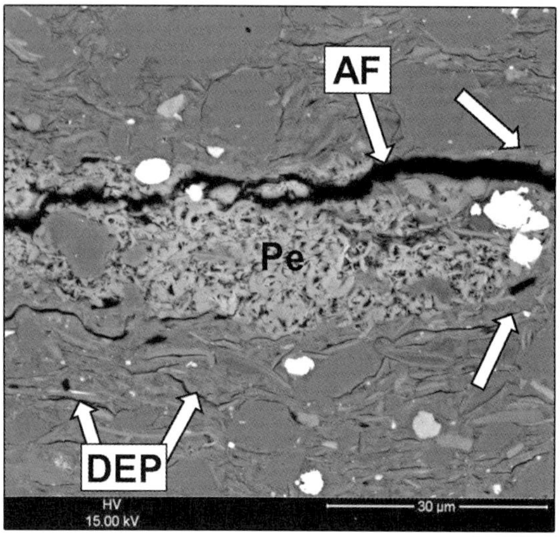

F.71.Eagle Ford
Eagle Ford Formation (Late Cretaceous)
East Texas Basin, Texas
Depth: 6700–6800 ft (2042–2073 m)
Ar-ion milled, BSE image
Thermal maturity: N/A
Submitted by D. Jennings, Core Laboratories
Scattered biogenic aggregates or fecal pellets (Pe) are composed of calcareous coccolith and foraminifer fragments. Notice the differential compaction of clay-rich material at the edge of the aggregate (white arrows). Horizontal artificial microfractures (AF) that likely developed during sample preparation roughly follow pellet boundaries. Desiccation-enhanced pores (DEP) are also common in clay-rich samples, which are more susceptible to drying during the sample preparation process.

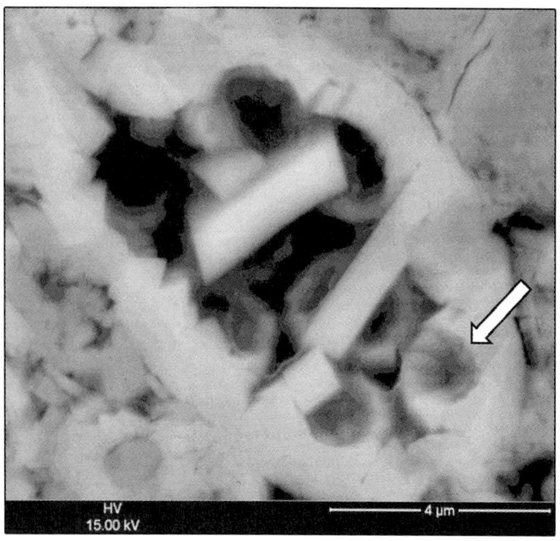

F.72.Eagle Ford
Eagle Ford Formation (Late Cretaceous)
East Texas Basin, Texas
Depth: 7500–7600 ft (2286–2316 m)
Ar-ion milled, BSE image
Thermal maturity: 1.2–1.3%R_o
Submitted by D. Jennings, Core Laboratories
Individual plates (i.e., arrow) are visible along the walls of this intact coccosphere.

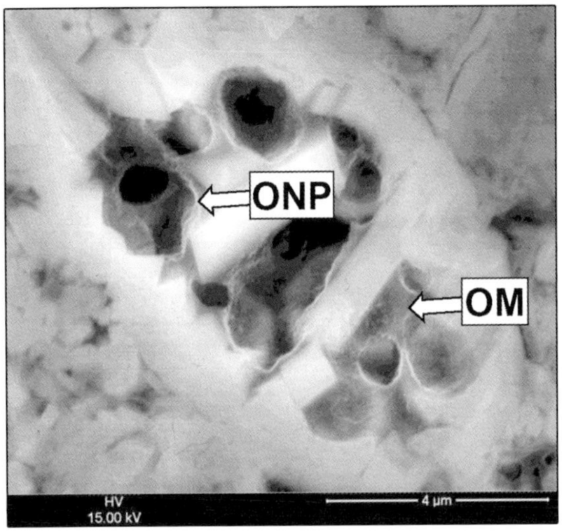

F.73.Eagle Ford
Eagle Ford Formation (Late Cretaceous)
East Texas Basin, Texas
Depth: 7500–7600 ft (2286–2316 m)
Ar-ion milled, SE image
Thermal maturity: 1.2–1.3%R_o
Submitted by D. Jennings, Core Laboratories
A secondary electron image demonstrates that organic matter (OM) within the coccosphere contains organic nanopores that are highlighted by light-color rims (ONP).

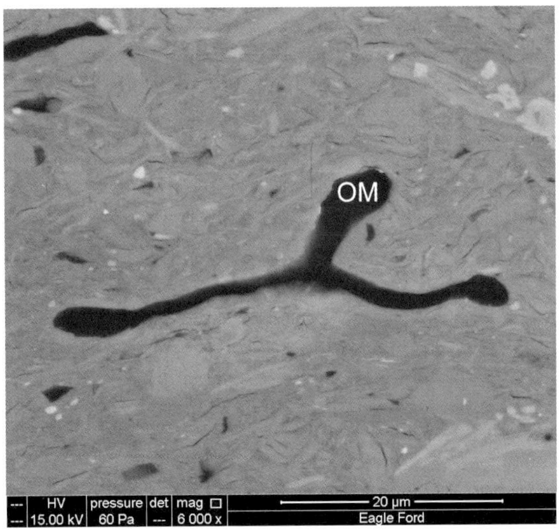

F.74.Eagle Ford
Eagle Ford Formation (Late Cretaceous)
Western Gulf of Mexico Basin, Texas
Ar-ion milled, BSE image
Thermal maturity: 0.55%
Submitted by B. Wawak and T. Gentzis, Core Laboratories
Irregularly shaped, silt-size, discrete particle of organic matter (OM) in a matrix that is predominantly disorganized phyllosilicate minerals and some quartz. This particle has a smooth, nonporous surface texture.

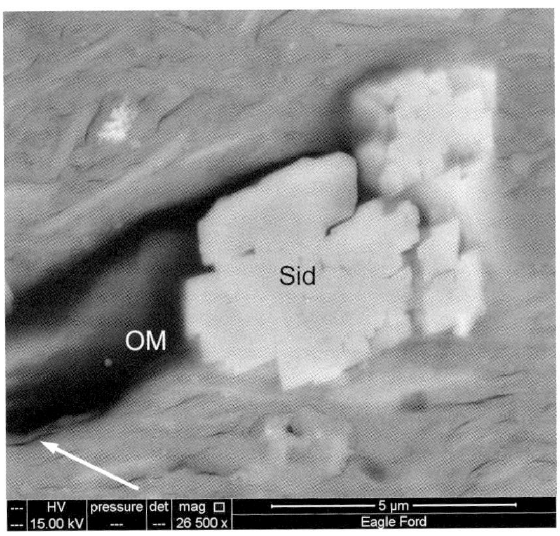

F.75.Eagle Ford
Eagle Ford Formation (Late Cretaceous)
Western Gulf of Mexico Basin, Texas
Ar-ion milled, BSE image
Thermal maturity: 0.55%
Submitted by B. Wawak and T. Gentzis, Core Laboratories
Relatively large (silt-size) particle of organic matter (OM) and siderite (Sid). Note elongated pore along the edge of the organic matter particle (arrow). Most of adjacent matrix is phyllosilicate minerals that contain some interparticle pores.

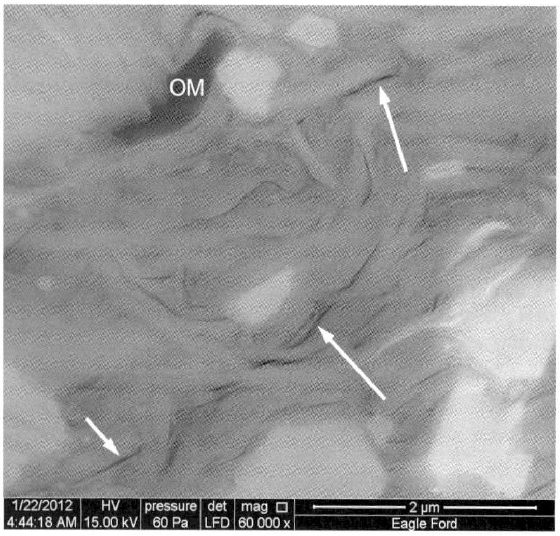

F.76.Eagle Ford

Eagle Ford Formation (Late Cretaceous)
Western Gulf of Mexico Basin, Texas
Ar-ion milled, BSE image
Thermal maturity: 0.55%
Submitted by B. Wawak and T. Gentzis, Core Laboratories
Phyllosilicate matrix in this part of the sample consists primarily of clay minerals. The morphology of the clays suggests they are primarily mixed-layer illite/smectite. Note also occurrence of interparticle pores associated with the clays (arrows) and small clay-size particle of organic matter (OM).

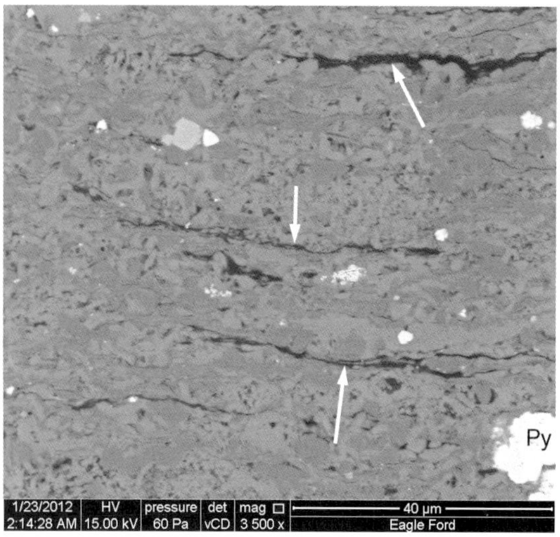

F.77.Eagle Ford

Eagle Ford Formation (Late Cretaceous)
Western Gulf of Mexico Basin, Texas
Ar-ion milled, BSE image
Thermal maturity: 0.58%
Submitted by B. Wawak and T. Gentzis, Core Laboratories
This image shows a calcareous mudstone that has an oriented fabric. The fabric is characterized by discontinuous laminations that are elevated concentrations of organic matter (arrows). These organic-rich laminations occur within a matrix that is predominantly pieces of coccoliths (lighter gray) and clays (darker gray). There are also scattered particles of pyrite (Py).

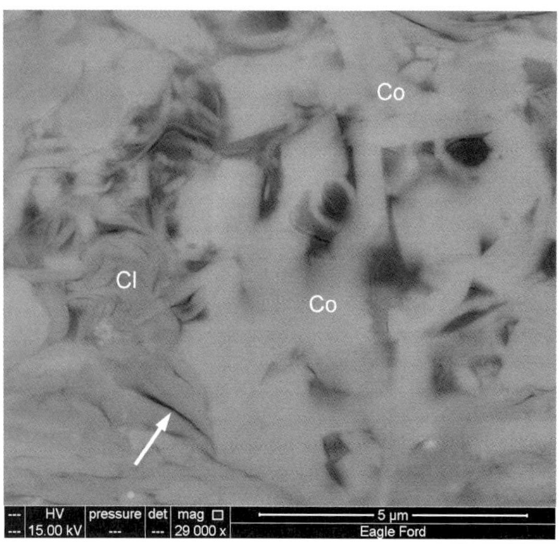

F.78.Eagle Ford

Eagle Ford Formation (Late Cretaceous)
Western Gulf of Mexico Basin, Texas
Ar-ion milled, BSE image
Thermal maturity: 0.58%
Submitted by B. Wawak and T. Gentzis, Core Laboratories
Accumulation of coccolith fragments (Co) and clays (Cl) in a calcareous mudstone. Most of the interparticle pores are filled with organic matter (dark color). Note also interparticle pores associated with clays (arrow).

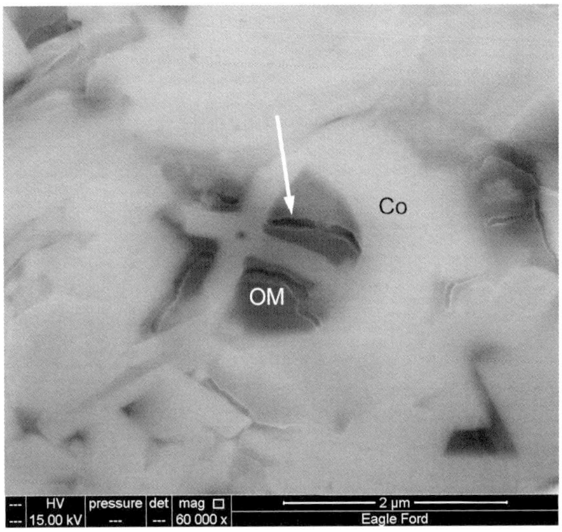

F.79.Eagle Ford
Eagle Ford Formation (Late Cretaceous)
Western Gulf of Mexico Basin, Texas
Ar-ion milled, BSE image
Thermal maturity: 0.58%
Submitted by B. Wawak and T. Gentzis, Core Laboratories
Organic matter (OM) partially filling intraparticle pores associated with coccoliths (Co). Note also fractures within the organic matter (arrow).

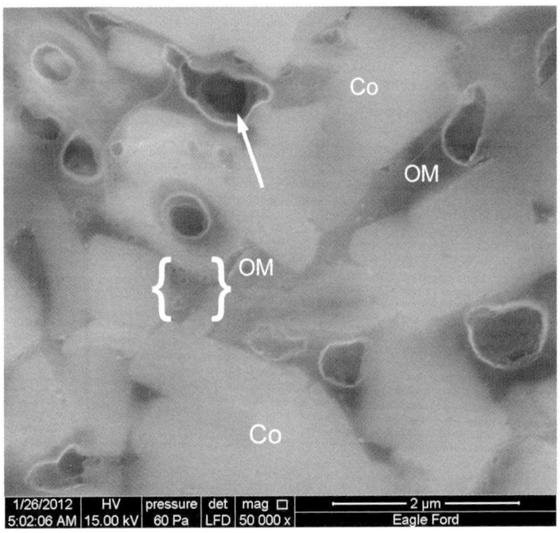

F.80.Eagle Ford
Eagle Ford Formation (Late Cretaceous)
Western Gulf of Mexico Basin, Texas
Ar-ion milled, BSE image
Thermal maturity: 2.20%
Submitted by B. Wawak and T. Gentzis, Core Laboratories
Submicron pores in organic matter (OM) that fills interparticle pores in-between pieces of disarticulated coccoliths (Co). The long dimension of the pore indicated by the arrow is 923 nm. This sample has a bimodal distribution of pore sizes that are associated with organic matter. The bracket indicates an area showing the smaller-size pores and is enlarged in Figure 81.

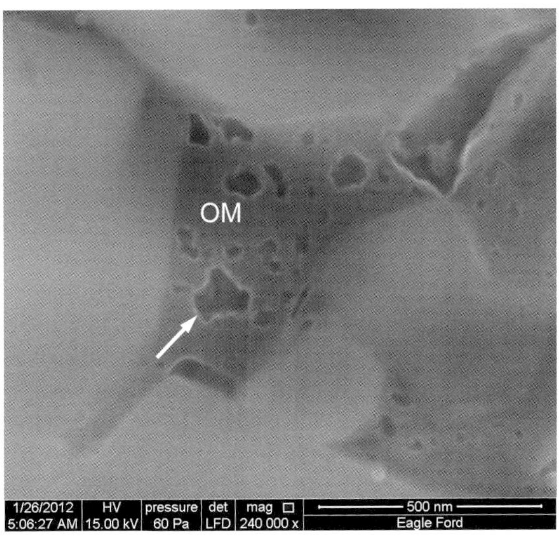

F.81.Eagle Ford
Eagle Ford Formation (Late Cretaceous)
Western Gulf of Mexico Basin, Texas
Ar-ion milled, BSE image
Thermal maturity: 2.20%
Submitted by B. Wawak and T. Gentzis, Core Laboratories
This is an enlargement of the area indicated by the bracket in Figure 80 that shows the smaller pore size population in this sample. The smaller pores are shown here and have long dimensions that are less than 100 nm. The long dimension of the pore indicated by the arrow is 94 nm.

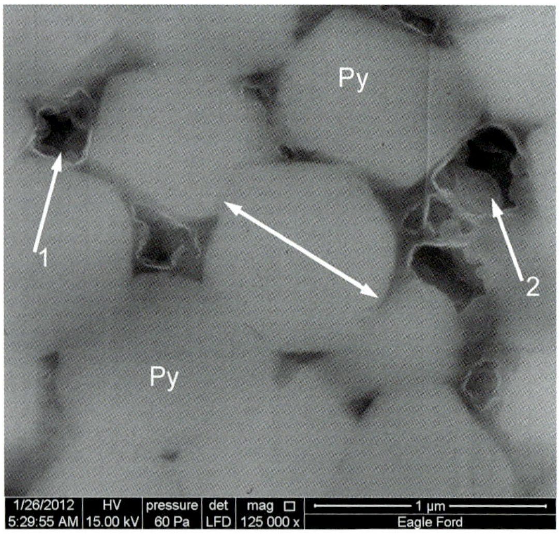

F.82.Eagle Ford
Eagle Ford Formation (Late Cretaceous)
Western Gulf of Mexico Basin, Texas
Ar-ion milled, BSE image
Thermal maturity: 2.20%
Submitted by B. Wawak and T. Gentzis, Core Laboratories
Relatively large nanopores associated with a pyrite framboid (Py). The sizes of the long dimension of the measured pores are 230 nm (1) and 342 nm (2). The width of the measured pyrite crystal (arrow) is 790 nm. The groove seen in the right half of the image is an artifact caused by argon-ion milling.

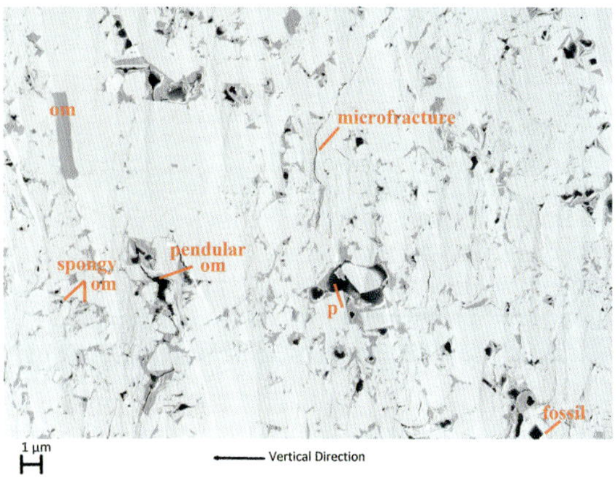

F.83.Eagle Ford
Eagle Ford Formation (Late Cretaceous)
Maverick Basin, Texas
Depth: 10500–11000 ft (3200–3353 m)
Ar-ion milled, SE2 image
Thermal maturity: N/A
Submitted by R. Aylsworth, Ingrain
Stratigraphic top to left (Upper Cretaceous). Porosity (p) is associated with both spongy and pendular organic matter. Interparticle porosity within fossils. Solid organic matter (om) and microfractures (fra) are also present. Total porosity 2.7%, organic matter 10.1%. Values given were computed by digital rock physics.

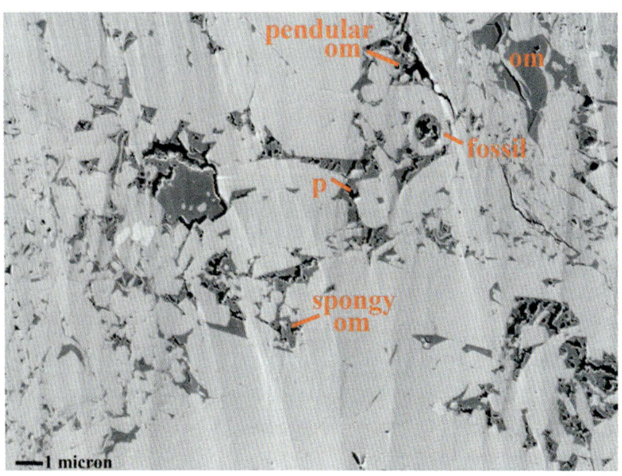

F.84.Eagle Ford
Eagle Ford Formation (Late Cretaceous)
Maverick Basin, Texas
Depth: 10500–11000 ft (3200–3353 m)
Ar-ion milled, SE2 image
Thermal maturity: N/A
Submitted by R. Aylsworth, Ingrain
Stratigraphic top to left. Porosity (p) is associated with both spongy and pendular organic matter. Interparticle porosity within fossils. Solid organic matter (om). Total porosity 3.8%, organic matter 14.0%. Values given were computed by digital rock physics.

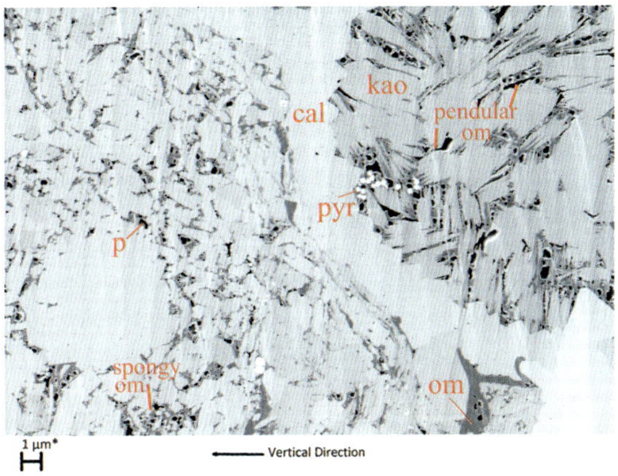

F.85. Eagle Ford
Eagle Ford Formation (Late Cretaceous)
Maverick Basin, La Salle County, Texas
Depth: 8000–8500 ft (2438–2591 m)
Ar-ion milled, SE2 image
Thermal maturity: N/A
Submitted by J. DeVito, Ingrain
Stratigraphic top to left. Kaolinite (kao) and porous organic matter fill internal chamber of a calcareous (cal) planktonic foraminifera (upper right). A small amount of pyrite (pyr) is located within the clay. Organic matter (om) texture is spongy and pendular. More than half of the porosity (p) is associated with organic matter. Total porosity is 2.6%, total organic matter is 9.6%, and high-density material is 0.2%. Values given were computed by digital rock physics.

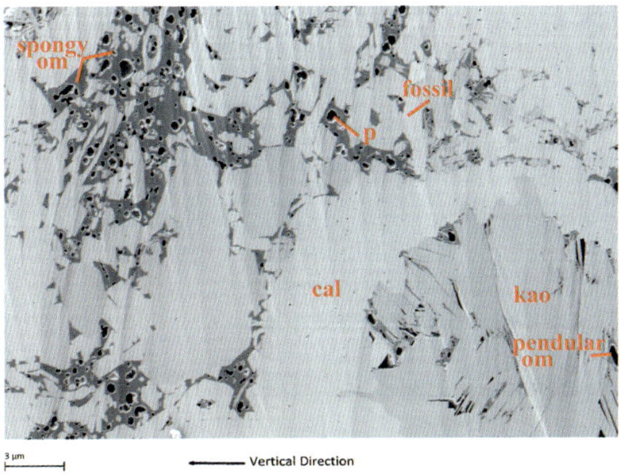

F.86. Eagle Ford
Eagle Ford Formation (Late Cretaceous)
Maverick Basin, Texas
Depth: 10000–10500 ft (3048–3200 m)
Ar-ion milled, SE2 image
Thermal maturity: N/A
Submitted by J. DeVito, Ingrain
Stratigraphic top to left. Pendular organic matter pores occur within clay matrix and between kaolinite (kao) within a planktonic foraminifer test chamber. Calcite (cal) is predominantly nannofossil fragments. Porosity (p) is associated with organic matter (om). Total porosity 2.1%, total organic matter 13.5%. Values given were computed by digital rock physics.

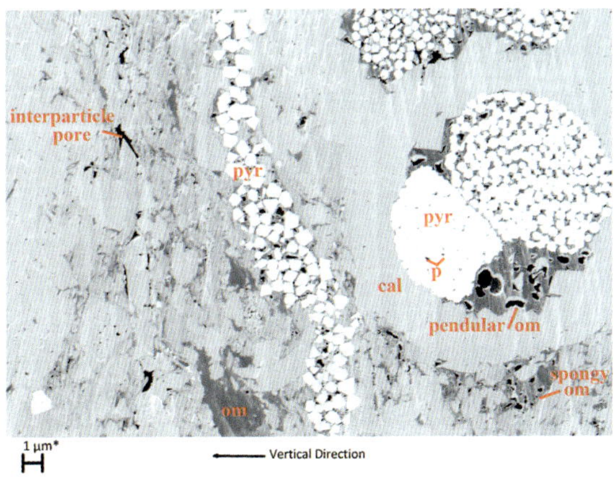

F.87. Eagle Ford
Eagle Ford Formation (Late Cretaceous)
Maverick Basin
La Salle County, Texas
Depth: 8000–8500 ft (2438–2591 m)
Ar-ion milled, SE2 image
Thermal maturity: N/A
Submitted by J. DeVito, Ingrain
Stratigraphic top to left. Pyrite framboids (pyr) with intraparticle porosity (p). Pendular organic matter borders the pyrite framboid within the calcareous foraminifer cavity. An elongated pyrite cluster (pyr) is aligned parallel to bedding. Interparticle pores and solid organic matter (om) are located at top. Total porosity 1.7%, organic matter 12.3%, high-density material 16.3%. Values given were computed by digital rock physics.

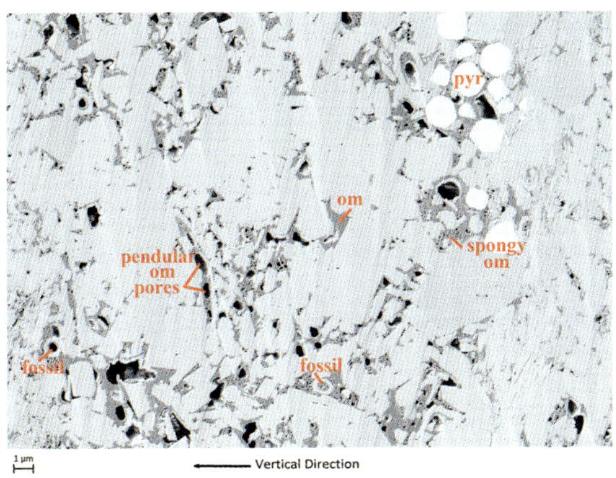

F.88. Eagle Ford
Eagle Ford Formation (Late Cretaceous)
Maverick Basin, Texas
Depth: 10000–10500 ft (3048–3200 m)
Ar-ion milled, SE2 image
Thermal maturity: N/A
Submitted by R. Aylsworth, Ingrain
Stratigraphic top to left. Porosity (black) is associated with both spongy and pendular organic matter (dark gray). Pyrite (pyr) is approximately 1 µm in size. Circular microfossils are present throughout the image. Total porosity 4.4%, organic matter 3.2%. Values given were computed by digital rock physics.

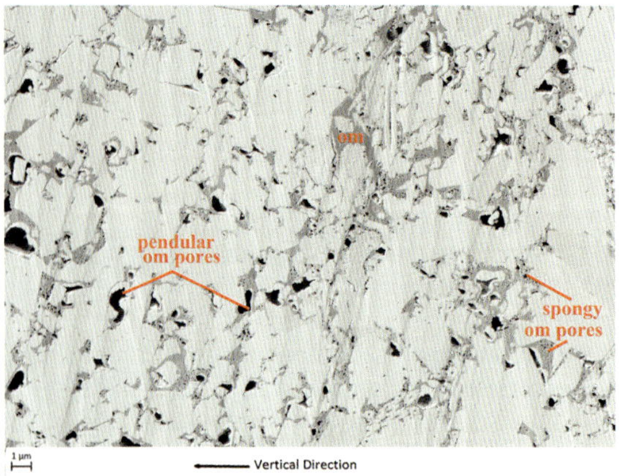

F.89. Eagle Ford

Eagle Ford Formation (Late Cretaceous)
Maverick Basin, Texas
Depth: 10000–10500 ft (3048–3200 m)
Ar-ion milled, SE2 image
Thermal maturity: N/A
Submitted by J. DeVito, Ingrain
Stratigraphic top to left. Porosity (black) is associated with both spongy and pendular organic matter (om). Intraparticle pores are associated with the organic matter toward the top. Total porosity 3.2%, organic matter 9.5%. Values given were computed by digital rock physics.

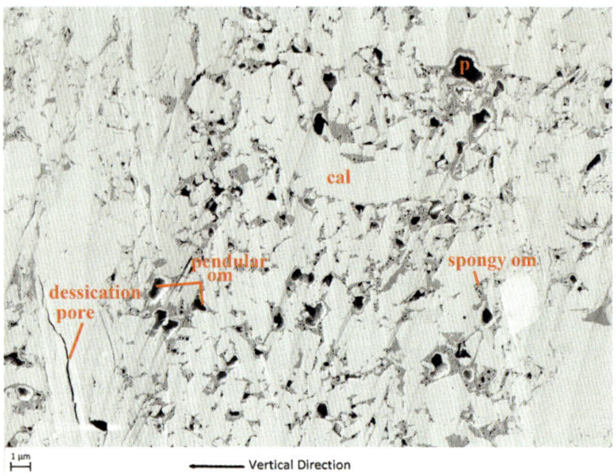

F.90. Eagle Ford

Eagle Ford Formation (Late Cretaceous)
Maverick Basin, Texas
Depth: 10000–10500 ft (3048–3200 m)
Ar-ion milled, SE2 image
Thermal maturity: N/A
Submitted by J. DeVito, Ingrain
Stratigraphic top to left. Porosity (p) is associated with both the spongy and pendular organic matter textures (dark gray). The matrix contains calcareous fossils (cal). Interpreted desiccation pore is likely induced. Total porosity 5.3%, organic matter 3.8%. Values given were computed by digital rock physics.

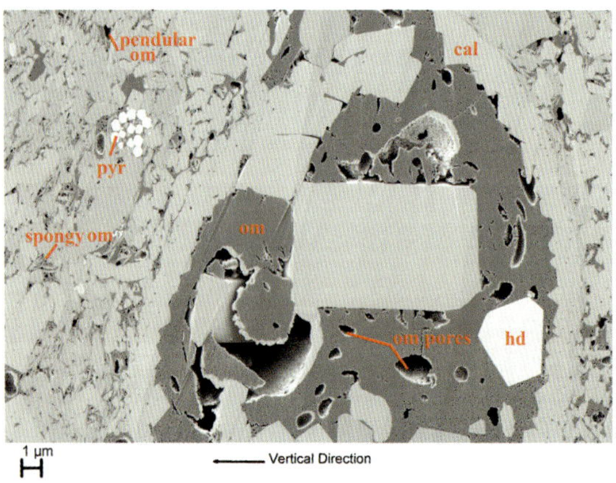

F.91. Eagle Ford

Eagle Ford Formation (Late Cretaceous)
Maverick Basin, Texas
Depth: 10000–10500 ft (3048–3200 m)
Ar-ion milled, SE2 image
Thermal maturity: N/A
Submitted by J. DeVito, Ingrain
Stratigraphic top to left. Spongy and pendular organic matter (om) surrounds the outer calcite (cal) rim of a microfossil. Large intraparticle mold pores sit within the organic matter. A pyrite framboid (pyr) occurs outside the fossil. Hexagonal (highly dense, white), rectangular, rhombic, and euhedral crystals (hd) are associated with the organic matter. Total porosity 6.2%, organic matter 29.3%, high-density material 2.01%. Values given were computed by digital rock physics.

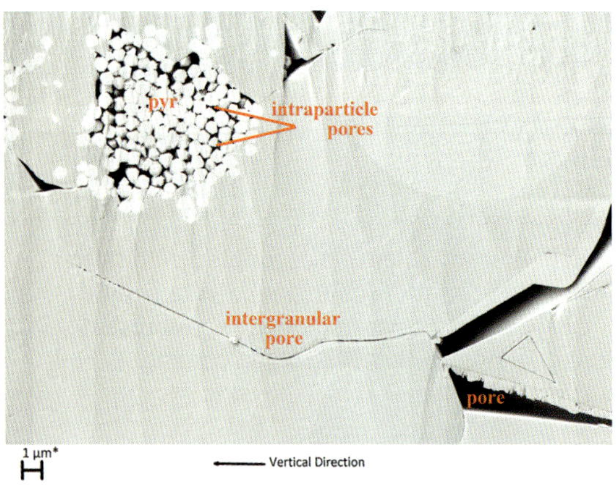

F.92. Eagle Ford

Eagle Ford Formation (Late Cretaceous)
Western Gulf of Mexico Basin, Gonzales County, Texas
Depth: 11000–11500 ft (3352–3505 m)
Ar-ion milled, SE2 image
Thermal maturity: N/A
Submitted by J. DeVito, Ingrain
Stratigraphic top to left. A large intergranular pore (pore) surrounds an angular mineral. An intergranular pore highlights a grain boundary. The pyrite framboid (pyr) contains intercrystalline porosity. Porosity 3.9%, pyrite 7.9%. Values given were computed by digital rock physics.

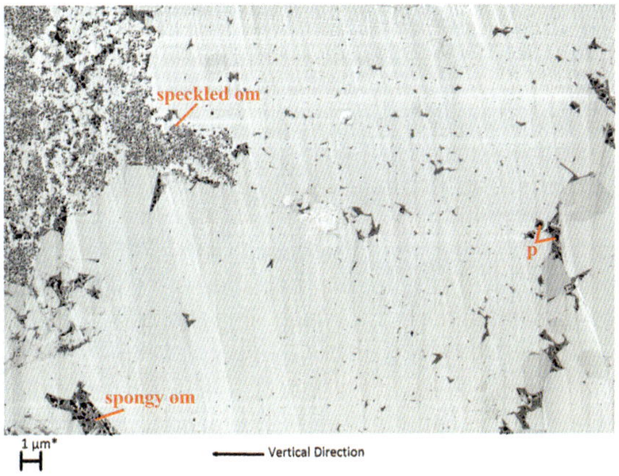

F.93.Eagle Ford
Eagle Ford Formation (Late Cretaceous)
Maverick Basin, La Salle County, Texas
Depth: 7500–8000 ft (2286–2438 m)
Ar-ion milled, SE2 image
Thermal maturity: N/A
Submitted by J. DeVito, Ingrain
Stratigraphic top to left. A line of spongy organic matter porosity (p) is organized parallel to bedding. Isolated organic matter and porosity exist within the middle of the matrix. "Speckled" nanometer-scale organic matter (om) and porosity are mixed with spongy organic matter surrounded by clay. Total porosity 1.8%, total organic matter 4.9%. Values given were computed by digital rock physics.

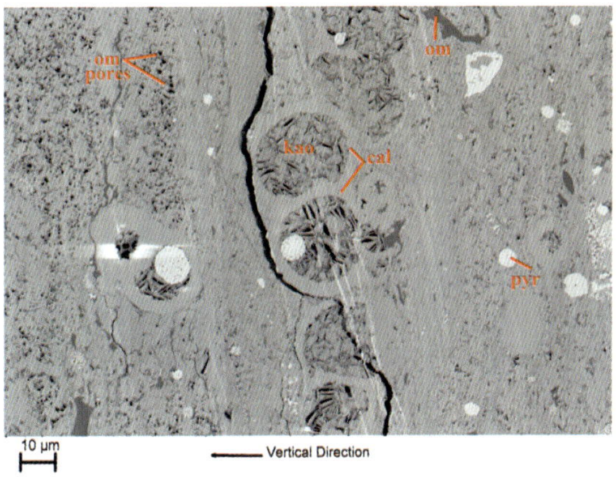

F.94.Eagle Ford
Eagle Ford Formation (Late Cretaceous)
Maverick Basin, Texas
Depth: 10500–11000 ft (3200–3353 m)
Ar-ion milled, SE2 image
Thermal maturity: N/A
Submitted by J. DeVito, Ingrain
Stratigraphic top to left. Kaolinite (kao) and porous organic matter fill the internal chambers of calcareous foraminifera microfossils (cal) arranged parallel to the bedding. Pyrite framboids (pyr) occur within the foraminifer cavity and the matrix. A mixture of pendular, spongy, and solid organic matter texture (om) is to the left of the fracture. Pores are in black (p). Total porosity 6.21%, organic matter 12.83%, high-density material 1.33%. Values given were computed with digital rock physics.

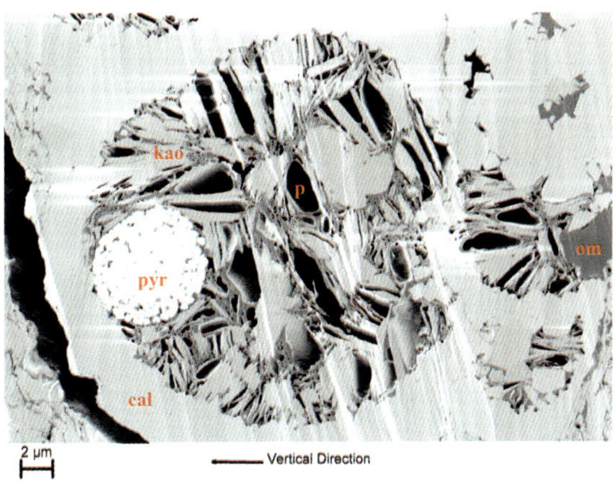

F.95.Eagle Ford
Eagle Ford Formation (Late Cretaceous)
Maverick Basin, Texas
Depth: 10500–11000 ft (3200–3353 m)
Ar-ion milled, SE2 image
Thermal maturity; N/A
Submitted by J. DeVito, Ingrain
Stratigraphic top to left. Kaolinite (kao) and porosity (p) fill the internal chambers of a Calcareous (cal) foraminifer microfossil. Organic matter (om) is spongy and pendular in texture. A pyrite framboid (pyr) sits within the clay within the foraminifer chamber. Pores are in black (p). Total porosity 14.52%, organic matter 6.08%, high-density material 3.83%. Values given were computed by digital rock physics.

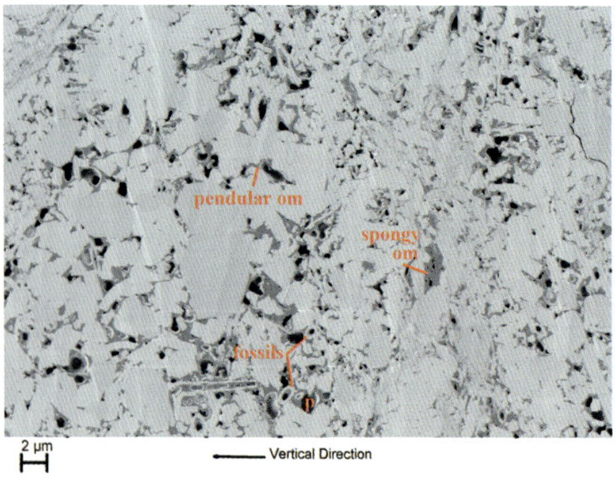

F.96.Eagle Ford
Eagle Ford Formation (Late Cretaceous)
Maverick Basin, Texas
Depth: 10500–11000 ft (3200–3353 m)
Ar-ion milled, SE2 image
Thermal maturity: N/A
Submitted by J. DeVito, Ingrain
Stratigraphic top to left. Porosity (p) is associated with both spongy and pendular organic matter (om). Circular and irregular-shape microfossils are present throughout the image. Total porosity 5.15%, organic matter 11.83%. Values given were computed by digital image analysis.

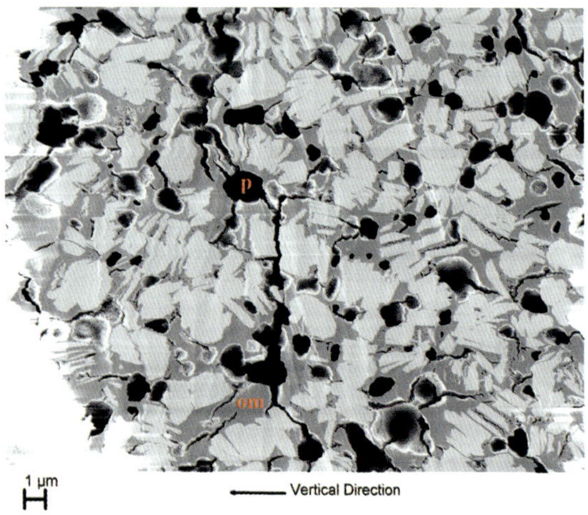

F.97.Eagle Ford
Eagle Ford Formation (Late Cretaceous)
Maverick Basin, Texas
Depth: 9500–10000 ft (2896–3048 m)
Ar-ion milled, SE2 image
Thermal maturity: N/A
Submitted by J. DeVito, Ingrain
Stratigraphic top to left. Image shows the chamber of a calcareous microfossil. Pendular and fractured organic matter textures (om) fill the internal chamber. Pores are in black (p). Total porosity 18.9%, organic matter 29.4%. Values given were computed by digital rock physics.

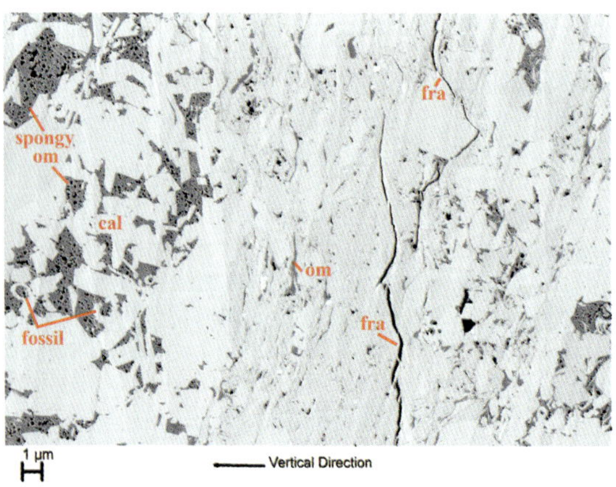

F.98.Eagle Ford
Eagle Ford Formation (Late Cretaceous)
Maverick Basin, Texas
Depth: Depth: 9500–10000 ft (2896–3048 m)
Ar-ion milled, SE2 image
Thermal maturity: N/A
Submitted by J. DeVito, Ingrain
Stratigraphic top to left. Spongy organic matter (om) is associated with a calcareous matrix (cal). The calcite is predominantly fossil fragments. Elongated organic matter (om) is perpendicular to the bedding within a clay matrix. Intergranular porosity is present near microfractures (fra). Total porosity 4.2%, organic matter 10.3%. Values given were computed with digital rock physics.

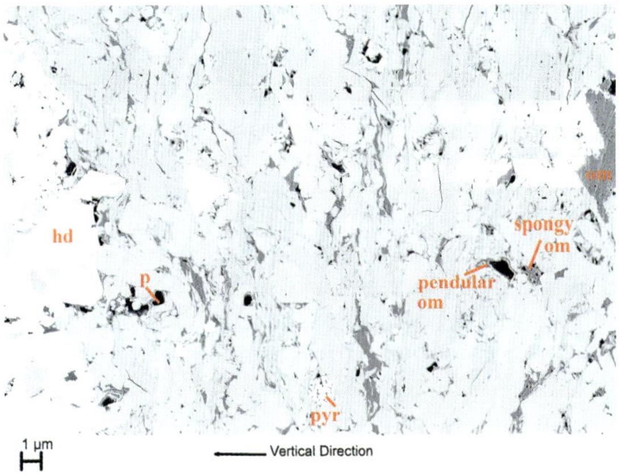

F.99.Eagle Ford
Eagle Ford Formation (Late Cretaceous)
Maverick Basin, Texas
Depth: 10500–11000 ft (3200–3353 m)
Ar-ion milled, SE2 image
Thermal maturity: N/A
Submitted by J. DeVito, Ingrain
Stratigraphic top to left. Porosity (p) is associated with both spongy and pendular organic matter (om). Pyrite (pyr), highly dense minerals (hd), and microfractures lie within the clay matrix. Total porosity 1.9%, organic matter 5.7%, high-density material 11.6%. Values given were computed by digital rock physics.

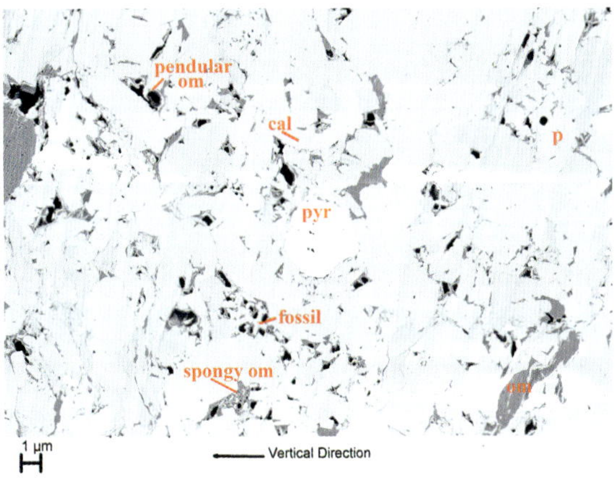

F.100.Eagle Ford
Eagle Ford Formation (Late Cretaceous)
Maverick Basin, Texas
Depth: 10500–11000 ft (3200–3353 m)
Ar-ion milled, SE2 image
Thermal maturity: N/A
Submitted by J. DeVito, Ingrain
Stratigraphic top to left. Porosity (p) is associated with both spongy and pendular organic matter. Interparticle porosity is within calcareous nannofossils and pyrite framboids (pyr). Solid organic matter (om) texture is present. Total porosity 3.2%, organic matter 6.1%, high-density 1.7%. Values given were computed by digital rock physics.

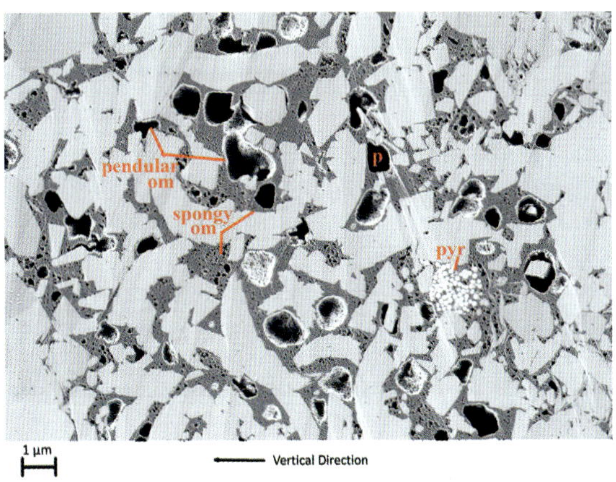

F.101.Eagle Ford
Eagle Ford Formation (Late Cretaceous)
Maverick Basin, Texas
Depth: 10500–11000 ft (3200–3353 m)
Ar-ion milled, SE2 image
Thermal maturity: N/A
Submitted by J. DeVito, Ingrain
Stratigraphic top to left. Porosity (p) is associated with both spongy and pendular organic matter (om) textures occurring within a predominantly calcareous matrix. High-density pyrite minerals (pyr) are in white. Total porosity 13.1%, organic matter 23.5%, pyrite 0.4% . Values presented computed by digital rock physics.

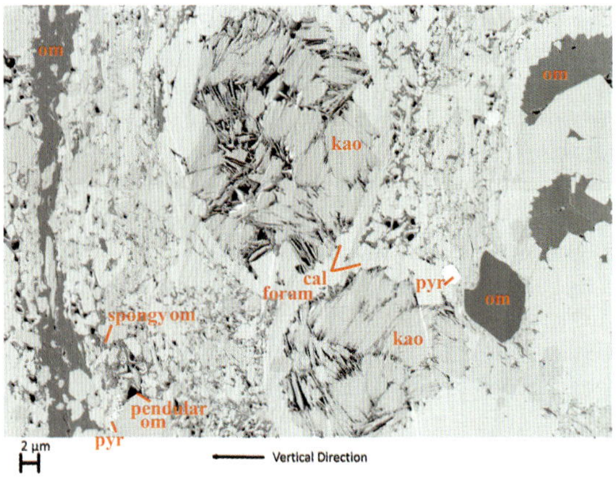

F.102.Eagle Ford
Eagle Ford Formation (Late Cretaceous)
Maverick Basin, Texas
Depth: 10500–11000 ft (3200–3353 m)
Ar-ion milled, SE2 image
Thermal maturity: N/A
Submitted by J. DeVito, Ingrain
Stratigraphic top to left. Organic matter (om) is parallel to bedding and contains nanometer-size spongy porosity. Two foraminifera fossils (foram) are filled with kaolinite (kao) and rimmed with calcite (cal). Porosity (black) and organic matter (dark gray) occur within the clay. Three organic matter textures are observed: solid organic matter (om) sits within a calcareous (cal) fossil; spongy and pendular occur between grains. Total porosity 3.9%, organic matter 12.6%. Values presented computed by digital rock physics.

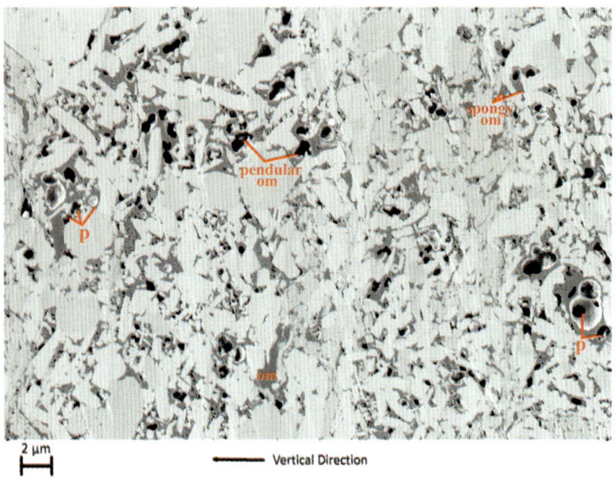

F.103.Eagle Ford
Eagle Ford Formation (Late Cretaceous)
Maverick Basin, Texas
Depth: 10500–11000 ft (3200–3353 m)
Ar-ion milled, SE2 image
Thermal maturity: N/A
Submitted by J. DeVito and L. Paquette, Ingrain
Stratigraphic top to left. Porosity (p) is associated with both spongy and pendular organic matter (om). Matrix is predominantly calcareous with clay. Total porosity 6.5%, organic matter 17.1%. Values presented computed by digital image analysis.

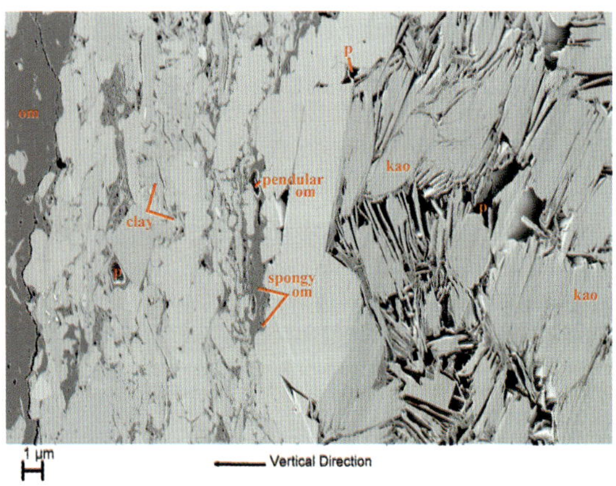

F.104.Eagle Ford
Eagle Ford Formation (Late Cretaceous)
Maverick Basin, Texas
Depth: 10500–11000 ft (3200–3353 m)
Ar-ion milled, SE2 image
Thermal maturity: N/A
Submitted by J. DeVito, Ingrain
Stratigraphic top to left. Porosity (p) occurs within the organic matter and between clays. A calcareous foraminifer fossil is filled with kaolinite (kao). Clay and organic matter (om) are laminated parallel to the bedding and contain nanometer-size spongy porosity. Total porosity 8.9%, organic matter 14.5%. Values presented computed by digital image analysis.

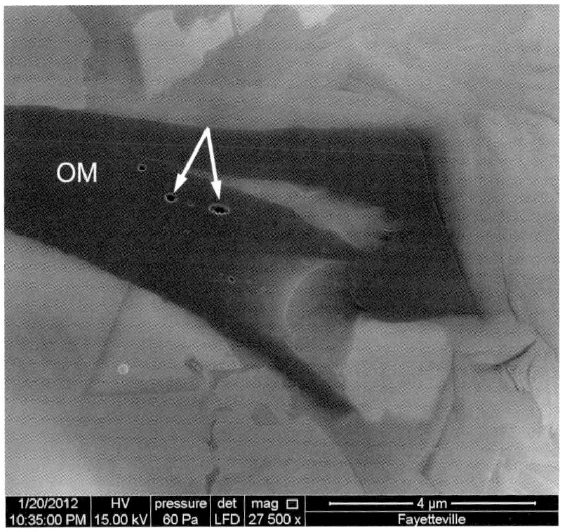

F.105.Fayetteville
Fayetteville Formation (Mississippian)
Arkoma Basin, Arkansas
Ar-ion milled, SE image
Thermal Maturity: 4.55%R_o
Submitted by B. Wawak and T. Gentzis, Core Laboratories
Scattered ovoid nanopores (arrows) in a relatively large (silt-size) discrete particle of organic matter (OM).

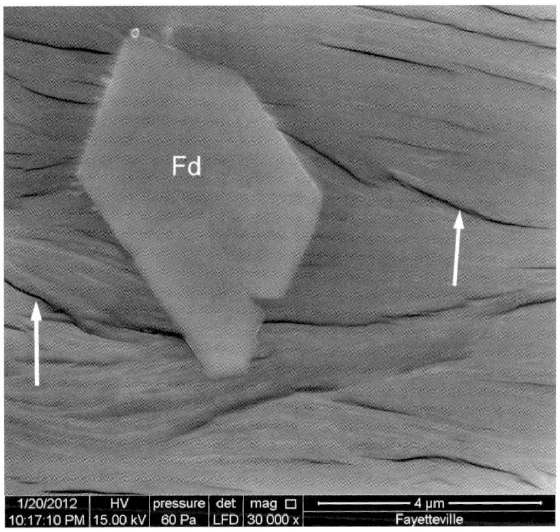

F.106.Fayetteville
Fayetteville Formation (Mississippian)
Arkoma Basin, Arkansas
Ar-ion milled, SE image
Thermal Maturity: 4.55%R_o
Submitted by B. Wawak and T. Gentzis, Core Laboratories
Clay size feldspar crystal (Fd) associated with phyllosilicate (predominantly clays) minerals. Note interparticle pores (arrows) associated with adjacent clays.

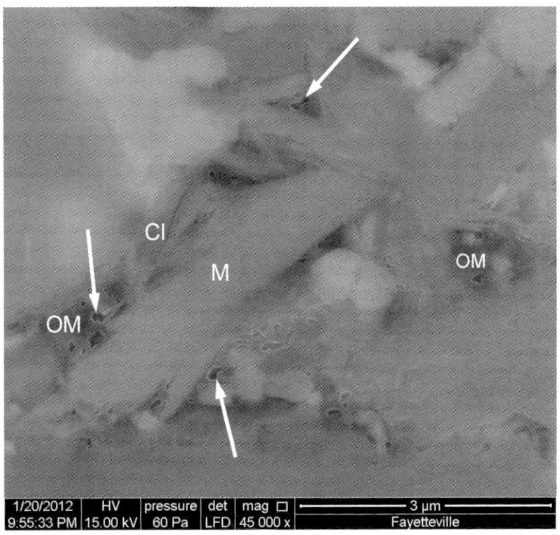

F.107.Fayetteville
Fayetteville Formation (Mississippian)
Arkoma Basin, Arkansas
Ar-ion milled, SE image
Thermal Maturity: 4.55%R_o
Submitted by B. Wawak and T. Gentzis, Core Laboratories
Scattered nanopores (arrows) associated with dispersed organic matter (OM) in interparticle pores associated with clays (Cl) and other phyllosilicate minerals (M).

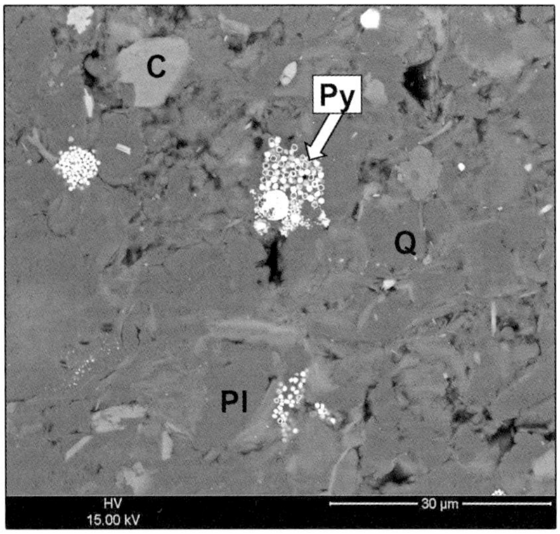

F.108.Floyd
Floyd Formation (Mississippian)
West-central Alabama
Depth: 7400–7500 ft (2456–2286 m)
Ar-ion milled, BSE image
Thermal maturity: N/A
Submitted by D. Jennings, Core Laboratories
Intercrystalline and intracrystalline pores are abundant in clusters of hollow, degraded pyrite framboids (Py). Quartz (Q), plagioclase (Pl), mica, and carbonate grains (C) are the predominant framework grains.

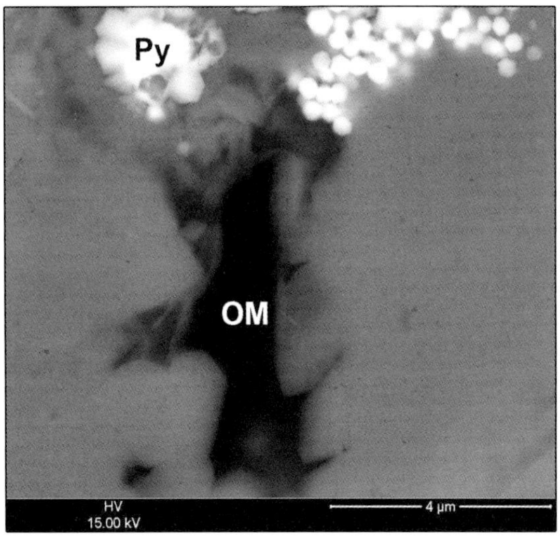

F.109.Floyd
Floyd Formation (Mississippian)
West-central Alabama
Depth: 7400–7500 ft (2456–2286 m)
Ar-ion milled, BSE image
Thermal maturity: N/A
Submitted by D. Jennings, Core Laboratories
Organic matter (OM) is commonly present within scattered nanometer- to micron-scale pores. Quartz overgrowths are evident at the edge of the pore. Note the degradation evident in the pyrite framboid (Py).

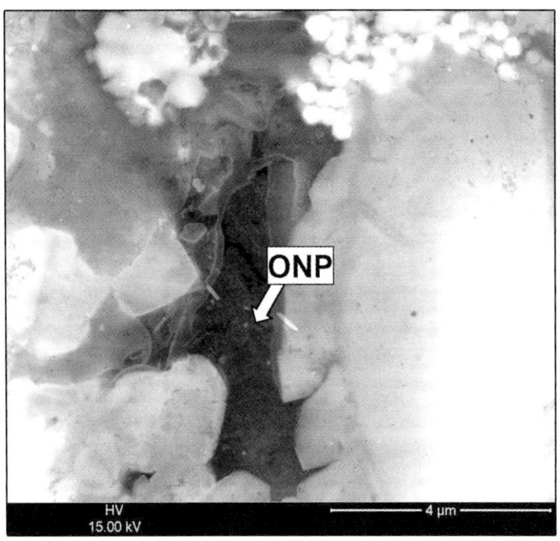

F.110.Floyd
Floyd Formation (Mississippian)
West-central Alabama
Depth: 7400–7500 ft (2456–2286 m)
Ar-ion milled, SE image
Thermal maturity: N/A
Submitted by D. Jennings, Core Laboratories
Organic matter in this sample displays very small, internal organic nanopores (ONP) as evidenced by the secondary electron image.

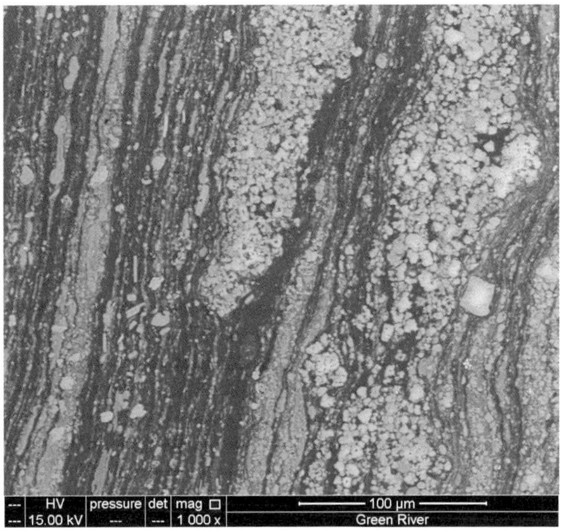

F.111.Green River
Green River Formation (Eocene)
Piceance Basin, Colorado
Ar-ion milled, BSE/SE image
Thermal maturity: 0.32%R_o
Submitted by B. Wawak and T. Gentzis, Core Laboratories
Relatively low-magnification view showing laminated fabric of the sample. The fabric is characterized by concentrations of light-color crystals that are separated by darker laminations that contain elevated amounts of organic matter. Minerals identified in the lighter-color laminations include dolomite, albite, nahcolite, and other minerals.

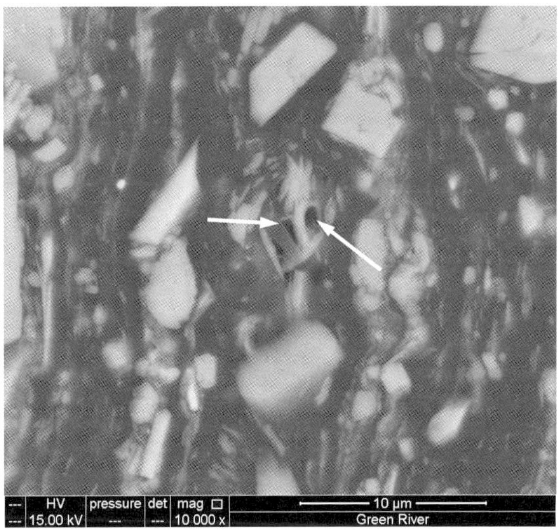

F.112.Green River
Green River Formation (Eocene)
Piceance Basin, Colorado
Ar-ion milled, BSE/SE image
Thermal maturity: 0.32%R_o
Submitted by B. Wawak and T. Gentzis, Core Laboratories
Open pores are rare in the sample, and the ones that are identified typically occur around the edges of the inorganic particles (arrows).

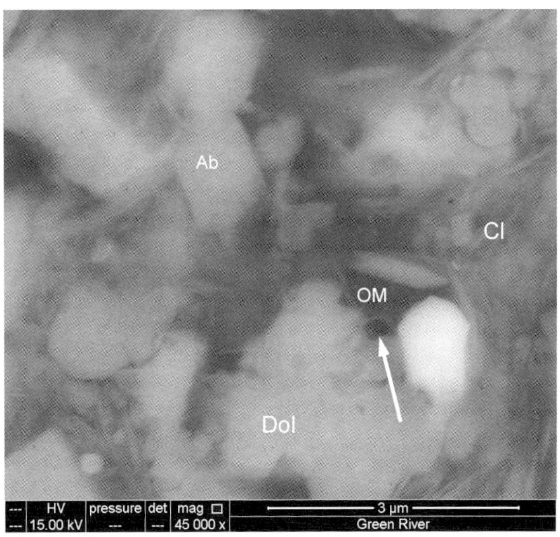

F.113.Green River
Green River Formation (Eocene)
Piceance Basin, Colorado
Ar-ion milled, BSE/SE image
Thermal maturity: 0.32%R_o
Submitted by B. Wawak and T. Gentzis, Core Laboratories
Small nanopore (arrow) associated with organic matter (OM) that occurs at the edge of a dolomite (Dol) crystal. Other inorganic particles seen in this view include phyllosilicate minerals (Cl) and albite (Ab).

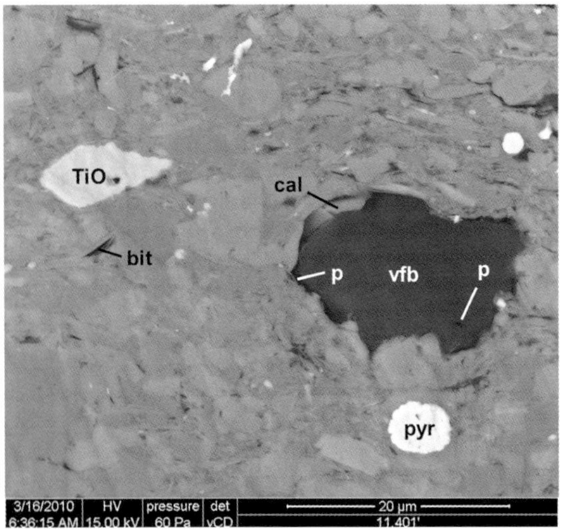

F.114.Haynesville

Haynesville Formation (Late Jurassic)
East Texas Basin, Harrison County, Texas
Depth: 11401 ft (3475 m)
Ar-ion milled, BSE image
Thermal maturity: 1.4–1.5%R_o
Submitted by W. Camp, Anadarko Petroleum
Void-filling bitumen (vfb) postdates euhedral calcite cement (calc). Porosity (p) locally developed along outer rim (shrinkage crack?) and within the bitumen fill. Elongate pores associated with clay minerals are also filled with bitumen (bit). Framboidal pyrite (pyr) and rutile (TiO) also present.

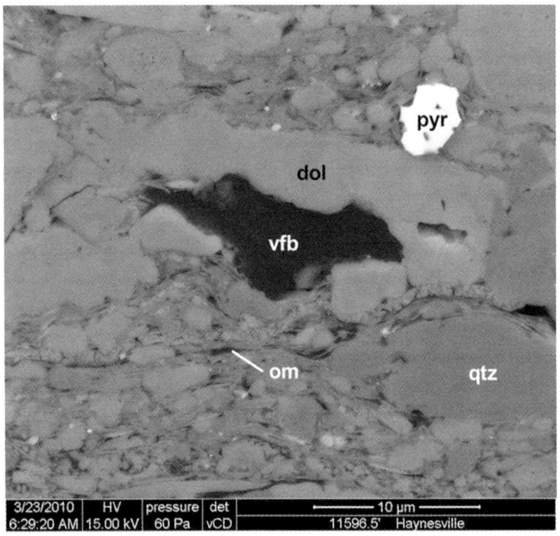

F.115.Haynesville

Haynesville Formation (Late Jurassic)
East Texas Basin, Harrison County, Texas
Depth: 11596.5 ft (3534 m)
Ar-ion milled, BSE image
Thermal maturity: 1.6–1.7%R_o
Submitted by W. Camp, Anadarko Petroleum
Irregular void below dolomitized fossil fragment (dol) filled with bitumen (vfb). Subparallel fabric created by clay grains is locally deformed around quartz grains (qtz) accentuated by interlaminated organic matter (om). Framboidal pyrite (pyr) also present.

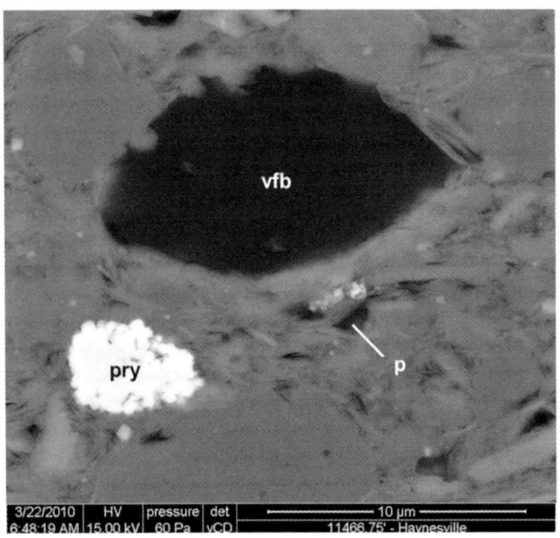

F.116.Haynesville

Haynesville Formation (Late Jurassic)
East Texas Basin, Harrison County, Texas
Depth: 11466.75 ft (3495 m)
Ar-ion milled, BSE image
Thermal maturity: 1.5%R_o
Submitted by W. Camp, Anadarko Petroleum
Elliptical void (burrow?) filled with bitumen (vfb). Note localized porosity (p) associated with bridging of platy and fibrous clay minerals (mica and illite). Framboidal pyrite (pyr) also present.

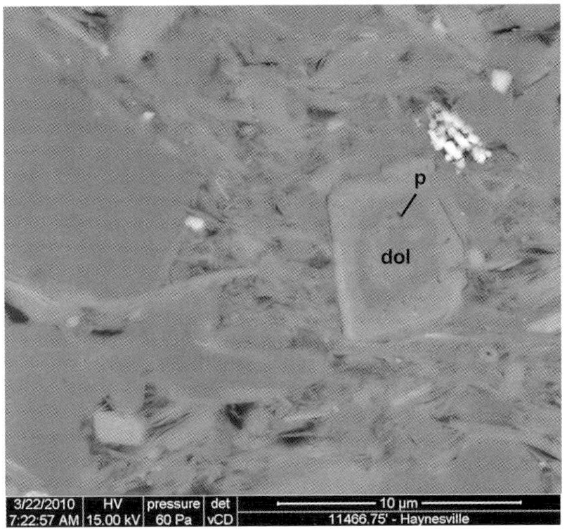

F.117.Haynesville
Haynesville Formation (Late Jurassic)
East Texas Basin, Harrison County, Texas
Depth: 11466.75 ft (3495 m)
Ar-ion milled, BSE image
Thermal maturity: 1.5%R_o
Submitted by W. Camp, Anadarko Petroleum
Authigenic dolomite rhomb (dol) exhibits compositional banding. Lighter shades of dolomite are iron rich based on energy-dispersive x-ray spectrometry analysis. Minor intracrystalline porosity (p) appears to be concentrated within iron-rich bands.

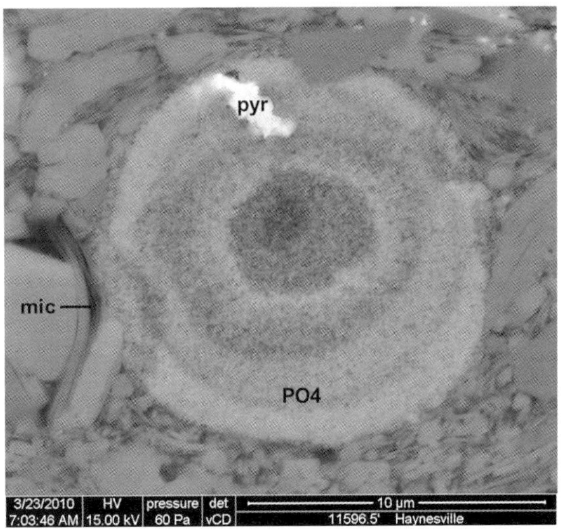

F.118.Haynesville
Haynesville Formation (Late Jurassic)
East Texas Basin, Harrison County, Texas
Depth: 11596.5 ft (3534 m)
Ar-ion milled, BSE image
Thermal maturity: 1.6–1.7%Ro
Submitted by W. Camp, Anadarko Petroleum
Phosphatic ooid (PO4). Note concentric layers of ooid and ductile mica grains (mic) locally deformed by more rigid grains due to compaction. Pyrite (pyr) is also present.

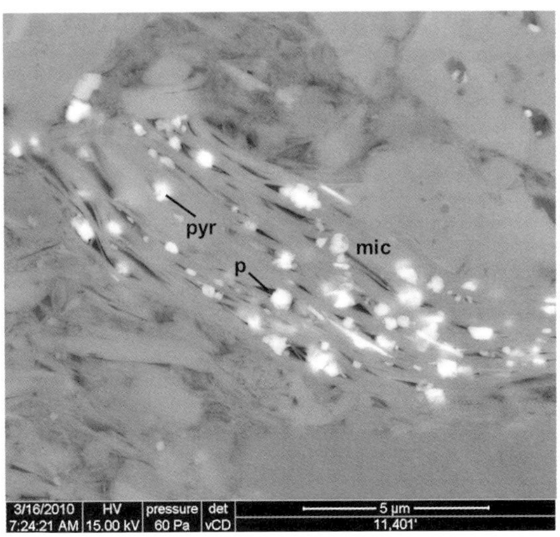

F.119.Haynesville
Haynesville Formation (Late Jurassic)
East Texas Basin, Harrison County, Texas
Depth: 11401 ft (3475 m)
Ar-ion milled, BSE image
Thermal maturity: 1.4–1.5%R_o
Submitted by W. Camp, Anadarko Petroleum
Exfoliated mica (mic) grain with pyrite (pyr) intercalated within interstitial pore spaces (p).

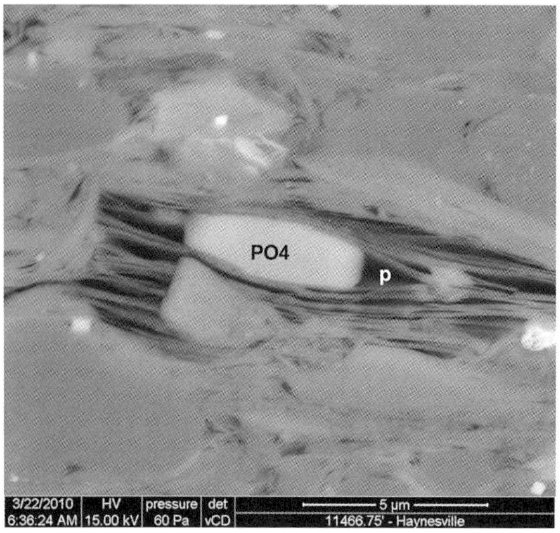

F.120.Haynesville
Haynesville Formation (Late Jurassic)
East Texas Basin, Harrison County, Texas
Depth: 11466.75 ft (3495 m)
Ar-ion milled, BSE image
Thermal maturity: 1.5%R_o
Submitted by W. Camp, Anadarko Petroleum
Exfoliated mica grain with apatite (PO4) intercalated within interstitial pore spaces (p).

F.121.Haynesville
Haynesville Formation (Late Jurassic)
East Texas Basin, Harrison County, Texas
Depth: 11596.5 ft (3534 m)
Ar-ion milled, BSE image
Thermal maturity: 1.6–1.7%R_o
Submitted by W. Camp, Anadarko Petroleum
Interstitial pores developed within exfoliated mica grain filled with pyrite (pyr) and void-filling bitumen (vfb).

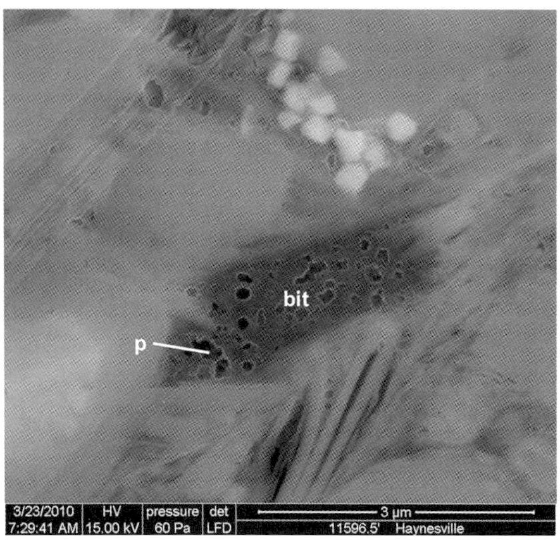

F.122.Haynesville
Haynesville Formation (Late Jurassic)
East Texas Basin, Harrison County, Texas
Depth: 11596.5 ft (3534 m)
Ar-ion milled, SE image
Thermal maturity: 1.6–1.7%R_o
Submitted by W. Camp, Anadarko Petroleum
Sponge-like texture of porous pyrobitumen (bit) filling interparticle pores within clay matrix. Pores (p) within pyrobitumen presumably formed during thermal cracking of oil to gas during burial.

F.123.Haynesville
Haynesville Formation (Late Jurassic)
East Texas Basin, Harrison County, Texas
Depth: 11401 ft (3475 m)
Ar-ion milled, BSE image
Thermal maturity: 1.4–1.5%R_o
Submitted by W. Camp, Anadarko Petroleum
Bedding plane parting (part). Low backscatter intensity indicates parting is filled with bitumen and thus may represent a hydrocarbon migration feature. Bitumen (bit) also fills larger pore spaces such as interstitial pores between exfoliated mica grains. Partings may create mechanical discontinuities that may respond to hydraulic fracture stimulation and facilitate hydrocarbon flow.

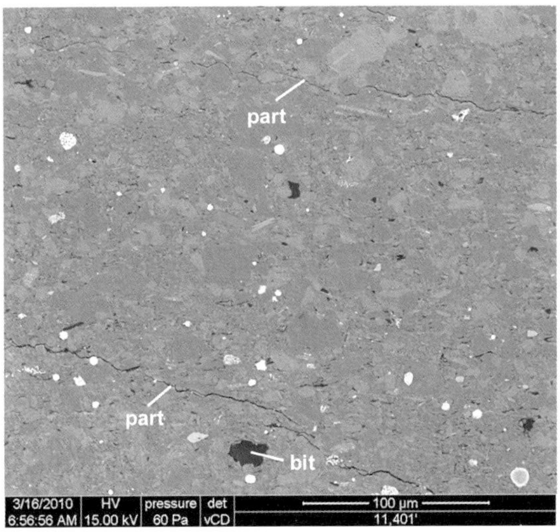

F.124.Haynesville
Haynesville Formation (Late Jurassic)
East Texas Basin, Harrison County, Texas
Depth: 11401 ft (3475 m)
Ar-ion milled, BSE image
Thermal maturity: 1.4–1.5%R_o
Submitted by W. Camp, Anadarko Petroleum
Bedding plane parting (part). Low backscatter intensity indicates parting is filled with bitumen and thus may represent a hydrocarbon migration feature. Bitumen (bit) also fills larger pore spaces. Partings may create mechanical discontinuities that may respond to hydraulic fracture stimulation and facilitate hydrocarbon flow.

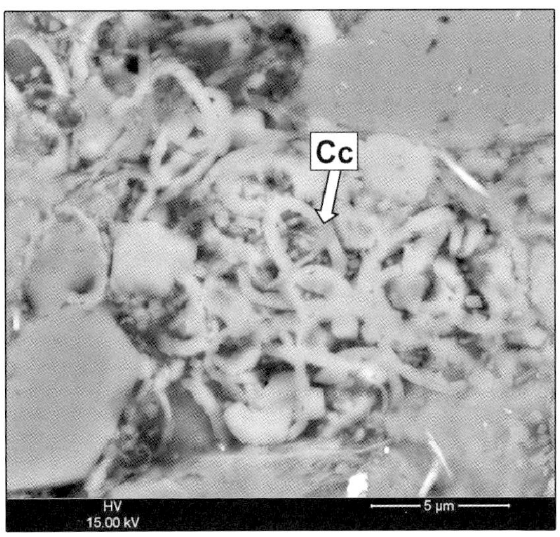

F.125.Haynesville
Haynesville Formation (Jurassic)
East Texas Basin, Texas
Depth: 13900–14000 ft (4237–4267 m)
Ar-ion milled, BSE image
Thermal maturity: 1.3–1.4%R_o
Submitted by D. Jennings, Core Laboratories
Organic matter is commonly present within interparticle pores between coccolith fragments (Cc) in this silty shale.

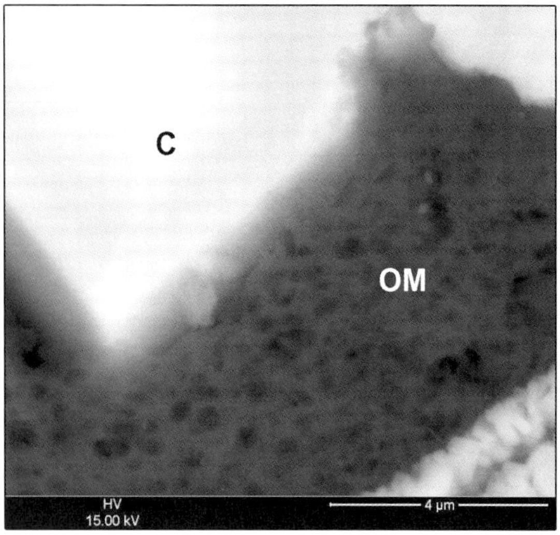

F.126.Haynesville
Haynesville Formation (Jurassic)
East Texas Basin, Texas
Depth: 13900–14000 ft (4237–4267 m)
Ar-ion milled, BSE image
Thermal maturity: 1.3–1.4%R_o
Submitted by D. Jennings, Core Laboratories
Authigenic calcite (C) and organic matter (OM) are present within a globose foraminifer chamber. The organic matter displays a spongy appearance.

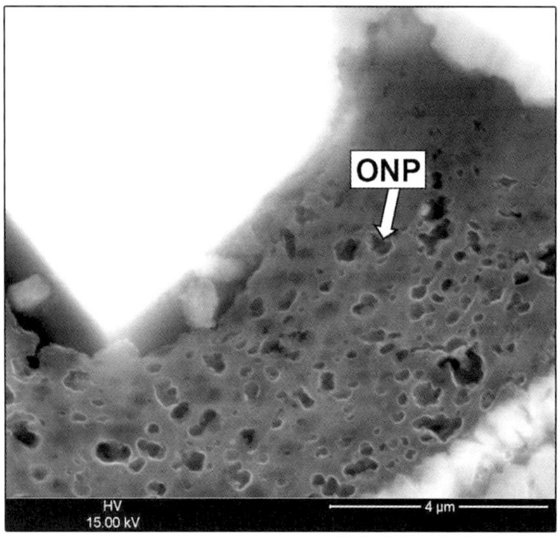

F.127.Haynesville
Haynesville Formation (Jurassic)
East Texas Basin, Texas
Depth: 13900–14000 ft (4237–4267 m)
Ar-ion milled, SE image
Thermal maturity: 1.3–1.4%R_o
Submitted by D. Jennings, Core Laboratories
The secondary electron image of organic matter within a foraminifer chamber reveals that well-developed, organic nanopores (ONP) are abundant.

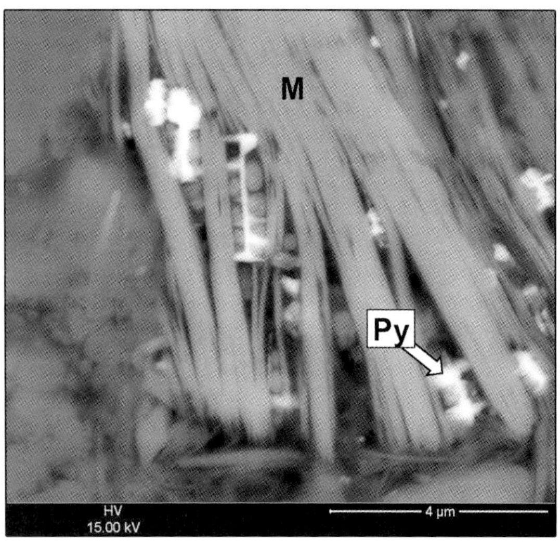

F.128.Haynesville
Haynesville Formation (Jurassic)
East Texas Basin, Texas
Depth: 13900–14000 ft (4237–4267 m)
Ar-ion milled, BSE image
Thermal maturity: 1.3–1.4%R_o
Submitted by D. Jennings, Core Laboratories
Pyrite (Py) and marcasite is present between layers of iron-rich mica (M). The authigenic clay particles associated with pyrite grains (medium gray) suggest that pyrite and authigenic clay formed penecontemporaneously as the mica degraded.

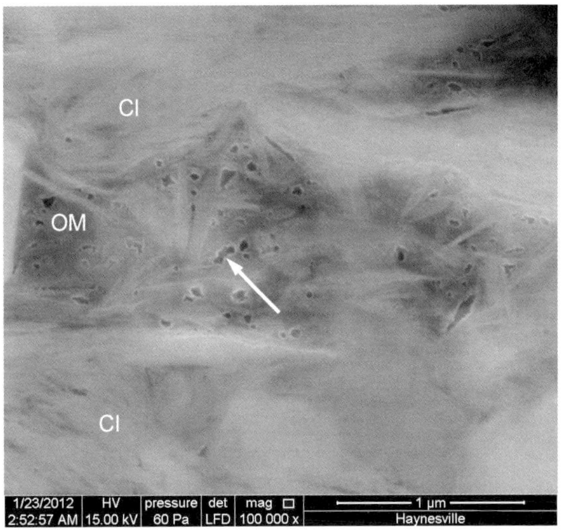

F.129.Haynesville
Haynesville Formation (Late Jurassic)
Northern Gulf of Mexico Basin, Louisiana
Ar-ion milled, SE image
Thermal maturity: 1.76%R_o
Submitted by B. Wawak and T. Gentzis, Core Laboratories
Nanopores associated with dispersed organic matter (OM). Organic matter fills interparticle areas associated with platy phyllosilicate minerals (Cl). The long dimensions of selected measured pores range from 24 to 103 nm. The long dimension of the pore indicated by the arrow is 103 nm.

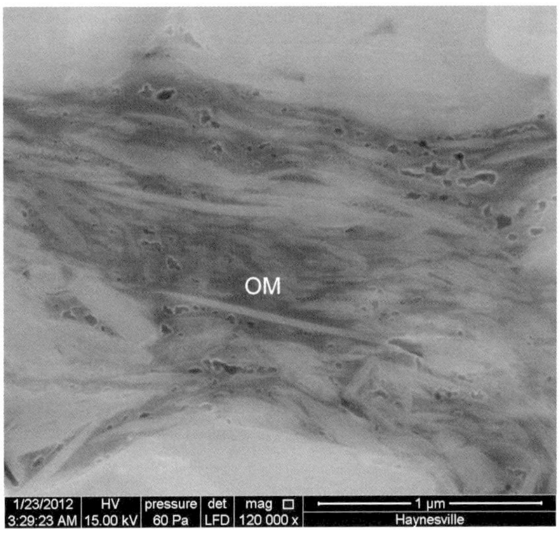

F.130.Haynesville
Haynesville Formation (Late Jurassic)
Northern Gulf of Mexico Basin, Louisiana
Ar-ion milled, SE image
Thermal maturity: 1.76%R_o
Submitted by B. Wawak and T. Gentzis, Core Laboratories
Nanopores associated with dispersed organic matter (OM). Organic matter fills interparticle areas associated with platy phyllosilicate minerals.

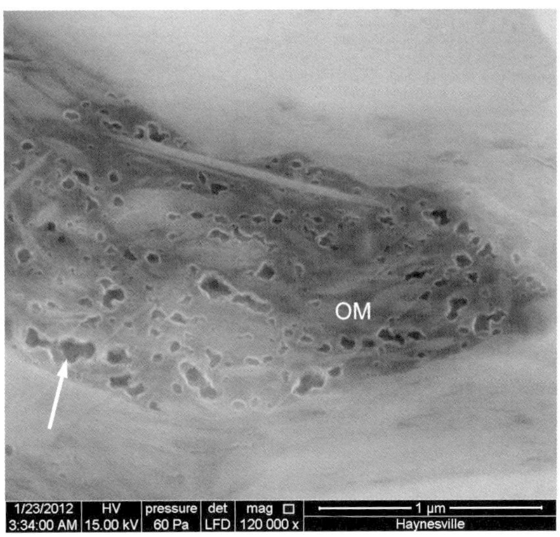

F.131.Haynesville
Haynesville Formation (Late Jurassic)
Northern Gulf of Mexico Basin, Louisiana
Ar-ion milled, SE image
Thermal maturity: 1.76%R_o
Submitted by B. Wawak and T. Gentzis, Core Laboratories
Nanopores in what appears to be a discrete particle of organic matter (OM). The long dimensions of selected measured pores range from 26 to 165 nm. Some of the larger pores are the product of the coalescing of several smaller pores (arrow).

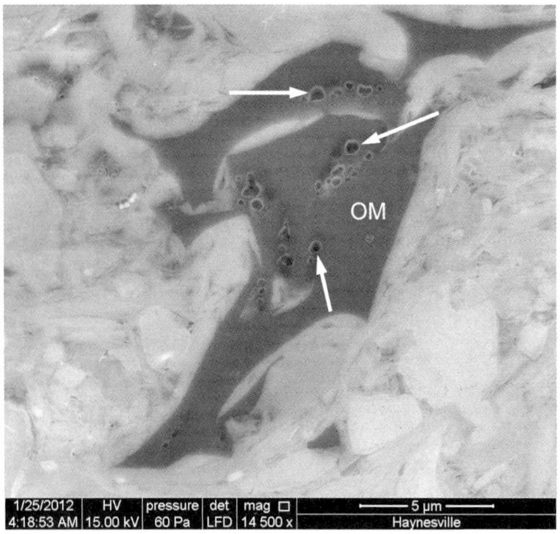

F.132.Haynesville
Haynesville Formation (Late Jurassic)
Northern Gulf of Mexico Basin, Louisiana
Ar-ion milled, SE image
Thermal maturity: 1.85%R_o
Submitted by B. Wawak and T. Gentzis, Core Laboratories
Relatively large (long dimension is hundreds of nanometers) pores (arrows) associated with silt-size particles of organic matter (OM).

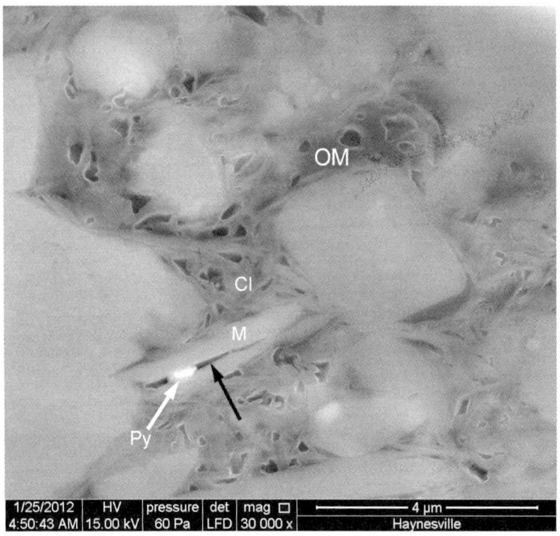

F.133.Haynesville
Haynesville Formation (Late Jurassic)
Northern Gulf of Mexico Basin, Louisiana
Ar-ion milled, SE image
Thermal maturity: 1.85%R_o
Submitted by B. Wawak and T. Gentzis, Core Laboratories
Nanopores in dispersed organic matter (OM) that fills interparticle pores associated with phyllosilicate minerals (Cl, clays; M, mica) and other inorganic particles. Note pyrite (Py) located within intraparticle pore (black arrow) associated with an expanded mica grain.

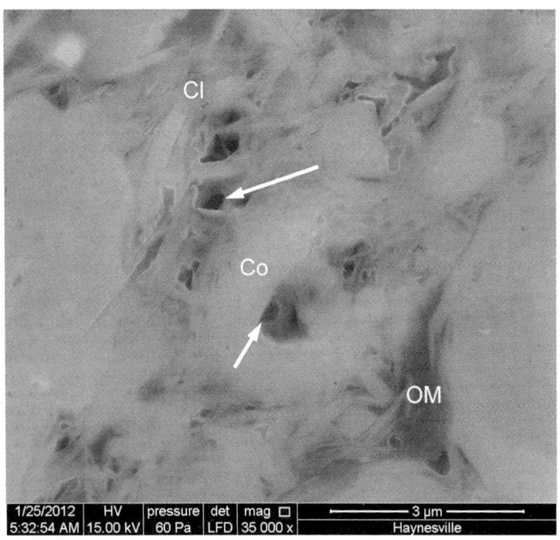

F.134.Haynesville
Haynesville Formation (Late Jurassic)
Northern Gulf of Mexico Basin, Louisiana
Ar-ion milled, SE image
Thermal maturity: 1.85%R_o
Submitted by B. Wawak and T. Gentzis, Core Laboratories
Nanopores (arrows) in organic matter filling interparticle pores associated with phyllosilicate minerals (Cl) and intraparticle pore space associated with a calcareous nannofossil (Co).

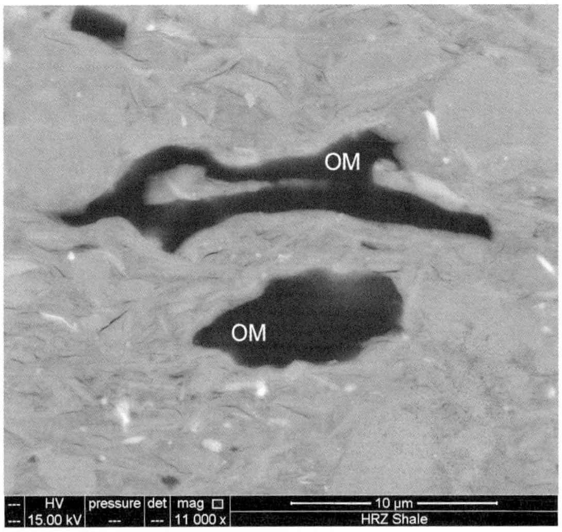

F.135.HRZ
HRZ Shale (Late Cretaceous)
North Slope, Alaska
Ar-ion milled, BSE/SE image
Thermal maturity: 0.76%R_o
Submitted by B. Wawak and T. Gentzis, Core Laboratories
Irregularly shaped, silt-size particles of organic matter (OM). These particles have a smooth, nonporous surface texture and occur in a surround matrix of phyllosilicate minerals and quartz particles.

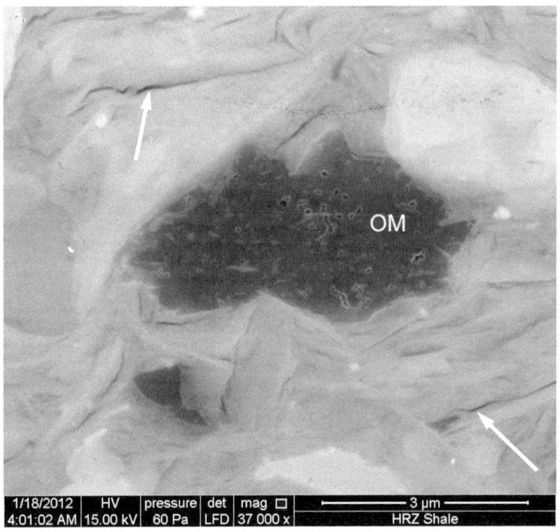

F.136.HRZ
HRZ Shale (Late Cretaceous)
North Slope, Alaska
Ar-ion milled, SE image
Thermal maturity: 0.76%R_o
Submitted by B. Wawak and T. Gentzis, Core Laboratories
Clay-size particle of organic matter (OM) that contains numerous nanometer-size pores. Note also scattered interparticle pores associated with clays (arrows).

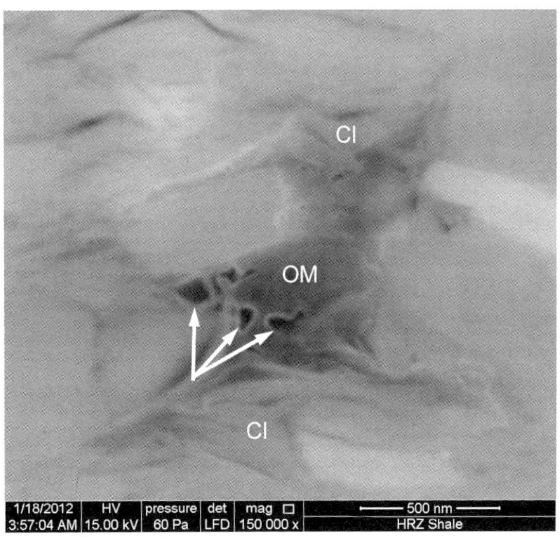

F.137.HRZ
HRZ Shale (Late Cretaceous)
North Slope, Alaska
Ar-ion milled, SE image
Thermal maturity: 0.76%R_o
Submitted by B. Wawak and T. Gentzis, Core Laboratories
Dispersed organic matter (OM) filling interparticle pores associated with clays (Cl) and other inorganic particles. The long dimensions of the larger pores (arrows) range from 70 to 111 nm.

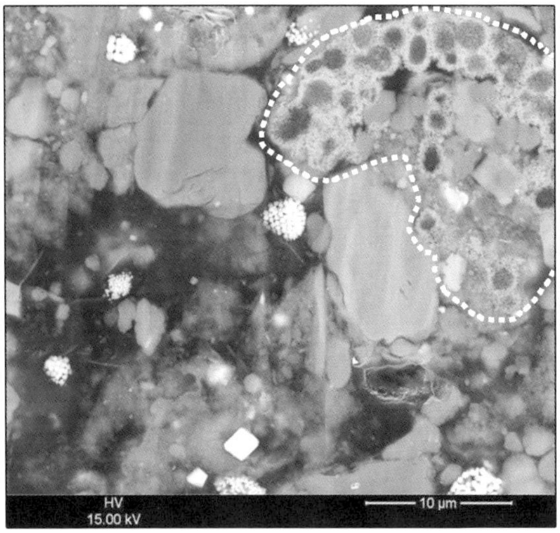

F.138.Manning Canyon
Manning Canyon Formation (Mississippian)
Uinta Basin, Utah
Depth: 7800–7900 ft (2347–2377 m)
Ar-ion milled, BSE image
Thermal maturity: N/A
Submitted by D. Jennings, Core Laboratories
A cluster of authigenic phosphate structures (outlined by white dashed line) include matrix material and occur adjacent to a micron-scale pore filled with organic matter (black).

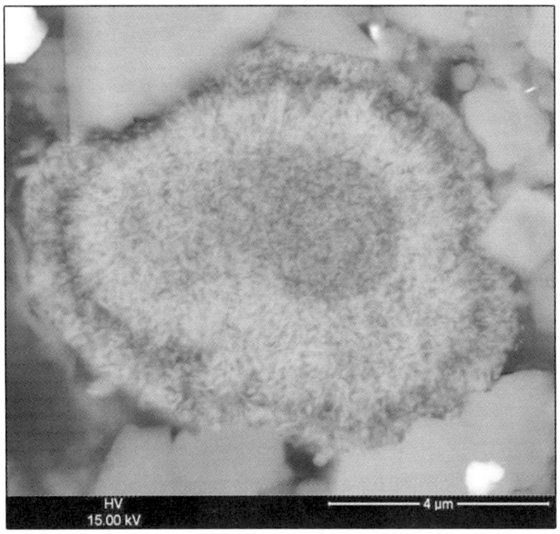

F.139.Manning Canyon
Manning Canyon Formation (Mississippian)
Uinta Basin, Utah
Depth: 7800–7900 ft (2347–2377 m)
Ar-ion milled, BSE image
Thermal maturity: N/A
Submitted by D. Jennings, Core Laboratories
High-magnification photomicrograph of accretionary, authigenic phosphate nodule. Layers of phosphate suggest that alternating geochemical conditions occurred during early burial.

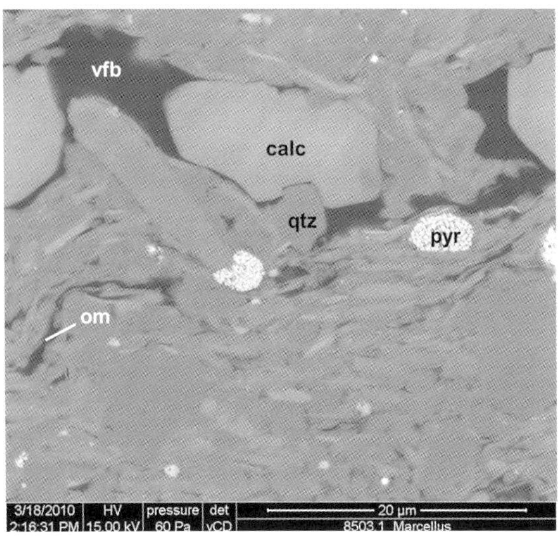

F.140.Marcellus
Marcellus Formation (Middle Devonian)
Appalachian Basin, Clinton County, Pennsylvania
Depth: 8503.1 ft (2592 m)
Ar-ion milled, BSE image
Thermal maturity: 2.6–2.7%R_o
Submitted by W. Camp, Anadarko Petroleum
Irregular-shape void partially filled with euhedral calcite cement (cal) and bitumen (vfb). Organic matter (om) interlaminated with clay mineral matrix (mica, illite, and chlorite) appears black in BSE images, accentuating the subparallel laminations. Framboidal pyrite (pry) and quartz grains (qtz) also present.

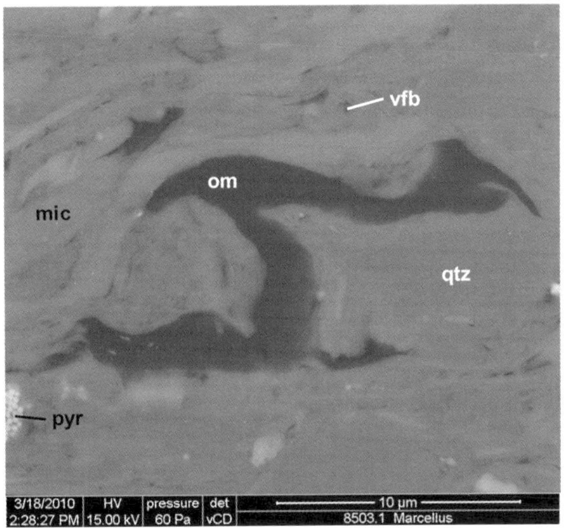

F.141.Marcellus
Marcellus Formation (Middle Devonian)
Appalachian Basin, Clinton County, Pennsylvania
Depth: 8503.1 ft (2592 m)
Ar-ion milled, BSE image
Thermal maturity: 2.6–2.7%R_o
Submitted by W. Camp, Anadarko Petroleum
Contorted organic matter (om). Irregular shape suggests the organic matter is a remnant of a structured kerogen maceral (possibly an algal *Tasmanites* remnant), instead of void-filling bitumen. Origin of organic matter associated with clays is less certain, but some void-filling bitumen (vfb) is apparent filling interstitial pore spaces. Mica (mic), quartz (qtz), and pyrite (pyr) are also present.

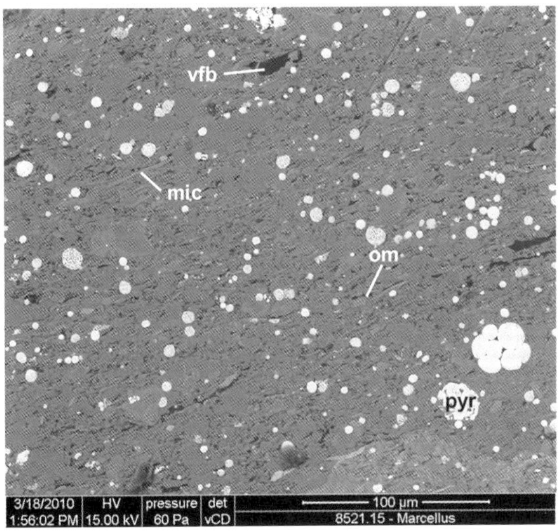

F.142.Marcellus
Marcellus Formation (Middle Devonian)
Appalachian Basin, Clinton County, Pennsylvania
Depth: 8521.15 ft (2597 m)
Ar-ion milled, BSE image
Thermal maturity: 2.6–2.7%R_o
Submitted by W. Camp, Anadarko Petroleum
Pyrite framboids (pyr) of various sizes are scattered throughout this sample. Most occur as single framboids, but some have formed clustered aggregates (e.g., lower right of image). Subparallel alignment of clay grains, including mica (mic) are accentuated by contrasting interlaminated organic matter (om). Void-filling bitumen (vfb) is also identifiable within larger pores.

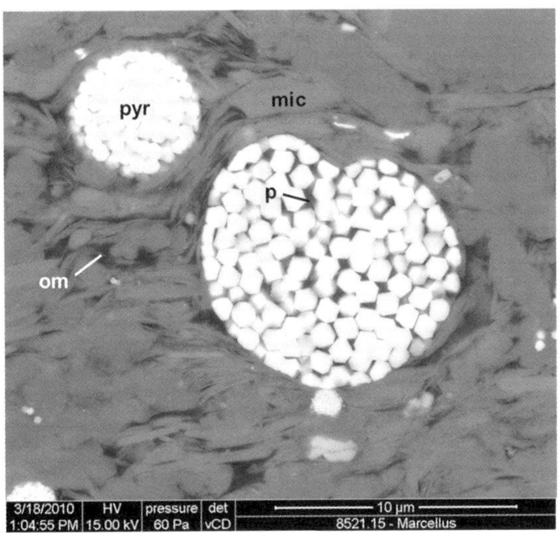

F.143.Marcellus
Marcellus Formation (Middle Devonian)
Appalachian Basin, Clinton County, Pennsylvania
Depth: 8521.15 ft (2597 m)
Ar-ion milled, BSE image
Thermal maturity: 2.6–2.7%R_o
Submitted by W. Camp, Anadarko Petroleum
Close-up view of pyrite framboids (pyr) illustrating clays, including mica (mic), deformed due to differential compaction. Note intercrystalline pore space (p) between individual hexagonal pyrite crystals and organic matter (om) distributed with clay grains.

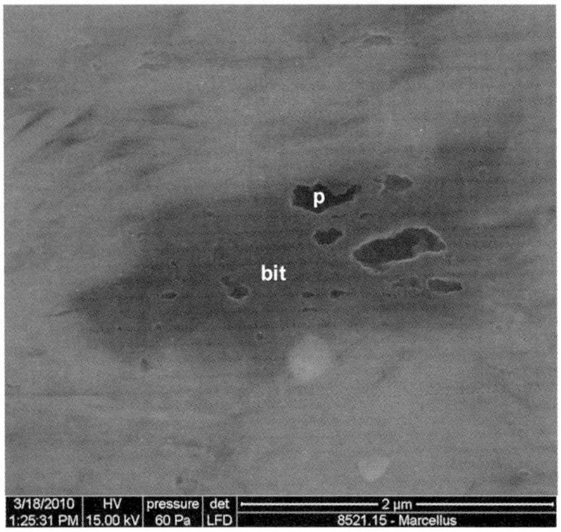

F.144.Marcellus
Marcellus Formation (Middle Devonian)
Appalachian Basin, Clinton County, Pennsylvania
Depth: 8521.15 ft (2597 m)
Ar-ion milled, SE image
Thermal maturity: 2.6–2.7%R_o
Submitted by W. Camp, Anadarko Petroleum
Sponge-like texture of porous pyrobitumen (bit) filling interparticle pores within clay matrix. Pores (p) within pyrobitumen presumably formed during thermal cracking of oil to gas during burial.

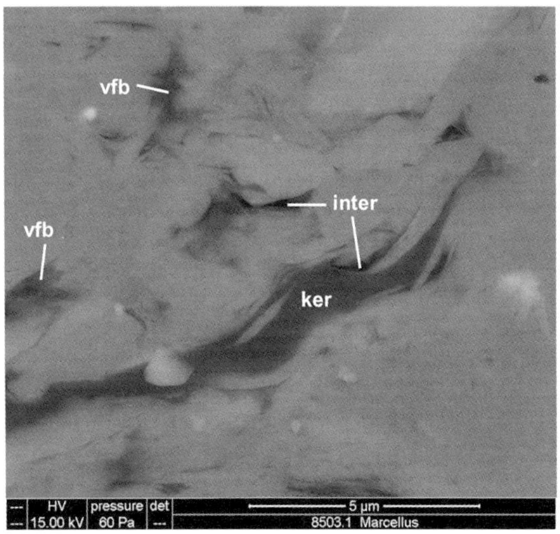

F.145.Marcellus
Marcellus Formation (Middle Devonian)
Appalachian Basin, Clinton County, Pennsylvania
Depth: 8503.1 ft (2592 m)
Ar-ion milled, superimposed SE and BSE image
Thermal maturity: 2.6–2.7%R_o
Submitted by W. Camp, Anadarko Petroleum
Superimposed SE and BSE image combines textural and compositional information within a single image. Angular intercrystalline pores (inter) locally preserved between clay and other mineral grains. Porous void-filling bitumen (vfb) exhibits sponge-like texture in the SE image. Note lack of organic matter pores within interpreted algal kerogen (ker) maceral (*Tasmanites*?).

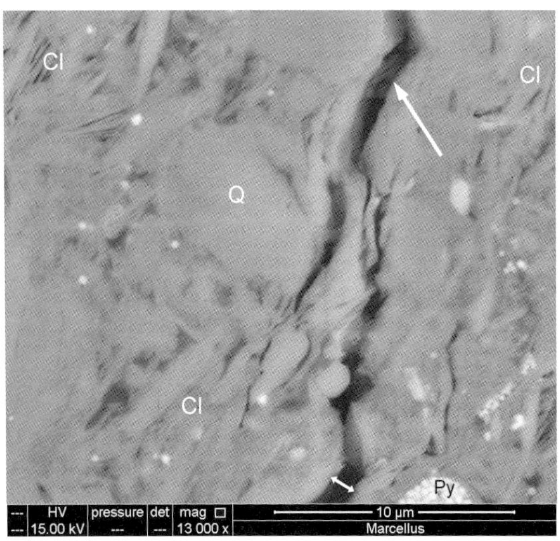

F.146.Marcellus
Marcellus Shale (Middle Devonian)
Appalachian Basin, Pennsylvania
Ar-ion milled, BSE/SE image
Thermal maturity: 2.47%R_o
Submitted by B. Wawak and T. Gentzis, Core Laboratories
Submicron microfracture that is partially filled with organic matter (arrow). Width of fracture (double-headed arrow, bottom of photo) is 987 nm. Most of inorganic particles in field of view are phyllosilicate minerals (Cl), quartz (Q), and pyrite (Py).

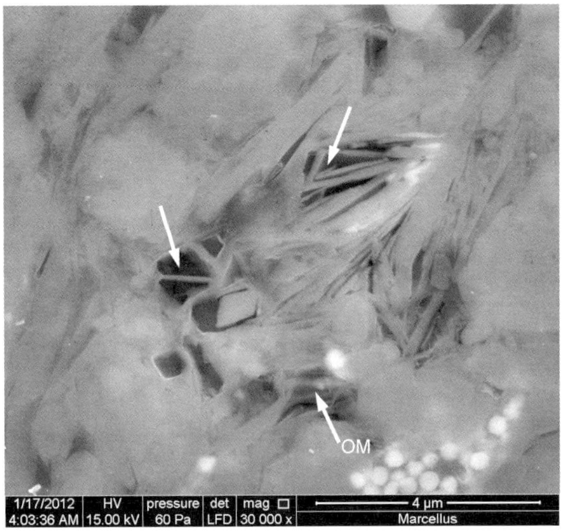

F.147.Marcellus
Marcellus Shale (Middle Devonian)
Appalachian Basin, Pennsylvania
Ar-ion milled, SE image
Thermal maturity: 2.47%R_o
Submitted by B. Wawak and T. Gentzis, Core Laboratories
Interparticle pores associated with phyllosilicate minerals (arrows). Some of these pores are lined with organic matter (OM). Note cluster of pyrite crystals at bottom of photo (white) and intercalated within expanded mica grains (white, center right part of photo).

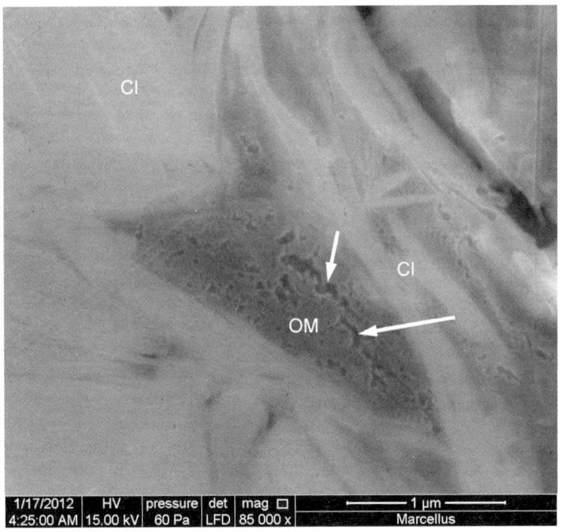

F.148.Marcellus
Marcellus Shale (Middle Devonian)
Appalachian Basin, Pennsylvania
Ar-ion milled, SE image
Thermal maturity: 2.47%R_o
Submitted by B. Wawak and T. Gentzis, Core Laboratories
Nanopores associated with organic matter (OM) that occurs in the interparticle pores associated with phyllosilicate minerals (Cl). Note some of the smaller pores have coalesced, forming relatively larger, linear pores (arrows).

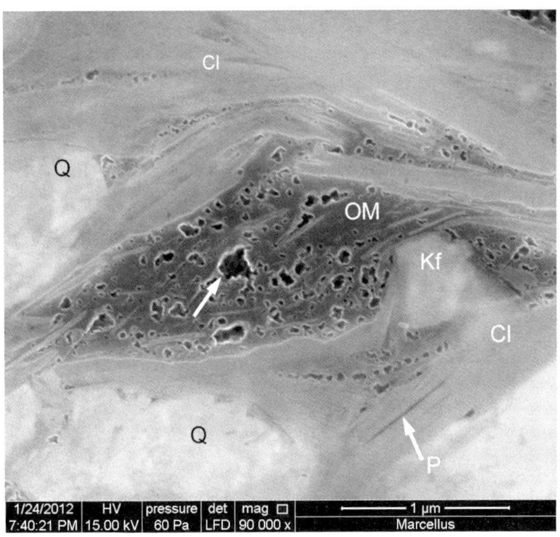

F.149.Marcellus
Marcellus Shale (Middle Devonian)
Appalachian Basin, West Virginia
Ar-ion milled, SE image
Thermal maturity: 1.25%R_o
Submitted by B. Wawak and T. Gentzis, Core Laboratories
Nanopores associated with dispersed organic matter (OM) that fills interparticle pores associated with a variety of inorganic particles, including quartz (Q), potassium feldspar (Kf), and phyllosilicate minerals (Cl). Note also interparticle pores (P) associated with phyllosilicate minerals. Long dimension of larger nanopore indicated by arrow is 192 nm.

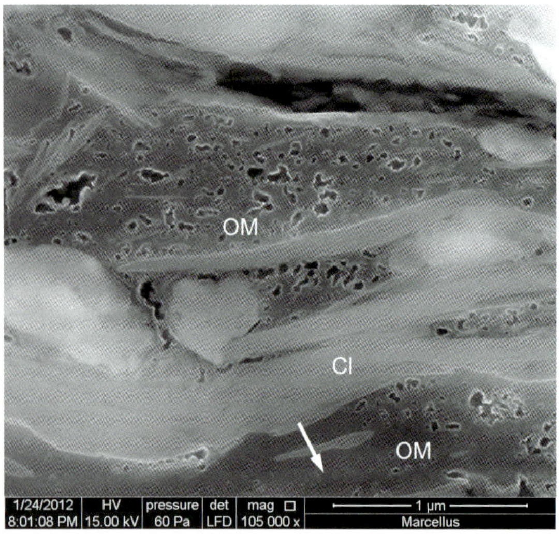

F.150.Marcellus
Marcellus Shale (Middle Devonian)
Appalachian Basin, West Virginia
Ar-ion milled, SE image
Thermal maturity: 1.25%R_o
Submitted by B. Wawak and T. Gentzis, Core Laboratories
Nanopores associated with dispersed organic matter (OM) that fills interparticle space associated with phyllosilicate minerals (Cl). Note organic matter at bottom of photomicrograph (arrow) is less porous than the organic in the center of the photo.

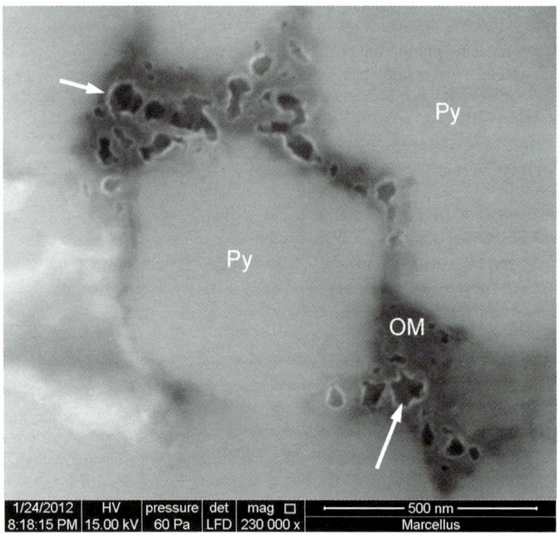

F.151.Marcellus
Marcellus Shale (Middle Devonian)
Appalachian Basin, West Virginia
Ar-ion milled, SE image
Thermal maturity: 1.25%R_o
Submitted by B. Wawak and T. Gentzis, Core Laboratories
Nanopores in organic matter (OM) that fill intraparticle pores within a pyrite (Py) framboid. The long dimensions of the pores indicated by the arrows are 79 (top) and 73 (bottom) nm.

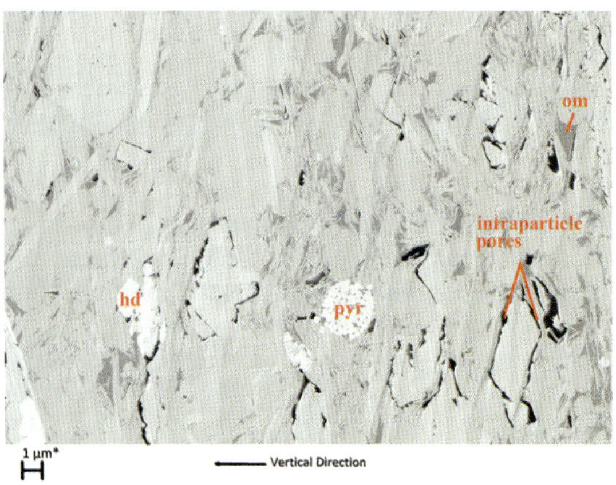

F.152.Marcellus
Marcellus Formation (Middle Devonian)
Appalachian Basin, Clearfield County, Pennsylvania
Depth: 7000–7500 ft (2134–2286 m)
Ar-ion milled, SE2 image
Thermal maturity: N/A
Submitted by J. DeVito, Ingrain
Stratigraphic top to left. Intraparticle pores (black) occur along crystal boundaries. Solid organic matter (om) encompasses the majority of the image. Pyrite (pyr) and highly dense minerals (hd) lie within the matrix. Total porosity 2.3%, organic matter 5.8%, high-density material 1.0%. Values given were computed by digital rock physics.

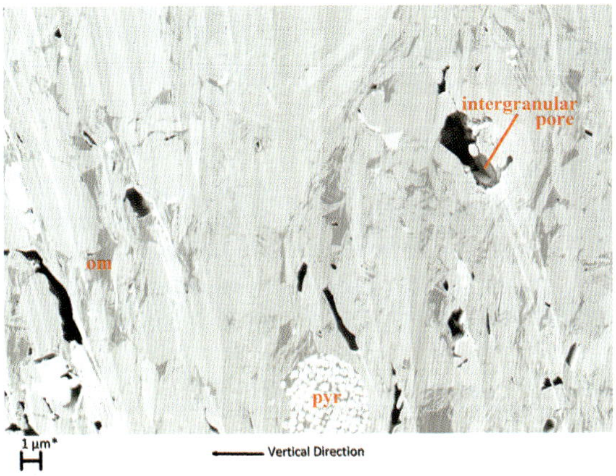

F.153.Marcellus
Marcellus Formation (Middle Devonian)
Appalachian Basin, Clearfield County, Pennsylvania
Depth: 7000–7500 ft (2134–2286 m)
Ar-ion milled, SE2 image
Thermal maturity: N/A
Submitted by J. DeVito, Ingrain
Stratigraphic top to left. Intergranular porosity (black), solid organic matter (om), and pyrite (pyr) occur within matrix. Total porosity 2.6%, organic matter 5.8%, highly dense material 1.8%. Values given were computed by digital rock physics.

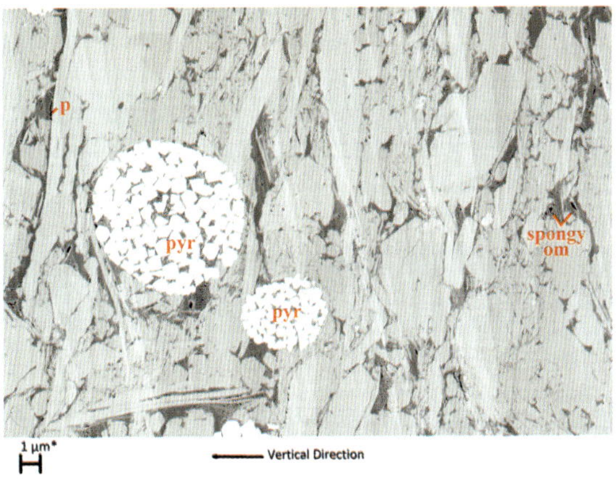

F.154.Marcellus
Marcellus Formation (Middle Devonian)
Appalachian Basin
Depth: 9500–10000 ft (2896–3048 m)
Ar-ion milled, SE2 image
Thermal maturity: N/A
Submitted by J. DeVito, Ingrain
Stratigraphic top to left. A mixture of solid and spongy textured organic matter (om) in a predominantly clay matrix. Pyrite framboids are in white (pyr), and porosity is black (p). Organic matter and nanopores fill voids within pyrite. Total porosity 0.7%, organic matter 12.3%, pyrite 8.7%. Values given were computed by digital rock physics.

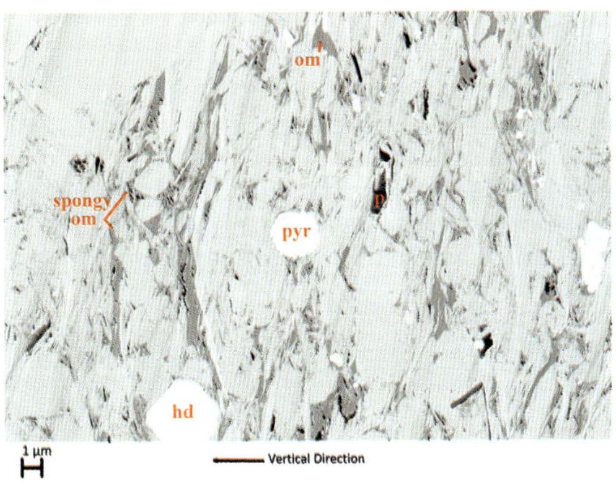

F.155.Marcellus
Marcellus Formation (Middle Devonian)
Appalachian Basin, Washington County, Pennsylvania
Depth: 6000–6500 ft (1829–1981 m)
Ar-ion milled, SE2 image
Thermal maturity: N/A
Submitted by J. DeVito, Ingrain
Stratigraphic top to left. Both spongy and solid organic matter textures (om) are present within the matrix. A pyrite framboid (pyr) 2 µm in size occurs within the center of the image. Interparticle porosity (p) is present within the matrix. Highly dense minerals (hd) are in white. Total porosity 3.6%, organic matter 7.0%, high-density material 2.7%. Values given were computed with digital rock physics.

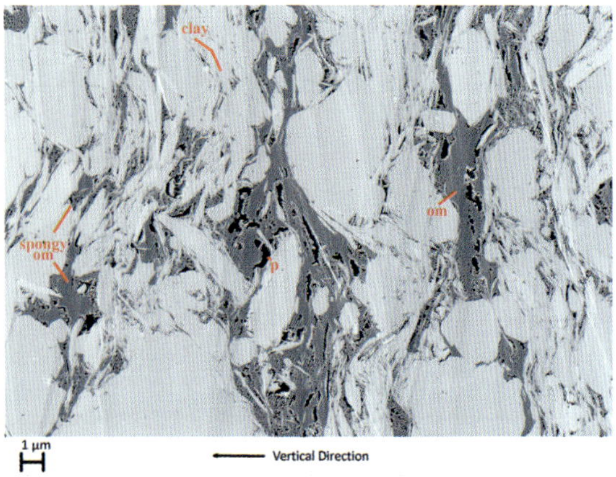

F.156.Marcellus
Marcellus Formation (Middle Devonian)
Depth: 6000–6500 ft (1829–1981 m)
Ar-ion milled, SE2 image
Thermal maturity: N/A
Submitted by J. DeVito, Ingrain
Stratigraphic top to left. Organic matter (om) and clay are organized in layers parallel to the bedding. Pores (p) are associated with spongy and pendular organic matter textures. Total porosity 4.7%, organic matter 17.6%. Values presented computed by digital image analysis.

F.157.Monterey
Monterey Formation—diatomite (Miocene)
San Joaquin Basin, California
SE image
Thermal maturity N/A
Submitted by M. Nelis and J. Nolting, Core Laboratories
Diatomite with a chaotic fabric composed of very fine silt-size to submicron fragments of siliceous microfossils. Interparticle pores (P) range from 1 to 3 μm, and intraparticle pores (arrow) are relatively smaller in size.

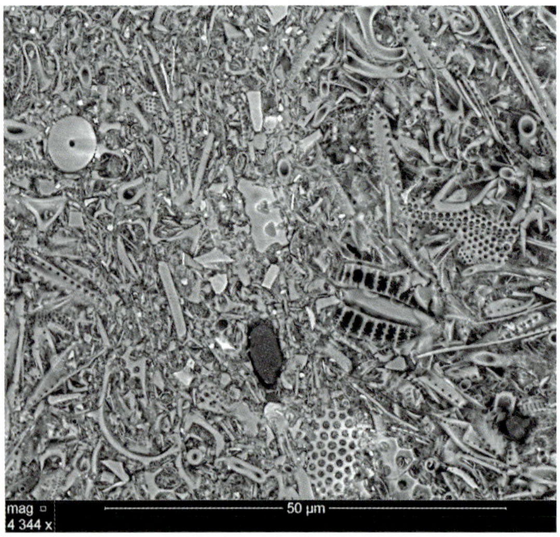

F.158.Monterey
Monterey Formation—diatomite (Miocene)
San Joaquin Basin, California
Ar-ion milled, SE image
Thermal maturity N/A
Submitted by M. Nelis and J. Nolting, Core Laboratories
This sample shows the sharp contact between highly microporous diatomite (right) composed of randomly oriented siliceous microfossil fragments and less-porous diatomite (left) composed of finely fragmented microfossils generally aligned parallel to the contact. Diatomite reservoir characteristics appear to vary at the microscale.

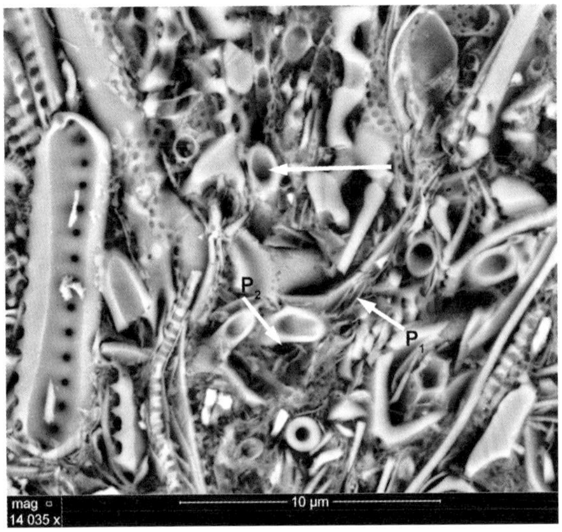

F.159.Monterey
Monterey Formation—diatomite (Miocene)
San Joaquin Basin, California
Ar-ion milled, SE image
Thermal maturity N/A
Submitted by M. Nelis and J. Nolting, Core Laboratories
Diatomite with a slightly aligned fabric composed primarily of medium silt-size to submicron fragments of siliceous microfossils. Trace amounts of phyllosilicate minerals (P_1) are observed within the interparticle pore spaces (P_2). Intraparticle pores (arrow) are abundant and readily identifiable due to Ar-ion milling.

F.160.Mowry
Mowry Formation (Late Cretaceous)
Powder River Basin, Wyoming
Depth: 10100–11200 ft (30788–3109 m)
Ar-ion milled, BSE image
Thermal maturity: N/A
Submitted by D. Jennings, Core Laboratories
Devitrified, volcanic glass shards in this sample are composed of mixed-layer illite-smectite, which is estimated to contain 25–35% smectite layers. Desiccation cracks are abundant in the smectite-rich volcanic shards (Sm). Nanometer-scale, interparticle, and intercrystalline pores are isolated to poorly connected. Authigenic quartz (Q) and sodium plagioclase (Pl) are common matrix minerals likely sourced from the alteration of volcaniclastic material. Pyrite pseudomorphs after plagioclase (Py) are also present.

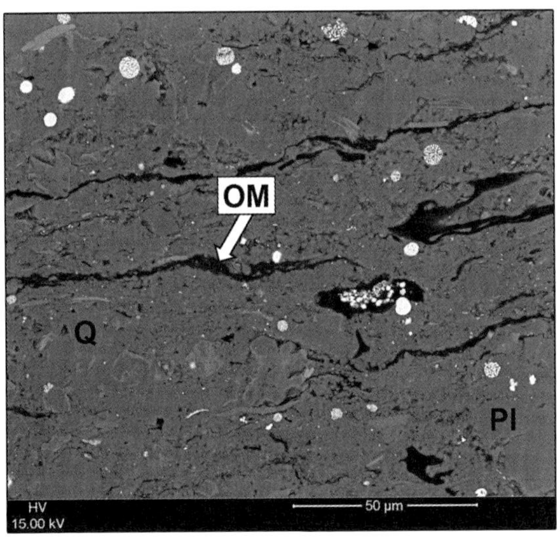

F.161.Mowry
Mowry Formation (Late Cretaceous)
Powder River Basin, Wyoming
Depth: 10100–10200 ft (3078–3109 m)
Ar-ion milled, BSE image
Thermal maturity: N/A
Submitted by D. Jennings, Core Laboratories
Scattered micrometer- and nanometer-scale interparticle pores are visible in this laminated silty shale. Strands of organic matter (OM) predominantly lie parallel to bedding planes. Rounded to subrounded quartz grains (Q), pyrite framboids (bright white), and plagioclase (Pl) are the most abundant constituents. Quartz is also present as an authigenic matrix constituent.

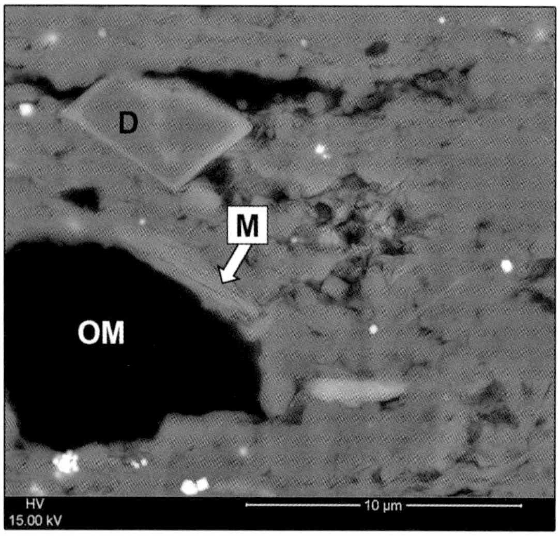

F.162.Mowry
Mowry Formation (Late Cretaceous)
Powder River Basin, Wyoming
Depth: 10200–10300 ft (3019–3139 m)
Ar-ion milled, BSE image
Thermal maturity: N/A
Submitted by D. Jennings, Core Laboratories
Tabular flakes of mica (M) and other matrix constituents are aligned around micron-scale pores filled with organic matter (OM). Dolomite rhombs (D) have ferroan dolomite rims (light gray).

F.163.Mulky
Mulky Shale (Pennsylvanian)
Cherokee Basin, Kansas
Ar-ion milled, BSE/SE image
Thermal maturity: 0.71%R_o
Submitted by B. Wawak and T. Gentzis, Core Laboratories
Relatively low magnification view showing the organized fabric of the sample. The fabric is characterized by the subparallel orientation of the elongated particles of organic matter (black particles) and platy/elongated inorganic grains.

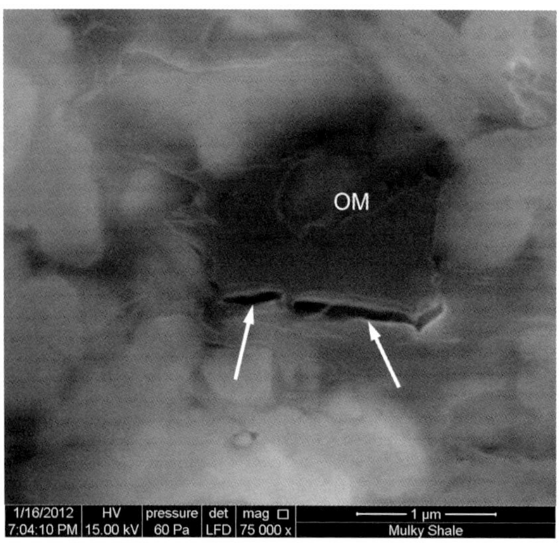

F.164.Mulky
Mulky Shale (Pennsylvanian)
Cherokee Basin, Kansas
Ar-ion milled, SE image
Thermal maturity: 0.71%R_o
Submitted by B. Wawak and T. Gentzis, Core Laboratories
Elongated pore (arrows) associated with the edge of small particle of organic matter (OM).

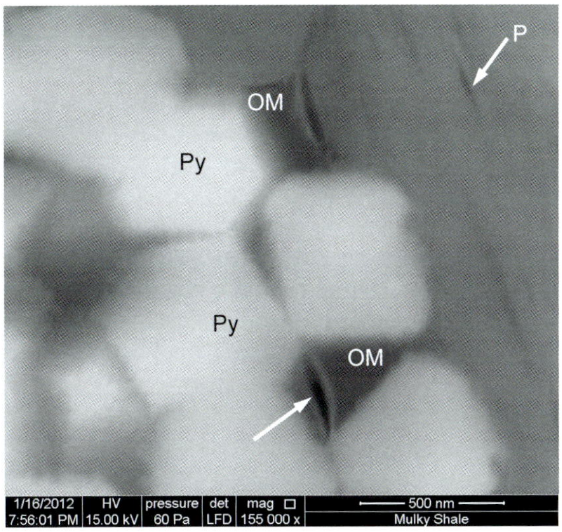

F.165.Mulky
Mulky Shale (Pennsylvanian)
Cherokee Basin, Kansas
Ar-ion milled, SE image
Thermal maturity: 0.71%R_o
Submitted by B. Wawak and T. Gentzis, Core Laboratories
Elongated pores (arrow) associated with organic matter (OM) that occurs at the outer edge of a pyrite framboid (Py) and interparticle pores associated with phyllosilicate minerals (P). Note the pores associated with the organic matter occur at or near the edges of the organic matter.

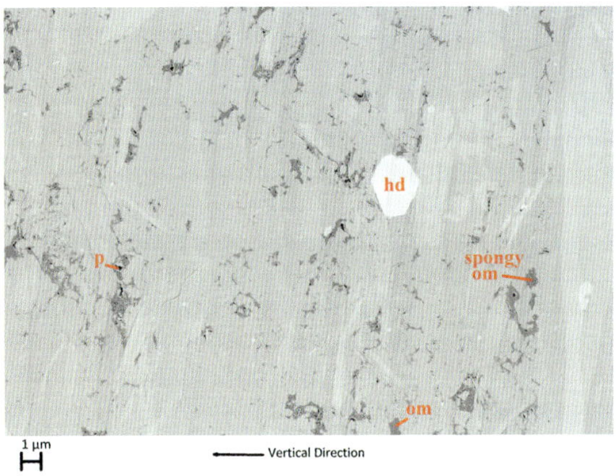

F.166.N.European
Formation: N/A
Northern Europe
Depth: 11500–12000 ft (3505–3658 m)
Ar-ion milled, SE2 image
Thermal maturity: N/A
Submitted by J. DeVito, Ingrain
Stratigraphic top to left. Organic matter (om) is solid and spongy. All porosity (p) is associated with organic matter. A hexagonal, highly dense mineral (hd) occurs within the matrix. Total porosity 0.5%, organic matter 4.2%, high-density material 0.8%. Values presented computed by digital image analysis.

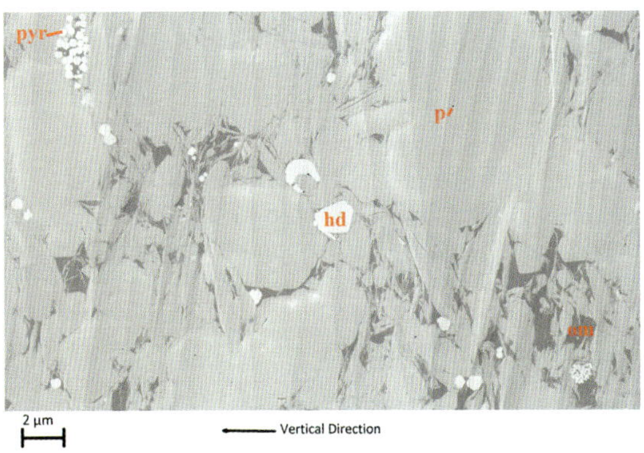

F.167.N.European
Formation: N/A
Northern Europe
Depth: 11500–12000 ft (3505–3658 m)
Ar-ion milled, SE2 image
Thermal maturity: N/A
Submitted by J. DeVito, Ingrain
Stratigraphic top to left. Pyrite framboids (pyr) and highly dense minerals (hd) are associated with solid organic matter (om). Organic matter 7.7%, high-density material 1.6%. Values given were computed by digital rock physics.

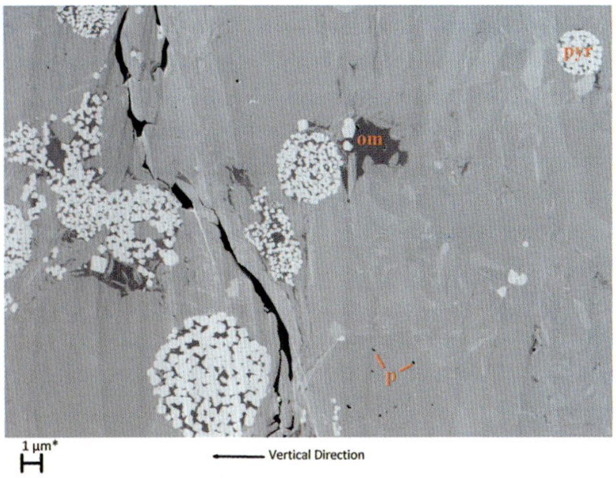

F.168.N.European
Formation: N/A
Northern Europe
Depth: 11000–11500 ft (3352–3505 m)
Ar-ion milled, SE2 image
Thermal maturity: N/A
Submitted by J. DeVito, Ingrain
Stratigraphic top to left. Intergranular porosity (p) and pyrite framboids (pyr) occur within the matrix. Organic matter (om) is mostly solid in texture. An open fracture (black) cuts across the image parallel to bedding. Total porosity 1.0%, organic matter 3.15%, high-density material 9.54%. Values given were computed by digital rock physics.

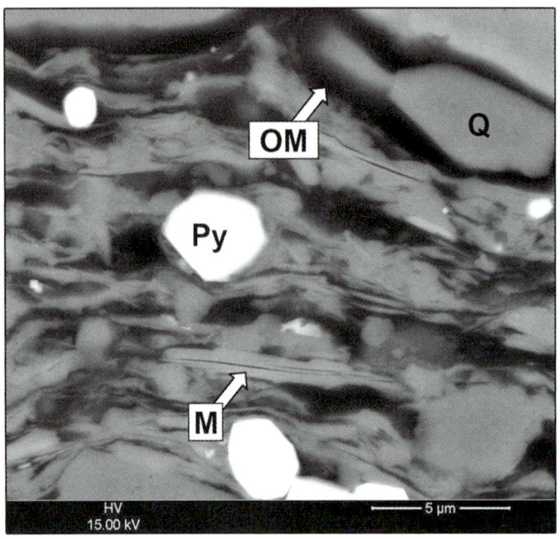

F.169.NewAlbany
New Albany Formation-Grassy Creek Member (Late Devonian)
Illinois Basin, Kentucky
Depth: 3600–3700 ft (1097–1128 m)
Ar-ion milled, BSE image
Thermal maturity: N/A
Submitted by D. Jennings, Core Laboratories
The overall laminated fabric of this silty shale is highlighted by strands of organic material (OM) and tabular flakes of mica (M) that lie parallel to bedding. Pyrite (Py), quartz (Q), and carbonate grains (bottom, right) are the most abundant constituents in this silty shale. Interparticle microporosity is abundant and includes common interparticle pressure shadows at the edge of resistant grains. Micropores are usually filled with organic material (black). Euhedral pyrite grains (Py) are abundant.

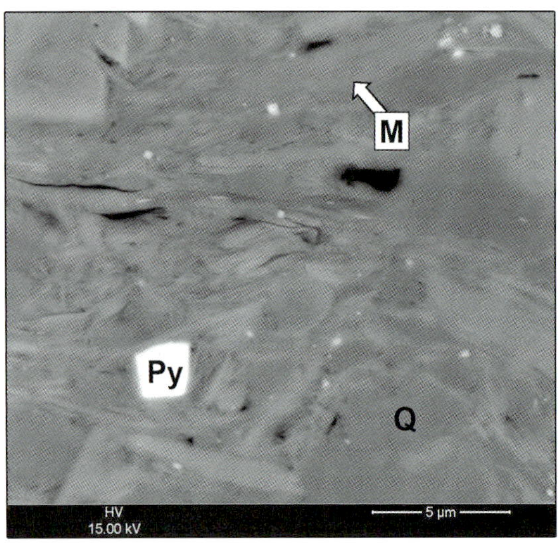

F.170.New Albany
New Albany Formation-Selmier Member (Late Devonian)
Illinois Basin, Kentucky
Depth: 3700–3800 ft (1128–1158 m)
Ar-ion milled, BSE image
Thermal maturity: N/A
Submitted by D. Jennings, Core Laboratories
Scattered micrometer- and nanometer-scale interparticle pores are visible in this microlaminated, silty shale. Rounded to subrounded quartz grains (Q) and tabular mica (M) are the most abundant constituents. Euhedral pyrite (Py) is also present..

F.171.Niobrara
Niobrara Formation (Late Cretaceous)
Denver Basin, Colorado
Ar-ion milled, BSE/SE image
Thermal maturity: 0.82%R_o
Submitted by B. Wawak and T. Gentzis, Core Laboratories
This photomicrograph shows the pelletal texture of this sample. The composition of the "pellet" is predominantly broken pieces of coccoliths; the surrounding part of the sample contains elevated amounts of clays (arrows). The interparticle area within the pellet is filled with dispersed organic matter (black). Pyrite occurs as individual crystals (white, top) and spherical framboids (white, right)

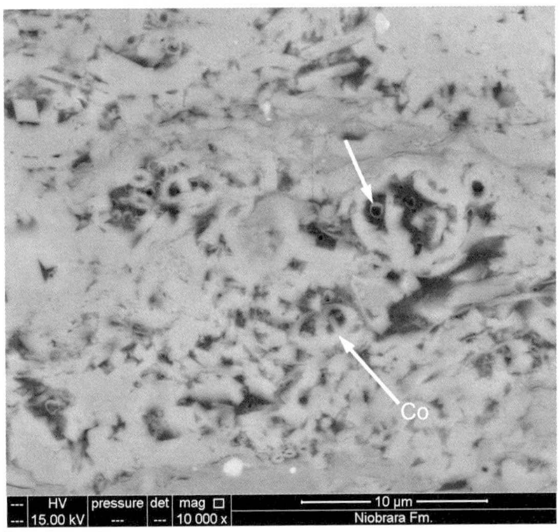

F.172.Niobrara
Niobrara Formation (Late Cretaceous)
Denver Basin, Colorado
Ar-ion milled, BSE/SE image
Thermal maturity: 0.82%R_o
Submitted by B. Wawak and T. Gentzis, Core Laboratories
Enlargement of part of calcareous pellet seen in Figure 172 that shows the texture of pellet. Most of particles in this field of view are pieces of coccoliths (Co). The interparticle and intraparticle areas associated with the coccolith pieces are partially filled with organic matter, which in turn contains open pores (arrow).

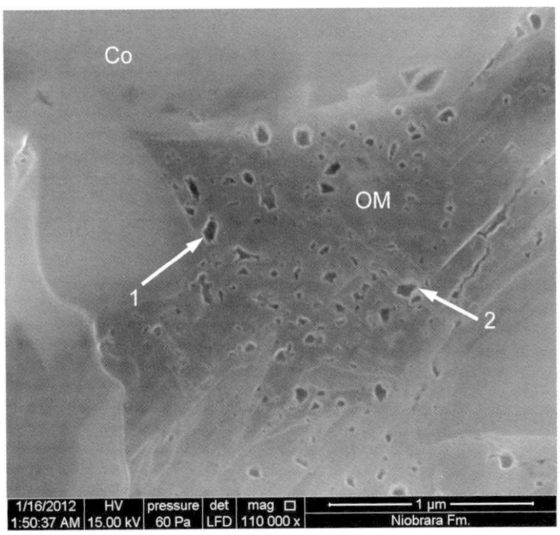

F.173.Niobrara
Niobrara Formation (Late Cretaceous)
Denver Basin, Colorado
Ar-ion milled, SE image
Thermal maturity: 0.82%R_o
Submitted by B. Wawak and T. Gentzis, Core Laboratories
This is an enlargement of some of the organic matter (OM) that fills the interparticle areas within the calcareous pellets seen in Figures 172 and 173. Particles at the top of the photomicrograph are pieces of coccoliths (Co). The long dimensions of the pores indicated by arrows are 82 nm (2) and 87 nm (1).

F.174.Niobrara
Niobrara Formation (Late Cretaceous)
Denver Basin, Colorado
Ar-ion milled, SE image
Thermal maturity: 0.82%R_o
Submitted by B. Wawak and T. Gentzis, Core Laboratories
Organic matter fills most of the intraparticle pores associated with this pyrite framboid . The organic matter contains abundant nanopores. Area indicated by brackets is enlarged in Figure 175.

F.175.Niobrara
Niobrara Formation (Late Cretaceous)
Denver Basin, Colorado
Ar-ion milled, SE image
Thermal maturity: 0.82%R_o
Submitted by B. Wawak and T. Gentzis, Core Laboratories
This image is an enlargement of the area indicated by brackets on Figure 174 showing the nanopores within organic matter that fills intraparticle space within a pyrite framboid. Measured width of indicated pyrite crystal (arrow) is 1.3 µm. Area indicated by brackets is enlarged in Figure 176.

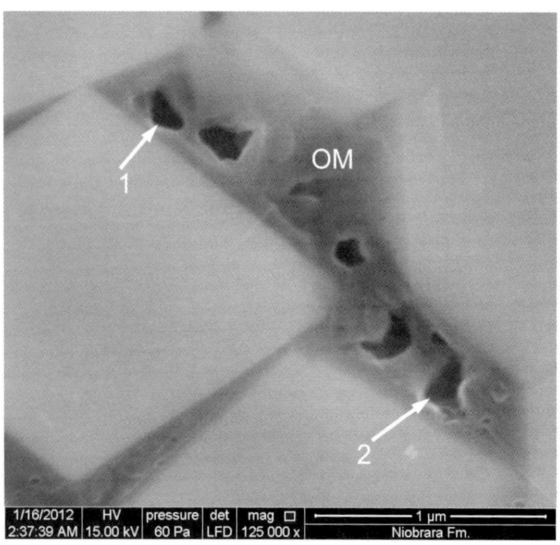

F.176.Niobrara
Niobrara Formation (Late Cretaceous)
Denver Basin, Colorado
Ar-ion milled, SE image
Thermal maturity: 0.82%R_o
Submitted by B. Wawak and T. Gentzis, Core Laboratories
This image is an enlargement of the area indicated by the brackets on Figure 175 showing the nanopores that are associated with organic matter (OM) within a pyrite framboid. Measured long dimensions of indicated pores are 172 nm (1) and 190 nm (2).

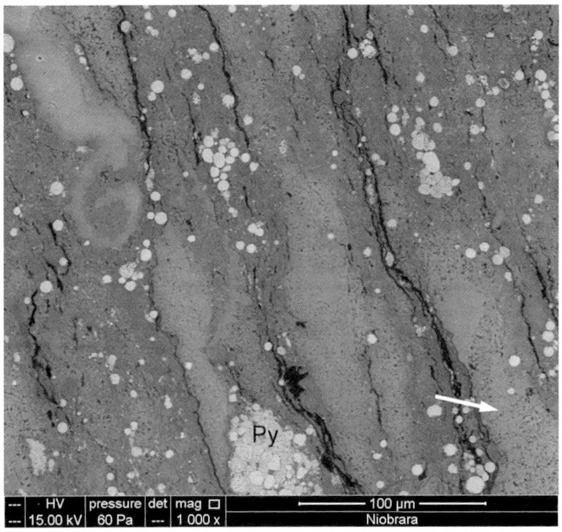

F.177.Niobrara
Niobrara Formation (Late Cretaceous)
Denver Basin, Colorado
Ar-ion milled, BSE/SE image
Thermal maturity: 0.69%R_o
Submitted by B. Wawak and T. Gentzis, Core Laboratories
Organized fabric of sample composed of linear accumulations of organic matter (black) and elongated patches of "clean" (virtually no clay) bioclasts and calcite cement (arrow) intercalated with areas composed of elevated amounts of clays and bioclasts. Note also abundance of clusters of pyrite framboids (Py). The patches of clean bioclasts are composed predominantly of broken pieces of coccoliths.

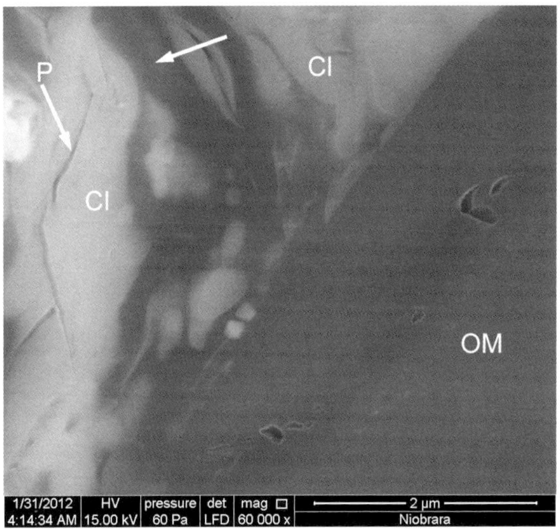

F.178.Niobrara
Niobrara Formation (Late Cretaceous)
Denver Basin, Colorado
Ar-ion milled, SE image
Thermal maturity: 0.69%R_o
Submitted by B. Wawak and T. Gentzis, Core Laboratories
Nanopores along the edge of relatively large (silt-size) discrete particle of organic matter (OM). Note also nonporous organic matter (arrow) filling interparticle pores associated with clays (CL) and scattered interparticle pores (P) that are not filled with organic matter.

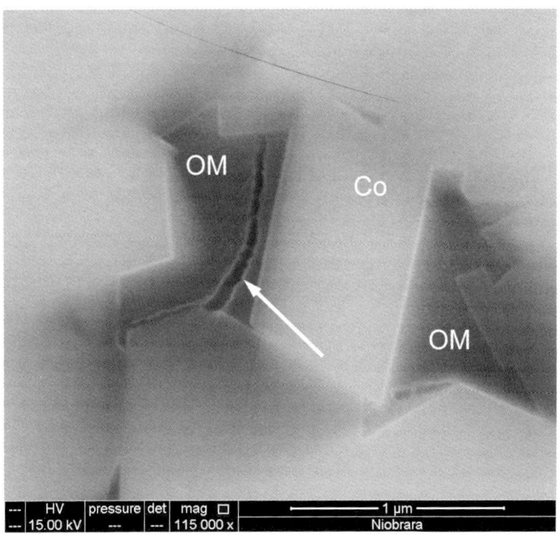

F.179.Niobrara
Niobrara Formation (Late Cretaceous)
Denver Basin, Colorado
Ar-ion milled, BSE/SE image
Thermal maturity: 0.69%R_o
Submitted by B. Wawak and T. Gentzis, Core Laboratories
Elongated pore (arrow) in organic matter (OM) that fills interparticle pores associated with disarticulated pieces of coccoliths (Co).

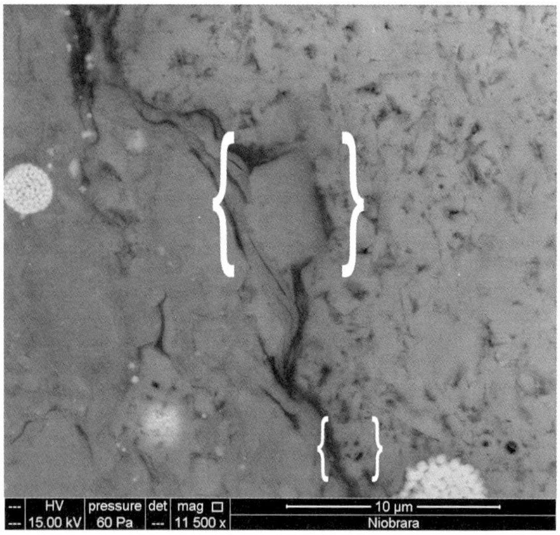

F.180.Niobrara
Niobrara Formation (Late Cretaceous)
Denver Basin, Colorado
Ar-ion milled, BSE/SE image
Thermal maturity: 0.69%R_o
Submitted by B. Wawak and T. Gentzis, Core Laboratories
Organic matter (black) concentrated in dissolution seam between argillaceous lithology (left side) and calcareous lithology (right side). Argillaceous lithology includes quartz, feldspar, clays, and bioclasts. Calcareous lithology is predominantly broken pieces of coccoliths. Organic matter partially fills interparticle and intraparticle pores in the calcareous lithology. Pyrite framboids are found in both lithologies (white). Large brackets indicate area enlarged in Figure 181, and small brackets indicate area enlarged in Figure 182.

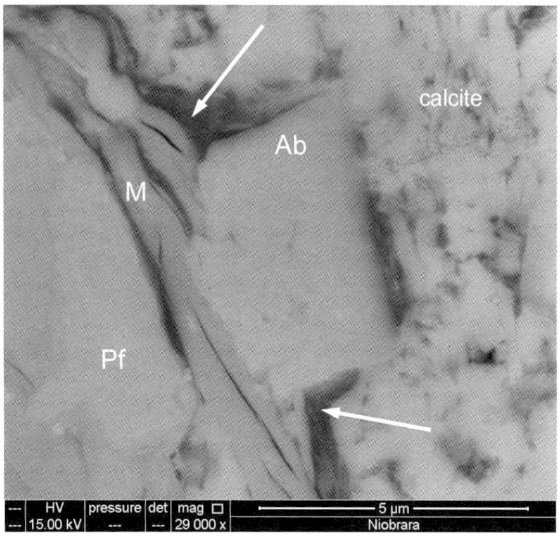

F.181.Niobrara
Niobrara Formation (Late Cretaceous)
Denver Basin, Colorado
Ar-ion milled, BSE/SE image
Thermal maturity: 0.69%R_o
Submitted by B. Wawak, Core Laboratories
Higher-magnification view of area indicated by large brackets in Figure 180 showing various inorganic particles identified in this sample, including albite (Ab), phyllosilicate minerals (M), and plagioclase feldspar (Pf). Note concentration of organic matter (arrow) at boundary between detrital grains and calcite grains.

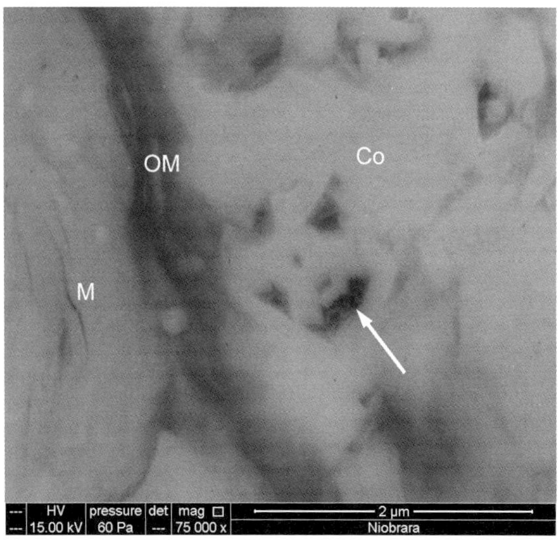

F.182.Niobrara
Niobrara Formation (Late Cretaceous)
Denver Basin, Colorado
Ar-ion milled, BSE/SE image
Thermal maturity: 0.69%R_o
Submitted by B. Wawak and T. Gentzis, Core Laboratories
High-magnification view of area indicated by small brackets in Figure 180 showing organic matter associated with dissolution seam. Particles in field of view include phyllosilicate minerals (M) and coccoliths (Co). The organic matter is associated with the boundary (OM) between the two lithologies and also partially fills intraparticle pores associated with coccoliths (arrow).

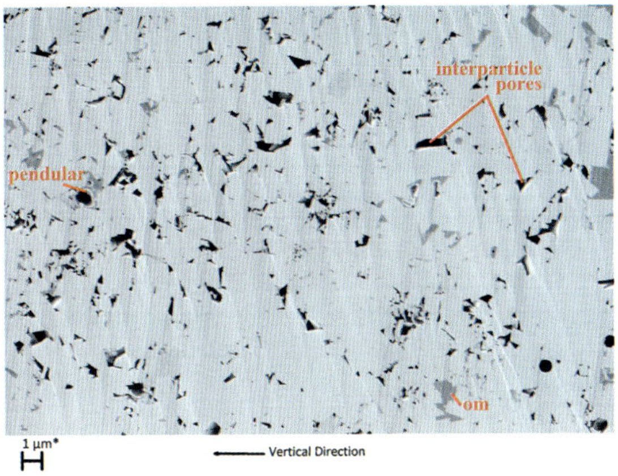

F.183.Niobrara
Niobrara Formation (Late Cretaceous)
Denver-Julesburg Basin, Weld County, CO
Depth: 7000–7500 ft (2134–2286 m)
Ar-ion milled, SE2 image
Thermal maturity: N/A
Submitted by J. DeVito, Ingrain
Stratigraphic top to left. Interparticle porosity (black) with solid and pendular organic matter (om). Grain (light gray) is predominantly calcareous fossil fragments. Total porosity 4.4%, organic matter 3.2%. Values given were computed by digital rock physics.

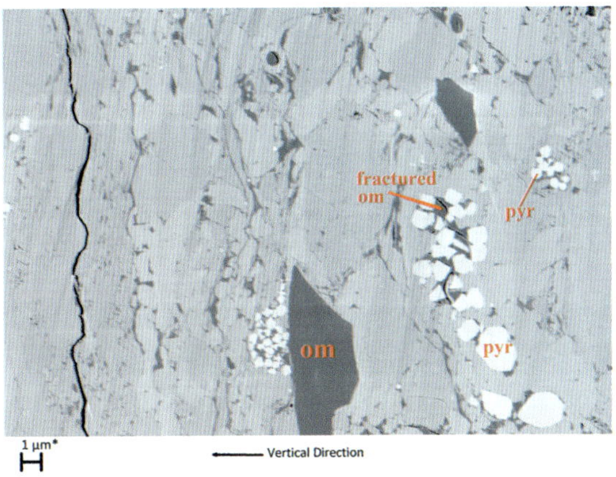

F.184.Niobrara
Niobrara Formation (Late Cretaceous)
Denver-Julesburg Basin, Weld County, CO
Depth: 7000–7500 ft (2134–2286 m)
Ar-ion milled, SE2 image
Thermal maturity: N/A
Submitted by J. DeVito, Ingrain
Stratigraphic top to left (Upper Cretaceous). Solid organic matter (om) and pyrite clusters (pyr) are organized parallel to bedding. An induced fracture cuts across the image parallel to bedding. Fractured organic matter pores (black) are between pyrite framboids and within the matrix. Total porosity 0.7%, organic matter 9.4%, highly dense material 3.5%. Values given were computed by digital rock physics.

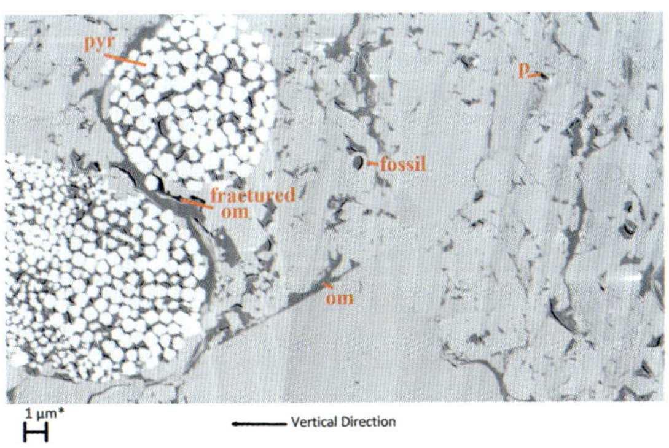

F.185.Niobrara
Niobrara Formation (Late Cretaceous)
Denver-Julesburg Basin, Weld County, Colorado
Depth: 7000–7500 ft (2134–2286 m)
Ar-ion milled, SE2 image
Thermal maturity: N/A
Submitted by J. DeVito, Ingrain
Stratigraphic top to left. Organic matter fractures are associated with two pyrite framboids (pyr). Within the matrix, fractured, solid organic matter (om) and intergranular porosity (p) are present. Organic matter is also contained within fossils. Total porosity 1.5%, organic matter 10.5%, pyrite 10.6%. Values given were computed by digital rock physics.

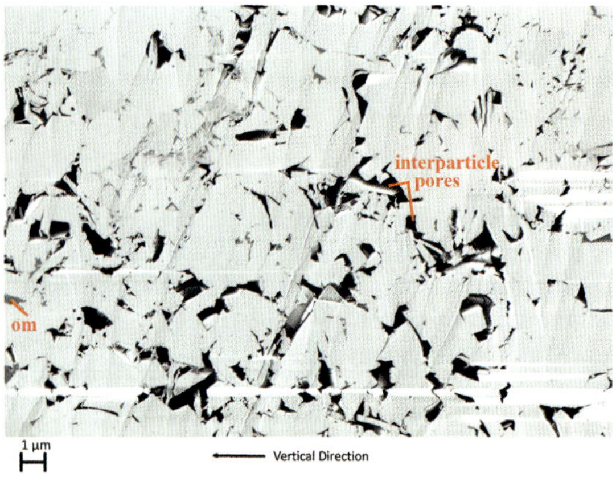

F.186.Niobrara
Niobrara Formation (Late Cretaceous)
Denver-Julesburg Basin, Weld County, Colorado
Depth: 6000–6500 ft (1829–1981 m)
Ar-ion milled, SE2 image
Thermal maturity: N/A
Submitted by J. DeVito, Ingrain
Stratigraphic top to left. Interparticle porosity (black) with solid organic matter (om). Grains (light gray) are predominantly calcareous fossil fragments. Total porosity 11.6%, organic matter 0.9%. Values given were computed with digital rock physics.
Submitted by J. DeVito, Ingrain

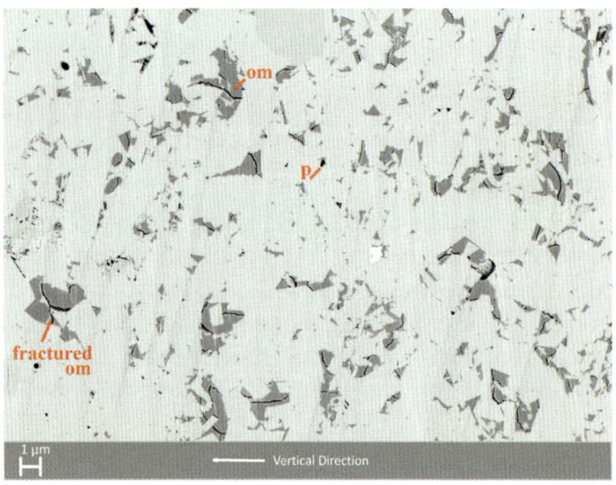

F.187.Niobrara
Niobrara Formation (Late Cretaceous)
Denver-Julesburg Basin, Weld County, Colorado
Depth: 6000–6500 ft (1829–1981 m)
Ar-ion milled, SE2 image
Thermal maturity: N/A
Submitted by J. DeVito, Ingrain
Stratigraphic top to left. Fractured organic matter (om) within a predominantly calcite matrix. Porosity is black (p). Total porosity 2.1%, organic matter 8.3%. Values given were computed with digital rock physics.

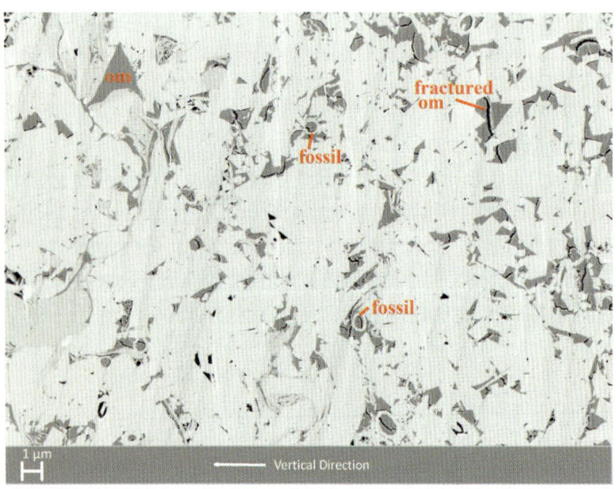

F.188.Niobrara
Niobrara Formation (Late Cretaceous)
Denver-Julesburg Basin, Weld County, Colorado
Depth: 6000–6500 ft (1829–1981 m)
Ar-ion milled, SE2 image
Thermal maturity: N/A
Submitted by J. DeVito, Ingrain
Stratigraphic top to left. Fractured and solid organic matter (om) within a predominantly calcite matrix. Interparticle pores are present (black). Total porosity 3.6%, organic matter 9.0%. Values given were computed by digital rock physics.

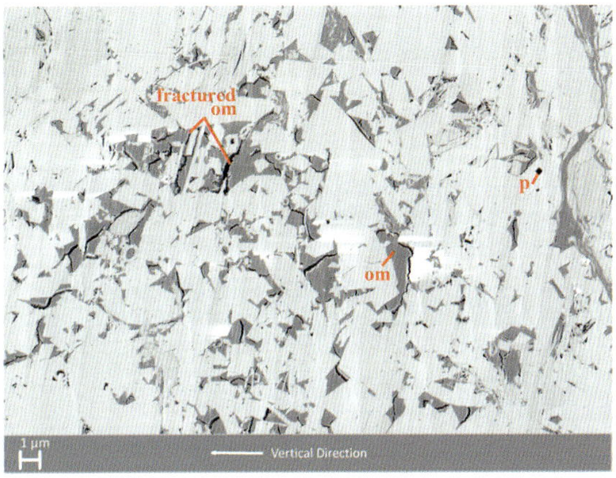

F.189.Niobrara
Niobrara Formation (Late Cretaceous)
Denver-Julesburg Basin, Weld County, Colorado
Depth: 6000–6500 ft (1829–1981 m)
Ar-ion milled, SE2 image
Thermal maturity: N/A
Submitted by J. DeVito, Ingrain
Stratigraphic top to left. Organic matter (om) texture is solid and fractured. Intergranular porosity (p) and microfossils occur within the calcareous matrix. Total porosity 3.5%, organic matter 11.9%. Values given were computed by digital rock physics.

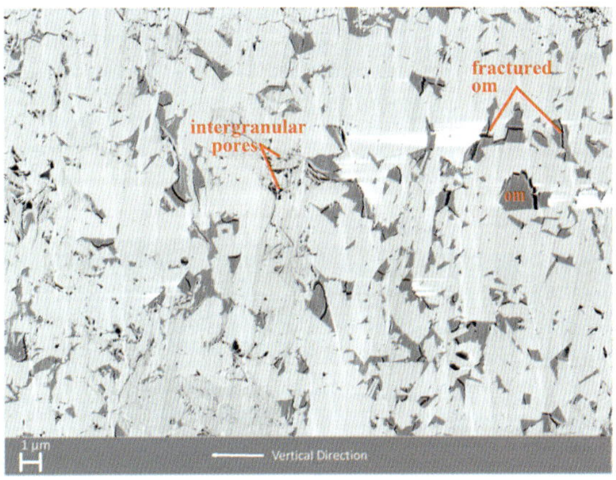

F.190.Niobrara
Niobrara Formation (Late Cretaceous)
Denver-Julesburg Basin, Weld County, Colorado
Depth: 6000–6500 ft (1829–1981 m)
Ar-ion milled, SE2 image
Thermal maturity: N/A
Submitted by J. DeVito, Ingrain
Stratigraphic top to left. Organic matter (om) texture is solid and fractured. Intergranular porosity (black) and microfossils occur within the calcareous matrix. Total porosity 3.4%, organic matter 9.6%. Values given were computed by digital rock physics.

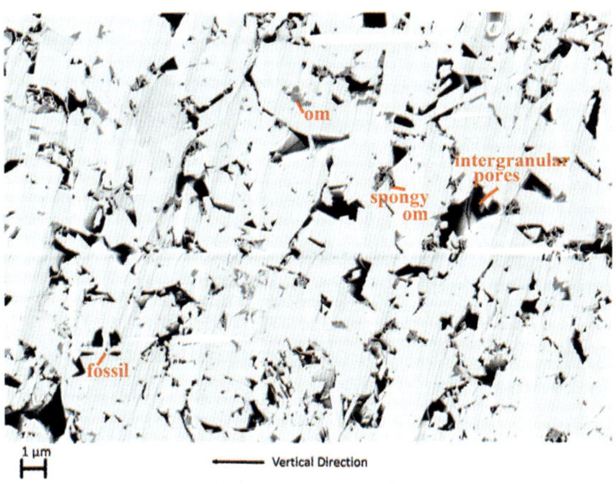

F.191.Niobrara
Niobrara Formation (Late Cretaceous)
Denver-Julesburg Basin, Weld County, Colorado
Depth: 6000–6500 ft (1829–1981 m)
Ar-ion milled, SE2 image
Thermal maturity: N/A
Submitted by J. DeVito, Ingrain
Stratigraphic top to left. Interparticle porosity (black) with solid organic matter (om). Organic matter texture is pendular and spongy. Grain (light gray) is predominantly calcareous fossil fragments. Total porosity 12.8%, organic matter 1.8%. Values given were computed by digital rock physics.

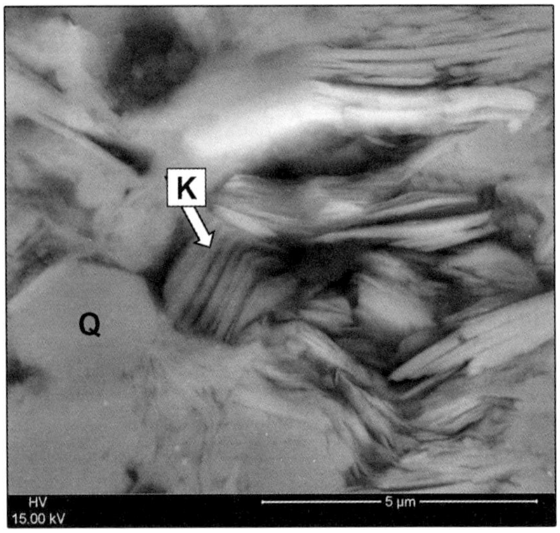

F.192.Ohio
Ohio Formation-Cleveland Member (Late Devonian)
Illinois Basin, Kentucky
Depth: 3400–3500 ft (1036–1067 m)
Ar-ion milled, BSE image
Thermal maturity: N/A
Submitted by D. Jennings, Core Laboratories
Authigenic clays occlude pores, restricting pore connectivity in this silty shale. Books of authigenic kaolinite (K) occur adjacent to degraded detrital mica. Authigenic quartz (Q) is also present.

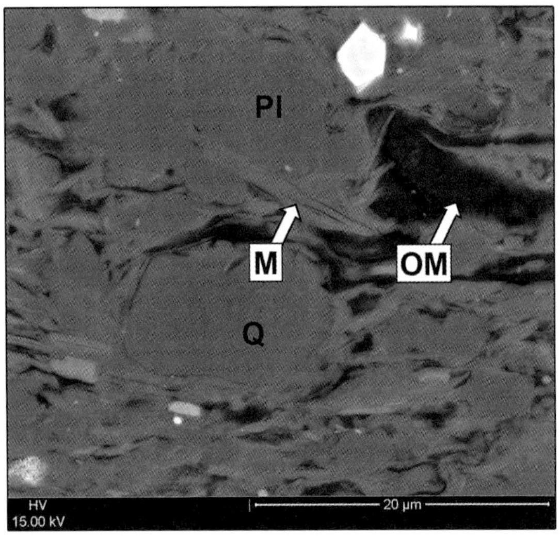

F.193.Ohio
Ohio Formation-Three Lick Member (Late Devonian)
Illinois Basin, Kentucky
Depth: 3500–3600 ft (1067–1097 m)
Ar-ion milled, BSE image
Thermal maturity: N/A
Submitted by D. Jennings, Core Laboratories
Detrital quartz silt (Q), plagioclase (Pl), and tabular flakes of mica (M) are the predominant framework grains in this microlaminated, silty shale. Interparticle pores are abundant and appear to be relatively well connected. Pressure shadow pores or phyllosilicate framework pores are locally present. Organic matter (OM) is scattered throughout the sample.

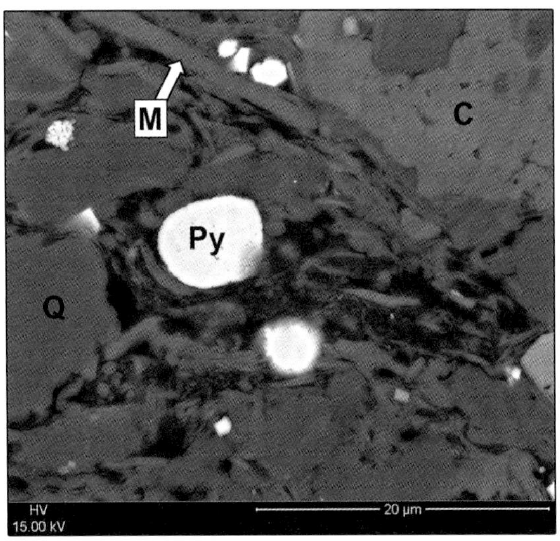

F.194.Ohio
Ohio Formation-Middle Huron Member (Late Devonian)
Illinois Basin, Kentucky
Depth: 3700–3800 ft (1067–1097 m)
Ar-ion milled, BSE image
Thermal maturity: N/A
Submitted by D. Jennings, Core Laboratories
Finely crystalline pyrite (Py), detrital quartz grains (Q), patches of calcite cement (C), and tabular flakes of mica are abundant in this silty shale. The matrix is predominantly composed of clay and organic matter (black). Interparticle pores are the most common pore type in this sample. Detrital clays commonly form bridged pores.

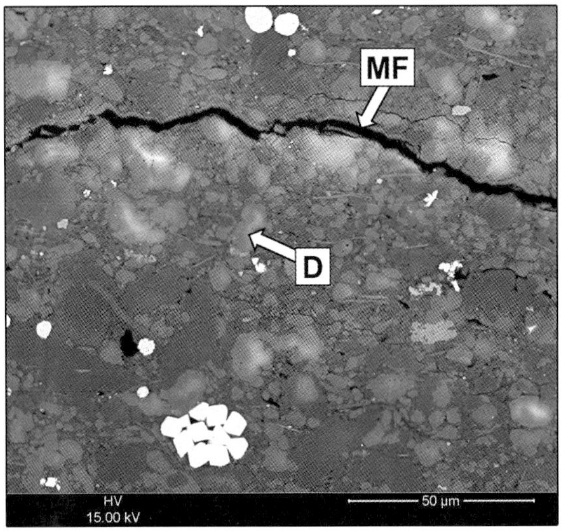

F.195.Paradox
Paradox Formation-Desert Creek Member (Pennsylvanian)
Paradox Basin, Colorado
Depth: 5800–5900 ft (1768–1798 m)
Ar-ion milled, BSE image
Thermal maturity: N/A
Submitted by D. Jennings, Core Laboratories
Ferroan dolomite rhombs (D) and detrital quartz silt (medium-gray grains) are the most abundant constituents in this dolomitic, silty shale. Euhedral pyrite is scattered throughout the sample. Bedding plane parallel, microfractures (MF) are common.

F.196.Paradox
Paradox Formation-Desert Creek Member (Pennsylvanian)
Paradox Basin, Colorado
Depth: 5800–5900 ft (1768–1798 m)
Ar-ion milled, BSE image
Thermal maturity: N/A
Submitted by D. Jennings, Core Laboratories
Micron-scale, marcasite-filled, horizontal microburrows are common in this sample. Note the matrix constituents that are compacted around the burrow, suggesting that this burrow was filled with marcasite early in the burial process. Authigenic clay particles occlude intracrystalline pores between the marcasite crystals.

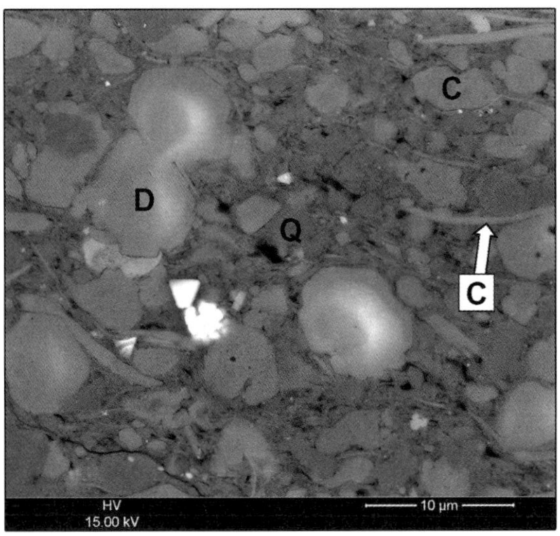

F.197.Paradox
Paradox Formation-Desert Creek Member (Pennsylvanian)
Paradox Basin, Colorado
Depth: 5800–5900 ft (1768–1798 m)
Ar-ion milled, BSE image
Thermal maturity: N/A
Submitted by D. Jennings, Core Laboratories
Light zones within dolomite grains (D) are composed of ferroan dolomite. Thin skeletal fragments composed of calcite (C, white box) emphasize the microlaminated fabric of this sample. Calcite grains (C) and quartz silt (Q) are also present.

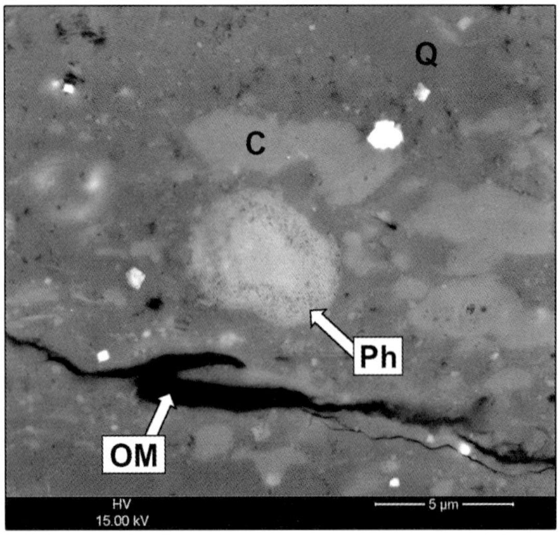

F.198.Pearsall (Early Cretaceous)
Pearsall Formation
Gulf Coast Basin, Texas
Depth: 11000–11100 ft (1768–1798 m)
Ar-ion milled, BSE image
Thermal maturity: N/A
Submitted by D. Jennings, Core Laboratories
Detrital quartz silt (Q) and calcite (C) are scattered throughout the clay-rich matrix of this silty shale. Micron-scale strands of organic matter (OM) lie parallel to bedding. Phosphate peloids (Ph) exhibit solid centers and do not appear to be compacted.

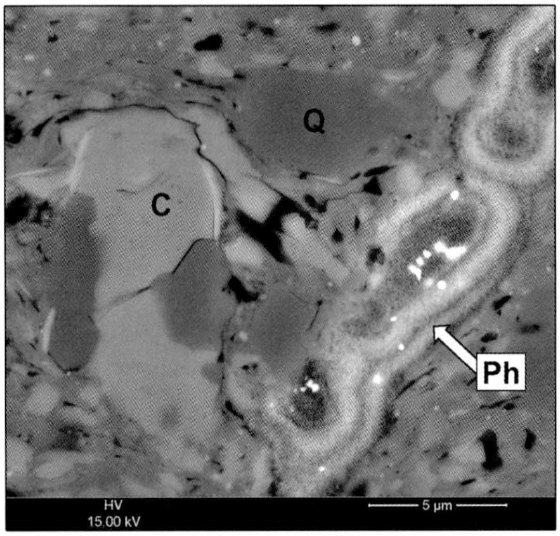

F.199.Pearsall
Pearsall Formation (Early Cretaceous)
Gulf Coast Basin, Texas
Depth: 11200–11300 ft (3414–3444 m)
Ar-ion milled, BSE image
Thermal maturity: N/A
Submitted by D. Jennings, Core Laboratories
Accretionary phosphate masses (Ph) that possess dark, organic-rich centers are common in this calcareous, silty shale. Calcite grains (C) that are partially replaced by quartz (medium gray) and quartz silt (Q) are the predominant framework grains. Interparticle pores are the most abundant pore type. Although most pores are nanometer scale, micron-scale pores are locally present.

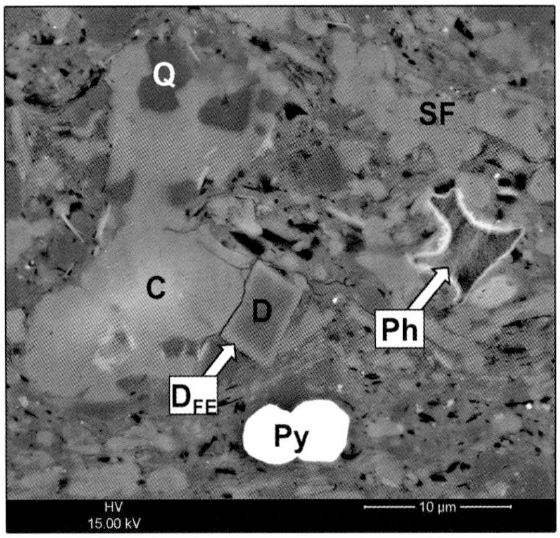

F.200.Pearsall
Pearsall Formation (Early Cretaceous)
Gulf Coast Basin, Texas
Depth: 11200–11300 ft (3414–3444 m)
Ar-ion milled, BSE image
Thermal maturity: N/A
Submitted by D. Jennings, Core Laboratories
The fabric of this calcareous, silty shale is microlaminated. Nanometer-scale, interparticle pores (black) are abundant. Calcite is present as calcite grains (C) that have been partially replaced by quartz (Q) and scattered skeletal fragments (SF). Dolomite rhombs (D) with ferroan dolomite rims (D_{FE}), euhedral pyrite (Py), and patches of fine-grain phosphate (Ph) are locally present.

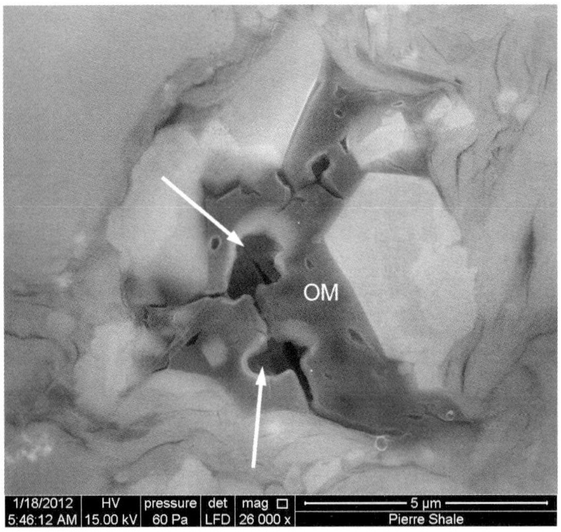

F.201.Pierre
Pierre Shale (Late Cretaceous)
Middle Park Basin, Colorado
Ar-ion milled, SE image
Thermal maturity: 0.58%R_o
Submitted by B. Wawak and T. Gentzis, Core Laboratories
Dispersed organic matter (OM) containing micrometer-size pores (arrows) that are connected with microfractures. Note also some interparticle pores associated with phyllosilicate minerals (top right corner).

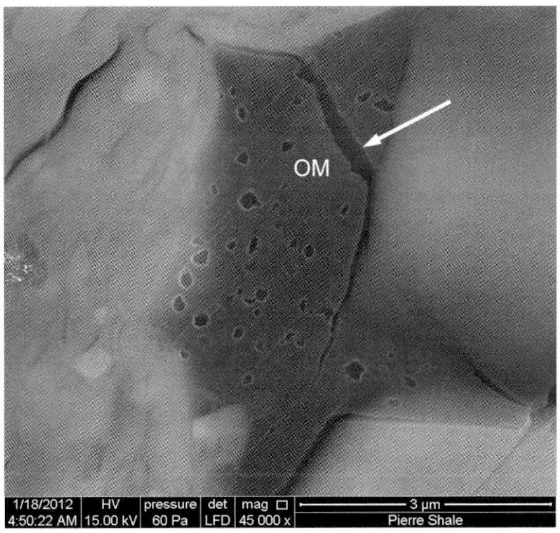

F.202.Pierre
Pierre Shale (Late Cretaceous)
Middle Park Basin, Colorado
Ar-ion milled, SE image
Thermal maturity: 0.58%R_o
Submitted by B. Wawak and T. Gentzis, Core Laboratories
Organic matter (OM) with abundant nanopores and a microfracture (arrow). The widest dimension of selected circular nanopores ranges from 60 to 190 nm.

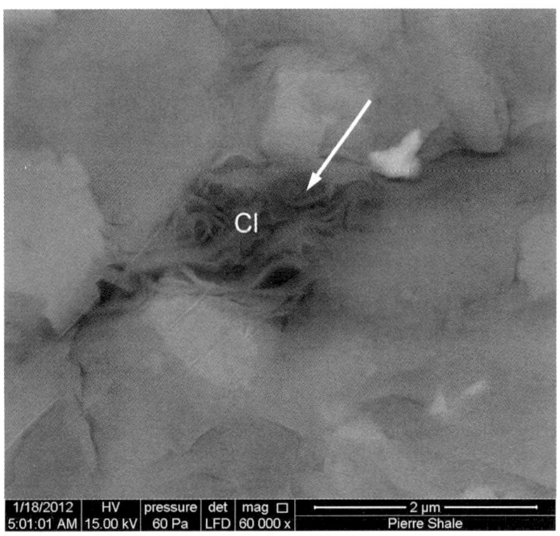

F.203.Pierre
Pierre Shale (Late Cretaceous)
Middle Park Basin, Colorado
Ar-ion milled, SE image
Thermal maturity: 0.58%R_o
Submitted by B. Wawak and T. Gentzis, Core Laboratories
Organic matter (arrow) and interparticle pores associated with phyllosilicate minerals (Cl). The morphology of the phyllosilicate minerals suggests they are clays.

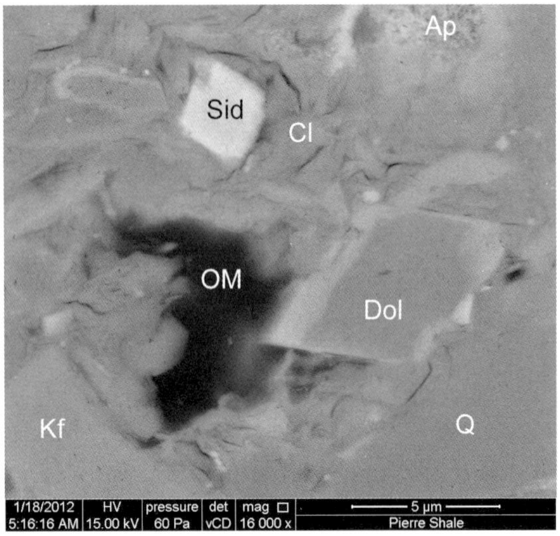

F.204.Pierre
Pierre Shale (Late Cretaceous)
Middle Park Basin, Colorado
Ar-ion milled, BSE image
Thermal maturity: 0.58%R_o
Submitted by B. Wawak and T. Gentzis, Core Laboratories
Nonporous particle of organic matter (OM) associated with various inorganic particles. Inorganic particles include apatite (Ap), siderite (Sid), dolomite (Dol), quartz (Q), phyllosilicate minerals (Cl, clays) and potassium feldspar (Kf). Most of the dolomite crystals have a distinctive ferroan rim.

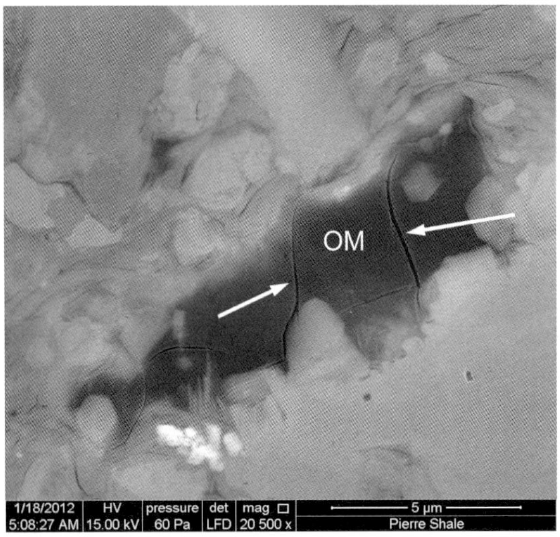

F.205.Pierre
Pierre Shale (Late Cretaceous)
Middle Park Basin, Colorado
Ar-ion milled, SE image
Thermal maturity: 0.58%R_o
Submitted by B. Wawak and T. Gentzis, Core Laboratories
Silt-size discrete particle of organic matter (OM) that contains microfractures that extend across the body of the particle (arrows) and microfractures that are perpendicular or near perpendicular to the "cross-particle" fractures.

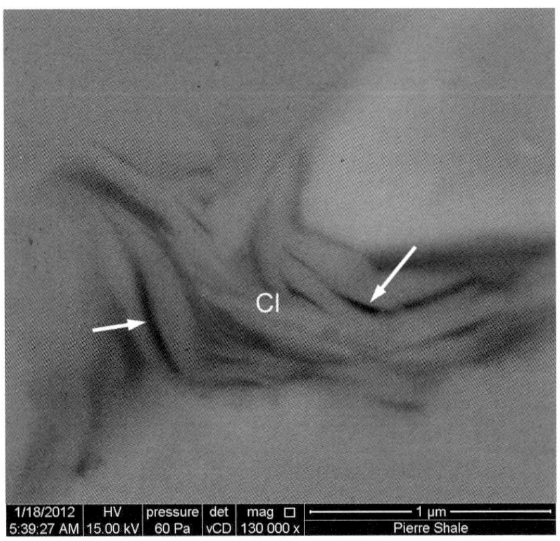

F.206.Pierre
Pierre Shale (Late Cretaceous)
Middle Park Basin, Colorado
Ar-ion milled, BSE/SE image
Thermal maturity: 0.58%R_o
Submitted by B. Wawak and T. Gentzis, Core Laboratories
Interparticle pores (arrows) associated with phyllosilicate minerals (Cl).

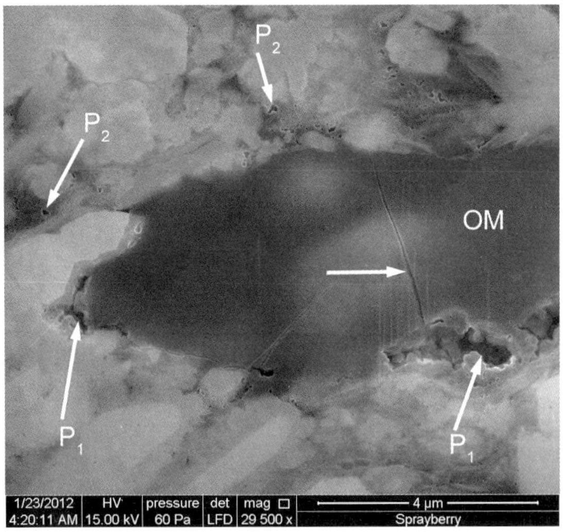

F.207.Spraberry
Spraberry Formation (Early Permian)
Permian Basin, Texas
Ar-ion milled, SE image
Thermal maturity: 1.08%R_o
Submitted by B. Wawak and T. Gentzis, Core Laboratories
Irregularly shaped pores (P_1) occurring along the edge of a silt-size discrete particle of organic matter (OM). Note also the microfracture extending across the body of the particle (arrow). There are also nanopores (P_2) associated with dispersed organic matter in the adjacent parts of the sample.

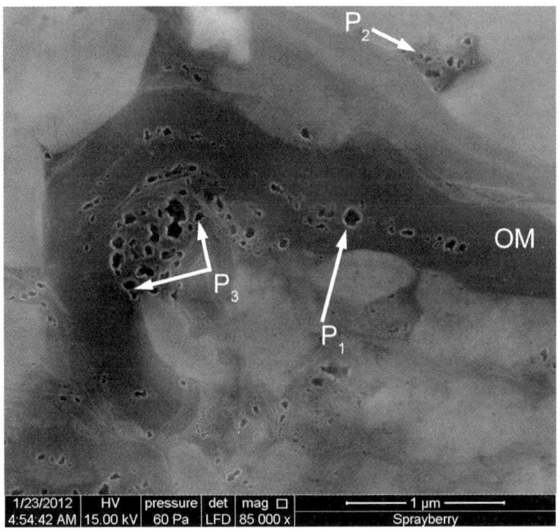

F.208.Spraberry
Spraberry Formation (Early Permian)
Permian Basin, Texas
Ar-ion milled, SE image
Thermal maturity: 1.08%R_o
Submitted by B. Wawak and T. Gentzis, Core Laboratories
Nanopores (P_1) occurring within relatively large, irregularly shaped discrete particle of organic matter (OM). Nanopores also occur within dispersed organic matter (P_2) in adjacent areas of the sample. The nanopores indicated by the arrows (P_3) appear to be porous dispersed organic matter.

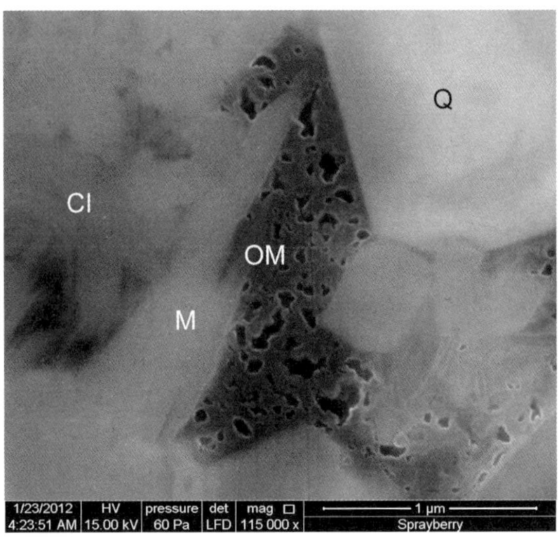

F.209.Spraberry
Spraberry Formation (Early Permian)
Permian Basin, Texas
Ar-ion milled, SE image
Thermal maturity: 1.08%R_o
Submitted by B. Wawak and T. Gentzis, Core Laboratories
Highly porous dispersed organic matter (OM) that fills interparticle pores associated with phyllosilicate minerals (Cl, clays; M, mica) and quartz (Q).

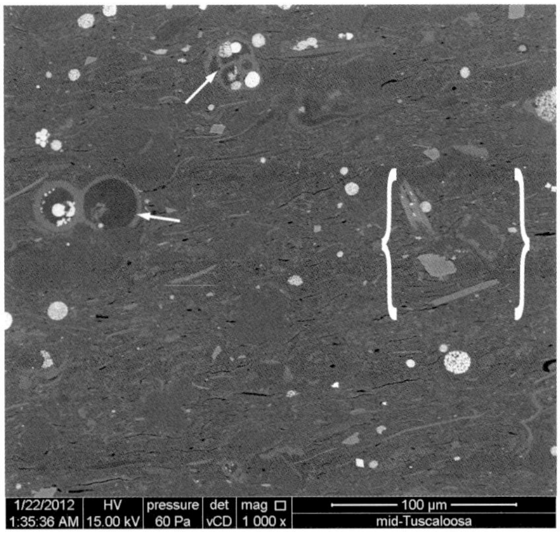

F.210.Tuscaloosa
Tuscaloosa Formation (Late Cretaceous)
Northern Gulf of Mexico Basin, Mississippi
Ar-ion milled, BSE image
Thermal maturity: 0.58%R_o
Submitted by B. Wawak and T. Gentzis, Core Laboratories
Relatively low-magnification view showing texture of sample and occurrence of a variety of larger (silt- to very fine sand-size) particles. Whole (or nearly whole) foraminifer tests (arrows) are scattered throughout the sample. The foraminifer tests are filled will kaolinite and pyrite (white). Brackets indicate area that is enlarged in Figure 211.

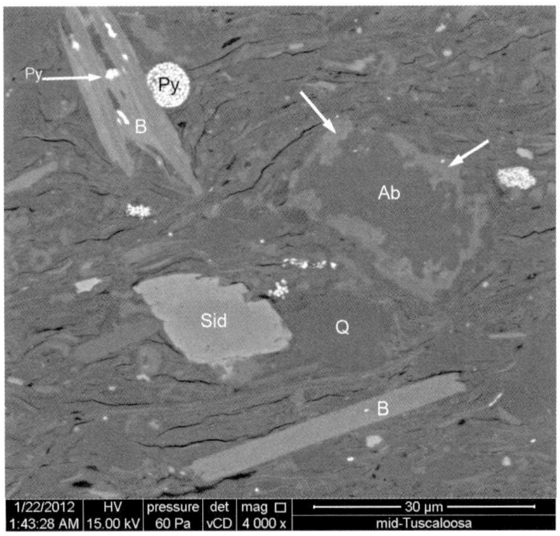

F.211.Tuscaloosa
Tuscaloosa Formation (Late Cretaceous)
Northern Gulf of Mexico Basin, Mississippi
Ar-ion milled, BSE image
Thermal maturity: 0.58%R_o
Submitted by B. Wawak and T. Gentzis, Core Laboratories
Enlargement of area indicated by brackets in Figure 210 showing the variety of silt-size inorganic particles. Inorganic particles identified in this sample include biotite (B), pyrite framboids (Py), albite (Ab), siderite (Sid), and quartz (Q). Note small pyrite crystals (Py arrow) within a biotite grain and calcite (light-color rim of irregular thickness, arrows) alteration around the edges of the albite grain.

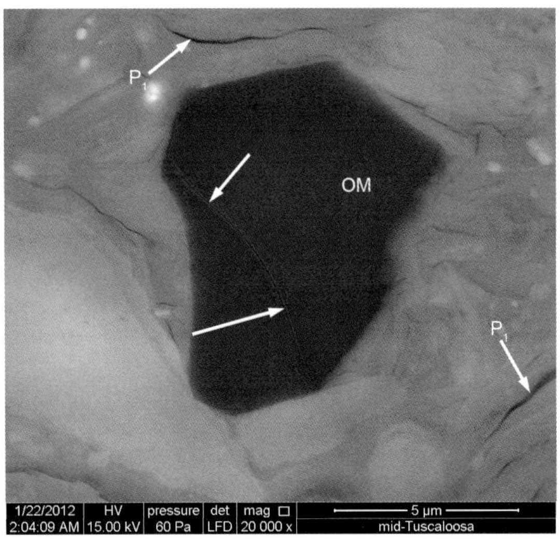

F.212.Tuscaloosa
Tuscaloosa Formation (Late Cretaceous)
Northern Gulf of Mexico Basin, Mississippi
Ar-ion milled, SE image
Thermal maturity: 0.58%R_o
Submitted by B. Wawak and T. Gentzis, Core Laboratories
Silt-size, discrete particle of organic matter (OM) that has a smooth (nonporous) surface texture and contains a fracture running across the body of the particle (arrows). Note also nearby interparticle pores associated with clays (P_1).

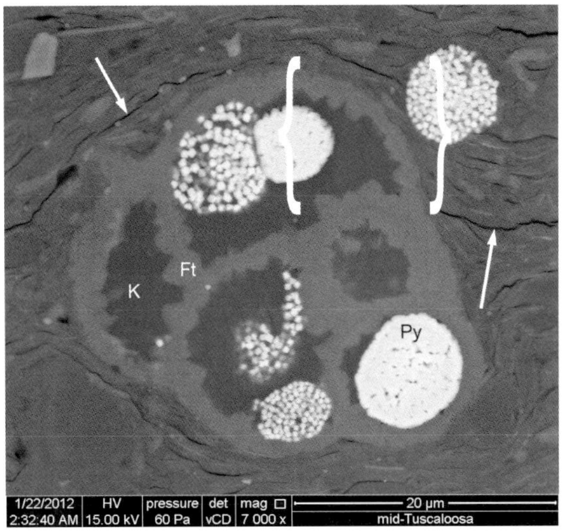

F.213.Tuscaloosa
Tuscaloosa Formation (Late Cretaceous)
Northern Gulf of Mexico Basin, Mississippi
Ar-ion milled, BSE image
Thermal maturity: 0.58%R_o
Submitted by B. Wawak and T. Gentzis, Core Laboratories
Enlargement of foraminifer test seen in Figure 210. The individual chambers of the foraminifer tests (Ft) are filled with kaolinite (K) and pyrite (Py). Note also the interparticle pores (arrows) associated with phyllosilicate minerals in the areas adjacent to the foraminifer test. Brackets indicate area that is enlarged in Figure 214.

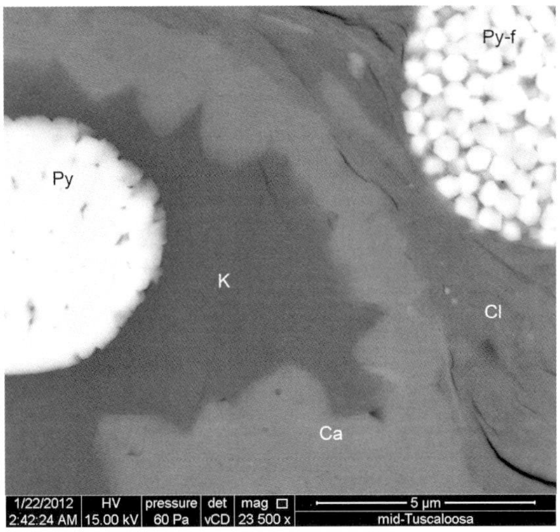

F.214.Tuscaloosa
Tuscaloosa Formation (Late Cretaceous)
Northern Gulf of Mexico Basin, Mississippi
Ar-ion milled, BSE image
Thermal maturity: 0.58%R_o
Submitted by B. Wawak and T. Gentzis, Core Laboratories
This image is a higher-magnification view of part of foraminifer test indicated by the brackets in Figure 213. The test is composed of calcite (Ca), and the test chambers are filled with kaolinite (K) and pyrite (Py). The matrix outside the foraminifer test is predominantly phyllosilicate minerals (Cl) and also includes pyrite framboids (Py-f). Note also some interparticle pores associated with the phyllosilicate minerals (bottom right corner) and intraparticle pores associated with the pyrite framboids (top right corner).

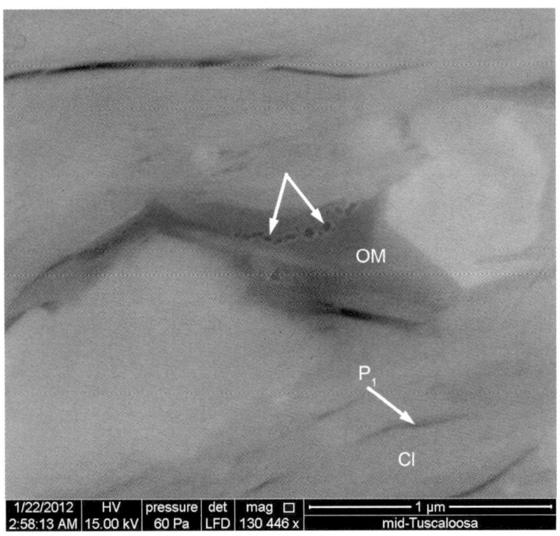

F.215.Tuscaloosa
Tuscaloosa Formation (Late Cretaceous)
Northern Gulf of Mexico Basin, Mississippi
Ar-ion milled, SE image
Thermal maturity: 0.58%R_o
Submitted by B. Wawak and T. Gentzis, Core Laboratories
Nanopores (arrows) associated with organic matter (OM). Note also interparticle pores (P_1) associated with nearby phyllosilicate minerals (Cl).

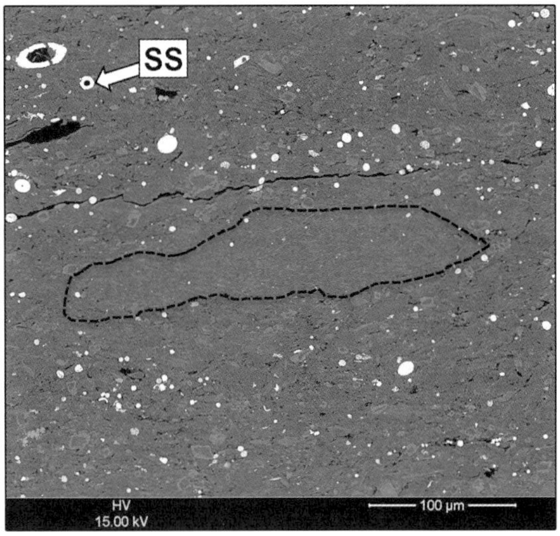

F.216.U.Devonian
Formation N/A (Late Devonian)
West-central, Alabama
Depth: 8200–8300 ft (2499–2530 m)
Ar-ion milled, BSE image
Thermal maturity: N/A
Submitted by D. Jennings, Core Laboratories
Pyrite-replaced sponge spicules (SS), zoned dolomite rhombs, quartz silt, and detrital flakes of mica are abundant in this silty shale. Micron-scale clay aggregates are also common (outlined with black dashed line). Relatively poorly connected interparticle pores (black) are moderately abundant.

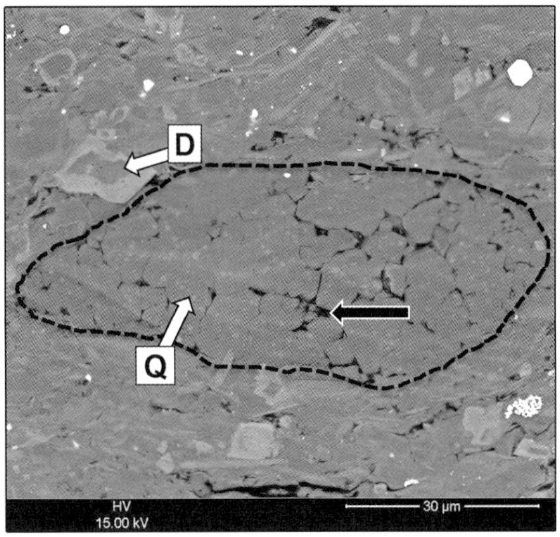

F.217.U.Devonian
Formation N/A (Late Devonian)
West-central, Alabama
Depth: 8200–8300 ft (2499–2530 m)
Ar-ion milled, BSE image
Thermal maturity: N/A
Submitted by D. Jennings, Core Laboratories
Zoned dolomite rhombs (D) typically display lighter-color, ferroan dolomite rims. Authigenic quartz (Q) fills a compacted, oval feature (likely a horizontal microburrow). Matrix material is compacted around the microburrow. Intrinsic pores are preserved between the quartz crystals (black arrow).

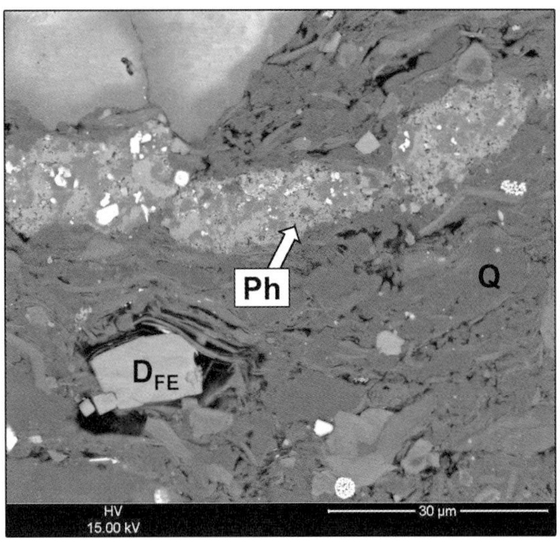

F.218.U.Devonian
Formation N/A (Late Devonian)
West-central, Alabama
Depth: 8200–8300 ft (2499–2530 m)
Ar-ion milled, BSE image
Thermal maturity: N/A
Submitted by D. Jennings, Core Laboratories
Scattered micrometer- and nanometer-scale interparticle pores (black) are abundant in this microlaminated, silty shale. Rounded to subrounded quartz grains (Q), pyrite grains (bright white), and ferroan dolomite (D_{FE}) are common. The fine-grain, porous fabric of phosphate peloids (Ph) suggests that they are early authigenic, microbially mediated features.

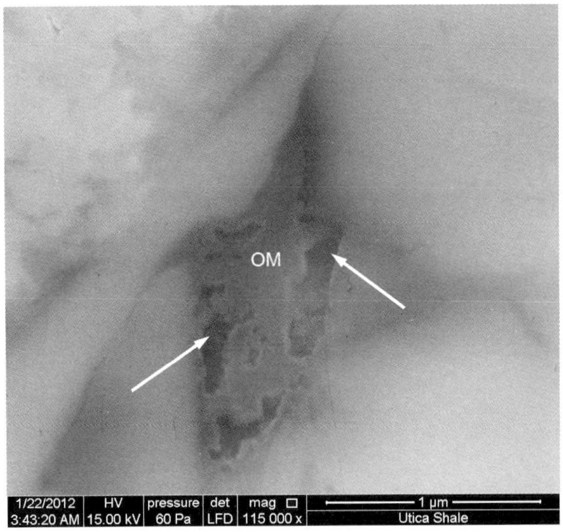

F.219.Utica
Utica Shale (Middle Ordovician)
Appalachian Basin, Ohio
Ar-ion milled, SE image
Thermal maturity: 0.88%R_o
Submitted by B. Wawak and T. Gentzis, Core Laboratories
Submicron pores (arrows) associated with dispersed organic matter (OM).

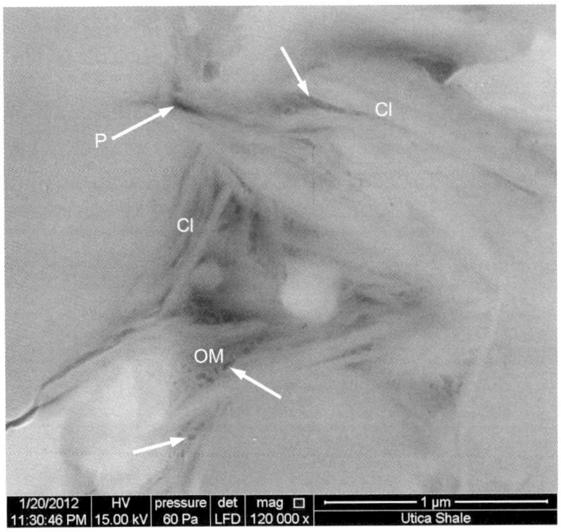

F.220.Utica
Utica Shale (Middle Ordovician)
Appalachian Basin, Ohio
Ar-ion milled, SE image
Thermal maturity: 0.88%R_o
Submitted by B. Wawak and T. Gentzis, Core Laboratories
The interparticle pores associated with these phyllosilicate minerals (Cl) is partially filled with dispersed organic matter (OM) that contains numerous nanopores (arrows). Note also interparticle pores associated with phyllosilicate minerals (P).

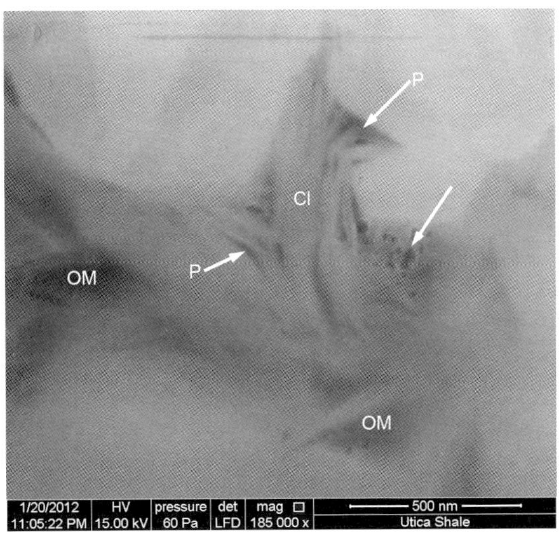

F.221.Utica
Utica Shale (Middle Ordovician)
Appalachian Basin, Ohio
Ar-ion milled, SE image
Thermal maturity: 0.88%R_o
Submitted by B. Wawak and T. Gentzis, Core Laboratories
Nanopores (arrow) associated with dispersed organic matter (OM) that partially fills interparticle pores associated with phyllosilicate minerals (Cl). Note also occurrence of interparticle pores associated with phyllosilicate minerals (P).

F.222.Wolfcamp
Wolfcamp (Early Permian)
Delaware Basin, Loving County, Texas
Depth: 12649.5 (3855 m)
Ar-ion milled, BSE image
Thermal maturity: 1.4%R_o
Submitted by W. Camp, Anadarko Petroleum
Several partially rounded (detrital?) dolomite (dol) grains rimmed by distinctive bright, ferroan-rich rims producing composite euhedral crystals.

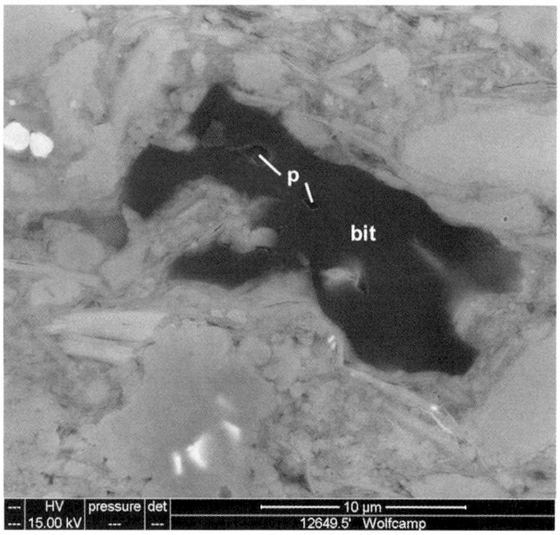

F.223.Wolfcamp
Wolfcamp (Early Permian)
Delaware Basin, Loving County, Texas
Depth: 12649.5 (3855 m)
Ar-ion milled, composite SE and BSE image
Thermal maturity: 1.4%R_o
Submitted by W. Camp, Anadarko Petroleum
Pyrobitumen (bit) exhibiting few intraparticle pores. Pores (p) within pyrobitumen presumably formed during thermal cracking of oil to gas during burial.

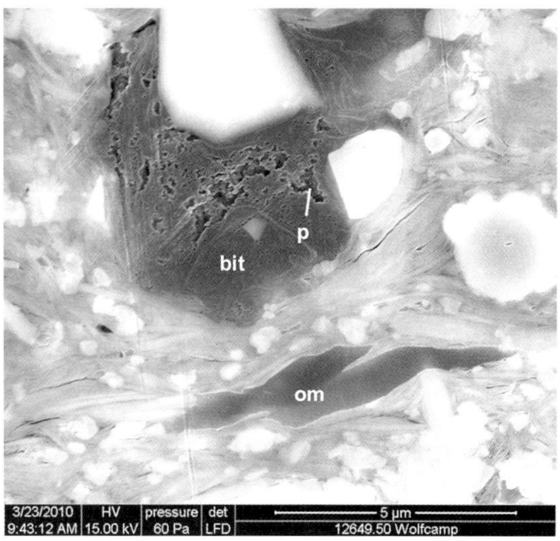

F.224.Wolfcamp
Wolfcamp (Early Permian)
Delaware Basin, Loving County, Texas
Depth: 12649.5 (3855 m)
Ar-ion milled, SE image
Thermal maturity: 1.4%R_o
Submitted by W. Camp, Anadarko Petroleum
Sponge-like texture of porous pyrobitumen (bit) within interparticle pores in clay matrix. Pores (p) within pyrobitumen presumably formed during thermal cracking of oil to gas during burial. Note lack of pores within structured organic matter (om) that appears to be a kerogen maceral.

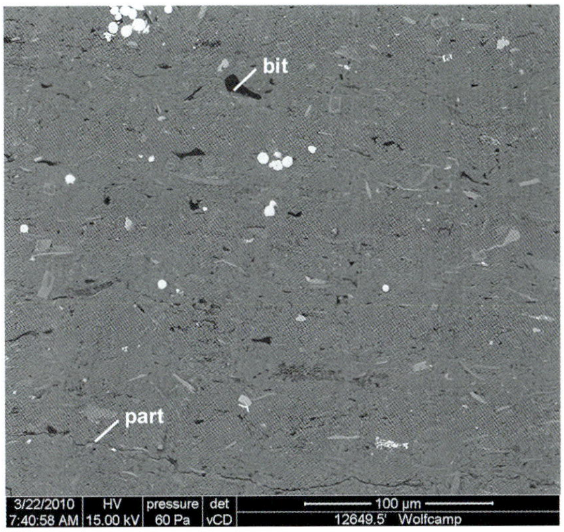

F.225.Wolfcamp
Wolfcamp (Early Permian)
Delaware Basin, Loving County, Texas
Depth: 12649.5 (3855 m)
Ar-ion milled, BSE image
Thermal maturity: 1.4%R_o
Submitted by W. Camp, Anadarko Petroleum
Bedding plane parting (part). Low backscatter intensity indicates parting is filled with bitumen and thus may represent a hydrocarbon migration feature. Bitumen (bit) also fills larger pore spaces. Partings may create mechanical discontinuities that may respond to hydraulic fracture stimulation and facilitate hydrocarbon flow.

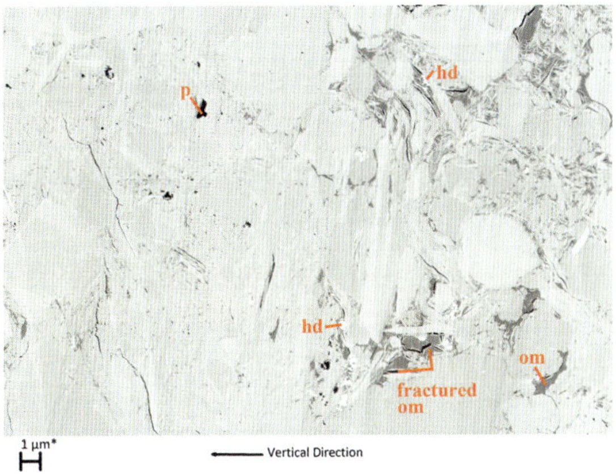

F.226.Wolfcamp
Wolfcamp Formation (Early Permian)
Permian Basin, Crockett County, Texas
Depth: 5000–5500 ft (1524–1676 m)
Ar-ion milled, SE2 image
Thermal maturity: N/A
Submitted by J. DeVito, Ingrain
Stratigraphic top to left. Thin, highly dense layers (hd) are associated with organic matter (om). Fractured organic matter porosity is within the clay matrix. Interparticle porosity (p) and solid organic matter (om) are present throughout the image. Thin desiccation cracks are parallel to the bedding. Total porosity 0.7%, organic matter 3.5%. Values given were computed by digital rock physics.

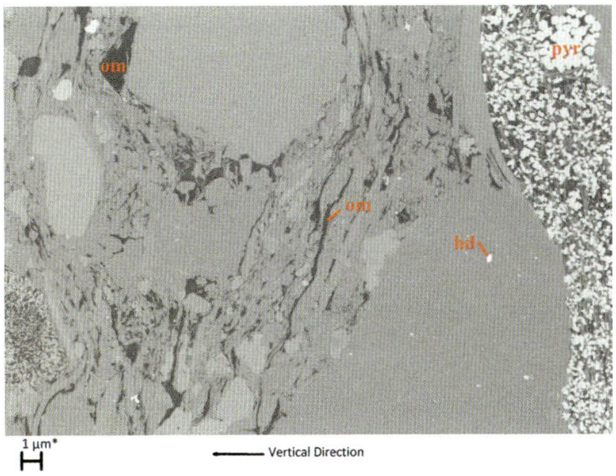

F.227.Wolfcamp
Wolfcamp Formation (Early Permian)
Permian Basin, Crockett County, Texas
Depth: 7000–7500 ft (2134–2286 m)
Ar-ion milled, BSE image
Thermal maturity: N/A
Submitted by J. DeVito, Ingrain
Stratigraphic top to left (Permian). Individual pyrite crystals and framboids (pyr) are organized parallel to bedding. Elongated solid organic matter (om) is located in the clay-rich layer above pyrite. A mixture of highly dense material (hd) and organic matter is within the matrix.

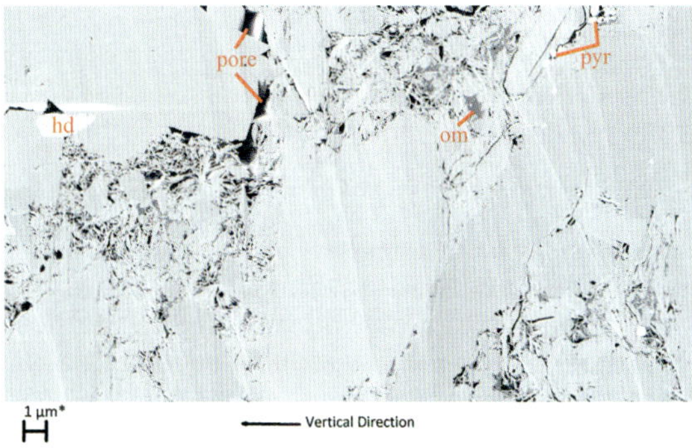

F.228.Wolfcamp
Wolfcamp Formation (Early Permian)
Permian Basin, Crockett County, Texas
Depth: 5000–5500 ft (1524–1676 m)
Ar-ion milled, SE2 image
Thermal maturity: N/A
Submitted by J. DeVito, Ingrain
Stratigraphic top to left. Angular interparticle porosity (black). Solid organic matter (om) is mixed with smaller spongy organic matter texture. Microfractures, pyrite (pyr), and other highly dense minerals (hd) are within the grain. Total porosity 4.2%, organic matter 2.1%. Values given were computed by digital rock physics.

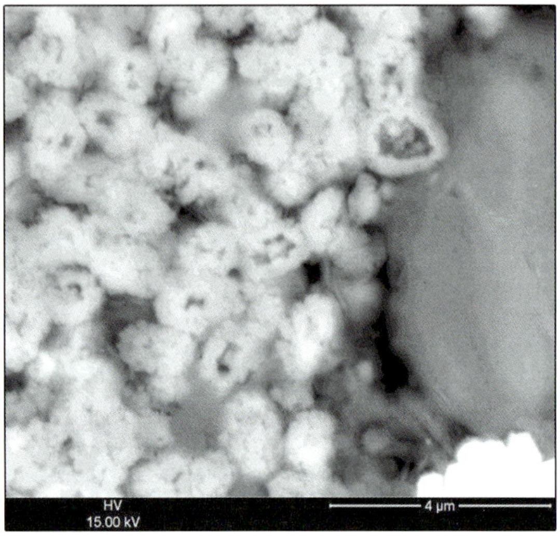

F.229.Woodford
Woodford Formation (Late Devonian)
Arkoma Basin, Oklahoma
Depth: 3900–4000 ft (1189–1219 m)
Ar-ion milled, BSE image
Thermal maturity: 1.5–1.6%R_o
Submitted by D. Jennings, Core Laboratories
Hollow spheres composed of clusters of phosphatic coccoids. Nanometer-scale pores are common within and between the spheres.

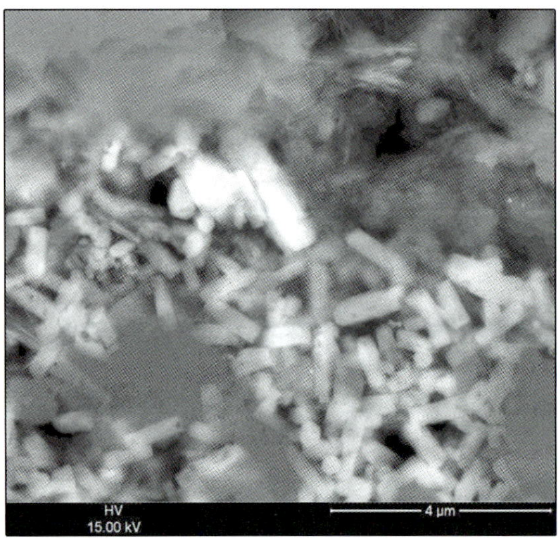

F.230.Woodford
Woodford Formation (Late Devonian)
Arkoma Basin, Oklahoma
Depth: 2000–2100 ft (610–640 m)
Ar-ion milled, BSE image
Thermal maturity: 1.0–1.1%R_o
Submitted by D. Jennings, Core Laboratories
Randomly oriented, hollow tubes of phosphate are present in a clay-rich matrix. Note the interparticle pores between the tubes.

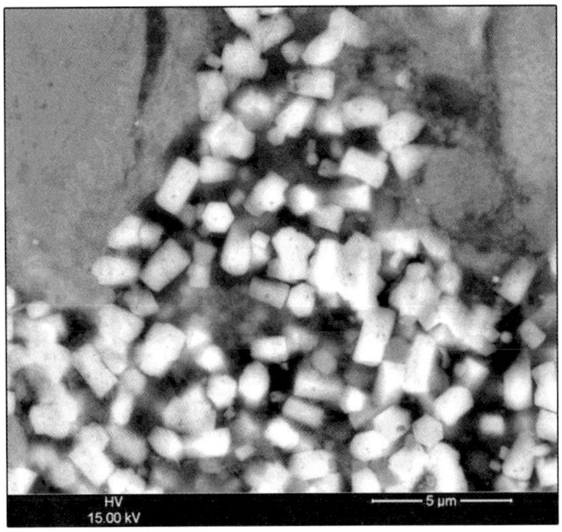

F.231.Woodford
Woodford Formation (Late Devonian)
Arkoma Basin, Oklahoma
Depth: 9600–9700 ft (2926–2957 m)
Ar-ion milled, BSE image
Thermal maturity: 1.1–1.2%R_o
Submitted by D. Jennings, Core Laboratories
Intercrystalline pores between euhedral phosphate crystals appear to be well connected. However, authigenic clay particles (medium gray) between phosphate crystals decrease porosity.

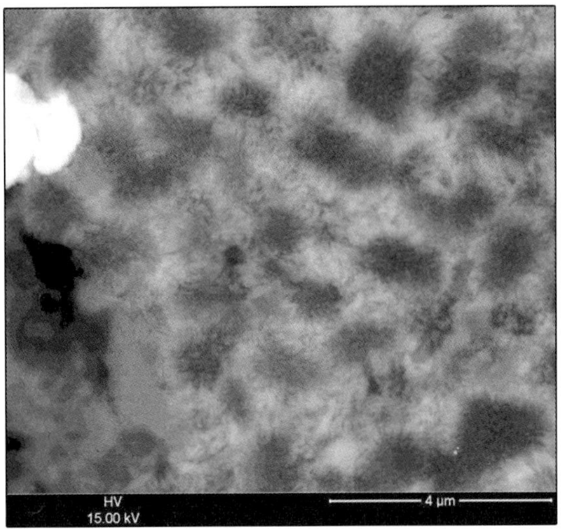

F.232.Woodford
Woodford Formation (Late Devonian)
Arkoma Basin, Oklahoma
Depth: 8600–8700 ft (2621–2652 m)
Ar-ion milled, BSE image
Thermal maturity: N/A
Submitted by D. Jennings, Core Laboratories
Fibrous phosphate (white) radiating from organic-rich centers (dark gray to black) occludes micron-scale pores.

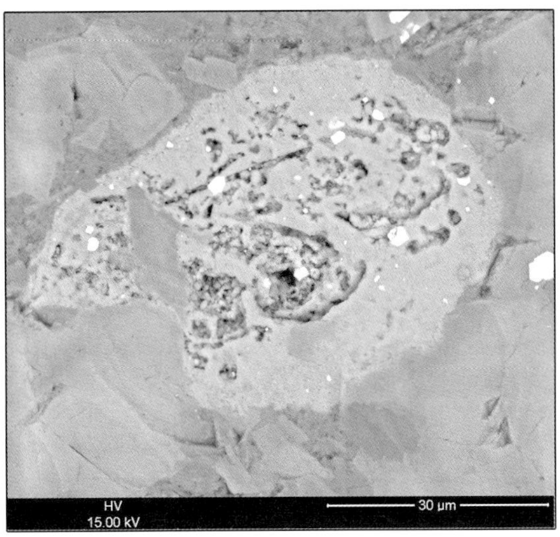

F.233.Woodford
Woodford Formation (Late Devonian)
Arkoma Basin, Oklahoma
Depth: 7400–7500 ft (2256–2286 m)
Ar-ion milled, BSE image
Thermal maturity: N/A
Submitted by D. Jennings, Core Laboratories
Phosphate peloids commonly display internal microporosity.

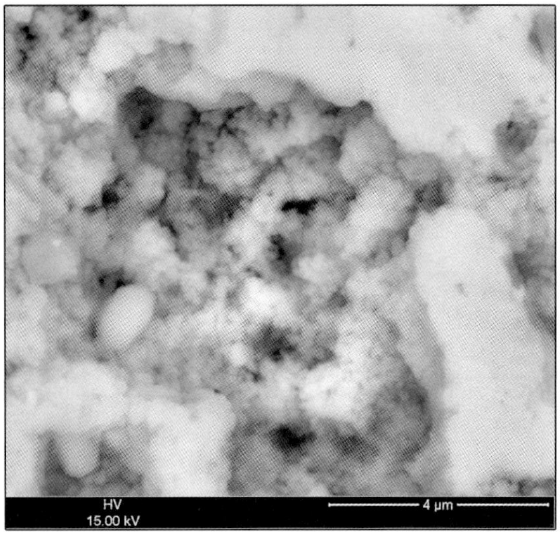

F.234.Woodford
Woodford Formation (Late Devonian)
Arkoma Basin, Oklahoma
Depth: 7400–7500 ft (2256–2286 m)
Ar-ion milled, BSE image
Thermal maturity: N/A
Submitted by D. Jennings, Core Laboratories
A high magnification of phosphate pellets reveals a network of clusters of coccoids. Tortuous intercrystalline pores (black) are abundant.

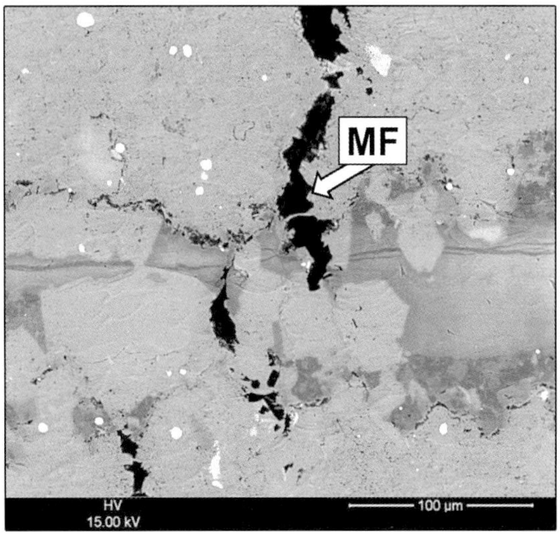

F.235.Woodford
Woodford Formation (Late Devonian)
Arkoma Basin, Oklahoma
Depth: 12200–12300 ft (3719–3749 m)
Ar-ion milled, BSE image
Thermal maturity: 1.7–1.8%R_o
Submitted by D. Jennings, Core Laboratories
Vertical microfractures (MF) infilled with bitumen (black). Microfractures occur in all lithologies and are locally abundant in chert beds.

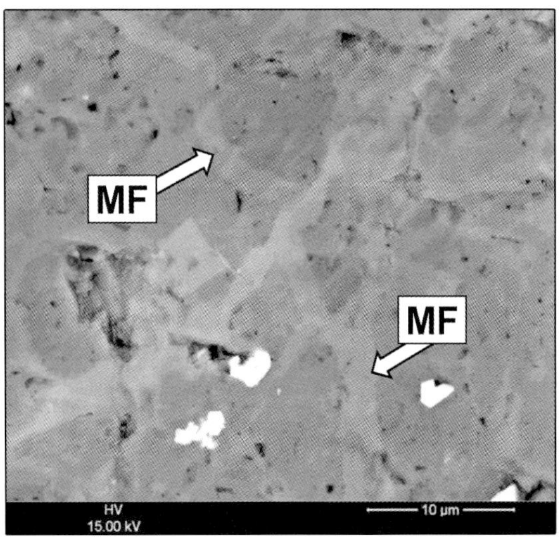

F.236.Woodford
Woodford Formation (Late Devonian)
Arkoma Basin, Oklahoma
Depth: 12600–12700 ft (3840–3871 m)
Ar-ion milled, BSE image
Thermal maturity: 1.7–1.8%R_o
Submitted by D. Jennings, Core Laboratories
Networks of micrometer-scale microfractures (MF) infilled with ferroan dolomite (light gray) are locally abundant in chert beds.

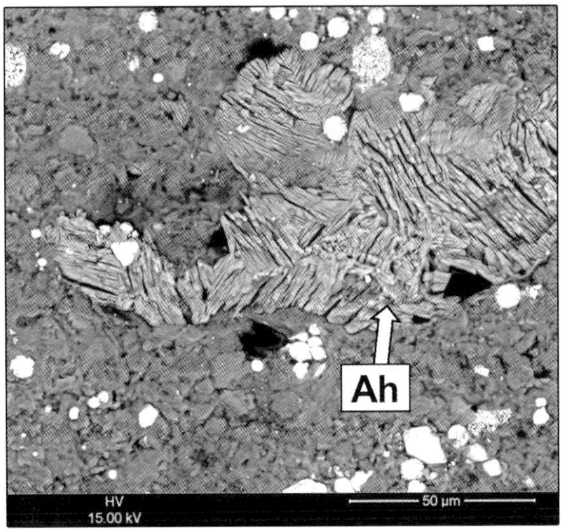

F.237.Woodford
Woodford Formation (Late Devonian)
Arkoma Basin, Oklahoma
Depth: 4200–4300 ft (1280–1311 m)
Ar-ion milled, BSE image
Thermal maturity: 1.7–1.8%R_o
Submitted by D. Jennings, Core Laboratories
Microfractures infilled with anhydrite (Ah) likely formed from gypsum that precipitated in microfractures during late-stage burial. Note that intercrystalline pores are present between individual anhydrite crystals.

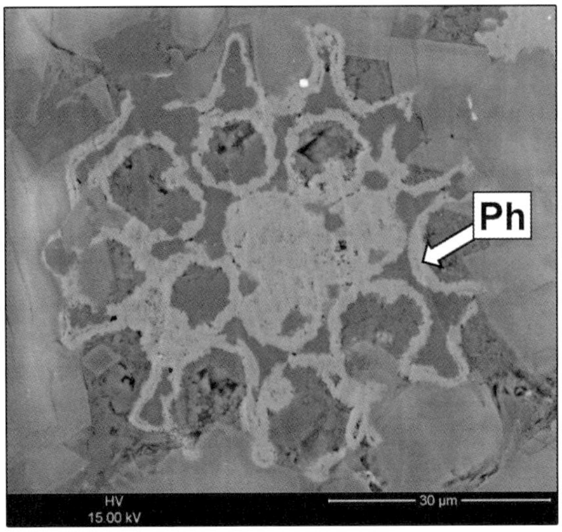

F.238.Woodford
Woodford Formation (Late Devonian)
Arkoma Basin, Oklahoma
Depth: 7400–7500 ft (2256–2286 m)
Ar-ion milled, BSE image
Thermal maturity: N/A
Submitted by D. Jennings, Core Laboratories
Phosphate (Ph) is present within the radiolaria central cavity and outlines the radiolaria. Siliceous radiolaria rays are partially preserved (dark gray).

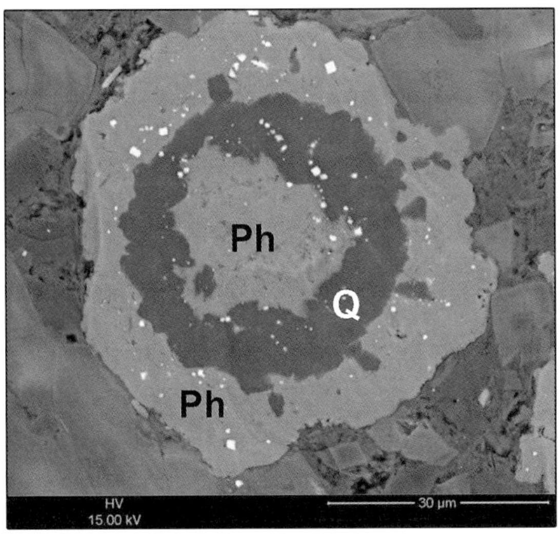

F.239.Woodford
Woodford Formation (Late Devonian)
Arkoma Basin, Oklahoma
Depth: 7400–7500 ft (2256–2286 m)
Ar-ion milled, BSE image
Thermal maturity: N/A
Submitted by D. Jennings, Core Laboratories
Phosphate (Ph) is present within the radiolaria central cavity and surrounds the entire siliceous radiolaria (Q). Siliceous radiolaria rays are partially preserved (dark gray).

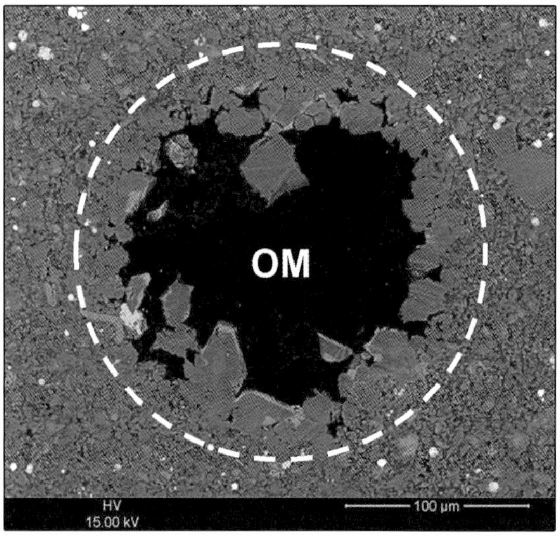

F.240.Woodford
Woodford Formation (Late Devonian)
Arkoma Basin, Oklahoma
Depth: 12800–12900 ft (3901–3932 m)
Ar-ion milled, BSE image
Thermal maturity: N/A
Submitted by D. Jennings, Core Laboratories
Organic matter (OM) is present within the central void of a remnant radiolaria. The outer rim of the radiolaria has been replaced by quartz (outer rim highlighted by white dashed line).

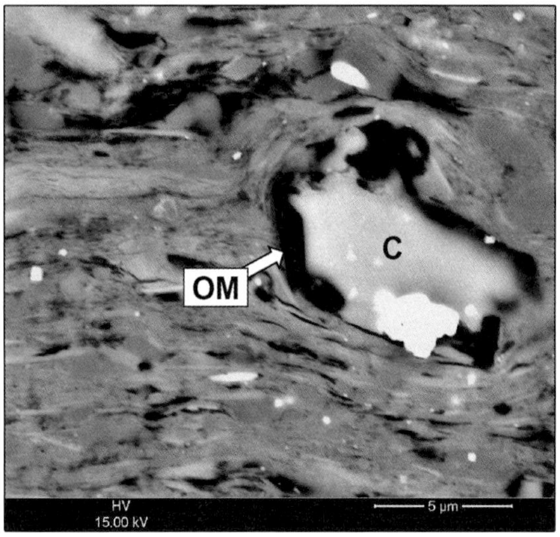

F.241.Woodford
Woodford Formation (Late Devonian)
Delaware Basin, Texas
Depth: 8200–8300 ft (2499–2530 m)
Ar-ion milled, BSE image
Thermal maturity: 1.2–1.3%R_o
Submitted by D. Jennings, Core Laboratories
Interparticle nanopores are abundant in this laminated, silty shale. Organic matter (OM) occurs around a partially dissolved carbonate grain (C). Note that clay is aligned around the organic matter, suggesting that this is a void of a dissolved framework grain.

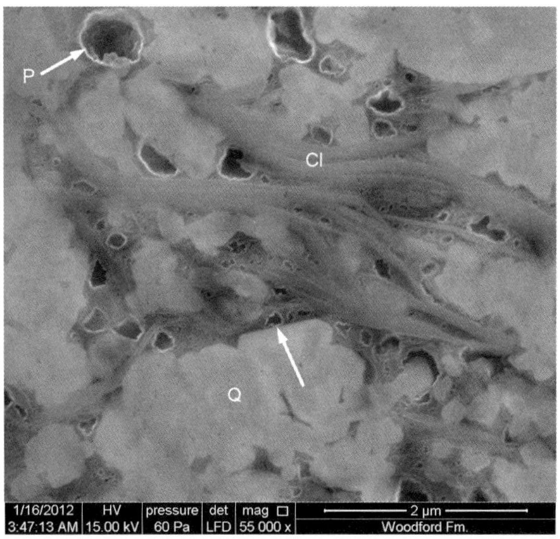

F.242.Woodford
Woodford Formation (Late Devonian)
Anadarko Basin, Oklahoma
Ar-ion milled, SE image
Thermal maturity: 1.15%R_o
Submitted by B. Wawak and T. Gentzis, Core Laboratories
Nanopores associated with dispersed organic matter that fills interparticle pores associated with phyllosilicate minerals (Cl) and quartz (Q). Pores in organic matter range from small, individual pores (arrow) to larger pores likely created by the coalescing of smaller pores (P).

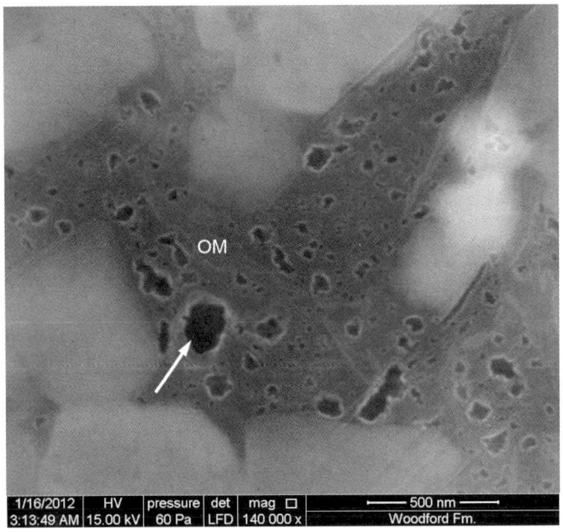

F.243.Woodford
Woodford Formation (Late Devonian)
Anadarko Basin, Oklahoma
Ar-ion milled, SE image
Thermal maturity: $1.15\%R_o$
Submitted by B. Wawak and T. Gentzis, Core Laboratories
Nanopores associated with dispersed organic matter (OM). The sizes of the pores range from approximately 2 nm (smallest resolved pores) to about 185 nm (long dimension of the pore indicated by arrow).

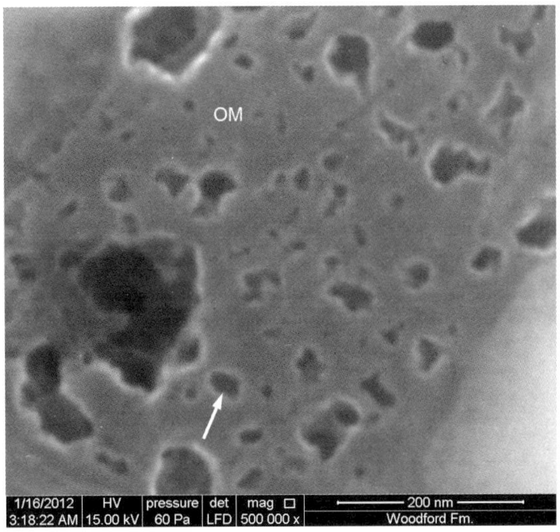

F.244.Woodford
Woodford Formation (Late Devonian)
Anadarko Basin, Oklahoma
Ar-ion milled, SE image
Thermal maturity: $1.15\%R_o$
Submitted by B. Wawak and T. Gentzis, Core Laboratories
Nanopores associated with dispersed organic matter (OM). The long dimension of the measured pore (indicated by arrow) is 27 nm.

F.245.Woodford
Woodford Formation (Late Devonian)
Anadarko Basin, Oklahoma
Ar-ion milled, BSE/SE image
Thermal maturity: $1.15\%R_o$
Submitted by B. Wawak and T. Gentzis, Core Laboratories
Organic matter and authigenic quartz filling intraparticle pores in a silicified fossil (Fs). Area indicated by bracket is enlarged in Figure 246. Note occurrence of irregularly shaped pyrite framboid and small cluster of pyrite crystals (arrows).

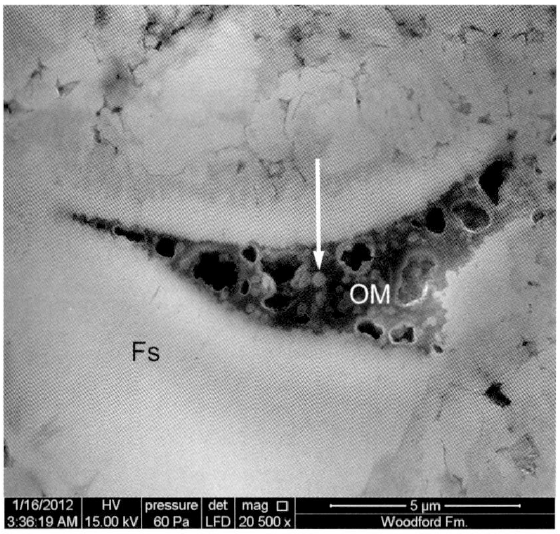

F.246.Woodford
Woodford Formation (Late Devonian)
Anadarko Basin, Oklahoma
Ar-ion milled, SE image
Thermal maturity: 1.15%R_o
Submitted by B. Wawak and T. Gentzis, Core Laboratories
Higher-magnification view of area indicated by brackets in Figure 245 showing intraparticle space associated with silicified fossil (Fs) that is filled with authigenic silica (arrow) and porous organic matter (OM). Long dimensions of pores range to about 1 μm.

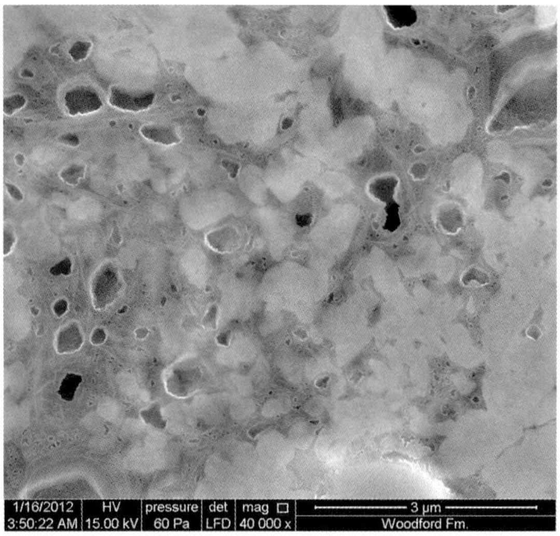

F.247.Woodford
Woodford Formation (Late Devonian)
Anadarko Basin, Oklahoma
Ar-ion milled, SE image
Thermal maturity: 1.15%R_o
Submitted by B. Wawak and T. Gentzis, Core Laboratories
This siliceous mudrock is composed of authigenic quartz (light gray) and phyllosilicate minerals. Highly porous dispersed organic matter (dark color) fills most of the interparticle pores. The longest dimensions of the larger pores are several hundred nanometers.